T0185567

Extreme Value Modeling and Risk Analysis

Methods and Applications

Extreme
Value Modeling
and Risk Analysis
Methods and Applications

Edited by

Dipak K. Dey
University of Connecticut
Storrs, USA

Jun Yan
University of Connecticut
Storrs, USA

CRC Press
Taylor & Francis Group
Boca Raton London New York

CRC Press is an imprint of the
Taylor & Francis Group, an **informa** business

A CHAPMAN & HALL BOOK

MATLAB® is a trademark of The MathWorks, Inc. and is used with permission. The MathWorks does not warrant the accuracy of the text or exercises in this book. This book's use or discussion of MATLAB® software or related products does not constitute endorsement or sponsorship by The MathWorks of a particular pedagogical approach or particular use of the MATLAB® software.

CRC Press
Taylor & Francis Group
6000 Broken Sound Parkway NW, Suite 300
Boca Raton, FL 33487-2742

First issued in paperback 2020

© 2016 by Taylor & Francis Group, LLC
CRC Press is an imprint of Taylor & Francis Group, an Informa business

No claim to original U.S. Government works

ISBN-13: 978-1-4987-0129-7 (hbk)
ISBN-13: 978-0-367-73739-9 (pbk)

Visit the Taylor & Francis Web site at
http://www.taylorandfrancis.com

and the CRC Press Web site at
http://www.crcpress.com

To my late parents with love and affection

D. K. Dey

To the memory of my mother and
To my father

J. Yan

Contents

Preface

Recent advances in risk analyses related to extreme events have given rise to challenging research problems that require combined expertise from statisticians and domain experts in fields such as climatology, hydrology, finance, insurance, sports, to name a few. The interdisciplinary aspects of these issues were addressed in the opening workshop for the SAMSI program on *Risk Analysis, Extreme Events and Decision Theory* which was held during September 16–19, 2007. The workshop aimed to take advantage of community input in the development of modeling extreme events in various applications. The workshop engaged a broadly representative segment of the statistical sciences, applied mathematics, decision analysis, and operations research communities in research on this subject. The program of the workshop covered a wide range of topics in both theory and application, with a theme to connect needed statistical/mathematical research with critical decision and risk assessment/management applications. The seed of this book was partially rooted in this workshop.

Although the literature on extreme value theory is enormous, applications are often sparse, scattered in journals in a variety of fields. We have not seen a book that provides various aspects in one place, especially advanced topics such as multivariate extremes. Journal articles rarely provide the conceptual background necessary for non-experts to understand and apply the approaches to their own problems. Researchers often find it difficult to go through the massive amount of literature. Moreover, the cost of journal subscriptions has skyrocketed recently. This book presents an up-to-date overview of statistical modeling of extreme events with related methods, computing, and various applications. A unique feature is that for many chapters the authors contributed their computer code, which is organized in a repository available from the editors' website, in an attempt to facilitate understanding of the topics and their applications to readers' own problems. We hope that this book will stimulate more collaboration between statisticians and substantive experts dealing with extreme events.

The book consists of 25 chapters. We start with two review chapters, one on univariate extreme value analysis and the other one on multivariate extremes. Chapter 3 considers univariate extreme value mixture modeling. Chapter 4 reviews threshold selection in extreme value analysis, followed by threshold modeling of nonstationary extremes in Chapter 5. Chapter 6 presents new results for block-maxima of vine copulas. Chapter 7 develops time series of extremes with applications from climatology. Chapter 8 focuses on max-autoregressive and moving maxima models for extremes. Extending to the spatial setting, spatial extremes and max-stable processes are considered in Chapter 9. Chapter 10 and Chapter 11 are concerned with simulation and conditional simulation of max-stable processes. Switching gears to inference methodologies, composite likelihood for extreme value analysis is considered in Chapter 12. Chapter 13 is devoted to Bayesian inference for extreme value modeling. Chapter 14 deals with approximate Bayesian computation for extremes modeling. Chapter 15 is a review of estimation of extreme conditional quantiles. The next three chapters are on extreme dependence. Chapter 16 presents parametric models for extremal dependence and their estimation. Then Chapter 17 focuses on nonparametric estimation of extremal dependence. An overview of nonparametric tests of extreme-value dependence is summarized in Chapter 18. Next are a few novel applications of extreme value modeling. Chapter 19 models multivariate extremes risks of financial investments. Chapter 20 presents some new results with bivariate regular variation motivated from insurance and financial risk management. Chapter 21 considers novel application of extreme value theory in weather and climate disasters. Chapter 22 discusses the extreme value analysis of safety data from clinical trials. Chapter 23 is an analysis of bivariate survival data based on copulas with log generalized extreme value marginal. Chapter 24 considers an application of extreme value theory in sports statistics through a change point analysis of top baseball batting averages. Finally in Chapter 25, computing software for extreme value analysis is reviewed.

We would like to thank all the authors for their contributions and peer reviewing with dedication, which made our job relatively simple. We also thank Jim Albert, Brian Bader, Forrest Crawford, and Stanislav Volgushev, for helping us as referees. As the project evolved over time, the patience of all the contributors was greatly appreciated.

We thankfully acknowledge the support rendered by David Grubbs, Acquisition Editor, Statistics, CRC Press/Taylor & Francis Group, who always stood behind us and always helped us with our sometimes unusual queries.

We express our indebtedness to each and every one who was associated with us directly or indirectly while the work was in progress. The list is certainly too lengthy to be exhaustive but we would like to give special mention to Brian Bader and Wenjie Wang. Brian helped convert a chapter in Word to LaTeX. Wenjie provided tremendous help in typesetting the whole book with LaTeX and setting up the computer code repository.

Contributors

Boris Beranger
Pierre and Marie Curie University
Paris, France

University of New South Wales
Sydney, Australia

Axel Bücher
Ruhr-Universität Bochum
Bochum, Germany

Frederico Caeiro
Universidade Nova de Lisboa
Lisbon, Portugal

Sy Han Chiou
Harvard University
Cambridge, Massachusetts

Claudia Czado
Technische Universität München
Munich, Germany

Dipak Dey
University of Connecticut
Storrs, Connecticut

Clément Dombry
Université de Franche-Comté
Besançon, France

Robert Erhardt
Wake Forest University
Winston-Salem, North Carolina

Eric Gilleland
National Center for Atmospheric
   Research
Boulder, Colorado

M. Ivette Gomes
Universidade de Lisboa
Lisbon, Portugal

Timothy Idowu
University of Wisconsin at Madison
Madison, Wisconsin

Yujing Jiang
University of Connecticut
Storrs, Connecticut

Philip Jonathan
Shell Research Ltd.
Chester, United Kingdom

Sangwook Kang
Yonsei University
Seoul, South Korea

Richard W. Katz
National Center for Atmospheric
   Research
Boulder, Colorado

Matthias Killiches
Technische Universität München
Munich, Germany

Anna Kiriliouk
Université Catholique de Louvain
Louvain-la-Neuve, Belgium

Ivan Kojadinovic
University of Pau
Pau, France

Deyuan Li
Fudan University
Shanghai, China

Ye Liu
JBA Risk Management
Broughton, United Kingdom

Paul J. Northrop
University College London
London, United Kingdom

Marco Oesting
University of Siegen
Siegen, Germany

Simone Padoan
Bocconi University
Milan, Italy

Ioannis Papastathopoulos
University of Edinburgh and Maxwell
    Institute
Edinburgh, United Kingdom

Liang Peng
Georgia State University
Atlanta, Georgia

David Randell
Shell Research Ltd.
Chester, United Kingdom

Brian J. Reich
North Carolina State University
Raleigh, North Caralina

Mathieu Ribatet
Université Montpellier 2
Montpellier, France

Université Lyon 1
Lyon, France

Dooti Roy
University of Connecticut
Storrs, Connecticut

Vivekananda Roy
Iowa State University
Ames, Iowa

Huiyan Sang
Texas A&M University
College Station, Texas

Carl Scarrott
University of Canterbury
Christchurch, New Zealand

Johan Segers
Université Catholique de Louvain
Louvain-la-Neuve, Belgium

Benjamin A. Shaby
Pennsylvania State University
State College, Pennsylvania

Scott A. Sisson
University of New South Wales
Sydney, Australia

Harry Southworth
Data Clarity Consulting Ltd.
Stockport, United Kingdom

Alec Stephenson
CSIRO
Melbourne, Australia

Qihe Tang
University of Iowa
Iowa City, Iowa

Jonathan A. Tawn
Lancaster University
Lancaster, United Kingdom

Huixia Judy Wang
George Washington University
Washington, District of Columbia

Michał Warchoł
Université Catholique de Louvain
Louvain-la-Neuve, Belgium

Jun Yan
University of Connecticut
Storrs, Connecticut

Zhongyi Yuan
Pennsylvania State University
State College, Pennsylvania

Zhengjun Zhang
University of Wisconsin at Madison
Madison, Wisconsin

1

Univariate Extreme Value Analysis

Dipak Dey
University of Connecticut, United States

Dooti Roy
University of Connecticut, United States

Jun Yan
University of Connecticut, United States

Abstract

This chapter reviews the basics of univariate extreme value analysis. The generalized extreme value (GEV) distribution is introduced as the limit distribution of sample maxima with appropriate standardization. The domains of attraction are reviewed and illustrated numerically with simulation for distributions that have different tail behaviors. Common properties of the GEV distribution are summarized. The connection to the generalized Pareto distribution is established by taking the limit of the conditional distribution over a threshold. Statistical inferences are grouped by the type of available data: the block maxima method, r largest order statistics method, the peaks-over-threshold method, and the point process method. For block maxima, the maximum product spacing method, which has been overlooked in extreme value applications, is worth highlighting for its numerical stability and competitive small sample performance in comparison with the likelihood method and the L-moments method. The efficiency gain of using r largest order statistics relative to using block maxima only is shown in a small simulation study. Compared to existing reviews on univariate extreme value analysis, this chapter provides more statistical flavor through numerical studies.

1.1 Generalized Extreme Value Distribution

Extreme value theory (EVT) is about modeling rare events, which occur with very small probability, and measuring their risks. It is widely used in risk management, finance, insurance, economics, biomedicine, hydrology, climate change, material sciences, telecommunications, sports, and many other fields wherever extreme events are of interest. The foundation of EVT is the generalized extreme value (GEV) distribution, which arises as the limiting distributions of sample maximum of independent, identically distributed (i.i.d.) random variables, as the sample size goes to infinity. As distributions of sample minima can be obtained from the sample maxima of negated variables, only maxima are discussed.

1.1.1 Limiting Distribution of Sample Maximum

Suppose that $Y_1, \ldots, Y_n$ is a sequence of i.i.d. random variables from distribution function F. Let $M_n = \max\{Y_1, Y_2, \ldots, Y_n\}$. The exact distribution of M_n is F^n. As in the case of central limit theorem, where the sample mean after appropriate standardization converges in distribution to a standard normal variable, we consider the limiting distribution of M_n with appropriate standardization. EVT seeks a non-degenerate distribution function $G(x)$, and sequences $a_n > 0$ and $b_n \in \mathbb{R}$, such that $a_n^{-1}(M_n - b_n)$ converges in distribution to G:

$$\lim_{n \to \infty} \Pr\left(\frac{M_n - b_n}{a_n} \leqslant x\right) = \lim_{n \to \infty} F^n(a_n x + b_n) = G(x) \qquad (1.1)$$

on all points in the continuity set of G. Note that by the convergence of types theorem (e.g., Billingsley, 2008, Theorem 14.2), G is unique up to a location-scale transformation. If (1.1) holds with a non-degenerate G, then F is said to be in the (maximum) domain of attraction of G, denoted as $F \in \mathcal{D}(G)$. Intuitively, the limit in (1.1) is possible because

$$F^n(a_n x + b_n) = \left\{1 - \frac{n[1 - F(a_n x + b_n)]}{n}\right\}^n \approx \exp\{-n[1 - F(a_n x + b_n)]\}$$

if $\lim_{n \to \infty} n[1 - F(a_n x + b_n)]$ exists.

The only possible limit G, up to location-scale transformation, is the GEV distribution (e.g., de Haan and Ferreira, 2007, Theorem 1.1.3)

$$G_\xi(x) = \begin{cases} \exp[-(1 + \xi x)_+^{-1/\xi}] & \xi \neq 0, \\ \exp[-\exp(-x)] & \xi = 0, \end{cases} \qquad (1.2)$$

where $v_+ = \max(v, 0)$. The distribution is Gumbel when $\xi = 0$, in which case, $G_0(x)$ is in fact $\lim_{\xi \to 0} G_\xi(x)$. In the sequel, expressions involving ξ are always to be understood as the limit as $\xi \to 0$ when $\xi = 0$. The GEV distribution is characterized by shape parameter ξ, which is known as the extreme value index (EVI).

The GEV distribution unifies the three limiting distributions studied historically (Fisher and Tippett, 1928; Gnedenko, 1943). The three distributions are Gumbel, Fréchet, and Weibull, also known as type I, II, and III extreme value distributions, respectively: The Gumbel distribution is $\Lambda(x) = G_0(x)$, $x \in \mathbb{R}$. The Fréchet distribution is Φ_α with $\alpha > 0$,

$$\Phi_\alpha(x) = \begin{cases} 0 & x \leq 0, \\ \exp(-x^{-\alpha}) & x > 0. \end{cases}$$

It is $G_\xi((x-1)/\xi)$ with $\xi = 1/\alpha > 0$. The Weibull distribution is Ψ_α with $\alpha > 0$,

$$\Psi_\alpha(x) = \begin{cases} \exp[-(-x)^\alpha] & x \leq 0, \\ 1 & x > 0. \end{cases}$$

It is $G_\xi(-(x+1)/\xi)$ with $\xi = -1/\alpha < 0$. This distribution is the reflection of the usual Weibull distribution from $\mathbb{R}_+$ to $\mathbb{R}_-$, hence also known as reversed-Weibull.

The three distributions are closely connected. For $X > 0$ and $\alpha > 0$, it is easy to see that the following statements are equivalent:
1. X follows distribution Φ_α.
2. $\alpha \log X$ follows distribution Λ.
3. $-1/X$ follows distribution Ψ_α.

1.1.2 Domain of Attraction

The EVI ξ determines the tail of the GEV distribution. The Fréchet distribution ($\xi > 0$) has a heavy tail with $1 - G_\xi(x) \sim (\xi x)^{-1/\xi}$ as $x \to \infty$, and, hence, moments only exist for order less than $1/\xi$. It fits well data from precipitation, stream flow, and economic impacts. The reversed-Weibull distribution has a finite right endpoint $-1/\xi$, with $1 - G_\xi(-1/\xi - x) \sim (-\xi x)^{-1/\xi}$ as $x \downarrow 0$. It can be used for extremes with an upper limit such as temperature, wind speed, sea level, and athletic record. The Gumbel distribution has right endpoint ∞ but the tail is light with $1 - G_0(x) \sim \exp(x)$ as $x \to \infty$. Figure 1.1 shows the density functions for the three distributions with $\xi \in (0.25, 0, 0.25)$, as well as the tail of their corresponding survival function $1 - G_\xi(\cdot)$.

For a given distribution F with support (x_*, x^*), it is of interest to check which domain of attraction G_ξ it is in, $F \in \mathcal{D}(G_\xi)$, and to find the standardizing constants $a_n > 0$ and b_n. Delicate conditions are required on the tail of F. A unified sufficient and necessary condition for $F \in \mathcal{D}(G_\xi)$ is, for all x with $1 + \xi x > 0$, there exists some positive function f such that

$$\lim_{t \uparrow x^*} \frac{1 - F[t + xf(t)]}{1 - F(t)} = (1 + \xi x)^{-1/\xi}, \tag{1.3}$$

(e.g., de Haan and Ferreira, 2007, Theorem 1.2.5). If (1.3) holds for some f, then

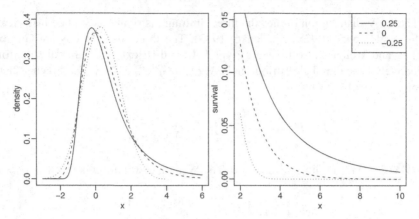

FIGURE 1.1
Density function and the survival tail of three GEV distributions with $\xi \in (0.25, 0, -0.25)$.

one possible f is

$$f(t) = \begin{cases} \xi t, & \xi > 0, \\ -\xi(x^* - t), & \xi < 0, \\ \frac{\int_t^{x^*} [1 - F(x)]\mathrm{d}x}{1 - F(t)}, & \xi = 0. \end{cases}$$

The tail quantile function plays an important role. For any nondecreasing function h, define its inverse function as $h^{-1}(y) = \inf\{x : h(x) \geq y\}$. The tail quantile function U is the inverse of $1/(1 - F)$. That is, $F(x) = 1 - 1/y$ and $U(y) = x$ are equivalent. It is nondecreasing on $[1, \infty)$ with $U(1) = x_*$ and $U(\infty) = x^*$. Quantity $U(n)$ is known as the return level for return period n, the level that is exceeded once every n periods on average, which is often a target of inference in fields such as hydrology or climatology. In financial risk management, $U(1/\alpha)$ of the loss distribution is a risk measure known as the α value at risk (VaR), which is the threshold loss value such that the probability that the loss exceeds this value in a given time horizon is α. For the Fréchet and Weibull distributions, the domain of attraction conditions have simpler forms in terms of U. Specifically, the condition is

$$\lim_{t \to \infty} \frac{U(tx)}{U(x)} = x^\xi$$

for Fréchet ($\xi > 0$) and

$$\lim_{t \to \infty} \frac{x^* - U(tx)}{x^* - U(t)} = x^{1/\xi}$$

with $x^* < \infty$ for Weibull ($\xi < 0$) (e.g., de Haan and Ferreira, 2007, Corollary 1.2.10).

If $F \in \mathcal{D}(G_\xi)$, then the standardizing constants $a_n > 0$ and b_n in $F^n(a_n x + b_n) \to G_\xi(x)$ can be found for three subfamilies in terms of U or/and f. For $\xi > 0$,

TABLE 1.1
Empirical rejection percentage of the KS goodness-of-fit test from 1000 replicates for m sample maxima of n i.i.d. observations from the distributions in the domain of attraction (DoA) of Fréchet, Gumbel, and Weibull using their limit distribution.

DoA	m				block size n				
		50	100	200	400	50	100	200	400
			Pareto $\alpha = 1$				Pareto $\alpha = 1.5$		
Φ_α	100	5.1	4.5	4.0	3.5	11.6	7.6	6.2	5.7
	200	4.2	3.7	5.0	4.9	19.8	10.6	7.5	4.9
	400	3.3	5.9	5.3	3.8	37.9	15.2	9.8	8.1
			Gamma $\Gamma(0.5, 1)$				Normal $N(0, 1)$		
Λ	100	6.0	5.7	6.3	5.1	15.9	17.0	12.8	12.1
	200	14.3	9.0	9.6	8.9	36.2	31.3	27.9	22.5
	400	36.7	22.5	17.5	13.3	70.2	61.7	56.4	48.9
			Beta$(3, b)$ with $b = 2$				Beta$(3, b)$ with $b = 3$		
Ψ_b	100	7.1	7.3	4.5	4.4	30.4	21.1	13.1	9.8
	200	11.3	8.8	4.2	5.4	58.0	39.8	26.2	17.6
	400	17.8	11.4	8.6	7.1	87.6	69.6	49.8	29.3

choosing $a_n = U(n)$ and $b_n = 0$ leads to limiting Fréchet distribution $\Phi_{1/\xi}$; for $\xi < 0$, choosing $a_n = x^* - U(n)$ and $b_n = x^*$ leads to limiting Weibull distribution $\Psi_{-1/\xi}$; for $\xi = 0$, choosing $a_n = f(U(n))$ and $b_n = U(n)$ leads to limiting Gumbel distribution G_0 (e.g., de Haan and Ferreira, 2007, Corollary 1.2.4). The Fréchet domain includes distributions such as the Pareto type, Burr, inverse gamma, log-gamma, Student t, F, and Fréchet; the Weibull domain includes distributions such as beta, power-law with an upper bound, reversed-Burr, and reversed-Weibull; the Gumbel domain includes many commonly used distributions such as exponential, Weibull, gamma, logistic, normal, and log-normal (e.g., Beirlant et al., 2004, Table 2.1–2.3).

How big an n is needed for $F^n(a_n x + b_n)$ to be well approximated by $G_\xi(x)$ if $F \in \mathcal{D}(G)$? With appropriately chosen $a_n > 0$ and b_n, the two distribution functions can be checked visually for an array of n (e.g., Beirlant et al., 2004, Ch.2). It is perhaps more interesting to apply goodness-of-fit test of the limiting distribution G_ξ to sample maxima data with finite n and the right sequence of a_n and b_n. Two distributions with different tail properties from each domain of attraction were selected in a simulation study, with the standardizing constants listed in Embrechts et al. (1997, Table 3.4.2–3.4.4). The Pareto distributions with shape $\alpha \in \{1, 1.5\}$ are used for the domain of attraction of Fréchet $\mathcal{D}(\Phi_\alpha)$. The survival function is $S(x) = 1/(1+x)^\alpha$, $x > 0$. The gamma distribution $\Gamma(a, 1)$ and normal distribution are used for the domain of attraction of $\mathcal{D}(\Lambda)$. The beta distributions Beta$(3, b)$ with $b \in \{1, 3\}$ are used for the domain of attraction of Weibull $\mathcal{D}(\Psi_b)$. The R code is available in the companion code repository. Table 1.1 reports the empirical rejection percentage of the

Kolmogorov–Smirnov (KS) test of the limiting distribution with significance level 0.05 from 1000 replicates; each replicate had a sample of size m of maxima of n i.i.d. observations. That is, n is the block size over which the maxima are taken, and m is the sample size of the block maxima.

Convergence to the limit distribution is indicated by close agreement of the empirical rejection percentage to the nominal level of 5%. For a given m, the agreement improves as n increases. For a given n, a larger m means higher power to detect the difference between the sampling distribution and the limit distribution, and, hence, the rejection rate increases. It is observed that the convergence of the sample maxima to their limit distribution is faster for heavier tailed distributions. The distribution on the left in the same domain has heavier right tail than the distribution on the right. For example, the gamma distribution has heavier tail than the normal distribution, and its sample maxima converges in distribution to Gumbel faster than the sample maxima of the normal distribution. It is also interesting to note that, for the most familiar normal distribution, the convergence is very slow. In an additional experiment with $n = 10,000$ and $m = 500$, the rejection percentage was still 37.3%.

1.2   Properties of the GEV Distribution

With a location parameter μ and scale parameter $\sigma > 0$, the distribution function in (1.2) becomes

$$G(x|\theta) = \begin{cases} \exp\left[-\left[1 + \xi\left(\frac{x-\mu}{\sigma}\right)\right]_+^{-1/\xi}\right] & \xi \neq 0, \\ \exp\left[-\frac{x-\mu}{\sigma}\right] & \xi = 0 \end{cases} \tag{1.4}$$

where $\theta = (\mu, \sigma, \xi)$. Closed-form quantile function is obtained as

$$Q(u|\theta) = \begin{cases} \mu + \frac{\sigma}{\xi}\left[(-\log u)^{-\xi} - 1\right] & \xi \neq 0, \\ \mu - \sigma\log(-\log u) & \xi = 0, \end{cases} \tag{1.5}$$

which can be used to generate random variables from the GEV distribution. The corresponding density function is

$$f(x|\theta) = \frac{1}{\sigma}t^{\xi+1}(x|\theta)\exp[-t(x|\theta)], \tag{1.6}$$

where

$$t(x|\theta) = \begin{cases} \left[1 + \xi\left(\frac{x-\mu}{\sigma}\right)\right]_+^{-1/\xi} & \xi \neq 0, \\ \exp\left(\frac{x-\mu}{\sigma}\right) & \xi = 0. \end{cases}$$

Note the double appearances of $t(x|\theta)$, which should be exploited in evaluating the density. The behavior of these functions (distribution, quantile, and density) can have numerical problems for $|\xi|$ close to zero if these expressions are implemented as is naively. An accurate evaluation of $\log(1 + u)$ for $|u| \to 0$ can be exploited for this purpose; see function `log1p` in R. A careful implementation is available in R package `texmex` (Southworth and Heffernan, 2013).

The median of the distribution is

$$Q(0.5|\theta) = \mu + \sigma \frac{(\log 2)^{-\xi} - 1}{\xi}$$

with limit $\mu - \sigma \log \log 2$ as $\xi \to 0$. The mode is

$$M_X = \mu + \sigma \frac{(1 + \xi)^{-\xi} - 1}{\xi}$$

with limit μ as $\xi \to 0$. As mentioned earlier, moments only exist for order less than $1/\xi$ because $1 - G_\xi(x) \sim x^{-1/\xi}$ in the heavy tail case $\xi > 0$. With a change of variable

$$y = \left[1 + \xi \left(\frac{x - \mu}{\sigma} \right) \right]^{-1/\xi},$$

it is easy to find

$$E(X) = \mu + \frac{\sigma}{\xi} [\Gamma(1 - \xi) - 1], \qquad \xi < 1,$$

with limit $\mu + \sigma\gamma$ as $\xi \to 0$, where γ is the Euler's constant (≈ 0.57721, which equals $-\Gamma'(1)$, where Γ' is digamma function or derivative of the gamma function). The same variable change can be used to find higher-order moments. The variance can be shown to be

$$\text{Var}(X) = \frac{\sigma^2 [\Gamma(1 - 2\xi) - \Gamma^2(1 - \xi)]}{\xi^2}, \qquad \xi < 1/2,$$

with limit $\sigma^2 \pi^2 / 6$ as $\xi \to 0$.

The skewness of the GEV exists for $\xi < 1/3$,

$$\text{sign}(\xi) \frac{g_3 - 3g_1 g_2 + 2g_1^3}{\text{Var}^{3/2}(X)},$$

with limit $12\sqrt{6}\zeta(3)/\pi^3 \approx 1.14$ as $\xi \to 0$, where $g_k = \Gamma(1 - k\xi)$, $k = 1, 2, 3$, and ζ is the Riemann zeta function. The skewness increases with ξ, and the ξ value at which G_ξ has zero skewness is about -0.2776. An alternative skewness measure with respect to mode defined by Arnold and Groenveld (1995) only requires existence of mode. For a distribution F, this skewness measure is $\gamma_M = 1 - 2F(M_F)$, where M_F is the mode of F. For the GEV distribution, $\gamma_M = 2\exp[-(1+\xi)]-1$ if $\xi > -1$; otherwise, the mode does not exist. The GEV distribution has negative skewness for $\xi < \log 2 - 1$ and is positively skewed for $\xi > \log 2 - 1$. The flexibility of the GEV skewness can be utilized to construct asymmetric link functions for binary data analysis (Wang and Dey, 2010).

A log-GEV distribution can be constructed from the GEV distribution and was shown to provide very flexible shapes that may be used in survival data analysis (Roy et al., 2013). See also Chapter 23 for a bivariate survival analysis with log-GEV marginals and Clayton copula.

1.3   Exceedances over a Threshold

Often an event is regarded as an extreme event only if it exceeds some high threshold value u. The conditional distribution of the exceedances converges to the generalized Pareto distribution (GPD) as $u \to \infty$ under the same conditions as those for the convergence of sample maxima in distribution (1.1) (Pickand, 1975). Consider random variable Y with distribution F. The interest is the limit of

$$\lim_{u \to \infty} \Pr(Y > u + y | Y > u) = \frac{1 - F(u + y)}{1 - F(u)}, \ y > 0.$$

A heuristic derivation is as follows.

Since (1.1) is assumed to hold, there is some θ such that, for large n,

$$F^n(z) \approx \exp\left\{ -\left[1 + \xi \left(\frac{z - \mu}{\sigma} \right) \right]^{-\frac{1}{\xi}} \right\}$$

with some parameter $\theta = (\mu, \sigma, \xi)$, or equivalently

$$n \log F(z) \approx -\left[1 + \xi \left(\frac{z - \mu}{\sigma} \right) \right]^{-\frac{1}{\xi}}.$$

As for large z, $F(z) \to 1$, and $\log F(z) \sim F(z) - 1$,

$$1 - F(z) \sim \frac{1}{n} \left[1 + \xi \left(\frac{z - \mu}{\sigma} \right) \right]^{-\frac{1}{\xi}}. \tag{1.7}$$

Evaluating z at u and $u + y$, $y > 0$ gives

$$\Pr(X > u + y \,|\, X > u) \sim \left[\frac{1 + \xi(u + y - \mu)/\sigma}{1 + \xi(u - \mu)/\sigma} \right]^{-1/\xi} = \left[1 + \frac{\xi y}{\sigma_u} \right]^{-1/\xi}, \tag{1.8}$$

which is the survival function of a GPD with shape parameter ξ and scale parameter $\sigma_u = \sigma + \xi(u - \mu)$.

The rigorous connection between the GPD and EVT was first established by Pickand (1975) albeit the GPD has long been used to model heavy tailed data. The shape parameter ξ of the GPD and the GEV distribution are the same, and the scale parameter σ_u of the GPD depends on the location parameter μ and scale parameter σ of the GEV distribution and the threshold value u. In practice, the exceedances over a threshold enables usage of more information from the data than just the block maxima; see 1.4.3 for application in parameter estimation.

1.4 Parameter Estimation

Parameter estimation can be based on different kinds of data available: block maxima, blocks of largest order statistics, exceedances over a threshold, or the whole historical records. The rationale of using more data beyond block maxima is that they contain additional information about the extremes upper tail.

1.4.1 Block Maxima

The block maxima approach assumes that maxima taken from independent blocks are available, and that they can be fitted with a GEV distribution. Larger blocks mean fewer maxima and larger variance in estimation, while smaller blocks may not fit the GEV distribution and lead to bias. Often block size determined from the pragmatic perspective or data availability, leading to data such as annual, monthly, or daily maxima. Suppose that $Y_1, \ldots, Y_m$ are maxima from m independent blocks. Let $Y_{1:m} \leq Y_{2:m} \leq \cdots \leq Y_{m:m}$ be the ordered sample.

1.4.1.1 Maximum Likelihood

The loglikelihood function is

$$L(\theta) = \sum_{i=1}^{m} \log f(X_i|\theta),$$

where $f(\cdot|\theta)$ is the GEV density in (1.6). The support of the GEV distribution depends on the parameter values: $\mu - \sigma/\xi$ is the upper endpoint for $\xi < 0$ and lower endpoint for $\xi > 0$; the support is $\mathbb{R}$ if $\xi = 0$. Therefore, the regularity conditions of the usual asymptotics of the maximum likelihood estimator (MLE) are violated. In fact, Smith (1985) showed that

1. when $\xi > -0.5$, the MLE has the usual asymptotic properties;
2. when $-1 < \xi \leq -0.5$, the MLE exists but does not have the standard asymptotic properties (especially for μ);
3. when $\xi \leq -1$, the MLE does not exist.

The cases where $\xi \leq -0.5$ correspond to distributions with bounded upper endpoint, which is not as frequently observed as the cases with $\xi > -0.5$ in practice.

The score function of the loglikelihood and the Fisher information have been derived by Prescott and Walden (1980), which can be useful in devising iterative algorithms to find the MLE and estimating its variance. The scores can be obtained from the derivatives of the log density

$$\frac{\partial \log f}{\partial \mu} = \frac{1}{\sigma w}(1 + \xi - v),$$

$$\frac{\partial \log f}{\partial \sigma} = \frac{1}{\sigma}\left[-1 + \frac{1}{\sigma w}(1 + \xi - v)(y - \mu)\right],$$

$$\frac{\partial \log f}{\partial \xi} = \frac{1}{\xi^2} \log(w)(1 - v) - \frac{1}{\sigma w}(1 + \xi - v)(y - \mu)\xi,$$

where $w = 1 + \xi(y - \mu)/\sigma$, and $v = w^{-1/\xi}$. The Fisher information is mI, and I has entries

$$I_{\mu\mu} = \frac{1}{\sigma^2}p,$$

$$I_{\sigma\sigma} = \frac{1}{\sigma^2\xi^2}\left[1 - 2\Gamma(2 + \xi) + p\right],$$

$$I_{\xi\xi} = \frac{1}{\xi^2}\left[\frac{\pi^2}{6} + \left(1 - \gamma + \frac{1}{\xi}\right)^2 - \frac{2q}{\xi} + \frac{p}{\xi^2}\right],$$

$$I_{\mu\sigma} = -\frac{1}{\sigma^2\xi}\left[p - \Gamma(2 + \xi)\right],$$

$$I_{\mu\xi} = \frac{1}{\sigma\xi}\left(q - \frac{p}{\xi}\right),$$

$$I_{\sigma\xi} = \frac{1}{\sigma\xi^2}\left[1 - \gamma + \frac{1 - \Gamma(2 + \xi)}{\xi} - q + \frac{p}{\xi}\right],$$

where

$$p = (1 + \xi)^2\Gamma(1 + 2\xi),$$

$$q = \Gamma(2 + \xi)\left[\Gamma'(1 + \xi) + \frac{1}{\xi} + 1\right],$$

$\Gamma'(r) = d \log \Gamma(r)/dr$ is the digamma function, and γ is Euler's constant $-\Gamma'(1)$. The information is only valid for $\xi > -0.5$, in which case, the asymptotic variance of the MLE is I^{-1}/m.

The MLE $\hat{\theta}$ is asymptotically consistent, normally distributed, and most efficient when $\xi > -0.5$, as the usual MLE theory states. Covariates can be easily incorporated into any of the three parameters. Nonetheless, for small-to-moderate sample sizes, the MLE is not robust in that it can be difficult to find with numerical issues in iterative algorithms, and if it can be found, the variance can be large relative to the probability weighted moments approach (Hosking et al., 1985).

Return level estimation is often of primary interest. The T-period return level R_T is estimated as $\hat{R}_T = Q(1 - 1/T | \hat{\theta})$ with the quantile function (1.5). Confidence intervals can be obtained in principle with the delta method using the asymptotic variance of $\hat{\theta}$. Their performance, however, may be poor in terms of empirical coverage probability when the sample size is not large enough, especially for return levels with large return periods T. An alternative approach is the profile likelihood method. Partition θ into (θ_1, θ_2), where θ_1 contains a subvector of dimension k of current interest. The profile loglikelihood for θ_1 is

$$L_p(\theta_1) = \max_{\theta_2} L\big((\theta_1, \theta_2)\big).$$

Under suitable regularity conditions, the deviance

$$D_p(\theta_1) = 2[L(\hat{\theta}) - L_p(\theta_1)]$$

evaluated at the true parameter value θ_1 follows approximately χ_k^2, chi-squared distribution with k degrees of freedom (e.g., Coles, 2001, Theorem 2.6). This result can be used to construct a $1 - \alpha$ confidence region for θ_1 with

$$C_\alpha = \{\theta_1 : D_p(\theta_1) \leq c_\alpha\}$$

where c_α is the upper α quantile of χ_k^2. For any component in θ, this approach is straightforward. For R_T, one can reparametrize the GEV distribution in terms of R_T using (1.5), and then apply the profile likelihood method to R_T.

1.4.1.2 Probability Weighted Moments or L-Moments

The probability weighted moments (PWM) of a random variable X with distribution function F are defined as

$$M_{p,r,s} = E\{X^p F^r(X)[1 - F(X)]^s\}$$

for real numbers p, r, and s (Greenwood et al., 1979). These quantities are easily computed when the quantile function Q of F has a closed form,

$$M_{p,r,s} = \int_0^1 Q^p(u) u^r (1 - u)^s du.$$

The quantity $M_{p,0,0}$ is the pth moment of X, which only exists when $\xi < 1/\xi$; the restriction makes the method of moments for GEV parameter estimation unattractive. Hosking et al. (1985) investigated quantities $\beta_r = M_{1,r,0}, r = 0, 1, 2, \ldots$, and matched them with their sample counterpart to estimate the parameters, where only the existence of the mean is required ($\xi < 1$).

With the closed-form quantile function (1.5), it can be shown that Hosking et al. (1985)

$$\beta_r = \frac{1}{r+1}\left[\mu - \frac{\sigma}{\xi}\{1 - (r+1)^\xi \Gamma(1-\xi)\}\right], \qquad \xi < 1.$$

With $r \in \{0, 1, 2\}$, this leads to

$$\beta_0 = \mu - \frac{\sigma}{\xi}(1 - \Gamma(1-\xi)) \tag{1.9}$$

$$2\beta_1 - \beta_0 = \frac{\sigma}{\xi}\Gamma(1-\xi)(2^\xi - 1) \tag{1.10}$$

$$\frac{3\beta_2 - \beta_0}{2\beta_1 - \beta_0} = \frac{3^\xi - 1}{2^\xi - 1} \tag{1.11}$$

Estimation of β_r can be conveniently based on the order statistics $Y_{1:m}, \ldots, Y_{m:m}$. An unbiased estimator of β_r is

$$b_r = \frac{1}{m}\sum_{j=1}^m \left(\prod_{l=1}^r \frac{(j-l)}{(m-l)}\right) Y_{j:m}$$

or an asymptotically equivalent version

$$\widehat{\beta}_r = \frac{1}{m} \sum_{j=1}^{m} \left(\frac{j}{m+1} \right)^r Y_{j:m}.$$

The PWM estimator is, after replacing β_r with b_r or $\widehat{\beta}_r$, the solution to the Equations (1.9)–(1.11). Hosking et al. (1985) showed that when β_r are substituted by b_r, the PWM estimate satisfies a feasibility criterion: $\widehat{\xi} < 1$ and $\widehat{\sigma} > 0$ almost surely, a clearly desirable property.

An equivalent approach is known as the L-moments approach which has been widely used in hydrology and climatology (Hosking and Wallis, 2005). The L-moments of a random variable X with distribution F are defined as

$$\lambda_r = \frac{1}{r} \sum_{k=0}^{r-1} (-1)^k \binom{r-1}{k} EX_{(r-k):r}, \qquad r = 1, 2, \ldots,$$

where $X_{(r-k):r}$ is the $(r-k)$th order statistics of a sample of size r. Clearly, λ_r is a linear function of the expected order statistics, and, hence, the L in "L-moments." The expectation of order statistics $X_{j:r}$ is

$$EX_{j:r} = \frac{r}{(j-1)!(r-j)!} \int xF^{j-1}(x)[1-F(x)]^{r-j} dF(x).$$

L-moments λ_r, $r = 1, 2, \ldots$, exist for any random variable X if and only if X has finite mean, in which case, the distribution is characterized by its L-moments (Hosking, 1990, Theorem 1). The first few L-moments are

$$\lambda_1 = EX = \int_0^1 Q(u)du,$$

$$\lambda_2 = \frac{1}{2}E(X_{2:2} - X_{1:2}) = \int_0^1 Q(u)(2u-1)du,$$

$$\lambda_3 = \frac{1}{3}E(X_{3:3} - 2X_{2:3} + X_{1:3}) = \int_0^1 Q(u)(6u^2 - 6u + 1)du.$$

The connection to the PWMs is clear — the PWMs can be expressed as linear functions of L-moments. Therefore, procedures based on PWMs and L-moments are equivalent. It is more convenient with L-moments as they have direct interpretations as scale or shape of F.

The second L-moment λ_2 is a scale measure that gives less weight to larger differences in comparison with the familiar standard deviation. Also known as L-scale, it is half mean difference of two random observations. L-moment ratios are L-moments standardized by λ_2:

$$\tau_r = \lambda_r / \lambda_2, \qquad r = 3, 4, \ldots.$$

Specifically, τ_3 and τ_4 are L-skewness and L-kurtosis, respectively, the counterparts of conventional skewness and kurtosis. The first few L-moments (or L-moments ratio) of the GEV distribution are (Hosking, 1990, Table 1)

$$\lambda_1 = \mu - \frac{\sigma}{\xi}(1 - \Gamma(1 - \xi)),$$

$$\lambda_2 = \frac{\sigma}{\xi}\Gamma(1 - \xi)(2^\xi - 1),$$

$$\tau_3 = 2\frac{3^\xi - 1}{2^\xi - 1} - 3.$$

The L-moments estimator is obtained by solving these equations numerically with λ_r replaced by their sample version, which is summarized from all r-element subsets of the sample as

$$l_r = \binom{m}{r}^{-1} \sum_{1 \le i_1 \le i_2 \le \cdots \le i_r} \frac{1}{r} \sum_{k=0}^{r-1} (-1)^k \binom{r-1}{k} X_{i_{r-k}:m}.$$

The PWM or L-moments estimator is consistent and asymptotically normal. The asymptotic variance matrix can be derived with the delta method from the variance of the sample PWMs provided $\xi < 0.5$. For small-to-moderate samples, the PWM estimator is more robust numerically and more efficient than the MLE (Hosking et al., 1985). The superb performance for small samples relative to the MLE is explained by its implicit constraint $\xi < 1$ (Katz et al., 2002). When a similar constraint is imposed in the likelihood approach, the performance of the resulting penalized MLE becomes competitive too for small samples (Coles and Dixon, 1999), but the penalty needs to be tuned. It was recently shown theoretically that the block maxima approach with PWM estimator is a rather efficient method in comparison to the peaks-over-threshold approach in Section 1.4.3 under usual practical conditions (Ferreira and de Haan, 2015).

1.4.1.3 Maximum Product Spacing

The maximum product spacing (MPS) is a general estimation method for univariate models (Cheng and Amin, 1983; Ranneby, 1984). The method allows efficient estimation in non-regular cases where the MLE may not exist, or more robust estimation when MLE exists. This is especially relevant to the GEV distribution, as the MLE does not exist when $\xi \le -1$ and exists but does not have the usual properties when $-1 < \xi \le -0.5$ (Smith, 1985). The idea of the MPS method is, on the uniform scale after the probability integral transformation, to choose parameter values that make the observed data as uniform as possible. This is achieved by maximizing the geometric mean of the spacing between consecutively ordered observations. Let $U_i(\theta) = G(Y_{i:m}|\theta)$, $i = 1, \ldots, m$, where G is the GEV distribution function (1.4). Define

$$D_i(\theta) = U_i(\theta) - U_{i-1}(\theta) \qquad i = 1, \ldots, m+1,$$

with $U_0(\theta) = 0$ and $U_{m+1}(\theta) = 1$. The MPS estimator $\breve{\theta}_n$ minimizes

$$M(\theta) = -\sum_{i=1}^{m+1} \log D_i(\theta).$$

A byproduct of the MPS method is that $M(\theta)$ evaluated at the estimator is the Moran's statistic for goodness-of-fit test (Cheng and Stephens, 1989).

For GEV modeling, the asymptotic properties of the MPS estimator are the same as those of the MLE when $\xi > -1$; when $\xi < -1$, the MPS estimator of μ is $n^{-\xi}$ consistent, and the MPS estimator for (σ, ξ) is $n^{1/2}$ consistent, asymptotically normal, and efficient (Wong and Li, 2006, Theorem 3.2). In the numerical study of Wong and Li (2006), the MPS estimator showed very competitive performance in comparison to the MLE when the latter can be calculated, and in comparison to the L-moments estimator for sample size as small as 10.

1.4.2 Largest Order Statistics

Instead of just using the maxima from each block, it is possible to use the r largest order statistics from each block which may lead to more efficient estimation if the distribution of the r largest order statistics from the blocks is justified. As an extension of the limiting distribution of the block maxima, as block size $n \to \infty$, for a finite r, the joint limiting distribution of the r largest order statistics $M_n^{(1)} > \cdots > M_n^{(r)}$ can be standardized with the same normalizing constants $a_n > 0$ and b_n as those for the block maxima (Weissman, 1978). Under similar conditions for the convergence of the block maxima,

$$\Pr\left(\frac{M_n^{(1)} - b_n}{a_n} \leq x_1, \ldots, \frac{M_n^{(r)} - b_n}{a_n} \leq x_r\right) \to G_r(x_1, \ldots, x_r)$$

for a joint distribution G_r, where $x_1 > \cdots > x_r$.

The joint distribution G_r has density

$$f_r(x_1, x_2, \ldots, x_r | \mu, \sigma, \xi) = \sigma^{-r} \exp\left\{-(1 + \xi z_r)^{-\frac{1}{\xi}} - \left(\frac{1}{\xi} + 1\right) \sum_{j=1}^{r} \log(1 + \xi z_j)\right\}$$

(1.12)

for some location parameter μ, scale parameter $\sigma > 0$ and shape parameter ξ, where $x_1 > \cdots > x_r$, $z_j = (x_j - \mu)/\sigma$, and $1 + \xi z_j > 0$ for $j = 1, \ldots, r$ (e.g., Tawn, 1988). The parameters $\theta = (\mu, \sigma, \xi)$ are the same as those in the limiting GEV distribution of the block maxima $M_n^{(1)}$. For $\xi = 0$, the density is the limit of (1.12) as $\xi \to 0$, which is the Gumbel case (Smith, 1986). When $r = 1$, G_1 is the GEV distribution (1.6). Suppose that $(X_1, \ldots, X_r)$ follows distribution G_r. From the density (1.12), it is easy to verify that the conditional distribution of X_k given $(X_1, \ldots, X_{k-1})$, $k = 2, \ldots, r$, is the GEV distribution right truncated at X_{k-1}. This makes it straightforward to simulate from G_r (Bader et al., 2015).

TABLE 1.2
Summaries of the MLE using r largest order statistics based on 1000 replicates generated from G_5 with $\xi = 0.5$. ESE: empirical standard error. ASE: average of standard errors estimated from inverting the Fisher information.

r	bias			ESE			ASE		
	$\widehat{\mu}$	$\widehat{\sigma}$	$\widehat{\xi}$	$\widehat{\mu}$	$\widehat{\sigma}$	$\widehat{\xi}$	$\widehat{\mu}$	$\widehat{\sigma}$	$\widehat{\xi}$
				$m = 50$					
1	0.001	−0.022	0.012	0.158	0.157	0.158	0.161	0.156	0.153
2	−0.006	−0.010	0.003	0.129	0.139	0.120	0.130	0.135	0.116
3	−0.004	−0.003	0.002	0.123	0.144	0.104	0.123	0.133	0.099
4	−0.003	−0.002	−0.000	0.119	0.136	0.090	0.119	0.132	0.088
5	−0.001	0.001	−0.001	0.124	0.151	0.086	0.118	0.131	0.081
				$m = 100$					
1	0.005	−0.012	0.011	0.118	0.110	0.102	0.114	0.110	0.103
2	0.000	−0.002	0.004	0.095	0.097	0.078	0.093	0.096	0.080
3	−0.001	0.001	0.004	0.087	0.094	0.070	0.087	0.094	0.069
4	−0.001	0.002	0.004	0.085	0.093	0.063	0.085	0.094	0.062
5	0.002	0.006	0.004	0.103	0.132	0.060	0.084	0.093	0.057
				$m = 200$					
1	0.001	−0.008	0.002	0.079	0.078	0.069	0.080	0.077	0.071
2	−0.001	−0.004	0.001	0.067	0.068	0.054	0.065	0.067	0.056
3	−0.001	−0.002	0.000	0.063	0.066	0.048	0.061	0.066	0.048
4	−0.001	−0.001	−0.001	0.061	0.066	0.043	0.060	0.065	0.043
5	−0.001	−0.001	−0.000	0.061	0.066	0.040	0.059	0.065	0.040

Suppose that the r largest order statistics are available from m independent blocks:

$$\{(Y_{i1}, \cdots, Y_{ir}) : i = 1, \ldots, m\}.$$

The likelihood can be constructed from (1.12). The score function and the Fisher information matrix $mI(\theta)$ have been derived by Tawn (1988). The MLE $\widehat{\theta}_n$ behaves as usual when $\xi > −0.5$ (Smith, 1985); $\widehat{\theta}_n$ is consistent and asymptotically normal with variance $I^{-1}(\theta)/m$.

To illustrate the efficiency gain of using r largest order statistics relative to using block maxima only, a simulation study was carried out with random samples generated from G_5 with $\mu = 0$, $\sigma = 1$, and $\xi \in \{-0.25, 0.5, 1\}$. Three sample sizes were considered $m \in \{50, 100, 200\}$. Function rlarg.fit from R package ismev (Heffernan and Stephenson, 2012) was used to obtain the MLEs and their standard errors for $r \in \{1, \ldots, 5\}$. With 1000 replicates, summaries of the bias, empirical standard error (ESE), and average of the standard errors (ASE) based on the Fisher information are reported in Table 1.2 for $\xi = 0.5$.

From Table 1.2, the estimates are virtually unbiased. The empirical standard errors decrease as r increases, but the rate appears to slow down after $r = 3$ in this specific setting. The agreement between ESE and ASE, especially for the shape parameter, suggest that a large sample size ($m = 200$) is needed for the asymptotic variance of the MLE to take effect. The results for $\xi = -0.25$ are similar, but for $\xi = 1$, because the tail is rather heavy, a bigger sample size is needed for better agreement between ESE and ASE.

It is worth emphasizing, however, that the efficiency gain of larger r in the simulation study may not be realizable in practice because of the bias-variance tradeoff. A smaller r means less information and higher variance of the estimator, while a larger r means stronger distributional assumption which may not be supported by the data, and, hence, larger bias. Ideally, one would like to select r as large as possible subject to adequate model diagnostics such as goodness-of-fit test. Bader et al. (2015) proposed a sequential approach to select r based on model specification tests.

1.4.3 Peaks-over-Threshold (POT)

The peaks-over-threshold (POT) approach selects peaks over a higher threshold from the initial observations, and estimates the tail behavior using the conditional distribution of these exceedances. which is the GPD, as justified in Section 1.3. Suppose that there are k peaks over a threshold u, denoted as $X_1, \ldots, X_k$, from the initial n i.i.d. observations. Let $Y_i = X_i - u$, $i = 1, \ldots, k$, be the exceedances. The density of Y_i from the survival function (1.8) is

$$h(y|\sigma_u, \xi) = \begin{cases} \frac{1}{\sigma_u} \left(1 + \frac{\xi y}{\sigma_u}\right)_+^{-1/\xi} & \xi \neq 0, \\ \frac{1}{\sigma_u} \exp(-y/\sigma_u) & \xi = 0, \end{cases} \tag{1.13}$$

where $y \geq 0$ for $\xi \geq 0$ and $0 \leq y \leq -\sigma_u/\xi$ for $\xi < 0$. The parameters (σ_u, ξ) can be estimated by maximizing the likelihood constructed from (1.13). Similar to evaluation of the GEV density, this density needs to be evaluated with care for $|\xi| \to 0$ (e.g., using `log1p` in R).

The same problem with the MLE for block maxima retains. The usual properties of the MLE only hold when $\xi > -0.5$; the asymptotic variance does not hold when $-1 < \xi \leq -0.5$; and the MLE does not exist when $\xi < -1$ (Smith, 1985). The PWM or L-moments method can be used with the implicit restriction $\xi < 1$ (Hosking, 1990; Hosking et al., 1985). The MPS method also has the same appealing properties as in the case of block maxima (Wong and Li, 2006, Theorem 3.2). Castillo and Hadi (1997) proposed an elemental percentile method, which consists of two steps: first, some elemental estimates are obtained by solving equations relating the distribution function to their percentiles for some elemental subsets of the observations; second, the elemental estimates are combined in a suitable way. This method is easy to compute, giving unique, well-defined estimates in cases where other classical estimators often fail.

Estimation of (σ_u, ξ) is only half way toward estimating the tail behaviors such as return level or VaR, because the GPD is conditional on exceeding the threshold. From (1.8),

$$\Pr(X > u + y) = \eta_u \left(1 + \frac{\xi y}{\sigma_u}\right)^{-1/\xi},$$

where $\eta_u = \Pr(X > u)$ can be estimated by $\widehat{\eta}_u = k/n$. Suppose that there are b_s observations per block or time period (e.g., year), the T-period return level x_T is the solution to

$$\eta_u \left[1 + \frac{\xi(x_T - u)}{\sigma_u}\right]^{-1/\xi} = \frac{1}{Tb_s}.$$

For $\xi > -0.5$, inferences about x_T can be done by combining inferences of η_u and (σ_u, ξ) with the delta method. Interval estimators constructed from appropriate profile likelihood are again preferred for their precision, and a simple approximation ignoring the uncertainty in $\widehat{\eta}_u$ can be obtained by parametrizing σ_u in terms of x_T (Coles, 2001, p.83).

A critical step in the POT method is the selection of threshold u, which again is a compromise between bias and variance. Too low a threshold may violate the asymptotic assumptions of the model while a higher threshold gives fewer extreme events with which the estimator will have greater variance. One needs to choose as low a threshold as possible subject to that the limiting model provides a valid approximation of the data. A widely used exploratory method is the mean excess plot. When $\xi < 1$, the mean of the GPD in (1.8) is

$$\frac{\sigma_u}{1-\xi} = \frac{\sigma + \xi(u - \mu)}{1-\xi},$$

which is a linear function of u. This is the mean of the excess of the threshold u for the initial i.i.d. observations X, $E(X - u|X > u)$. By plotting the sample mean of the excesses over u against u, along with their confidence intervals, one can select the point u_0 beyond which the curve appears to be linear. A similar method is to plot the estimates $(\widehat{\sigma}_u, \widehat{\xi})$ over a grid of u values, and select the point u_0 beyond which $\widehat{\xi}$ is stable and $\widehat{\sigma}_u$ is linear.

A recent review on threshold selection is Scarrott and MacDonald (2012). Chapter 4 provides a comprehensive treatment on this subject in this volume.

1.4.4 Point Process

The point process method bears much resemblance to the POT method. It also works with the initial observations and a high threshold u. It is based on the point process characterization of the extremes, a nonhomogeneous Poisson process (Pickands, 1971). Consider i.i.d. variables $X_1, \ldots, X_n$ from distribution F such that (1.1) holds for some $a_n > 0$ and b_n with limiting GEV distribution G in (1.4) for some $\mu, \sigma > 0$, and ξ. Define the point process

$$P_n = \left\{ \left(\frac{i}{n+1}, \frac{X_i - b_n}{a_n} \right) : i = 1, \ldots, n \right\}$$

on $\mathbf{R}^2$. The probability of each point in P_n falling into region $A_u = (0,1) \times (u, \infty)$ is, from (1.7),

$$p_n \sim \frac{1}{n} \left[1 + \xi\left(\frac{u - \mu}{\sigma}\right)\right]^{-1/\xi}.$$

Let $N_n(A)$ be the number of points of P_n in region A. Because X_i's are i.i.d., $N_n(A_u)$ follows a binomial distribution $\text{Bin}(n, p_n)$. As $n \to \infty$, the binomial distribution converges to a Poisson distribution with rate

$$\Lambda(A_u) = \left[1 + \xi\left(\frac{u - \mu}{\sigma}\right)\right]^{-1/\xi}.$$

Note that the rate is homogeneous in the time direction. Therefore, this is exactly the characterizing property of a nonhomogeneous Poisson process with intensity function

$$\lambda(t, v) = \frac{1}{\sigma}\left[1 + \xi\left(\frac{u - \mu}{\sigma}\right)\right]^{-1/\xi - 1}. \tag{1.14}$$

Smith (1989) proposed statistical inferences with the point process characterization (for possibly nonstationary extremes). Absorbing a_n and b_n into μ and σ, the point process

$$\left\{\left(\frac{i}{n + 1}, X_i\right) : i = 1, \dots, n,\right\}$$

is approximately a nonhomogeneous Poisson process over A_u with intensity function (1.14) for large u. Selection of u is the same problem as in the POT method. For a suitably selected u, the likelihood of the Poisson process is

$$L_u(\theta) = \exp\{-n_b\Lambda(A_u)\} \prod_{X_i > u} n_b\lambda\left(\frac{i}{n + 1}, X_i\right), \tag{1.15}$$

where n_b is the number of blocks or periods (e.g., years) to which the inference is of interest. For $\xi > -0.5$, the usual likelihood inference can be carried out to obtain MLE of θ and their standard errors. Confidence intervals can be constructed for θ with the profile likelihood method, but for return levels, it is impractical because of no explicit reparametrization in terms of return level. Approximate confidence intervals can be obtained from simulation (Coles, 2001, Section 7.8). Covariates can be incorporated into, for example, the location parameter μ to model nonstationary extremes.

The point process method is closely linked to the POT method. The POT method is connected to the point process method through
1. the scale parameter $\sigma_u = \sigma + \xi(u - \mu)$;
2. the probability of exceedances

$$\eta_u = \Pr(X > u) \sim \frac{1}{n}\left[1 + \xi\left(\frac{u - \mu}{\sigma}\right)\right]^{-\frac{1}{\xi}}.$$

Using these two connections, the point process likelihood can be shown to be the product of the GPD likelihood for exceedances and the binomial likelihood for the number of exceedances (e.g., Coles, 2001, Section 7.6). The two methods are therefore equivalent. As the binomial distribution converges to a Poisson distribution, the POT method is also known as the Poisson–GPD method. The point process method has the advantages that the model parameters are GEV parameters without dependence on threshold u, and that all the parameters are estimated in a single step. The invariance of all parameters to u makes it easy to allow varying threshold which is necessary for nonstationary series.

The point process characterization provides a general framework that unifies all the results presented in earlier sections. The limiting distribution of the sample maximum can be derived as

$$\Pr\left(\frac{M_n - b_n}{a_n} \leq x\right) \sim \Pr\left[N_n(A_x) = 0\right] \to \exp\left\{\left[1 + \xi\left(\frac{u - \mu}{\sigma}\right)\right]^{-1/\xi}\right\}.$$

The distribution of exceedances over a large threshold u can be derived, using the exponentially distributed waiting time of Poisson processes, as

$$\Pr\left(\frac{X_i - b_n}{a_n} > u + y \,\middle|\, \frac{X_i - b_n}{a_n} > u\right) = \frac{\Lambda(A_{u+y})}{\Lambda(A_u)} \to \left[1 + \frac{\xi y}{\sigma_u}\right]^{-1/\xi}.$$

The limiting distribution of the rth largest order statistic can be derived from

$$\Pr\left(\frac{M_n^{(r)} - b_n}{a_n} \leq x\right) \sim \Pr\left(N_n(A_x) \leqslant r - 1\right).$$

The joint distribution of the r largest order statistics can be derived from the likelihood (1.15) by setting the rth largest order statistic as u.

1.5 Further Reading

Model diagnosis can be done with quantile-quantile (QQ) plot visually or goodness-of-fit test for the GEV or GPD model. The Komolgorov–Smirnov test and the Anderson–Darling test can be used, with special tables developed in earlier works (e.g., Chandra et al., 1981; Laio, 2004; Stephens, 1977). As in any goodness-of-fit test procedure, the uncertainty in parameter estimation needs to be appropriately accounted for. Parametric bootstrap provides a general solution, and with modern computer power, it is not considered computing intensive as it used to be, especially for small-to-moderate sample sizes. A fast large sample alternative to parametric bootstrap is based on the multiplier central limit theorem (Kojadinovic and Yan, 2012).

The models and methods presented in this review are asymptotically justified based on i.i.d. assumption on the initial observations. In reality, the i.i.d. assumption is often violated as most data are collected sequentially over time and are temporally dependent. A review of EVT for stochastic processes is Leadbetter et al. (2012). Coles (2001, Chapter 5–6) gives an excellent, accessible treatment to EVT for dependent sequences and nonstationary sequences. Chapter 10 of Beirlant et al. (2004) and Chapter 7 of Embrechts et al. (1997) review extremes of time series.

In this volume, Chapter 4 reviews threshold selection in extreme value analysis; Chapter 5 treats the threshold as a modeling target and estimates it. Chapter 7 reviews statistical modeling for time series of extremes, while Chapter 8 summarizes the max-autoregressive and moving maxima models for extremes, the analog of the autoregressive moving average model. An important inference method, the Bayesian approach, is comprehensively treated in Chapter 13.

References

Arnold, B. C. and R. A. Groenveld (1995). Measuring skewness with respect to the mode. *The American Statistician 49*, 34–38.

Bader, B., J. Yan, and X. Zhang (2015). Selecting the number of largest order statistics in extreme value analysis. Technical Report 11, Department of Statistics, University of Connecticut.

Beirlant, J., Y. Goegebeur, J. Segers, and J. Teugels (2004). *Statistics of Extremes: Theory and Applications*. John Wiley & Sons.

Billingsley, P. (2008). *Probability and Measure*. John Wiley & Sons.

Castillo, E. and A. S. Hadi (1997). Fitting generalized Pareto distribution to data. *Journal of American Statistical Association 92*, 1609–1620.

Chandra, M., N. D. Singpurwalla, and M. A. Stephens (1981). Kolmogorov statistics for tests of fit for the extreme value and Weibull distributions. *Journal of American Statistical Association 76*, 729–731.

Cheng, R. C. H. and N. A. K. Amin (1983). Estimating parameters in continuous univariate distributions with a shifted origin. *Journal of the Royal Statistical Society: Series B (Statistical Methodology) 45*(3), 394–403.

Cheng, R. C. H. and M. A. Stephens (1989). A goodness-of-fit test using Moran's statistic with estimated parameters. *Biometrika 76*(2), 385–392.

Coles, S. (2001). *An Introduction to Statistical Modeling of Extreme Values*. Springer.

Coles, S. G. and M. J. Dixon (1999). Likelihood based inference for extreme value models. *Extremes 2*, 5–23.

de Haan, L. and A. Ferreira (2007). *Extreme Value Theory: An Introduction*. New York: Springer.

Embrechts, P., C. Klüppelberg, and T. Mikosch (1997). *Modelling Extremal Events*. Springer Science & Business Media.

Ferreira, A. and L. de Haan (2015). On the block maxima method in extreme value theory: PWM estimators. *The Annals of Statitics 43*(1), 276–298.

Fisher, R. A. and L. H. C. Tippett (1928). Limiting forms of the frequency distribution of the largest or smallest member of a sample. *Mathematical Proceedings of the Cambridge Philosophical Society 24*(2), 180–190.

Gnedenko, B. (1943). Sur la distribution limite du terme maximum d'une serie aleatoire. *Annals of Mathematics 44*, 423–453.

Greenwood, J. A., J. M. Landwehr, N. C. Matalas, and J. R. Wallis (1979). Probability weighted moments: Definition and relation to parameters of several distributions expressable in inverse form. *Water Resources Research 15*(5), 1049–1054.

Heffernan, J. E. and A. G. Stephenson (2012). *ismev: An Introduction to Statistical Modeling of Extreme Values*. R package version 1.39.

Hosking, J. R. M. (1990). *L*-moments: Analysis and estimation of distributions using linear combinations of order statistics. *Journal of the Royal Statistical Society: Series B (Statistical Methodology) 52*, 105–124.

Hosking, J. R. M. and J. R. Wallis (2005). *Regional Frequency Analysis: An Approach Based on L-Moments*. Cambridge University Press.

Hosking, J. R. M., J. R. Wallis, and E. F. Wood (1985). Estimation of the generalized extreme value distribution by the method of probability-weighted moments. *Technometrics 27*, 251–261.

Katz, R. W., M. B. Parlange, and P. Naveau (2002). Statistics of extremes in hydrology. *Advances in Water Resources 25*(8), 1287–1304.

Kojadinovic, I. and J. Yan (2012). Goodness-of-fit testing based on a weighted bootstrap: A fast large-sample alternative to the parametric bootstrap. *Canadian Journal of Statistics 40*(3), 480–500.

Laio, F. (2004). Cramer–von Mises and Anderson–Darling goodness of fit tests for extreme value distributions with unknown parameters. *Water Resources Research 40*, 1–10.

Leadbetter, M. R., G. Lindgren, and H. Rootzén (2012). *Extremes and Related Properties of Random Sequences and Processes*. Springer Science & Business Media.

Pickand, III, J. (1975). Statistical inference using extreme order statistics. *Annals of Statistics 3*, 119–131.

Pickands, III, J. (1971). The two-dimensional Poisson process and extremal processes. *Journal of Applied Probability 8*(4), 745–756.

Prescott, P. and A. T. Walden (1980). Maximum likelihood estimation of the parameters of the generalized extreme-value distribution. *Biometrika 67*, 723–724.

Ranneby, B. (1984). The maximum spacing method. An estimation method related to the maximum likelihood method. *Scandinavian Journal of Statistics 11*(2), 93–112.

Roy, D., V. Roy, and D. K. Dey (2013). Bayesian analysis of survival data under generalized extreme value distribution with application in cure rate model. Technical report, Department of Statistics, Iowa State University.

Scarrott, C. and A. MacDonald (2012). A review of extreme value threshold estimation and uncertainty quantification. *REVSTAT tatistical Journal 10*(1), 33–60.

Smith, R. L. (1985). Maximum likelihood estimation in a class of non–regular cases. *Biometrika 72*, 67–90.

Smith, R. L. (1986). Extreme value theory based on the r largest annual events. *Journal of Hydrology 86*(1), 27–43.

Smith, R. L. (1989). Extreme value analysis of environmental time series: An application to trend detection in ground-level ozone. *Statistical Science 4*(4), 367–377.

Southworth, H. and J. E. Heffernan (2013). *texmex: Threshold Exceedences and Multivariate Extremes*. R package version 2.1.

Stephens, M. A. (1977).   Goodness of fit for the extreme value distribution. *Biometrika 64*(3), 583–588.

Tawn, J. A. (1988).   An extreme-value theory model for dependent observations. *Journal of Hydrology 101*(1), 227–250.

Wang, X. and D. K. Dey (2010).   Generalized extreme value regression for binary response data: An application to B2B electronic payments system adoption. *The Annals of Applied Statistics 4*, 2000–2023.

Weissman, I. (1978).   Estimation of parameters and large quantiles based on the k largest observations. *Journal of the American Statistical Association 73*(364), 812–815.

Wong, T. S. T. and W. K. Li (2006).   A note on the estimation of extreme value distributions using maximum product of spacings.   In H.-C. Ho, C.-K. Ing, and T.-L. Lai (Eds.), *Time Series and Related Topics: In Memory of Ching-Zong Wei*. Institute of Mathematical Statistics.

2

Multivariate Extreme Value Analysis

Dipak Dey
University of Connecticut, United States

Yujing Jiang
University of Connecticut, United States

Jun Yan
University of Connecticut, United States

Abstract

The dependence among extreme observations is critical in decision making that involves multivariate extreme value analysis. Marginal generalized extreme value distributions coupled with a copula have been applied in practice, but the extreme dependence structure in such models may be limited if the copula is not an extreme value copula. Multivariate extreme value distributions are characterized jointly by marginal generalized extreme value distributions and an extreme-value copula. This chapter reviews multivariate extreme value distributions, extreme value copulas, measures of extremal dependence, multivariate generalized Pareto distributions, semiparemetric conditional dependence models, and their statistical inferences. The multivariate generalized Pareto distribution characterizes the multivariate exceedances over certain thresholds conditioning on that there is at least one marginal exceedance. In the conditional approach, each variable is taken in turn as a conditional variate, given a large value of which, the remaining variables are modeled by their limiting distribution. Parameter estimation and goodness-of-fit test are outlined with references to relevant chapters.

2.1   Multivariate Extreme Value Distribution

2.1.1   Limiting Standardization

A d-dimensional multivariate extreme value distribution (MEVD) arises as the limiting distribution of the componentwise maxima of a random sample from a d-dimensional multivariate distribution. Let $\{X_i = (X_{i,1}, \ldots, X_{i,d}) : i = 1, \ldots, n\}$ be a random sample of a d-dimensional random vector with joint distribution function F, marginal distribution functions $F_1, \ldots, F_d$, and copula C_F. Let $M_{n,j} = \max_{i=1}^{n} X_{ij}$, $j = 1, \ldots, d$, be the componentwise maxima. As in the univariate case, the distribution of component minima can be obtained from the distribution of the component maxima of the negated variables, so we focus on maxima in the sequel. For each margin $j = 1, \ldots, d$, assume suitable normalizing sequence $a_n = (a_{n,1}, \ldots, a_{n,d}) > 0$ and $b_n = (b_{n,1}, \ldots, b_{n,d})$ such that the limiting marginal distribution of $Z_{n,j} = (M_{n,j} - b_{n,j})/a_{n,j}$ is a non-degenerate generalized extreme value (GEV) distribution as $n \to \infty$. Our interest here is the limiting joint distribution G of $\boldsymbol{Z}_n = (Z_{n,1}, \ldots, Z_{n,d})$, which is an MEVD. Specifically, for $\boldsymbol{z} \in \mathbb{R}^d$,

$$\Pr(\boldsymbol{Z}_n \leq \boldsymbol{z}) = F^n(\boldsymbol{az} + \boldsymbol{b}) \to G(\boldsymbol{z}), \tag{2.1}$$

where and throughout the vector operations are componentwise. Distribution F is said to be in the domain of attraction of G, denoted by $F \in \mathcal{D}(G)$.

MEVDs can be characterized by max-stability. A d-dimensional distribution function G is max-stable if for every integer $k \geq 1$, there exist vectors $\boldsymbol{\alpha}_k > 0$ and $\boldsymbol{\beta}_k$ such that

$$G^k(\boldsymbol{\alpha}_k \boldsymbol{z} + \boldsymbol{\beta}_k) = G(\boldsymbol{z}) \tag{2.2}$$

for all $\boldsymbol{z} \in \mathbb{R}^d$. It is trivial to see that a max-stable distribution G is in its own domain of attraction and is an MEVD. The converse is also true which can be seen as follows. If G is an MEVD, then by definition there exists F, $\boldsymbol{a}_n$ and $\boldsymbol{b}_n$, such that $F^n(\boldsymbol{a}_n \boldsymbol{z} + \boldsymbol{b}_n) \to G(\boldsymbol{z})$. This gives $F^{nk}(\boldsymbol{a}_n \boldsymbol{z} + \boldsymbol{b}_n) \to G^k(\boldsymbol{z})$ and $F^{nk}(\boldsymbol{a}_{nk} \boldsymbol{z} + \boldsymbol{b}_{nk}) \to G(\boldsymbol{z})$ for some $\boldsymbol{a}_{nk}$ and $\boldsymbol{b}_{nk}$. Then choosing $\boldsymbol{\alpha}_k = \lim_{n \to \infty} \boldsymbol{a}_{nk}/\boldsymbol{a}_n$ and $\boldsymbol{\beta}_k = \lim_{n \to \infty} (\boldsymbol{b}_{nk} - \boldsymbol{b}_n)/\boldsymbol{a}_n$ will make (2.2) hold for every $k \geq 1$.

2.1.2   Extreme-Value Copula

The dependence structure of an MEVD G is characterized by its copula C_G, which has to satisfy certain properties due to the limiting argument in (2.1) — the copula C_G of an MEVD G has to be an extreme-value copula. A copula C is an extreme copula if there exists a copula C_F such that, as $n \to \infty$,

$$C_F^n(\boldsymbol{u}^{1/n}) \to C(u_1, \ldots, u_d), \qquad (u_1, \ldots, u_d) \in [0,1]^d.$$

In this case, C_F is said to be in the domain of attraction of C. From Sklar's theorem (Sklar, 1959), the copula C_n of $\boldsymbol{M}_n$ (and $\boldsymbol{Z}_n$, because $\boldsymbol{Z}_n$ is a monotone increasing transformation of M_n) is

$$C_n(u) = C_F^n(u^{1/n}) \to C_G(u) \qquad u \in [0,1]^d,$$

where C_G is an extreme-value copula. For example, the t-copula is in the domain of attraction of the t-EV copula with the same degrees of freedom (Demarta and McNeil, 2005). A simulation study with componentwise maxima from samples of multivariate t distribution showed that the convergence of the copula is acceptable for a sample size as small as 100 for a few lower degrees of freedom, while the convergence of the marginal maximum to Fréchet distribution needs a sample size of 500 or more for degrees of freedom 3, and much higher as the degrees of freedom increase (Shang et al., 2015, Table 3 in supplementary materials).

Extreme-value copulas are also characterized by max-stability. A copula C is max-stable if it satisfies

$$C(u) = C^k(u^{1/k}), \qquad u \in [0,1]^d, \tag{2.3}$$

for all integer $k \geq 1$. A max-stable copula is in the domain of attraction of its own. A copula is an extreme-value copula if and only if it is max-stable (e.g., Gudendorf and Segers, 2010, Theorem 2.1). By definition, the family of extreme-value copulas or max-stable copulas coincides with the set of copulas of MEVDs.

An extreme-value copula C can be expressed in terms of the Pickands dependence function (Pickands, 1981). Let Δ_{d-1} be the unit simplex in $\mathbb{R}^d$ defined as

$$\Delta_{d-1} = \{(w_1, \ldots, w_d) : w_i > 0, i = 1, \ldots, d, \sum_{i=1}^d w_i = 1\}.$$

Then C can be expressed as

$$C(u_1, \ldots, u_d) = \exp\left\{ \left(\sum_{j=1}^d \log u_j \right) A \left(\frac{\log u_1}{\sum_{j=1}^d \log u_j}, \cdots, \frac{\log u_d}{\sum_{j=1}^d \log u_j} \right) \right\}, \tag{2.4}$$

where A, known as the Pickands dependence function, is $\Delta_{d-1} \to [1/d, 1]$, convex and satisfies $\max(w) \leq A(w) \leq 1$ for all $w \in \Delta_{d-1}$. The conditions on A are necessary for C to be an extreme-value copula. For $d = 2$, they are sufficient and necessary (Tawn, 1990).

Marginal GEV distributions can be coupled to form a multivariate distribution with a non-extreme-value copula by Sklar's theorem. For example, Gaussian copulas have been used for dependence among extremes in hydrological applications (Renard and Lang, 2007, e.g.,). The tails of Gaussian copulas, however, are asymptotically independent, offering no extreme value dependence; see Section 2.2. This does not prevent non-extreme-value copulas from being useful. The copula C_n with $n < \infty$ can still be of interest; see Chapter 6 on block maxima of vine copulas.

2.1.3   Max-Stable Distribution with Unit Fréchet Margins

Representations of MEVDs are often stated with unit Fréchet margins whose distribution function is $G_1(z) = \exp(-1/z)$ for $z > 0$. A multivariate distribution G is an MEVD with unit Fréchet margins if and only if

$$G(z) = \exp(-V(z)), \quad z \in [0, \infty)^d \backslash \{0\}, \tag{2.5}$$

where

$$V(z) = \int_{\Delta_{d-1}} \max_{j=1,\ldots,d} \left(\frac{w_j}{z_j} \right) \mathrm{d}S(w_1, \ldots, w_d), \tag{2.6}$$

for a positive measure S defined on Δ_{d-1} satisfying $\int_{\Delta_{d-1}} w_j \mathrm{d}S(w_1, \ldots, w_d) = 1$, $j = 1, \ldots, d$ (de Haan and Resnick, 1977; Pickands, 1981). Measure S is called the spectral measure, which characterizes the dependence structure of G.

The max-stability (2.2) of an MEVD G implies that $G^{-1/k}$ is a distribution function for every positive integer k. This means that G is max-infinitely divisible (Balkema and Resnick, 1977). Then there exists a measure μ on $[0, \infty)^d \backslash \{0\}$ such that

$$V(z) = \mu((0, z]^c), \quad z \in [0, \infty)^d \backslash \{0\},$$

where B^c is the complement of a set B. This measure μ is known as the exponent measure and function V the exponent function. Transforming the max-stability of extreme-value copulas (2.3) to MEVDs with Fréchet margins gives $G^r(rz) = G(z)$ for all $r > 0$. Equivalently, $V(r \cdot) = r^{-1} V(\cdot)$ and $\mu(r \cdot) = r^{-1} \mu(\cdot)$; that is, both V and μ are homogeneous of order -1.

The stable tail dependence function $l : [0, \infty]^d \to [0, \infty)$ is a closely related concept defined as

$$l(v) = V(1/v), \quad v \in [0, \infty]^d, \tag{2.7}$$

It is easy to see that the stable tail dependence function l satisfies the following properties (e.g., Beirlant et al., 2004, p.257):

1. $l(r \cdot) = r \, l(\cdot)$ for $r > 0$ (homogeneous of order 1);
2. $l(e_j) = 1$ for $j = 1, \ldots, d$ with e_j being the jth unit vector in $\mathbb{R}^d$;
3. $\max_{i=1}^d v_i \le l(v) \le \sum_{i=1}^d v_i$ for $v \in [0, \infty)^d$;
4. l is convex.

Note that the lower and upper bounds in the 3rd property are themselves stable tail dependence functions corresponding to complete dependence and independence, respectively. With homogeneity of order 1, l is connected to the Pickands dependence function A, which is simply the restriction of l to the unit simplex, through

$$l(x) = \left(\sum_{i=1}^d x_i \right) A(w_1, \ldots, w_d), \quad x \in \mathbb{R}_+^d, \tag{2.8}$$

where $w_j = x_j / \sum_{i=1}^d x_i$.

A number of parametric families of MEVDs have appeared in the literature. Gudendorf and Segers (2010) reviewed a list of extreme value copulas. Commonly used examples are the logistic model (and its variants), the Hüsler–Reiss model, and the t-extreme-value copula. See Chapter 16 on extreme dependence models for details.

2.2 Measures of Extremal Dependence

The dependence structure of an MEVD is characterized by the spectral measure S, stable tail dependence function l, or the Pickands dependence function A, which have complex structures which cannot be easily inferred from data. Simple summaries of the dependence are useful. For the bivariate case, dependence measures such as Kendall's tau or Spearman's rho can be expressed, for example, in terms of A (Fougeres, 2004; Ghoudi et al., 1998):

$$\tau = \int_0^1 \frac{t(1-t)}{A(t)} \, dA'(t), \qquad \rho = 12 \int_0^1 \{A(t)+1\}^{-2} dt - 3.$$

They measure the overall dependence of the two margins. It is often more interesting to measure the dependence of extreme observations at the tails.

2.2.1 Upper Tail Dependence

For a bivariate uniform vector (U_1, U_2) with copula C, define

$$\chi(u) = \Pr(U_1 > u | U_2 > u) = \frac{\bar{C}(u, u)}{1 - u} = 2 - \frac{1 - C(u, u)}{1 - u}, \quad u \in [0, 1]$$

where $\bar{C}(u_1, u_2)$ is the survival copula of C defined by $C(u_1, u_2) = \Pr(U_1 > u_1, U_2 > u_2)$. Joe (1993) defined the upper tail dependence parameter of C as $\chi = \lim_{u \to 1} \chi(u)$, where $0 \leq \chi \leq 1$. Asymptotic independence is equivalent to $\chi = 0$. Asymptotic dependence is achieved if and only if $\chi > 0$. The upper tail dependence parameter of a distribution F is linked to the domain of attraction G of this distribution. Specifically, if $F \in \mathcal{D}(G)$, then the upper tail parameters associated to the distributions F and G are the same (Joe, 1997, Theorem 6.8).

Function $\chi(u)$ is a quantile-dependent measure of dependence. The sign of $\chi(u)$ indicates if the two variables are positively or negatively associated at the quantile level u. Its range is $2 - \log(2u - 1)/\log(u) \leqslant \chi(u) \leqslant 1$ where the lower bound is interpreted as $-\infty$ for $u \leqslant \frac{1}{2}$, and 0 for $u = 1$ (Coles et al., 1999). If G is a bivariate extreme value distribution, then $\chi(u)$ is constant and it is easy to show that $\chi(u) = 2 - l(1, 1)$, leading to $\chi = 2 - l(1, 1)$, where l is the stable tail dependence function defined in (2.7).

2.2.2 Complementary Upper Tail Dependence

One disadvantage of χ and $\chi(u)$ is that, for two asymptotically independent variables, the null value of χ does not provide any information on the strength of the dependence at the sub-asymptotic level, and $\chi(u)$ can be erroneously interpreted. Coles et al. (1999) proposed a measure of dependence $\bar{\chi}$ which is useful in the case of asymptotic independence:

$$\bar{\chi} = \lim_{u \to 1} \bar{\chi}(u) \equiv \lim_{u \to 1} \frac{2 \log(1-u)}{\log \bar{C}(u, u)} - 1,$$

where $-1 < \bar{\chi} \leqslant 1$, for all $u \in [0, 1]$. This measure complements the upper tail dependence χ in that asymptotic dependence is equivalent to $\bar{\chi} = 1$, and that for asymptotic independence, $\bar{\chi}$ provides a whole range of dependence measures whose magnitude increases with dependence strength. Together, $(\chi, \bar{\chi})$ provides a fuller summary of the extremal dependence: $(\chi > 0, \bar{\chi} = 1)$ means asymptotic dependence, in which case, χ measures the strength of the dependence; $(\chi = 0, \bar{\chi} < 1)$ means asymptotic independence, in which case $\bar{\chi}$ measures the strength of the dependence.

2.2.3   Extremal Coefficient

Extremal coefficient was first introduced by Buishand (1984). For the copula C of a bivariate random vector, define

$$\theta_C(u) = \frac{\log C(u, u)}{\log u} \qquad u \in [0, 1].$$

The extremal coefficient is $\theta = \lim_{u \to 1} \theta(u)$. From the definition, the connection to the upper tail dependence parameter χ is clear, $\theta = 2 - \chi$. For a bivariate MEVD G with unit Fréchet margins, θ is such that

$$G(z, z) = \exp(-\theta/z), \qquad z > 0.$$

The extremal coefficient has range $[1, 2]$, with the lower and upper ends indicating, respectively, independence and perfect dependence. This measure can be generalized to the d-dimensional multivariate case with range $[1, d]$ (Schlather and Tawn, 2003). It can be expressed in terms of the Pickands dependence function A as $dA(1/d, \ldots, 1/d)$ (e.g., Kojadinovic et al., 2015).

2.2.4   Coefficient of Tail Dependence

Similar to χ, the extremal coefficient provides no information about the dependence in the case of asymptotic independence. To overcome this, Ledford and Tawn (1996, 1997) defined a coefficient of tail dependence η. With (Z_1, Z_2) having unit Fréchet margins, the joint survival function satisfies the asymptotic condition

$$\Pr(Z_1 > z, Z_2 > z) \sim \mathcal{L}(z)[\Pr(Z_1 > z)]^{1/\eta} \sim \mathcal{L}(z) z^{-1/\eta}$$

for large z, where $\mathcal{L}$ is a slowly varying function (such that for all $a > 0$, $\lim_{z \to \infty} \mathcal{L}(az)/\mathcal{L}(z) = 1$) and $0 < \eta \leq 1$ is the coefficient of tail dependence. The strength of the dependence is determined through η and $\mathcal{L}(\cdot)$. The η parameter determines the decay rate of survival function for large z and characterizes the nature of the tail dependence. There are four classes of dependence (Ledford and Tawn, 1997):

1. asymptotic dependence: $\eta = 1$ and $\mathcal{L}(\cdot) \not\to 0$;
2. negative association: $0 < \eta < \frac{1}{2}$;
3. positive association: $\frac{1}{2} < \eta < 1$;
4. near independence: $\eta = \frac{1}{2}$, $\mathcal{L}(\cdot) \geqslant 1$ (independence corresponds to $\mathcal{L}(x) \equiv 1$).

The last three cases are all asymptotic independence, and η measures the strength of dependence in the asymptotic independence. The coefficient of tail dependence is connected to the complementary upper tail dependence $\bar{\chi}$ through $\bar{\chi} = 2\eta - 1$ (Coles et al., 1999, Section 4.1). Heffernan (2000) provided a directory of coefficient of tail dependence for a wide range of bivariate distributions. Many commonly used distributions have little flexibility in the range of limiting dependence structure.

2.3 Multivariate Generalized Pareto Distributions

The univariate generalized Pareto distribution (GPD) is the asymptotic conditional distribution of a GEV variable given that it exceeds a large threshold; see Chapter 1. To extend the GPD to the multivariate case, the first question one needs to consider is how to define the conditioning event. Various forms of extensions have been studied (e.g., Falk and Guillou, 2008; Ferreira and de Haan, 2014; Ledford and Tawn, 1996). The one presented here is from Rootzén and Tajvidi (2006), where a multivariate generalized Pareto distribution (MGPD) is defined conditional on that at least one component of the multivariate vector is large.

A d-dimensional distribution function H is an MGPD if

$$H(\boldsymbol{x}) = \frac{1}{-\log G(\boldsymbol{0})} \log \frac{G(\boldsymbol{x})}{G(\boldsymbol{x} \wedge \boldsymbol{0})}, \quad \boldsymbol{x} \in \mathbb{R}^d, \tag{2.9}$$

for some MEVD G with non-degenerate margins and with $0 < G(\boldsymbol{0}) < 1$, where $\boldsymbol{a} \wedge \boldsymbol{b}$ is the componentwise minimum of vectors $\boldsymbol{a}$ and $\boldsymbol{b}$. In particular, $H(\boldsymbol{x}) = 0$ for $\boldsymbol{x} < \boldsymbol{0}$ and $H(\boldsymbol{x}) = 1 - \log G(\boldsymbol{x}) / \log G(\boldsymbol{0})$ for $\boldsymbol{x} > \boldsymbol{0}$.

Rootzén and Tajvidi (2006) give two motivations of this definition. First, exceedances (of suitably coordinated thresholds) asymptotically have an MGPD if and only if the limiting distribution of componentwise sample maxima is MEVD (Rootzén and Tajvidi, 2006, Theorem 2.1). Specifically, let $\boldsymbol{X} = (X_1, \ldots, X_d)$ be a d-dimensional random vector with distribution function F and define $\bar{F} = 1 - F$ as the tail function of F. Let $\{\boldsymbol{u}(t) = (u_1(t), \ldots, u_d(t)) | t \in [1, \infty)\}$ be a d-dimensional curve starting at $\boldsymbol{u}(1) = \boldsymbol{0}$, and $\boldsymbol{\sigma}(u) = \boldsymbol{\sigma}(\boldsymbol{u}(t)) = (\sigma_1(u_1(t)), \ldots, \sigma_d(u_d(t))) > \boldsymbol{0}$ a function with values in $\mathbb{R}^d$. Consider the vector of normalized exceedances of the thresholds $\boldsymbol{u}$

$$\boldsymbol{X}_{\boldsymbol{u}(t)} = \left(\frac{X_1 - u_1(t)}{\sigma_1(u_1(t))}, \ldots, \frac{X_d - u_d(t)}{\sigma_d(u_d(t))} \right).$$

If $F \in \mathcal{D}(G)$ with $0 < G(\boldsymbol{0}) < 1$, then there exists an increasing continuous curve

u with $F\big(u(t)\big) \to 1$ as $t \to \infty$ and a function $\sigma(u) > 0$ such that

$$\Pr(X_u \leqslant x | X_u \not\leqslant 0) \to \frac{1}{-\log G(0)} \log \frac{G(x)}{G(x \wedge 0)}$$

as $t \to \infty$ for all $x \in \mathbb{R}^d$. The components of the curve $\{u(t)\}$ specify how the threshold increases "in a suitably coordinated way"; the asymptotic distributions can differ for different relations between the levels.

Conversely, suppose that there exists an increasing continuous curve u with $F(u(t)) \to 1$ as $t \to \infty$, and a function $\sigma(u) > 0$ such that $\Pr(X_u \leqslant x | X_u \not\leqslant 0) \to H(x)$ for some function H and for $x > 0$, where the marginals of H on $\mathbb{R}_+$ are non-degenerate. Then $\Pr(X_u \leqslant x | X_u \not\leqslant 0)$ converges to a limit $H(x)$ for all x and there is a unique MEVD G with $G(0) = e^{-1}$ such that

$$H(x) = \log \frac{G(x)}{G(x \wedge 0)},$$

$G(x) = e^{-\bar{H}(x)}$ for $x > 0$, and $F \in D(G)$.

The second motivation of the distribution H in (2.9) is that it is the only one which is preserved under (a suitably coordinated) change of exceedance levels (Rootzén and Tajvidi, 2006, Theorem 2.2). The theorem also has two directions. If X has an MGPD, then there exists an increasing continuous curve u with $F(u(t)) \to 1$ as $t \to \infty$, and a function $\sigma(u) > 0$ such that

$$\Pr(X_u \leqslant x | X_u \not\leqslant 0) \to \Pr(X \leq x), \tag{2.10}$$

for $t \in [1, \infty)$ and all x. Conversely, if there exists an increasing continuous curve u with $\Pr\big(X < u(t)\big) \to 1$ as $t \to \infty$, and a function $\sigma(u) > 0$ such that (2.10) holds for $x > 0$, and X has non-degenerate margins, then X has an MGPD.

Of note on lower-dimensional marginal distributions of the MGPD is that the marginal distribution of any single component of a random vector with an MGPD does not have a univariate GPD. This may seem undesired at first glance but it makes sense. The reason is that the marginal distribution of one component of H is the asymptotic conditional distribution of that component given that at least one of the d components is large. In contrast, a univariate GPD is the asymptotic distribution of a random variable given that itself is large.

An advantage of the MGPD in (2.9) is that it allows to utilize more of the data and better estimation of parameters because it is defined for all $x \not\leqslant 0$ — partially negative components in x also contribute to estimation. This is different from the MGPDs developed from the point process approach (Coles and Tawn, 1991; Joe et al., 1992) where only values of $x > 0$ are modeled parametrically. Further, the MGPD provides information for the "time structure" — whether the component maxima have occurred simultaneously or not — which is not available from multivariate block maxima. This distribution has been used in bivariate extremes (Rakonczai and Zempléni, 2012) and pairwise likelihood method for spatial extremes (Bacro and Gaetan, 2014). Ferreira and de Haan (2014) further extended the MGPD to the generalized Pareto process, which is yet to be applied to data from stochastic processes.

2.4 Semiparametric Conditional Dependence Model

To overcome the limited flexibility provided by parametric MEVDs and the restriction that all components become large at the same rate in existing approaches, Heffernan and Tawn (2004) proposed a semiparametric conditional approach that is different from all preexisting methods. The approach is based on the asymptotic structure of the conditional distributions arising from a d-dimensional random vector $X = (X_1, \ldots, X_d)$ with standard Gumbel margins. Let X_{-i} be X without the ith component. Consider the limiting conditional distribution of X_{-i}, with appropriate standardization, given $X_i = x_i$ for a large value x_i.

Assume that for any given i, there exists vector normalizing functions $a_{|i}$ and $b_{|i}$, both $R \to \mathbb{R}^{d-1}$, such that for all $y_i \geq u_i$ with a high threshold u_i, the standardized variables

$$Z_{|i} = \frac{X_{-i} - a_{|i}(y_i)}{b_{|i}(y_i)} \tag{2.11}$$

converge in distribution,

$$\lim_{x_i \to \infty} \Pr(Z_{|i} \leq z_{|i} | Y_i = y_i) = G_{|i}(z_{|i}), \tag{2.12}$$

for any sequence of y_i values such that $y_i \to \infty$, where all margins of the limiting distribution $G_{|i}$ are non-degenerate. This assumption implies that, conditionally on $Y_i > u_i$, as $u_i \to \infty$, the variables $Y_i - u_i$ and $Z_{|i}$ are independent in the limit with limiting marginal distributions being unit exponential and $G_{|i}$, respectively (Heffernan and Tawn, 2004, p.503).

The normalizing functions a, b, and the limiting distribution $G_{|i}$ characterize the joint tail behavior of X_{-i} given $X_j = x_j > u$ for large u. Heffernan and Tawn (2004) proposed parametric forms for the normalizing functions a and b:

$$a_{|i}(y) = \alpha_{|i} y + I_{\{\alpha_{|i}=0, \beta_{|i}<0\}} (c_{|i} - d_{|i} \log y),$$

$$b_{|i}(y) = y^{b_{|i}},$$

where $\alpha_{|i} \in [0,1]^{d-1}$, $\beta_{|i} \in (-\infty, 1)^{d-1}$, $c_{|i} \in \mathbb{R}^{d-1}$, and $d_{|i} \in [0,1]^{d-1}$ are vector parameters, and I is the indicator function. This form covers a wide range of known dependence structures including all the copula models studied in Heffernan (2000). The distribution $G_{|i}$ is unspecified, and can be estimated by the empirical distribution of $Z_{|i}$ after the parameters in the normalizing functions $a(\cdot)$ and $b(\cdot)$ are estimated.

The conditioning distribution in this approach is to be distinguished from the MGPD in Section 2.3. The MGPD is the conditional distribution of a random vector given that at least one component of the random vector is large. In contrast, the distribution in (2.12) is the limiting conditional distribution of the remaining components given a large value in a single component of a random vector. Both approaches allow more information from the data than merely block maxima for improved efficiency in inferences. An extension to allow hierarchical structured parameters with a financial application is in Chapter 19.

2.5 Statistical Inferences

Statistical inferences for multivariate extremes can be based on block maxima data or exceedance data. Block maxima data are assumed to follow an MEVD. For exceedance data, it is assumed that the full data come from a distribution in the domain of attraction of an MEVD which is the target of inference. The marginal distributions and the dependence structures are often separated in inferences in light of Sklar's theorem. The topics in this section are not necessarily mutually exclusive. Chapters dedicated to these topics will be referred to. More details are presented to the threshold approach based on composite likelihood and the conditional approach of Heffernan–Tawn.

2.5.1 Likelihood Approach with Block Maxima

If a random sample of block maxima is available and a parametric MEVD is specified, the maximum likelihood estimator is tempting. Nonetheless, the explicit joint densities of parametric MEVDs are hardly available for $d \geq 3$ because of the combinatorial explosion as the dimension d increases when differentiating the distribution function (2.5) when the exponent function V has a known expression; see also Chapter 14. It would be even less possible when V is not available explicitly.

To overcome the difficulty, the composite likelihood approach has been used, which composes pieces of lower dimensional likelihood information to form an objective function and maximize it to estimate the parameters. The composite likelihood method as a general methodology was first proposed by Lindsay (1988) and has found wide applications wherever full likelihood evaluation is unavailable; see Varin et al. (2011) for a recent review. The composite likelihood approach has been applied in the form of pairwise likelihood, constructed from bivariate marginal densities, for high-dimensional data in spatial extreme analysis (e.g., Davison and Gholamrezaee, 2012; Padoan et al., 2010; Smith and Stephenson, 2009). Chapter 12 provides a comprehensive treatment.

2.5.2 Threshold Approach

Extending the threshold approach to the multivariate case needs to determine how to define exceedances and what distribution they follow. Bacro and Gaetan (2014) considered two different threshold exceedances and their corresponding conditional distributions. In either approach, because the joint density is tractable only for bivariate marginals, pairwise likelihood is used for inferences. Consider a d-dimensional random vector $Y = (Y_1, \ldots, Y_d)$ with joint distribution F in the domain of attraction of G. To focus on the dependence modeling, suppose that the marginals have been transformed to the unit Fréchet or standard Gumbel scale.

The first approach defines a bivariate exceedance as both components exceeding a large threshold u (Ledford and Tawn, 1996). Suppose that (Y_i, Y_j) have unit

Fréchet margins. Their joint distribution F_{ij} is in the domain of attraction of the corresponding bivariate margin G_{ij} of G. Therefore, $F_{ij}(y_i, y_j)$ can be approximated on $(u, \infty)^2$ as

$$F_{ij}(y_i, y_j) \approx G_{ij}(y_i, y_j). \tag{2.13}$$

The second approach defines a bivariate exceedance as at least one component exceeding a large threshold, which is the setting of the MGPD of Rootzén and Tajvidi (2006) in Section 2.3. Suppose that (Y_i, Y_j) have standard Gumbel margins. There exist increasing continuous curves $(u_i(\cdot), u_j(\cdot))$ such that, for $t \to \infty$, $F_{ij}(u_i(t), u_j(t)) \to 1$ and positive functions $\sigma_i(\cdot)$ and $\sigma_j(\cdot)$ such that the conditional distribution of the standardized pairs given the exceedance converges to H in (2.9). Specifically, for large u,

$$\Pr\left(\frac{Y_i - u}{\sigma_u} \leqslant z_i, \frac{Y_j - u}{\sigma_u} \leqslant z_j \middle| Y_i > u \text{ or } Y_j > u \right) \approx H_{ij}(z_i, z_j) \tag{2.14}$$

for some σ_u, where H_{ij} is the (i, j)th margin of H.

Suppose that the observed data $Y_1, \ldots, Y_n$ are independent copies of the vector Y on the scale of unit Fréchet or standard Gumbel as appropriate for the method. Let $f_{ij}(\cdot, \cdot; \theta)$ be the bivariate density with parameter vector θ for the (i, j) pair. The pairwise likelihood is

$$\sum_{t=1}^{n} \sum_{i=1}^{d} \sum_{j>i}^{d} w_{ij} \log f_{ij}(Y_{ti}, Y_{tj}; \theta), \tag{2.15}$$

where $w_{ij} \geqslant 0$ is a user-specified weight, possibly data dependent. The resulting estimator is consistent and asymptotically normal with a variance that can be estimated by a sandwich estimator, even when the data are weakly dependent (Guyon, 1995, Theorem 3.4.7).

In the first approach with the simultaneous exceedances of two components, the approximation (2.13) is only valid on the region $(u, \infty)^2$ for high enough u. For pairs with at least one component below u, their contribution is accounted for with the censored approach (Ledford and Tawn, 1996). Assuming unit Fréchet margins, the likelihood contribution of each pair is

$$f_{ij}(Y_{ti}, Y_{tj}; \theta) = \begin{cases} \nabla_{ij} G_{ij}(Y_{ti}, Y_{tj}; \theta), & Y_{ti} \geqslant u, \ Y_{tj} \geqslant u, \\ \nabla_i G_{ij}(Y_{ti}, u; \theta), & Y_{ti} \geqslant u, \ Y_{tj} < u, \\ \nabla_j G_{ij}(u, Y_{tj}; \theta), & Y_{ti} < u, \ Y_{tj} \geqslant u, \\ G_{ij}(u, u; \theta), & Y_{ti} < u, \ Y_{tj} < u, \end{cases}$$

where ∇ is the partial differentiation operator.

In the second approach using (2.14), only pairs with at least one exceedance contribute to the likelihood, so the number of pairs in (2.15) is a random variable. The likelihood contribution for each such pair is $\nabla_{ij} H_{ij}$:

$$f_{ij}(Y_{ti}, Y_{tj}; \theta) = -\frac{1}{\log G_{ij}(u,u;\theta)} \left[\frac{\nabla_{ij} G_{ij}(Y_{ti}, Y_{tj}; \theta)}{G_{ij}^2(Y_{ti}, Y_{tj}; \theta)} - \frac{\nabla_i G_{ij}(Y_{ti}, Y_{tj}; \theta) \nabla_j G_{ij}(Y_{ti}, Y_{tj}; \theta)}{G_{ij}^2(Y_{ti}, Y_{tj}; \theta)} \right],$$

for $(Y_{ti}, Y_{tj}) \notin (-\infty, u]^2$. The parameter vector θ contains marginal parameter σ_u in addition to the dependence parameters.

The simulation results in Bacro and Gaetan (2014) showed that both approaches outperformed the approach based on block maxima in mean squared error, but they can have substantial bias due to the approximation errors of the tail of bivariate distribution (2.13) and (2.14). Shang et al. (2015) proposed a two-step approach where the marginal parameters are first estimated with marginal exceedance data while the dependence parameters are then estimated with block maxima. Efficiency gain in the first step may help to improve the estimator in the second step as shown by their simulation studies.

2.5.3   Bayesian Approach

Classical Bayesian inferences depend on the likelihood. The same challenge faced by likelihood approaches — evaluating the full likelihood — remains a bottleneck for the Bayesian approach. For lower-dimensional data such as bivariate extremes, the Bayesian approach for univariate extremes in Chapter 13 can be generalized when the joint density can be evaluated. In general, however, Bayesian approaches are not straightforward.

One solution is the approximate Bayesian computation (ABC). It assumes that given a candidate parameter vector, observations from the model can be efficiently generated; this random sample is compared to the observed data through some distance metric, which can be used as an importance weight for the candidate. The resulting weighted sample provides an approximate to the posterior distribution of the parameter without likelihood evaluation. See Chapter 14 for details with applications. Bayesian inference with composite likelihood in place of likelihood was proposed recently by Ribatet et al. (2012), with an application to spatial extremes. Adjustments to the composite likelihood based on asymptotic theory are necessary to ensure appropriate inference.

2.5.4   Conditional Approach of Heffernan–Tawn

Suppose that the observed data $Y_1, \ldots, Y_n$ is a random sample from some multivariate distribution F. Inferences about the tail behavior of F using the conditional approach of Heffernan and Tawn (2004) consist of three steps.

The first step is to transform every margin to standard Gumbel. This needs to estimate the marginal distribution function F_i for each margin $i = 1, \ldots, d$. A semiparametric model is specified for F_i, where exceedences over threshold v_i is characterized by a GPD with parameter vector ψ_i, and the body part below v_i is unspecified and to be estimated by the empirical distribution function. The marginal parameters $\boldsymbol{\psi} = (\psi_1^\top, \ldots, \psi_d^\top)^\top$ are estimated by maximizing an independence likelihood

$$\sum_{i=1}^{d} \sum_{k=1}^{n_{v_i}} \log f_{X_i}(x_{i|i,k}) \tag{2.16}$$

with respect to ψ, f_{X_i} is the semiparametric marginal density of F_i, n_{v_i} is the number of observations with ith component exceeding the marginal threshold v_i, and $x_{j|i,k}$: $j = 1, \ldots, d$, $k = 1, \ldots, n_{v_i}$, is the jth component of the kth such observation. The transformation is completed by applying the transformation $G^{-1}\widehat{F}_i(\cdot)$ to the ith margin, where G^{-1} is the quantile function of standard Gumbel. Let $X_1, \ldots, X_n$ be the transformed sample, from which the standardized version $Z_1, \ldots, Z_n$ can be calculated given parameter values in the normalizing constants.

The second step is to estimate the parameters $\theta_{|i} = (\alpha_{|i}^{\top}, \beta_{|i}^{\top}, c_{|i}^{\top}, d_{|i}^{\top})^{\top}$, $i = 1, \ldots, d$, in the conditional distribution given a large value of one component. Assuming that the mean $\nu_{|i}$ and variance $\delta_{|i}^2$ of the standardized variables $Z_{|i}$ exist, $\theta_{|i}$ can be estimated by a set of multivariate regression models for pairs of variables jointly with the nuisance parameter $\lambda_{|i} = (\nu_{|i}, \delta_{|i})$, $i = 1, \ldots, d$. Specifically, this can be done by maximizing a pseudolikelihood based on independent working normal distributions for each pair

$$Q_{|i}(\theta_{|i}, \lambda_i) = \sum_{j \neq i} \sum_{k=1}^{n_{u_i}} \log \phi(Z_{j|i,k}|\nu_{j|i}, \delta_{j|i}),$$

where $\phi(\cdot|\nu, \delta)$ is the density of $N(\nu, \delta^2)$, n_{u_i} is the number of observations with ith component exceeding threshold u_i, and $Z_{j|i,k}$ is the jth component of the kth such observation after standardization (2.11). All the conditional model parameters $\theta = (\theta_{|i}, \ldots, \theta_{|n})$ can be jointly estimated by maximizing

$$\sum_{i=1}^{d} Q_{|i}(\theta_{|i}, \lambda_{|i}).$$

Note that the normal density ϕ is used as an artifact only; no distributional assumption is imposed on $Z_{|i}$ beyond the first two moments.

In the third step, inferences about the tail behavior of F is based on Monte Carlo approximation using a large sample drawn from the fitted conditional model. Random samples from the conditional distribution of Y given $Y_i > v_i$ are drawn as the following:

1. generate X_i from the Gumbel distribution conditional on its exceeding u_i where u_i is the transformed threshold of v_i from the Y_i scale to the Gumbel scale.
2. generate $Z_{|i}$ from the empirical distribution $\widehat{G}_{|i}$ independently of X_i.
3. transform $Z_{|i}$ to the scale of X_{-i}: $X_{-i} = \widehat{a}_{|i}(X_i) + \widehat{b}_{|i}(X_i)Z_{|i}$.
4. transform (X_{-i}, X_i) to the original scale of F by the $Y_j = \widehat{F}_i^{-1}(X_j)$ for each $j = 1, \ldots, d$, and output $Y = (Y_1, \ldots, Y_d)$.

For example, $\Pr(Y \in C|Y_i > v_i)$ for some event C in the joint tail region can be approximated by the proportion of the generated sample that falls in to C.

This approach has several advantages. It admits a wide range of extremal dependence without specifying the distribution of the standardized variables. No marginal distributional assumption is needed except for the exceedances. Conditioning on a single component being large allows usage of more information from the data for more efficient inferences. The implementation of the estimation and sampling is simple. It applies to high dimensions with little difficulty. Uncertainty from various sources, such as parameter estimation from the marginal and conditional models and the nonparametric distribution of the standardized residuals, can be accounted for with a semiparametric bootstrap method that preserves the dependence structure of the data (Heffernan and Tawn, 2004, Section 5.4). The bootstrap method can be very useful beyond this context (e.g., Wang et al., 2014).

2.5.5   Goodness-of-Fit Test

Goodness-of-fit of multivariate extremes can be tested separately for the marginal GEV distributions and the extreme-value copulas.

When the marginal distributions have no shared parameters across margins, standard univariate goodness-of-fit tests can be applied; see Chapter 1. Often the marginal distributions have shared parameters; for example, in a spatial setting, the location parameter is often described as a function of the longitude and latitude. A goodness-of-fit test statistic such as the Kolmogorov–Smirnov (KS) statistic can be computed for each margin, and to construct an overall summary statistic, their p-values can be combined. Specifically, let $p_{i,n}$, $i = 1, \ldots, d$, be the p-value of the KS statistic for the ith margin computed from a random sample of size n from a d-dimensional distribution. Among others, the p-values can be combined with the classic method of Fisher (1932) and Tippett (1931) as

$$W_n = -2 \sum_{i=1}^{d} \log p_{i,n} \quad \text{and} \quad T_n = \min(p_{1,n}, \ldots, p_{d,n}).$$

To assess the significance of W_n and T_n, a parametric bootstrap can be devised which needs to account for the uncertainty from parameter estimation and dependence structure in the data. The procedure that preserves the dependence structure proposed in Heffernan and Tawn (2004, Section 5.4) becomes handy. Bootstrap samples from the fitted marginal models can be generated without parametric specification of the dependence structure. For each sample, the fitting process can be carried out and the combined p-values be computed to form a bootstrap sample of them. The p-values of the observed combined p-values is the proportion in the bootstrap sample that is at least as extreme as the observed. An advantage of this approach is that no dependence structure needs to be specified.

Goodness-of-fit for the dependence structure can be assessed at two levels: whether the copula is fitted by an extreme-value copula and, if yes, whether the copula is fitted by a specific parametric copula. Recent developments on tests of extreme-value dependence are reviewed in detail in Chapter 18. For bivariate extreme-value copulas, Genest et al. (2011) proposed a goodness-of-fit test based on the distance between parametric and nonparametric estimates of the Pickands dependence function. This test is competitive in comparison with the test for general copula based on the distance between the parametric estimate of the copula and the empirical copula (Genest et al., 2009; Kojadinovic et al., 2011). For higher-dimensional setting, composite likelihood can be used to obtain the parametric estimators and dependence measures such as extremal coefficient can be used to construct testing statistics (Kojadinovic et al., 2015).

2.6 Further Reading

The overview on multivariate extremes presented in this chapter may be biased by the authors' personal experiences and expertise. A number of reviews on this subject are available; for example, Chapter 8 of Coles (2001), Chapters 7–8 of Beirlant et al. (2004), Fougeres (2004), Part II of de Haan and Ferreira (2007), Bortot and Gaetan (2012), Bacro and Gaetan (2012), Ribatet and Sedki (2013), and others.

Some chapters later in this volume complement this review. Extension to spatial extremes and max-stable processes is covered in Chapter 9. Representations of multivariate extremes through spectral measure and estimation of the extremal dependence are reviewed in Chapters 16 and 17.

References

Bacro, J.-N. and C. Gaetan (2012). A review on spatial extreme modelling. In E. Porcu, J. Montero, and M. Schlather (Eds.), *Advances and Challenges in Space-time Modelling of Natural Events*, pp. 103–124. Springer.

Bacro, J.-N. and C. Gaetan (2014). Estimation of spatial max-stable models using threshold exceedances. *Statistics and Computing 24*(4), 651–662.

Balkema, A. A. and S. I. Resnick (1977). Max-infinite divisibility. *Journal of Applied Probability 14*(2), 309–319.

Beirlant, J., Y. Goegebeur, J. Segers, and J. Teugels (2004). *Statistics of Extremes: Theory and Applications*. John Wiley & Sons.

Bortot, P. and C. Gaetan (2012). *Multivariate Extremes* (2 ed.)., pp. 1685–1694. Chichester: John Wiley & Sons.

Buishand, T. A. (1984). Bivariate extreme-value data and the station-year method. *Journal of Hydrology 69*(1), 77–95.

Coles, S. G. (2001). *An Introduction to Statistical Modeling of Extreme Values*. London: Springer.

Coles, S. G., J. E. Heffernan, and J. A. Tawn (1999). Dependence measures for extreme value analyses. *Extremes 2*(4), 339–365.

Coles, S. G. and J. A. Tawn (1991). Modelling extreme multivariate events. *Journal of the Royal Statistical Society: Series B (Statistical Methodology) 53*(2), 377–392.

Davison, A. C. and M. M. Gholamrezaee (2012). Geostatistics of extremes. *Proceedings of the Royal Society A: Mathematical, Physical & Engineering Sciences 468*(2138), 581–608.

de Haan, L. and A. Ferreira (2007). *Extreme Value Theory: An Introduction.* New York: Springer.

de Haan, L. and S. Resnick, I (1977). Limit theory for multivariate sample extremes. *Probability Theory and Related Fields 40*(4), 317–337.

Demarta, S. and A. J. McNeil (2005). The t copula and related copulas. *International Statistical Review 73*(1), 111–129.

Falk, M. and A. Guillou (2008). Peaks-over-threshold stability of multivariate generalized Pareto distributions. *Journal of Multivariate Analysis 99*(4), 715–734.

Ferreira, A. and L. de Haan (2014). The generalized Pareto process; with a view towards application and simulation. *Bernoulli 20*, 1717–1737.

Fisher, R. A. (1932). *Statistical Methods for Research Workers.* London: Olivier and Boyd.

Fougeres, A.-L. (2004). Multivariate extremes. In B. Finkenstadt and H. Rootzén (Eds.), *Extreme Values in Finance, Telecommunications and the Environment*, pp. 373–388. Florida: Chapman & Hall/CRC.

Genest, C., I. Kojadinovic, J. Nešlehová, and J. Yan (2011). A goodness-of-fit test for bivariate extreme-value copulas. *Bernoulli 17*(1), 253–275.

Genest, C., B. Rémillard, and D. Beaudoin (2009). Goodness-of-fit tests for copulas: A review and a power study. *Insurance: Mathematics and Economics 44*, 199–213.

Ghoudi, K., A. Khoudraji, and L. P. Rivest (1998). Statistical properties of couples of bivariate extreme-value copulas. *Canadian Journal of Statistics 26*(1), 187–197.

Gudendorf, G. and J. Segers (2010). Extreme-value copulas. In P. Jaworski, F. Durante, W. K. Hárdle, and T. Rychlik (Eds.), *Copula Theory and Its Applications*, pp. 127–145. New York: Springer.

Guyon, X. (1995). *Random Fields on a Network: Modeling, Statistics, and Applications.* New York: Springer–Verlag.

Heffernan, J. E. (2000). A directory of coefficients of tail dependence. *Extremes 3*(3), 279–290.

Heffernan, J. E. and J. A. Tawn (2004). A conditional approach for multivariate extreme values (with discussion). *Journal of the Royal Statistical Society: Series B (Statistical Methodology) 66*(3), 497–546.

Joe, H. (1993). Parametric families of multivariate distributions with given margins. *Journal of Multivariate Analysis 46*(2), 262–282.

Joe, H. (1997). *Multivariate Models and Multivariate Dependence Concepts.* London, New York: Chapman & Hall/CRC.

Joe, H., R. L. Smith, and I. Weissman (1992). Bivariate threshold methods for extremes. *Journal of the Royal Statistical Society: Series B (Statistical Methodology) 54*(1), 171–183.

Kojadinovic, I., H. Shang, and J. Yan (2015). A class of goodness-of-fit tests for spatial extremes models based on max-stable processes. *Statistics and Its Interfaces 8*(1), 45–62.

Kojadinovic, I., J. Yan, and M. Holmes (2011). Fast large-sample goodness-of-fit for copulas. *Statistica Sinica 21*(2), 841–871.

Ledford, A. W. and J. A. Tawn (1996). Statistics for near independence in multivariate extreme values. *Biometrika 83*(1), 169–187.

Ledford, A. W. and J. A. Tawn (1997). Modelling dependence within joint tail regions. *Journal of the Royal Statistical Society: Series B (Statistical Methodology) 59*(2), 475–499.

Lindsay, B. G. (1988). Composite likelihood methods. *Contemporary Mathematics 80*(1), 221–239.

Padoan, S. A., M. Ribatet, and S. A. Sisson (2010). Likelihood-based inference for max-stable processes. *Journal of the American Statistical Association 105*(489), 263–277.

Pickands, III, J. (1981). Multivariate extreme value distributions. In *Proceedings 43rd Session International Statistical Institute*, Volume 2, pp. 859–878. Amsterdam: International Statistical Institute.

Rakonczai, P. and A. Zempléni (2012). Bivariate generalized Pareto distribution in practice: Models and estimation. *Environmetrics 23*(3), 219–227.

Renard, B. and M. Lang (2007). Use of a Gaussian copula for multivariate extreme value analysis: Some case studies in hydrology. *Advances in Water Resources 30*(4), 897–912.

Ribatet, M., D. Cooley, and A. C. Davison (2012). Bayesian inference from composite likelihoods, with an application to spatial extremes. *Statistica Sinica 22*, 813–845.

Ribatet, M. and M. Sedki (2013). Extreme value copulas and max-stable processes. *Journal de la Société Française de Statistique 154*(1), 138–150.

Rootzén, H. and N. Tajvidi (2006). Multivariate generalized Pareto distributions. *Bernoulli 12*(5), 917–930.

Schlather, M. and J. A. Tawn (2003). A dependence measure for multivariate and spatial extreme values: Properties and inference. *Biometrika 90*(1), 139–156.

Shang, H., J. Yan, and X. Zhang (2015). A two-step approach to model precipitation extremes in California based on max-stable and marginal point processes. *The Annals of Applied Statistics 9*(1), 452–473.

Sklar, A. (1959). Fonctions de répartition à n dimensions et leurs marges. *Publications de l'Institut de Statistique de l'Université de Paris 8*, 229–231.

Smith, E. L. and A. G. Stephenson (2009). An extended Gaussian max-stable process model for spatial extremes. *Journal of Statistical Planning and Inference 139*(4), 1266–1275.

Tawn, J. A. (1990). Modelling multivariate extreme value distributions. *Biometrika 77*(2), 245–253.

Tippett, L. H. C. (1931). *The Methods of Statistics*. London: Williams and Norgate.

Varin, C., N. M. Reid, and D. Firth (2011). An overview of composite likelihood methods. *Statistica Sinica 21*(1), 5–42.

Wang, Z., J. Yan, and X. Zhang (2014). Incorporating spatial dependence in regional frequency analysis. *Water Resources Research 50*(12), 9570–9585.

3

Univariate Extreme Value Mixture Modeling

Carl Scarrott

University of Canterbury, New Zealand

Abstract

A plethora of univariate extreme value mixture models have been developed, which combine a classic tail model with a component to describe the bulk of the distribution. The threshold which defines the tail model support is typically treated as a parameter to be inferred, thus permitting both estimation and uncertainty quantification. These models potentially provide a more objective approach to threshold choice than the traditional graphical diagnostics. This chapter summarises the key features of univariate extreme value mixture models and inference approaches for them. They are presented in a general framework with consistent notation to compare their properties, including a summary of their performance from recent comparative simulation studies. The R packages available on CRAN for implementing these mixture models are outlined. Some advice for developers and users is provided.

3.1 Introduction

The generalised Pareto distribution (GPD) is the classic extreme value model for the exceedances of a threshold u, justified by the Pickands–Balkema–de Haan theorem of Pickands (1975) and Balkema and de Haan (1974). It is an asymptotically motivated model for conditional exceedances of a sufficiently high threshold for independent and identically distributed random variables, under some mild regularity conditions. Similar justification exists for stationary sequences satisfying short range dependence conditions, e.g., the $D(u_n)$ condition of Leadbetter et al. (1983) or the asymptotic independence of maxima by Hsing (1987) and Hsing et al. (1988).

Let X be a random variable for an exceedance of a predefined threshold u which is assumed to follow a GPD with scale $\sigma_u > 0$ and shape ξ parameters. The cumulative distribution function (cdf) is given by:

$$G(x \mid \boldsymbol{\theta}_u) = \Pr(X \leqslant x \mid X > u) = \begin{cases} 1 - \left[1 + \xi\left(\dfrac{x-u}{\sigma_u}\right)\right]_+^{-1/\xi}, & \xi \neq 0, \\ 1 - \exp\left[-\left(\dfrac{x-u}{\sigma_u}\right)\right]_+, & \xi = 0, \end{cases} \tag{3.1}$$

where $x_+ = \max(x, 0)$ and the GPD parameters include the threshold $\boldsymbol{\theta}_u = (u, \sigma_u, \xi)$. The GPD scale notation σ_u makes the dependence on the threshold explicit. For $\xi \geqslant 0$ the support is $x > u$, but when $\xi < 0$ there is an upper end point so that $u < x < u - \sigma_u/\xi$. The special case of $\xi = 0$ is defined by continuity in the limit $\xi \to 0$ giving the exponential form. The conditional probability density function (pdf) of the GPD $g(x \mid \boldsymbol{\theta}_u)$ is defined by differentiation. Detailed properties of the GPD are described in Chapter 1. The threshold is referred to as the location parameter of the GPD, and has typically been pre-chosen in applications.

It should be remembered that to use the GPD for describing unconditional quantities that an implicit parameter $\phi_u = \Pr(X > u)$ is needed:

$$\Pr(X > x) = \phi_u \left[1 - \Pr(X \leqslant x \mid X > u)\right], \tag{3.2}$$

which is called the *"tail fraction"* or "threshold exceedance probability."

Practitioners apply the GPD as an approximation to the upper tail of the population distribution above a sufficiently high threshold. Hence, threshold selection is a key first step in tail modelling using the GPD. Threshold choice is a tradeoff between the bias and variance. It is desirable to choose as low a threshold as possible, to maximise the sample information for inference purposes and thus reduce the variance of the estimates. However, if the threshold is too low then the asymptotic theory underlying the GPD model will be invalid, which could lead to biased estimates.

A reasonably wide-ranging review of threshold estimation approaches in the literature is provided by Scarrott and MacDonald (2012). This chapter aims to consolidate the current literature on univariate extreme value mixture modelling, which typically includes threshold inference. A general framework which includes the majority of the extreme value mixture models, which are subsequently referred to as the "standard" models, is detailed in Section 3.2. A short bibliographic review of the standard models is given in Section 3.2.2, with a summary of various extensions in Section 3.2.1. Inference approaches for these standard models is covered in Section 3.2.3. Mixture models which do not fit within this general framework, but which include classical extreme value tail models, are then outlined and discussed in Section 3.3.

Relevant R (R Core Team, 2013) packages available on CRAN are discussed in Section 3.4. Supplementary materials provide examples of using the evmix package (Scarrott and Hu, 2015), as it has the most comprehensive coverage of extreme value mixture models. A more comprehensive user guide to the evmix package is provided by Hu and Scarrott (2013). Advice for developers and users of such models is provided in Section 3.5, with the aim to prevent some of the pitfalls of such models and their inference.

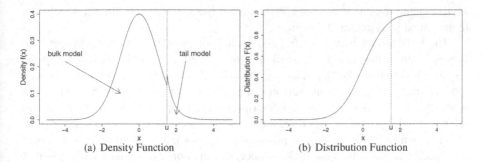

(a) Density Function   (b) Distribution Function

FIGURE 3.1
Density function and cumulative distribution functions of the parametric extreme
value mixture model, with normal bulk model with $\theta_b = (\mu = 0, \sigma = 1)$ and GPD
with $\theta_u = (u = 1.5, \sigma_u = 0.4, \xi = 0)$. Bulk model based tail fraction $\phi_u = 1 - \Phi(1.5) = 6.7\%$. Vertical dashed line indicates the threshold.

3.2   Standard Extreme Value Mixture Models

The simplest extreme value mixture model was proposed by Behrens et al. (2004)
and has been widely developed. It is also a natural extension of the classical extreme value tail modelling approach, represented by Equations 3.2 and 3.1. The "bulk
model" below the threshold is simply spliced together with the "tail model," , which
is commonly the GPD.

Denote the bulk model cdf as $H(x \mid \theta_b)$, where the subscript b denotes the bulk
model parameter vector. The cdf of the spliced mixture is then given by:

$$F(x \mid \theta_b, \theta_u) = \begin{cases} H(x \mid \theta_b) & \text{for } x \leqslant u, \\ H(u \mid \theta_b) + [1 - H(u \mid \theta_b)] \, G(x \mid \theta_u) & \text{for } x > u. \end{cases} \quad (3.3)$$

with pdf given by

$$f(x \mid \theta_b, \theta_u) = \begin{cases} h(x \mid \theta_b) & \text{for } x \leqslant u, \\ [1 - H(u \mid \theta_b)] \, g(x \mid \theta_u) & \text{for } x > u, \end{cases} \quad (3.4)$$

where $h(.)$ is the corresponding bulk model pdf. The density and distribution function
of this extreme value mixture model is depicted in Figure 3.1. By definition the cdf is
continuous at the threshold, but there is no such constraint on the pdf. The parameters
were chosen to demonstrate the potential discontinuity in the pdf at the threshold in
Figure 3.1(a).

The second line of Equation 3.4 shows that the tail fraction ϕ_u, from the classical
unconditional GPD in Equation 3.2, is the bulk model survival probability assuming
it continues beyond the threshold. Defining $\phi_u = 1 - H(u \mid \theta_b)$ is referred to as
the *bulk model based tail fraction*. The bulk density up to u integrates to $1 - \phi_u =$

$H(u \mid \boldsymbol{\theta}_b)$ and the unconditional GPD density beyond u integrates to ϕ_u, which sum to unity, and so Equation 3.4 is proper.

The bulk model based tail fraction scaling of the GPD ensures a proper pdf. It benefits from borrowing information from the bulk data, which is usually ample compared to that in the tail. This is, however, also a rather odd choice. The key motivation for splicing in the tail model is that we are interested in applications where the bulk model is not a good descriptor for the upper tail; therefore, assuming the bulk model provides a good estimator of the upper tail fraction is defeating the model's own purpose. Further, it potentially exposes tail estimation to bulk model mis-specification. If the bulk model is poor, or if its upper tail behaviour is different to that of the population, then the tail fraction estimate and thus upper tail fit may be compromised, which is discussed in Section 3.5.

Hu (2013a) demonstrated that protection from the bulk model mis-specification is provided by a reformulation close to the classical threshold modelling approach in Equation 3.2, where the tail fraction is an extra parameter ϕ_u. MacDonald et al. (2011) generalised Equation 3.3 to give a cdf:

$$F(x \mid \boldsymbol{\theta}_b, \boldsymbol{\theta}_u, \phi_u) = \begin{cases} \dfrac{(1 - \phi_u)}{H(u \mid \boldsymbol{\theta}_b)} H(x \mid \boldsymbol{\theta}_b) & \text{for } x \leqslant u, \\ (1 - \phi_u) + \phi_u\, G(x \mid \boldsymbol{\theta}_u) & \text{for } x > u. \end{cases} \tag{3.5}$$

The denominator $H(u \mid \boldsymbol{\theta}_b)$ on the first line renormalises the bulk model cdf to make it a proper truncated distribution with support up to the threshold. As such, the rescaled bulk model now integrates to $1 - \phi_u$ up to the threshold, with the GPD contributing the usual tail fraction ϕ_u above the threshold.

The tail fraction scaling ϕ_u gives a proper density:

$$f(x \mid \boldsymbol{\theta}_b, \boldsymbol{\theta}_u, \phi_u) = \begin{cases} \dfrac{(1 - \phi_u)}{H(u \mid \boldsymbol{\theta}_b)} h(x \mid \boldsymbol{\theta}_b) & \text{for } x \leqslant u, \\ \phi_u\, g(x \mid \boldsymbol{\theta}_u) & \text{for } x > u. \end{cases} \tag{3.6}$$

which will henceforth be described as the *standard model*. This generalisation is referred to as the *parameterised tail fraction* approach and includes the bulk model based tail fraction as a special case where $\phi_u = 1 - H(u \mid \boldsymbol{\theta}_b)$.

Section 3.2.3 shows that the maximum likelihood estimator of the parameterised tail fraction is the sample proportion of excesses $\widehat{\phi}_u = \frac{n_u}{n}$, where n_u is the number of excesses and n is the sample size. This intuitive estimator is used in all implementations of the parameterised tail fraction approach thus far (Cabras and Castellanos, 2011; Gonzalez et al., 2013; MacDonald et al., 2011; Scarrott and MacDonald, 2010) and the classical extreme value tail modelling approach, see Coles (2001).

Interestingly, Cabras and Castellanos (2011) use the bulk model based tail fraction for parametric bulk models and revert to a parameterised tail fraction when using a semiparametric density estimator for the bulk model. However, due to the mean preserving properties of the Poisson regression used for their semiparametric density estimator, the parameterised tail fraction and bulk model based tail fraction are the same in this special case.

Similar models have been published in the actuarial science literature, where they are referred to as "composite models." Cooray and Ananda (2005) considered a composite lognormal-Pareto, which has a lognormal bulk model and a Pareto (rather than GPD) tail model as they focus on heavy tailed populations. Unfortunately, their formulation did not include the tail fraction scaling, so ignores the fact that the Pareto is a conditional tail model. As such, their formulation has some restrictive properties and exhibits poor performance in wide applications Scollnik (2007). This oversight was corrected by Scollnik (2007) by inclusion of a parameterised tail fraction, as in Equation 3.6. These composite models are discussed in Sections 3.3 and 3.2.2.

An alternative presentation of the standard model uses a proper truncated density for the bulk model. The GPD component is already truncated at the threshold and proper. Denote a proper truncated pdf for the bulk model by

$$h_t(x \mid u, \boldsymbol{\theta}_b) = \frac{h(x \mid \boldsymbol{\theta}_b)}{H(u \mid \boldsymbol{\theta}_b)} \, I(x \leqslant u). \tag{3.7}$$

The standard model with parameterised tail has a very concise pdf given by

$$f(x \mid \boldsymbol{\theta}_b, \boldsymbol{\theta}_u, \phi_u) = \begin{cases} (1 - \phi_u) \, h_t(x \mid u, \boldsymbol{\theta}_b) & \text{for } x \leqslant u, \\ \phi_u \, g(x \mid \boldsymbol{\theta}_u) & \text{for } x > u. \end{cases} \tag{3.8}$$

The bulk and tail scaling factors, $1 - \phi_u$ and ϕ_u, respectively, sum to unity thus giving a proper density. There is an extensive literature on mixture models in the wider statistics literature, see McLachlan and Peel (2000), where the components usually have overlapping support. The standard model can be presented in a form commonly used in that literature as $f(x) = (1 - \phi_u) \, h_t(x) + \phi_u \, g(x)$, where the parameter dependence is ignored for clarity.

3.2.1 Continuity Constraints and Tail Model Extensions

The standard extreme value mixture models defined by Equations 3.3 and 3.5 have been extended by various authors to allow more flexibility or to apply physically justifiable constraints. This section details some of the key developments. The alternative formulations which do not fit within these standard model equations are outlined in Section 3.3.

A key drawback with the standard model is that, although the cdf is continuous by definition, the pdf is not necessarily continuous (or more generally smooth) at the threshold. In practice, inferred parameters from data will often provide a fitted model which is close to continuous but this cannot be guaranteed. This can present bias and uncertainty where the quantile (or similar characteristic) of interest is close to the threshold. Nonstationary extensions of such models can be particularly problematic with the extent of discontinuity varying along the threshold function.

The first formulations to include continuity constraints on the pdf at the threshold were the composite lognormal-Pareto of Cooray and Ananda (2005) and hybrid Pareto of Carreau and Bengio (2006). These formulations exclude the tail fraction scaling, which is equivalent to the tail models being treated as unconditional models.

The bulk model (lognormal and normal, respectively) is spliced with the tail model (Pareto and GPD, respectively) at the threshold, with no differential scaling of the two components. In essence, the tail fraction scaling ϕ_u on the second lines of Equation 3.3 is set to one, thus leading to domination by the tail model as the exceedance probability will be greater than 0.5. This oversight leads to performance issues in wide applications (Scollnik, 2007). As these formulations are not standard models as specified by Equations 3.3 or 3.5, they are detailed with the alternative models in Section 3.3. However, their general approach for imposing the continuity constraints have also been developed for the standard model, so are now detailed.

A simple but generally applicable approach to constrain the pdf to be continuous at the threshold is to restrict the GPD scale parameter σ_u. The GPD pdf evaluated in the limit approaching the threshold u from the right is simply $\lim_{x \to u^+} g(x \mid \boldsymbol{\theta}_u) = 1/\sigma_u$. Equating this result to the bulk model pdf evaluated at the threshold, for a bulk model based tail fraction, gives

$$\sigma_u = \frac{1 - H(u \mid \boldsymbol{\theta}_b)}{h(u \mid \boldsymbol{\theta}_b)}. \tag{3.9}$$

Similarly for the parameterised tail fraction gives the constraint

$$\sigma_u = \frac{H(u \mid \boldsymbol{\theta}_b)}{1 - \phi_u} \frac{\phi_u}{h(u \mid \boldsymbol{\theta}_b)}, \tag{3.10}$$

where the numerator ϕ_u replaces the upper tail fraction $1 - H(u \mid \boldsymbol{\theta}_b)$ in Equation 3.9. The left fraction is the reciprocal of the truncated bulk scaling factor in Equation 3.5. Both constraints are functions of the threshold and bulk model parameters, and for the parameterised tail fraction approach ϕ_u is also needed.

Scollnik (2007) uses the parameterised tail fraction model and constrains the tail fraction to achieve continuity of the pdf at the threshold using

$$\phi_u = \frac{h(u \mid \boldsymbol{\theta}_b)}{h(u \mid \boldsymbol{\theta}_b) + g(u \mid \boldsymbol{\theta}_u) \, H(u \mid \boldsymbol{\theta}_b)}, \tag{3.11}$$

which is a more complex constraint than Equation 3.10 as it depends on both the bulk and tail parameters $(\boldsymbol{\theta}_b, \boldsymbol{\theta}_u)$. The pdf from Equation 3.6 becomes

$$f(x \mid \boldsymbol{\theta}_b, \boldsymbol{\theta}_u) = \begin{cases} \dfrac{g(u \mid \boldsymbol{\theta}_u)}{h(u \mid \boldsymbol{\theta}_b) + g(u \mid \boldsymbol{\theta}_u) \, H(u \mid \boldsymbol{\theta}_b)} \, h(x \mid \boldsymbol{\theta}_b) & \text{for } x \leqslant u, \\[2ex] \dfrac{h(u \mid \boldsymbol{\theta}_b)}{h(u \mid \boldsymbol{\theta}_b) + g(u \mid \boldsymbol{\theta}_u) \, H(u \mid \boldsymbol{\theta}_b)} \, g(x \mid \boldsymbol{\theta}_u) & \text{for } x > u. \end{cases} \tag{3.12}$$

This formulation requires that the upper tail density $g(.)$ is a conditional or proper truncated form. Nadarajah and Baka (2014) further extend Scollnik's result to allow for an arbitrary upper tail density $g(.)$, which may not be truncated below the threshold, which is discussed in Sections 3.2.2 and 3.3.

There appears to be no general result for constraining the bulk model or threshold parameters, but specific conditions have been found in special cases. So and Chan

(2014) developed an extreme value mixture model with GPD for both the upper and lower tails with an arbitrary parametric bulk component, with the latter chosen to be a normal in their application and simulations, extending the previous work of Zhao et al. (2010) and Zhao et al. (2014). They implemented a constraint of continuity of the pdf at both thresholds by constraining the bulk model parameters (mean and variance of a normal).

As far as the author is aware, the comparative benefits of continuity constraints on the pdf implemented by restricting either the bulk, threshold, tail fraction, or tail model parameters have not been explored. Further smoothness constraints, e.g., continuity in first derivative, have been considered. The potential for increased lack of robustness of the upper tail fit to the bulk model/data due to such parameter constraints is discussed in Section 3.5.

The key extensions of the standard extreme value mixture models are:

- Scollnik (2007) considers the special case of the Pareto tail model, which suggests that alternative tail models are plausible, e.g., extended GPD proposed by Papastathopoulos and Tawn (2013). Nadarajah and Baka (2014) also consider any arbitrary distribution for the upper tail, not necessarily truncated below the threshold, with a focus on the Burr distribution.
- MacDonald et al. (2011), Oumow et al. (2012) and Northrop and Jonathan (2011) consider the point process representation of univariate independent and identically distributed extremes, which overcomes the dependence of the usual GPD scale parameter σ_u on the threshold u. However, Lee (2013) recently highlighted that the rate of occurrence component of the point process likelihood does not provide a good approximation and as such has a tendency to infer high thresholds. This is a subject of ongoing research. Tancredi et al. (2006) consider the GEV representation for the upper tail.
- Zhao et al. (2010), Zhao et al. (2014), Mendes and Lopes (2004), So and Chan (2014), Solari and Losada (2004), Nadarajah and Baka (2014), and MacDonald et al. (2011) consider extensions with both the lower and upper tails described by an extreme value tail model.
- The hierarchical model of Pigeon and Denuit (2011) includes a distribution of thresholds, which are integrated out in the resulting inference.
- Couturier and Victoria-Feser (2010) consider a zero-inflated GPD.

3.2.2 Bibliography of Standard Models

Scarrott and MacDonald (2012) review threshold estimation approaches, including references to all the known extreme value mixture models at the time of writing. A recent comparative simulation study between most of the extreme value mixture models was completed by Hu (2013a), which led to the development of the evmix package (Scarrott and Hu, 2015). Further comparisons of the mixture models and their implementation in the evmix package are also provided by Hu and Scarrott (2013). In this section, the key points from these comparative studies are provided, including an updated bibliography of such models in the literature.

Scarrott and MacDonald (2012) loosely categorised the existing extreme value mixture models into those with parametric, semiparametric, and nonparametric bulk component; a convention continued here. The following is a summary of the standard extreme value mixture models defined by Equations 3.3 and 3.5. Examples outside of this framework are given in Section 3.3.

The usual parametric bulk models have been considered by many authors:

- Normal — Cabras and Castellanos (2011) and Behrens et al. (2004).
- Lognormal — Cabras and Castellanos (2011), Scollnik (2007), and Nadarajah and Baka (2014).
- Gamma — Zheng et al. (2014), Teodorescu and Vernicu (2013), and Behrens et al. (2004).
- Weibull — Cabras and Castellanos (2011), Teodorescu and Vernicu (2013) and Behrens et al. (2004).
- Exponential — Wong and Li (2010) and Teodorescu and Vernicu (2009).
- Beta — MacDonald (2012).

Variants with tail models spliced for both upper and lower tails are

- Normal — Zhao et al. (2010), Zhao et al. (2014), So and Chan (2014), Oumow et al. (2012), and Mendes and Lopes (2004).
- Lognormal — Solari and Losada (2004) and Nadarajah and Baka (2014).

As discussed above, the non-standard variants which exclude the tail fraction scaling (e.g., the hybrid Pareto) will be discussed separately in Section 3.3.

The spectrum of semi and nonparametric bulk model are more difficult to classify. The semiparametric bulk models are loosely defined as those where the effective degrees of freedom of the density estimate is expected to grow slowly with sample size, whereas nonparametric bulk models grow more quickly. The standard models with semiparametric bulk components are:

- Finite mixtures of gammas by do Nascimento et al. (2012).
- Finite mixtures of exponentials by Lee et al. (2012).
- Dirichlet process mixture of gammas by Fuquene (2015), which permits the full spectrum of parametric to nonparametric estimation.
- Cabras and Castellanos (2011) proposed a semiparametric density estimator for the bulk, estimated by Poisson regression smoothing with a polynomial basis applied to histogram binned observations.

The semiparametric mixture of hybrid Paretos, which excludes the tail fraction will be discussed in Section 3.3.1.

The nonparametric density estimators for the bulk component are:

- Tancredi et al. (2006) were the first to propose a nonparametric bulk component, using the mixture of uniforms density estimator of Robert (1998) and Robert and Casella (2010) which provides a piecewise linear approximation to the bulk cdf. This nonparametric mixture is defined between the usual threshold and a pre-defined lower threshold that is definitely too low, which could be the lower bound on support.
- A kernel density estimator (KDE) for the bulk component was proposed by Scarrott and MacDonald (2010) and MacDonald et al. (2011), with a boundary corrected KDE for bounded support applications by MacDonald et al. (2013). MacDonald et al. (2011) and MacDonald et al. (2013) consider variants with a tail

model for both the upper and lower tails. A cross-validation likelihood is used so the kernel bandwidth and other mixture model parameters can be estimated simultaneously.

- Gonzalez et al. (2013) consider an arbitrary nonparametric density estimator for the bulk, with a two stage inference approach similar to that of Cabras and Castellanos (2011). The example of a nonparametric density estimator used in their applications is the same kernel density estimator of Scarrott and MacDonald (2010) and MacDonald et al. (2011).

3.2.3 Inference for Extreme Value Mixture Models

A wide range of inference approaches have been implemented for extreme value mixture models. Some researchers have been successful with maximisation of the complete likelihood of all the parameters, $\theta = (\theta_b, \theta_u)$ which may also include the tail fraction ϕ_u where appropriate, e.g., Carreau and Bengio (2009b), Scollnik (2007), Nadarajah and Baka (2014), and Vrac and Naveau (2007). Penalties to prevent overfitting (e.g., AIC and BIC) are commonly used. Wong and Li (2010) suggested constraining the maximum number of excesses to $\lfloor n/4 \rfloor$ which is equivalent to placing a lower bound on the threshold.

The complete likelihood for n i.i.d. observations $x = (x_1, \ldots, x_n)$ from the standard model in Equation 3.4, with a bulk model based tail fraction, is

$$
\begin{aligned}
L(\theta \mid x) &= \prod_{\{i; x_i \leqslant u\}} h(x_i \mid \theta_b) \prod_{\{i; x_i > u\}} [1 - H(u \mid \theta_b)] g(x_i \mid \theta_u). \qquad (3.13) \\
&= [1 - H(u \mid \theta_b)]^{n_u} \prod_{\{i; x_i \leqslant u\}} h(x_i \mid \theta_b) \prod_{\{i; x_i > u\}} g(x_i \mid \theta_u),
\end{aligned}
$$

where n_u is the number of excesses. The complete likelihood for standard model in Equation 3.6, with a parameterised tail fraction, is given by

$$
\begin{aligned}
L(\theta, \phi_u \mid x) &= \prod_{\{i; x_i \leqslant u\}} \frac{1 - \phi_u}{H(u \mid \theta_b)} h(x_i \mid \theta_b) \prod_{\{i; x_i > u\}} \phi_u\, g(x_i \mid \theta_u). \qquad (3.14) \\
&= (1 - \phi_u)^{n - n_u} \phi_u^{n_u} \prod_{\{i; x_i \leqslant u\}} h_t(x_i \mid u, \theta_b) \prod_{\{i; x_i > u\}} g(x_i \mid \theta_u),
\end{aligned}
$$

where the last line uses the proper truncated bulk model density from Equation 3.8 to make the scaling effects of the bulk/tail fractions clear.

Except in special cases, the complete likelihood is not separable between the bulk and tail model parameters. The threshold, which is a tail model parameter, appears with the bulk model parameters in $H(u \mid \theta_b)$ and $h_t(x_i \mid u, \theta_b)$ in Equations 3.13 and 3.14, respectively. However, if one conditions on a particular threshold u the corresponding conditional likelihoods are separable, so maximisation with respect to the bulk and tail parameters can then be carried out separately. In the case of the parameterised tail fraction in Equation 3.14, conditional on a threshold u, the maximum likelihood estimator of ϕ_u is the usual sample proportion of excesses. The separability property when conditioning on u, naturally leads to various multi-stage inference approaches.

However, the separability of the complete likelihood into bulk and tail components, conditional on u, requires that there is no relationship between the bulk and tail parameters. For example, constraints on the parameters to achieve continuity of the pdf at the threshold may induce dependence between the bulk and tail parameters, like those seen in Equations 3.9, 3.10, and 3.11.

In general, direct optimisation of the complete likelihood is problematic as it typically exhibits multiple local modes (Gonzalez et al., 2013; Lee et al., 2012). Figures 3 and 5b of Cabras and Castellanos (2011) are the marginal posterior estimates for the threshold, which also exhibits this multi-modal behaviour. This is an inherent feature of such threshold models. A fundamental feature of the GPD is that once a suitably high threshold has been reached then for all higher thresholds the GPD will be a suitable model. Further, if the bulk model is able to adequately approximate the tail then the two components will compete and will exhibit similar likelihoods at multiple thresholds. Various advanced optimisation tools could be used (e.g., multiple starting values or stochastic optimizers) to overcome this multi-modality.

One of the commonly used approaches to avoid the direct optimisation of the complete likelihood is profile likelihood estimation, e.g., Hu and Scarrott (2013), Lee et al. (2012), Gonzalez et al. (2013), Solari and Losada (2004), Bee (2012), and Cooray and Ananda (2005). Denote the parameter vector excluding the threshold $\boldsymbol{\theta}_{-u}$ and the complete likelihood using the partitioned parameters $L(\boldsymbol{\theta}_{-u}, u)$. The profile likelihood at u is given by

$$L_p(u) = \max_{\boldsymbol{\theta}_{-u}} L(\boldsymbol{\theta}_{-u}, u).$$

The profile likelihood over u often exhibits multi-modality, so black-box optimization is similarly problematic. A grid search over potential thresholds is typically used to find the value which maximises the profile likelihood. The profile likelihood estimation procedure is then:

- Define a suitable subset of potential thresholds $\boldsymbol{u}^* = (u_1, \ldots, u_k)$.
- Evaluate the profile likelihood at each potential threshold and choose the value that maximises it:

$$\widehat{u} = \arg\max_{u^* \in \boldsymbol{u}^*} L_p(u^*).$$

- Fix the threshold at its estimated value and maximise the conditional likelihood over the non-threshold parameters:

$$\widehat{\boldsymbol{\theta}}_{-u} = \arg\max_{\boldsymbol{\theta}_{-u}} L(\boldsymbol{\theta}_{-u}, u, |u = \widehat{u}).$$

From personal experience, the conditional likelihood in the final step does not exhibit such severe multi-modality. As such, black-box optimisation of the conditional likelihood can be used, which could include taking advantage of separability of the likelihood when conditioning on the threshold.

The general properties of profile likelihood estimation are not discussed here; the interested reader is referred to Pawitan (2001). The main drawback with the current profile likelihood implementations is that the threshold is fixed at step 2, prior to inferring the non-threshold parameters and as such the uncertainty associated with its selection are ignored. Of course, the commonplace approach of calculating uncertainty estimates using asymptotic Wald-type approximations for multi-modal likelihoods is also questionable.

Gonzalez et al. (2013) use a similar multi-stage inference approach. Firstly, they obtain a nonparametric density estimator for the entire dataset which is subsequently fixed/conditioned upon, when applying the above profile likelihood procedure for estimating the threshold and other GPD parameters. As they use a KDE for the bulk component, the first step is simply to fix the kernel bandwidth. In addition, they consider heuristics to determine the lowest threshold where the profile likelihood stabilises.

Cabras and Castellanos (2011) use Bayesian inference for their semiparametric mixture model, within a multi-step inference process which has some similarities with that of Gonzalez et al. (2013). For each threshold sampled from the posterior, the semiparametric bulk density estimator is fitted to all the observations below the threshold. The semiparametric density estimate for each observation below that threshold is then conditioned upon in the profile likelihood of the GPD parameters (including the threshold).

If one ignores the obvious differences in these two inference approaches, due to using Bayesian and likelihood inference for the GPD parameters, there is a more subtle difference. Gonzalez et al. (2013) fit the kernel density estimator to the entire dataset (so the likelihood contribution for a bulk observation does not vary with each potential threshold), whereas Cabras and Castellanos (2011) refit the semiparametric density estimator to only those observations below each threshold (so the likelihood contribution for a bulk observation potentially varies with each threshold). Clearly the approach of Gonzalez et al. (2013) will be more computationally efficient, as they only fit their nonparametric density estimator once. However, it is not obvious as to which approach will provide better tail estimates (or reduced uncertainty in those estimates). In both approaches the uncertainty associated with the bulk density estimate is ignored in the resulting inference. For Gonzalez et al. (2013) this can be overcome by using a combined cross-validation likelihood of MacDonald et al. (2011), but which comes with a high computational burden.

Mendes and Lopes (2004) apply a heuristic multi-stage inference approach with similarities to the usual profile likelihood approach, for a two-tailed mixture model. They use L-moment estimators for the upper and lower tail GPD parameters (not including the thresholds), and apply a grid search over potential pairs of thresholds with the likelihood (conditioning on those parameter estimates) as the objective function to be maximised. Before applying this procedure, they apply a robust location-scale rescaling of the data which is said to provide "well-defined tails," but it is not obvious that this can be achieved by the linear rescaling which cannot influence the tail behaviour.

Probability weighted moment estimation was compared to maximum likelihood estimation by Bee (2012) for variants of composite lognormal-Pareto and lognormal-GPD models.

As mentioned above, the parameters in the bulk and tail components are generally dependent on the threshold, so the full efficiency of the usual EM algorithm cannot be achieved for inference purposes (Frigessi et al., 2003). As pointed out above, the likelihood conditional on the threshold is separable (provided parameter constraints don't induce further dependencies) and so this efficiency can be regained if the EM algorithm is applied to the conditional likelihood. The threshold can then be chosen by a search over an appropriate objective function. However, there seems to be little gained by this approach, as the probability of being in each component estimated in the expectation step could just be the sample proportion of excesses and non-excesses.

Lee et al. (2012) used the EM algorithm. However, they defined the two exponential components used for the bulk model as the mixture components, rather than treating the bulk and tail components. For their special case of the mixture of exponentials bulk model with bulk model based tail fraction, they derived closed-form expressions for the exponential parameter updates in the maximization step. They applied the EM algorithm conditional on a threshold u, which was subsequently chosen by maximising the overall likelihood similar to the profile likelihood algorithm described above.

Bayesian inference has been used by many authors as a natural way to include prior information, using computational posterior sampling approaches like MCMC (Marin and Robert, 2007). For example, see Behrens et al. (2004), MacDonald et al. (2011), Scarrott and MacDonald (2010), do Nascimento et al. (2012), Zheng et al. (2014), Oumow et al. (2012), So and Chan (2014), and Cabras and Castellanos (2011). These computational posterior sampling schemes seemingly avoid the challenges associated with the multi-modal posterior. Specialised Bayesian inference schemes have been developed for the mixture of uniforms bulk model of Tancredi et al. (2006) and the Dirichlet process mixture of gammas bulk model by Fuquene (2015).

3.3   Alternative Extreme Value Mixture Models

A wide range of extreme value mixture models have been developed that do not fit within the general framework defined by Equations 3.3 or 3.5. This section gives an overview of the alternative formulations in the literature. The focus is on those which include a classical extreme value tail model, like the GPD or Pareto. Various mixture models which do not include such classical extreme value tail models have been developed to approximate extremal behaviour, which will not be discussed here. The notable exclusions are the gamma mixtures of Venturini et al. (2008) and Castillo et al. (2012) and the Dirichlet process mixture models for extremal point processes of Wang et al. (2014) and Kottas and Sanson (2007).

The inference approaches for these alternative mixture models are similar to those outlined in Section 3.2.3. Only those features in the inference procedures that are substantially different to those in Section 3.2.3 are mentioned when describing each alternative mixture model.

3.3.1 Hybrid Pareto

Arguably the closest alternative mixture model to the standard form in Equation 3.5 is the hybrid Pareto proposed by Carreau and Bengio (2006) and further developed by Carreau and Bengio (2009b), Carreau et al. (2009), and Carreau and Bengio (2009a). The hybrid Pareto splices the normal for the bulk model and GPD for upper tail, with no tail fraction scaling, giving a pdf

$$f_{hpd}(x \mid \mu, \sigma, \boldsymbol{\theta}_u) = \begin{cases} \frac{1}{\gamma}\phi(x \mid \mu, \sigma) & \text{for } x \leqslant u, \\ \frac{1}{\gamma}g(x \mid \boldsymbol{\theta}_u) & \text{for } x > u, \end{cases} \tag{3.15}$$

where the normalisation constant $\gamma = 1 + \Phi(\frac{u-\mu}{\sigma})$ ensures a proper density. The one in the normalisation constant is due to the unscaled GPD component and $\Phi(\frac{u-\mu}{\sigma})$ is the contribution from the unscaled normal bulk model. Continuity in the zeroth and first derivative is ensured by restricting the GPD scale σ_u and threshold u parameters; see Carreau and Bengio (2009b) for details. Carreau and Bengio (2009b) use likelihood inference and Carreau et al. (2009) and Carreau and Bengio (2009a) use neural network based inference.

The key difference in the hybrid Pareto to the standard model is the lack of tail fraction scaling, which leads to the GPD dominating over the bulk model. The implied threshold exceedance probability is given by

$$P(X > u) = 1 - F_{hpd}(u \mid \mu, \sigma, \boldsymbol{\theta}_u) = \frac{1}{\gamma} = \frac{1}{1 + \Phi\left(\dfrac{u - \mu}{\sigma}\right)}.$$

The normal cdf $\Phi(.) \in (0,1)$ for finite u and so the exceedance probability is between $(0.5, 1)$ and the GPD will dominate over the bulk model. The aforementioned continuity constraints applied by Carreau and Bengio (2009b) mean that the exceedance probability is not free to vary as it depends on ξ:

$$P(X > u) = \frac{1}{1 + \Phi\left\{\text{sign}(1 + \xi)\sqrt{\left[W\left(\dfrac{(1 + \xi)^2}{2\pi}\right)\right]}\right\}},$$

where $W(\cdot)$ is Lambert's W function.

The dominance of the GPD component explains why the hybrid Pareto tends to perform well for limited asymmetric and heavy tailed distributions. However, the lack of freedom in the tail fraction leads to poor performance of the hybrid Pareto in wider applications. To resolve the observed poor performance, Carreau and Bengio

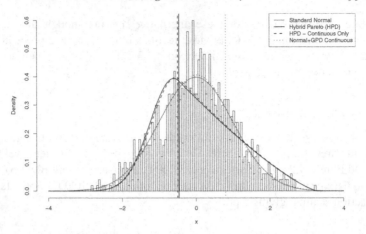

FIGURE 3.2
Density histogram of 1,000 simulated standard normal variates overlaid with the true standard normal density. The fitted (1) hybrid Pareto, (2) usual extreme value mixture model, and (3) hybrid Pareto with single continuity constraint are shown as solid, dashed, and dotted lines, respectively.

(2009b) proposed using a mixture of hybrid Paretos, which gives a mixture representation for the tail. Of course, the GPD component with the highest shape parameter will dominate the asymptotic tail behaviour, but the sub-asymptotic tail behaviour is less easy to define.

Hu (2013a) showed that the hybrid Pareto performance is improved by inclusion of the tail fraction scaling, as in the standard model in Equation 3.6. An example of the unexpected poor performance of the hybrid Pareto is demonstrated in Figure 3.2. A sample of 1,000 standard normal observations are simulated. Intuitively, one may expect the hybrid Pareto would be ideal, as the normal bulk model is the population distribution and the GPD can adequately approximate the upper tail. Figure 3.2 compares three models:

1. Hybrid Pareto as defined in Equation 3.15, with the parameter constraints by Carreau and Bengio (2009b) to ensure continuity in zeroth and first derivative of the pdf at the threshold.
2. Standard model with normal bulk and GPD tail, with bulk model based tail fraction. The results are unchanged if the parameterised tail fraction is used. The GPD scale parameter is constrained to ensure continuity of the pdf at the threshold as in Equation 3.9.
3. A less restrictive version of the hybrid Pareto, where only continuity of the pdf at the threshold is imposed by setting $\sigma_u = 1/\phi(u \mid \mu, \sigma)$.

Models (2) and (3) are directly comparable as they have similar constraints on σ_u and the only difference is the lack of tail fraction scaling in the latter. Profile likelihood estimation for the threshold is used for models (2) and (3), but this makes little difference compared to using the complete likelihood.

It is clear in Figure 3.2 that the hybrid Pareto variants perform poorly, due to the GPD domination. The hybrid Pareto with a single continuity constraint and standard model are directly comparable and clearly show the negative effect of ignoring the tail fraction scaling. The usual standard model, which includes the tail fraction scaling, is similar to the true normal density.

3.3.2 Composite Lognormal-GPD and Its Variants

A similar problem to that identified with the hybrid Pareto is also observed in the composite lognormal-Pareto model of Cooray and Ananda (2005), Preda and Ciumara (2006) and Teodorescu (2010) which has propagated to the Weibull-Pareto of Ciumara (2006) and Teodorescu and Panaitescu (2009) and exponential-Pareto of Teodorescu and Vernicu (2006). Scollnik (2007) highlighted the lack of differential scaling between the lognormal bulk and Pareto tail models and introduced a parameterised tail fraction to the composite lognormal-Pareto and lognormal-GPD as in Equation 3.5. The inclusion of the parameterised tail fraction then propagated to the exponential-Pareto by Teodorescu and Vernicu (2009), Weibull-Pareto and gamma-GPD by Teodorescu and Vernicu (2013). The latter paper also examined the properties of the general form of composite Pareto, with any bulk model.

The following exposition of these composite models focuses on the composite lognormal-GPD of Cooray and Ananda (2005), but the features are similar to the other variants mentioned. As with the hybrid Pareto, no tail fraction scaling was applied in this model, but the parameters were constrained to ensure continuity in the zeroth and first derivative of the pdf at the threshold. However, the lack of tail fraction scaling is even more restrictive for this particular composite model than the hybrid Pareto. Scollnik (2007) showed that the Pareto will dominate the upper tail as the implied tail fraction is fixed a priori at $P(X > u) \approx 0.61$, which does not vary with the threshold u.

Scollnik (2007) introduced the parameterised tail fraction to overcome this oversight, which makes the composite model equivalent to the standard model in Equation 3.6, so is not discussed in further detail. Scollnik (2007) also considered the combination of a lognormal bulk model with a GPD tail model, with both the aforementioned continuity constraints.

Nadarajah and Baka (2014) provided an extra generalisation to allow for both the bulk component to have support above the threshold and for the tail components to have support below the threshold

$$f(x \mid \boldsymbol{\theta}_b, \boldsymbol{\theta}_t, \phi_u) = \begin{cases} (1 - \phi_u)\, h_t(x \mid \boldsymbol{\theta}_b) & \text{for } x \leqslant u, \\ \phi_u\, g_t^*(x \mid \boldsymbol{\theta}_t) & \text{for } x > u. \end{cases} \quad (3.16)$$

The bulk component is truncated above the threshold as in Equation 3.7 and the upper component is truncated below the threshold

$$g_t^*(x \mid \boldsymbol{\theta}_t) = \frac{g^*(x \mid \boldsymbol{\theta}_t)}{1 - G^*(u \mid \boldsymbol{\theta}_t)}\, I(x > u), \quad (3.17)$$

where $\boldsymbol{\theta}_t$ is the parameter vector for upper tail component. Notice that even when the upper model is a truncated GPD then it will have a threshold $u^* \leqslant u$, and will reduce to the standard model in Equation 3.8 when the truncated GPD has $u^* = u$, i.e., when it is not truncated at u.

The parameter ϕ_u can be allowed to vary as an extra degree of freedom, but Nadarajah and Baka (2014) also consider restricting it to ensure continuity of the pdf at the threshold:

$$\phi_u = \frac{h(u)\,[1 - G^*(u)]}{h(u)\,[1 - G^*(u)] + g^*(u)\,H(u)} \tag{3.18}$$

where the parameters have been dropped for clarity. For the usual untruncated GPD this constraint reduces to the continuity constraint on ϕ_u given by Equation 3.11 above, as $G^*(u) = 0$.

When inference for these composite models is considered, maximum likelihood estimation is typically used with some penalties (e.g., AIC) for model comparison and to prevent overfitting.

3.3.3 Dynamically Weighted Mixture Models

Frigessi et al. (2003) were the first to propose a general transition function between the bulk and tail components, with the aim to ensure a smooth transition between them. Their dynamically weighted extreme value mixture model (DWM) pdf is given by

$$f_{dwm}(x \mid \boldsymbol{\theta}_b, \boldsymbol{\theta}_w, \boldsymbol{\theta}_{-u}) = \frac{1}{\gamma}\left\{[1 - p(x \mid \boldsymbol{\theta}_w)]\,h(x \mid \boldsymbol{\theta}_b) + p(x \mid \boldsymbol{\theta}_w)\,g(x \mid u = 0, \boldsymbol{\theta}_{-u})\right\}, \tag{3.19}$$

where $x \geqslant 0$ and the parameter vectors $\boldsymbol{\theta}_b$ and $\boldsymbol{\theta}_w$ are for the bulk model and weight function, respectively. The transition function $p(.)$ can be any function with domain $[0, 1]$, but is typically a strictly monotonically increasing function. The normalisation constant γ ensures the resultant density is proper and so is a function of the other parameters.

Notice that in the original definition of Frigessi et al. (2003) the DWM has no threshold, as the GPD threshold is prefixed at zero and the notation $\boldsymbol{\theta}_{-u}$ indicates the exclusion of the threshold from the GPD parameters. In practice, the data can be shifted to be bounded below by zero when applying this model, so the nonnegative support is not so restrictive. It is also possible to extend the lower bound on the support by suitable choice of the bulk model and transition function. However in this case, care has to be taken in specifying the GPD threshold and transition function to ensure the pole of the GPD at the threshold does not result in a spurious peak in the resulting density.

It should be noted that the choice to set the GPD threshold to zero in the original formulation in Equation 3.19 is entirely arbitrary. It has been set at an intuitively sensible value, but a priori choice of the threshold will now be shown to impact the implied tail fraction for this model and so some care has to be taken in its specification.

Frigessi et al. (2003) state that if the transition function is a Heaviside function with step at some threshold u^* given by:

$$p(x \mid u^*) = \begin{cases} 0 & \text{for } x \leqslant u^*, \\ 1 & \text{for } x > u^*, \end{cases} \qquad (3.20)$$

then the more usual extreme value mixture model of Behrens et al. (2004) in Equation 3.3 results as a special case. This is not quite as simple as it first seems, as can be seen from the resultant pdf

$$f(x \mid \boldsymbol{\theta}_b, u^*, \boldsymbol{\theta}_{-u}) = \begin{cases} \dfrac{h(x \mid \boldsymbol{\theta}_b)}{H(u^* \mid \boldsymbol{\theta}_b) + [1 - G(u^* \mid u = 0, \boldsymbol{\theta}_{-u})]} & \text{for } x \leqslant u^*, \\ \dfrac{g(x \mid u = 0, \boldsymbol{\theta}_{-u})}{H(u^* \mid \boldsymbol{\theta}_b) + [1 - G(u^* \mid u = 0, \boldsymbol{\theta}_{-u})]} & \text{for } x > u^*. \end{cases}$$

$$(3.21)$$

The Heaviside threshold has been denoted as u^* to differentiate it from the GPD threshold u. On first sight one gets the impression that this model falls into the same trap as that of the hybrid Pareto in Equation 3.15, by failing to apply a tail fraction scaling to the GPD tail model and the normalisation is the same for the bulk and tail components.

However, there is a subtle difference which means that a tail fraction scaling is implicit in this model, but care has to be taken in its specification. The threshold exceedance probability is given by

$$P(X > u^*) = 1 - F(u^* \mid \boldsymbol{\theta}_b, u^*, \boldsymbol{\theta}_{-u}) = \frac{1 - G(u^* \mid u = 0, \boldsymbol{\theta}_{-u})}{H(u^* \mid \boldsymbol{\theta}_b) + [1 - G(u^* \mid u = 0, \boldsymbol{\theta}_{-u})]}.$$

This form of tail fraction is free to vary over u^*, dependent on the bulk and tail model parameters. In particular, it is dependent on the a priori fixed threshold of the GPD at $u = 0$, so this must be pre-chosen with care.

The concept underlying the DWM is that a light upper tailed distribution is used for the bulk distribution, so that in combination with the smooth transition function the GPD can dominate the upper tail. Frigessi et al. (2003) and Vrac and Naveau (2007) use Weibull and gamma bulk models, respectively. In both cases, likelihood inference is employed using the complete likelihood. The example from experiment 1 of Frigessi et al. (2003) is depicted in Figure 3.3. A Weibull($\lambda = 1/\Gamma(1.5)$, $k = 2$) distribution is used for the bulk, GPD($\sigma_u = 1$, $\xi = 0.5$), and Cauchy($\mu = 1$, $\sigma = 1$) cdf for the transition function.

When developing dynamically weighted mixture models it should be noted that the tail behaviour is determined by all three components, as there is no guarantee that the GPD will dominate in general. A sufficiently light tailed bulk distribution and transition function which converges at a sufficiently rapid rate to one, are needed to ensure that the GPD dominates.

A feature of this model to be aware of is that the GPD reaches a peak at zero (or wherever the GPD lower bound is defined). Therefore, the transition function must decay to zero quickly enough towards this lower bound to ensure no spurious peak

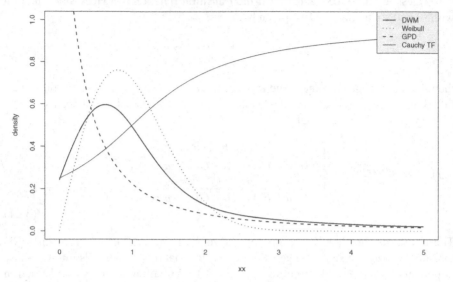

FIGURE 3.3
Example of dynamically weighted extreme value mixture model density with Weibull($\lambda = 1/\Gamma(1.5)$, $k = 2$) bulk pdf, GPD($\sigma_u = 1$, $\xi = 0.5$) pdf and Cauchy($\mu = 1$, $\sigma = 1$) cdf mixing function.

is created. This feature also means that the GPD parameter estimates are strongly influenced by not only the tail observations, but also the lower tail observations near lower bound.

3.3.4 Interval Transition Mixture Model

The dynamically weighted mixture model of Frigessi et al. (2003) is extended by Holden and Haug (2009) to transition between the bulk and tail models over an interval. The cdf of the interval transition mixture model is

$$F(x \,|\, \boldsymbol{\theta}_b, \boldsymbol{\theta}_u, \boldsymbol{\theta}_q, \boldsymbol{\theta}_p) = \frac{1}{\gamma} \left[H_t(q(x \,|\, \boldsymbol{\theta}_q) \,|\, \boldsymbol{\theta}_b) + G_t(p(x \,|\, \boldsymbol{\theta}_p) \,|\, \boldsymbol{\theta}_u) \right], \qquad (3.22)$$

where the normalisation constant γ ensures a proper density. The mixing functions are $p()$ and $q()$ provide a warping of the bulk and tail model cdf's, the properties of which are outlined below. The *bulk and tail models are truncated but are not necessarily proper*, with pdf's given by

$$h_t(x \,|\, \boldsymbol{\theta}_b) = \left\{ \begin{array}{ll} h(x \,|\, \boldsymbol{\theta}_b) & \text{for } x \leqslant u, \\ 0 & \text{for } x > u, \end{array} \right. \qquad g_t(x \,|\, \boldsymbol{\theta}_u) = \left\{ \begin{array}{ll} 0 & \text{for } x \leqslant u, \\ g(x \,|\, \boldsymbol{\theta}_u) & \text{for } x > u. \end{array} \right.$$
$$(3.23)$$

The normalisation constant is therefore $\gamma = H_t(\infty \,|\, \boldsymbol{\theta}_b) + G_t(\infty \,|\, \boldsymbol{\theta}_u)$. This is a slight abuse of notation, as in previous definitions the subscript t meant a proper truncated distribution, but in this section they may not be proper. If the bulk and tail models are truncated (and proper) then $\gamma = 2$.

The mixing functions apply over an interval $[u - \epsilon, u + \epsilon]$. The pdf defined over the three sub-intervals is given by

$$
f(x) = \begin{cases}
\frac{1}{\gamma} h(x) & \text{for } x \leqslant u - \epsilon, \\
\frac{1}{\gamma} \left[h(q(x)) \, q'(x) + g(p(x)) \, p'(x) \right] & \text{for } u - \epsilon < x \leqslant u + \epsilon \\
\frac{1}{\gamma} g(x) & \text{for } x > u + \epsilon,
\end{cases} \tag{3.24}
$$

where the parameters are dropped for clarity. Notice that this pdf formulation does not use truncated densities, as this is taken care of by the subintervals and that the mixing functions only enact within the mixing interval. The definition of the pdf makes it clear that the mixing functions are warping the bulk and tail model pdf's followed by re-weighting by their derivatives.

If there is an instant transition the $\epsilon = 0$ and $q(x \,|\, \theta_q) = p(x \,|\, \theta_p) = x$, $\forall \, x$, the transition interval disappears and the bulk and tail densities apply below and above the threshold, respectively.

We will now introduce the mixing functions properties. In general, within the transition interval the mixing function $q(.)$ spreads from the full interval $[u - \epsilon, u + \epsilon]$ to the smaller interval $[u - \epsilon, u]$ and similarly $p(.)$ maps the full interval to $[u, u + \epsilon]$, so that the bulk model will be spread beyond the threshold to $u + \epsilon$ and the tail model is spread below the threshold to $u - \epsilon$. No warping is applied below the interval for the bulk model and above the interval for the tail model, so the identity function applies:

- $q(x \,|\, \theta_q) = x$ for $x \leqslant u - \epsilon$ and
- $p(x \,|\, \theta_p) = x$ for $x \geqslant u + \epsilon$.

Further, the bulk cdf makes a consistent contribution above threshold and the tail cdf a constant contribution below the threshold so that:

- $q(x \,|\, \theta_q) \geqslant u$ for $x > u + \epsilon$ and
- $p(x \,|\, \theta_p) \leqslant u$ for $x < u - \epsilon$.

Holden and Haug (2009) set these to $x - \epsilon$ and $x + \epsilon$, respectively. Continuity of the pdf at the transition interval lower limit requires:

- $q(u - \epsilon \,|\, \theta_q) = u - \epsilon$ and $q'(u - \epsilon \,|\, \theta_q) = 1$; and
- $p'(u - \epsilon \,|\, \theta_p) = 0$,

and for upper transition interval limit:

- $q'(u + \epsilon \,|\, \theta_q) = 0$; and
- $p(u + \epsilon \,|\, \theta_p) = u + \epsilon$ and $p'(u + \epsilon \,|\, \theta_p) = 1$.

Continuity in first derivative of the pdf is further assured under the additional constraints of $q''(u - \epsilon \,|\, \theta_q) = p''(u + \epsilon \,|\, \theta_p) = 0$. Appropriate normalisation of the weights within the transition interval requires:

$$
q'(x \,|\, \theta_q) + p'(x \,|\, \theta_p) = 1 \quad \text{for } x \in [u - \epsilon, u + \epsilon].
$$

Although not required, it is physically sensible to have a symmetric contribution from the bulk and tail components at the threshold which gives $q'(u \,|\, \theta_q) = p'(u \,|\, \theta_p) = 0.5$.

Holden and Haug (2009) suggest the following mixing functions:

$$q(x \mid \boldsymbol{\theta}_q) = \begin{cases} x & \text{for } x \leqslant u - \epsilon, \\ \frac{1}{2}(x + u - \epsilon) + \frac{\epsilon}{\pi} \cos\left(\frac{\pi(x-u)}{2\epsilon}\right) & \text{for } u - \epsilon < x < u + \epsilon \\ x - \epsilon & \text{for } x \geqslant u + \epsilon, \end{cases} \qquad (3.25)$$

and

$$p(x \mid \boldsymbol{\theta}_q) = \begin{cases} x + \epsilon & \text{for } x \leqslant u - \epsilon, \\ \frac{1}{2}(x + u + \epsilon) - \frac{\epsilon}{\pi} \cos\left(\frac{\pi(x-u)}{2\epsilon}\right) & \text{for } u - \epsilon < x < u + \epsilon \\ x & \text{for } x \geqslant u + \epsilon, \end{cases} \qquad (3.26)$$

It is quickly checked that all the above requirements are met. Note that bulk model transition function $q(.)$ is not continuous in first derivative at the upper bound of the transition interval and similarly for the tail model transition function $p(.)$ at the lower bound of the transition interval. The bulk and tail model transition functions warp in a symmetric manner about the threshold, due to the cosine terms having alternate signs over the interval width 2ϵ.

Holden and Haug (2009) use maximum likelihood estimation over the complete likelihood. However, they note problems with lack of identifiability of the interval half-width ϵ, so they constrain it to be equal to the estimated value of normal bulk model standard deviation parameter.

This model formulation is interesting as it intuitively allows for a smooth transition between the two components, without suffering the problem of dealing with the pole of the GPD at the threshold as seen with the dynamically weighted mixture model described in the previous section. Unfortunately, it has the same flaw as that of the hybrid Pareto due to the lack of tail fraction scaling, which is most easily seen in the special case of $\epsilon = 0$ which gives pdf

$$f(x) = \begin{cases} \frac{1}{\gamma} h(x \mid \boldsymbol{\theta}_b) & \text{for } x \leqslant u, \\ \frac{1}{\gamma} g(x \mid \boldsymbol{\theta}_u) & \text{for } x > u. \end{cases} \qquad (3.27)$$

If the tail component is a GPD with support $x > u$ then $\gamma = H(u \mid \boldsymbol{\theta}_b) + 1$. Thus this mixture model reduces to exactly the hybrid Pareto in Equation 3.15 with the same features, including the lack of tail fraction scaling.

This oversight is rather more serious for this model, as the derivation suggests that truncated bulk and tail models could be used, which could also be proper. In the case of the $\epsilon = 0$ the normalisation constant becomes $\gamma = 2$ and as such the bulk and tail component will contribute equally to the density, which is a similar flaw to the a priori fixed tail fraction seen in the composite lognormal-GPD of Cooray and Ananda (2005).

The lack of tail fraction scaling could be overcome by a priori fixing the GPD threshold at some value v below that of the mixture model's threshold so that $v < u$. The tail model is then truncated below u and its contribution decreases with u. In this case, the model reduces to exactly the dynamical weight mixture model with a Heaviside function as in Equation 3.21. As such, it suffers from the same challenges associated with the a priori choice of GPD threshold and how this influences the tail fraction.

Holden and Haug (2009) comment that the standard model with a bulk model based tail fraction as in Equation 3.3 could be derived by setting $\epsilon = 0$ and choosing an appropriate c so that $c\,g(x\mid\boldsymbol{\theta}_b)$ gives $\gamma = 1$. This is exactly the parameterised tail fraction approach used in the standard model, but amounts to a priori defining the tail fraction, which is less than ideal. However, a slight adaptation of this suggestion to specify c using the threshold exceedance probability of the bulk model or as a parameter to be estimated, as in the standard model in Equations 3.4 and 3.6, provides a strategy for incorporating the tail fraction scaling to overcome this potential deficiency.

3.4 Implementation in R Packages

Various authors have either provided computer code in their publications or as supplementary materials. Most of the publicly available computer programs for extreme values mixture models are available for R (R Core Team, 2013), so this review will focus on these resources only. The key R packages for extreme value mixtures models are evmix (Scarrott and Hu, 2015), condmixt (Carreau, 2012), and CompLognormal (Nadarajah, 2013).

The compLognormal package results from the work of Nadarajah and Baka (2014) which provides a generalisation of the composite model with a lognormal bulk component below the threshold combined with essentially any model above the threshold. The upper model may or may not necessarily be an extreme value tail model and is not necessarily a priori truncated at the threshold. Only the most basic functionality is provided by the usual quartet (density, distribution, quantile, and random number generation functions). No model fitting or diagnostics are currently provided.

The condmixt package implements the hybrid Pareto discussed in Section 3.3.1, the mixture of hybrid Paretos and its many variants. A key variant replaces the normal bulk model with a mixture of normals, but retains the GPD tail model. Inference is provided by method of moments, maximum likelihood estimation for the usual hybrid Pareto, and neural networks for the more advanced models. Nonstationary (conditional) variants of all these models are provided.

The broadest collection of the extreme value mixture models (standard and alternative formulations) are provided in the evmix package, which was foreshadowed by the review of threshold estimation by Scarrott and MacDonald (2012). Hu (2013a) produced the original package, following an extensive simulation study comparing the performance of many such models, including the effects of the bulk or parameterised tail fractions and continuity constraints.

Examples of using the evmix package are provided in the supplementary materials. Further examples are provided in a short user guide by Hu (2013b) and a comprehensive description of the package features provided by Hu and Scarrott (2013), which are both distributed as part of the package help files.

3.5 Advice for Developers and Practitioners

A guiding principle in developing, or choosing, extreme value mixture models is to combine a suitable bulk model, or at least a flexible bulk model, with the tail model. If this is successfully achieved then these models and inference schemes can provide an automated and objective approach to threshold and tail estimation, including uncertainty quantification. One of the most serious concerns with applying these models is to ensure that the tail estimates are robust to the bulk data/model fit. The two components are not independent as they share some common information, particularly about the threshold. The following advice is borne out of personal experience, direct comparison of the model properties, and various simulation studies comparing their performance.

Hu (2013a) provides an extensive simulation study comparing the performance of many mixture models in approximating the tails of a variety of population distributions for samples of size 1,000 and 5,000. The performance in estimating the 10, 5, 1, and 0.1% upper tail quantiles (and lower tail in two-tailed models) is evaluated using the root mean squared error (RMSE).

There are some limitations in this simulation study, which will now be discussed. However, the following advice is unlikely to change as most points are rather intuitive. The key limitation is that the simulations used maximum likelihood estimation over the complete likelihood, optimised using BFGS algorithm, with the initial value for the threshold set as the population's 90% quantile (or 10% quantile for lower tail). As discussed in Section 3.2.3, profile likelihood estimation of the threshold is preferred to ameliorate the problems associated with the inherent multi-modal likelihood of these models. A direct comparison of likelihood and computational Bayesian inference based quantile estimation for some of these models was also carried out, with the same broad conclusions reached which gives confidence in the following advice.

Hu (2013a) measured the quantile estimation performance using the RMSE, which does not penalise for the degrees of freedom used in the model fit. As such, an advantage is provided to models which expend extra degrees of freedom, e.g., when using the parameterised tail fraction. The nonparametric and semiparametric extreme value mixture models may also expend more equivalent degrees of the freedom. However, the advantage they receive is likely to be small as these degrees of freedom are expended on the bulk model, rather than on the tail(s) where the performance is being evaluated.

1. A key insight from the model comparisons in this chapter is *not to forget about the tail fraction scaling*. This is a key feature of the classical extreme value tail modelling approach in Equation 3.2, but has been left out of some of the early

developments of these models, which has led to some performance issues in wider applications.

2. The majority of the mixture models use a bulk model based tail fraction, presuming it continues beyond the threshold. This approach has the benefit of making use of the ample data in the bulk of the distribution. However, Section 3.2 mentioned that this approach could be considered counterintuitive. A key reason for using a tail model is that we do not expect the bulk model to provide a good fit to the tail and so it may not be a good estimator of the tail fraction.

   The simulations of Hu (2013a) confirmed that, in general, it is better to use the parameterised tail fraction approach as this provides some robustness of the tail fit to that of the bulk. However, efficiency is gained if the bulk model is known to be correct (or sufficient data is available to determine that it is a good fit).

3. Parameter constraints to ensure continuity of the pdf at the threshold are physically sensible. However, Hu (2013a) found that with a well-specified mixture model (i.e., sensible bulk and tail components, and parameterised tail fraction) that little benefit is gained in practice. Further, inclusion of continuity constraints induces dependence between the bulk and tail parameters, which leads to a lack of robustness as discussed in item 2.

4. For well-specified mixture models the RMSE for extreme quantiles (e.g., 0.1% tail quantile for samples of size 1,000 and 5,000) are similar, whereas for the less extreme quantiles (e.g., 10 and 5% tail quantiles) the RMSE is much more sensitive to the mixture model specification. Essentially, these quantile estimates are closer to (and thus more sensitive to) the threshold and sensitive to the behaviour of density near it.

5. The usual consideration for choosing between parametric and semi/nonparametric estimators holds for these mixture models. If good prior information is available, then better are estimates are obtained with an appropriate parametric bulk model. However, in the more usual situation of an unknown population distribution the semi/nonparametric bulk models provide reliable tail estimation performance, similar to correctly specified parametric bulk models.

6. The complete likelihood typically exhibits multiple modes, e.g., as multiple thresholds can give similar likelihoods. As such, optimisation of the likelihood is challenging and profile likelihood estimation of the threshold is commonly used; see Section 3.2.3 for details.

References

Balkema, A. A. and L. de Haan (1974). Residual life time at great age. *Annals of Probability* 2(5), 792–804.

Bee, M. (2012). Statistical analysis of the lognormal-Pareto distribution using probability weighted moments and maximum likelihood. Technical Report 08/2012, University of Trento, Italy.

Behrens, C. N., H. F. Lopes, and D. Gamerman (2004). Bayesian analysis of extreme events with threshold estimation. *Statistical Modelling 4*(3), 227–244.

Cabras, S. and M. E. Castellanos (2011). A Bayesian approach for estimating extreme quantiles under a semiparametric mixture model. *ASTIN Bulletin 41*(1), 87–108.

Carreau, J. (2012). *condmixt: Conditional Density Estimation with Neural Network Conditional Mixtures*. R package version 1.0.

Carreau, J. and Y. Bengio (2006). A hybrid Pareto model for asymmetric fat-tail data. Technical Report 1283, University of Montreal.

Carreau, J. and Y. Bengio (2009a). A hybrid Pareto mixture for conditional asymmetric fat-tailed distributions. *IEEE Transaction on Neural Networks 20*(7), 1087–1101.

Carreau, J. and Y. Bengio (2009b). A hybrid Pareto model for asymmetric fat-tailed data: The univariate case. *Extremes 12*(1), 53–76.

Carreau, J., P. Naveau, and E. Sauquet (2009). A statistical rainfall-runoff mixture model with heavy-tailed components. *Water Resources Research 45*, W10437.

Castillo, J. D., J. Daoudi, and I. Serra (2012). The full-tails gamma distribution applied to model extreme values. arXiv:1211.0130.

Ciumara, R. (2006). An actuarial model based on the composite Weibull-Pareto distribution. *Mathematical Reports — Bucharest 8*(4), 401–414.

Coles, S. G. (2001). *An Introduction to Statistical Modelling of Extreme Values*. London: Springer-Verlag.

Cooray, K. and M. M. A. Ananda (2005). Modeling actuarial data with a composite lognormal-Pareto model. *Scandinavian Actuarial Journal 5*, 321–334.

Couturier, D. and M. Victoria-Feser (2010). Zero-inflated truncated generalized Pareto distribution for the analysis of radio audience data. *The Annals of Applied Statistics 4*(4), 1824–1846.

do Nascimento, F. F., D. Gamerman, and H. F. Lopes (2012). A semiparametric Bayesian approach to extreme value estimation. *Statistics and Computing 22*(2), 661–675.

Frigessi, A., O. Haug, and H. Rue (2003). A dynamic mixture model for unsupervised tail estimation without threshold selection. *Extremes 5*(3), 219–235.

Fuquene, J. (2015). A semiparametric Bayesian extreme value model using a Dirichlet process mixture of gamma densities. *Journal of Applied Statistics 42*, 267–280.

Gonzalez, J., D. Rodriguez, and M. Sued (2013). Threshold selection for extremes under a semiparametric model. *Statistical Methods and Applications 22*, 481–500.

Holden, L. and O. Haug (2009). A mixture model for unsupervised tail estimation. arXiv.0902.4137.

Hsing, T. (1987). On the characterization of certain point processes. *Stochastic Processes and Their Applications 26*(2), 297–316.

Hsing, T., J. Hüsler, and M. R. Leadbetter (1988). On the exceedance point process for a stationary sequence. *Probability Theory and Related Fields 78*(1), 97–110.

Hu, Y. (2013a). Extreme value mixture modeling with simulation study and applications in finance and insurance. MSc thesis, University of Canterbury, New

Zealand. Available at http://ir.canterbury.ac.nz/simple-search? query=extreme&submit=Go.

Hu, Y. (2013b). User's guide for evmix package in R. Available from http://www.math.canterbury.ac.nz/~c.scarrott/evmix.

Hu, Y. and C. J. Scarrott (2013). evmix: An R package for extreme value mixture modelling, threshold estimation and boundary corrected kernel density estimation. Available from http://www.math.canterbury.ac.nz/~c.scarrott/evmix.

Kottas, A. and B. Sanson (2007). Bayesian mixture modeling for spatial Poisson process intensities, with applications to extreme value analysis. *Journal of Statistical Planning and Inference 137*, 3151–3163.

Leadbetter, M. R., G. Lindgren, and H. Rootzén (1983). *Extremes and Related Properties of Random Sequences and Series*. London: Springer-Verlag.

Lee, D., W. K. Li, and T. S. T. Wong (2012). Modeling insurance claims via a mixture exponential model combined with peaks-over-threshold approach. *Insurance: Mathematics and Economic 51*(3), 538–550.

Lee, T. M. C. (2013). Modelling non-stationary extremes and threshold uncertainty. In *8th Conference on Extreme Value Analysis*. Available from http://people.math.gatech.edu/~peng/EVAFuDan/doc/EVA2013-ProgramAbstract.pdf.

MacDonald, A. (2012). *Extreme Value Mixture Modelling with Medical and Industrial Applications*. PhD thesis, University of Canterbury, New Zealand.

MacDonald, A., C. J. Scarrott, D. Lee, B. Darlow, M. Reale, and G. Russell (2011). A flexible extreme value mixture model. *Computational Statistics and Data Analysis 55*(6), 2137–2157.

MacDonald, A., C. J. Scarrott, and D. S. Lee (2013). Boundary correction, consistency and robustness of kernel densities using extreme value theory. Available from http://www.math.canterbury.ac.nz/~c.scarrott.

Marin, J. M. and C. P. Robert (2007). *Bayesian Core*. New York: Springer-Verlag.

McLachlan, G. and D. Peel (2000). *Finite Mixture Models*. Wiley.

Mendes, B. and H. F. Lopes (2004). Data driven estimates for mixtures. *Computational Statistics and Data Analysis 47*(3), 583–598.

Nadarajah, S. (2013). *CompLognormal: Functions for Actuarial Scientists*. R package version 3.0.

Nadarajah, S. and S. A. A. Baka (2014). New composite models for the Danish fire insurance data. *Scandinavian Actuarial Journal 2014*, 180–187.

Northrop, P. J. and P. Jonathan (2011). Threshold modelling of spatially dependent non-stationary extremes with application to hurricane-induced wave heights. *Environmetrics 22*(7), 799–809.

Oumow, B., M. de Carvalho, and A. C. Davison (2012). Bayesian P-spline mixture modeling of extreme forest temperatures. Submitted. Available from http://www.cma.fct.unl.pt/en/cma-9-2012.

Papastathopoulos, I. and J. A. Tawn (2013). Extended generalised Pareto models for tail estimation. *Journal of Statistical Planning and Inference 143*(1), 131–143.

Pawitan, Y. (2001). *In All Likelihood: Statistical Modelling and Inference Using Likelihood.* Oxford University Press, Oxford.

Pickands, III, J. (1975). Statistical inference using extreme order statistics. *Annals of Statistics 3*(1), 119–131.

Pigeon, M. and M. Denuit (2011). Composite lognormal-Pareto model with random threshold. *Scandinavian Actuarial Journal 2011*, 177–192.

Preda, S. and R. Ciumara (2006). On composite models: Weibull-Pareto and lognormal-Pareto — A comparative study. *Romanian Journal of Economic Forecasting 7*(2), 32–46.

R Core Team (2013). *R: A Language and Environment for Statistical Computing.* Vienna, Austria: R Foundation for Statistical Computing.

Robert, C. P. (1998). *Discretization and MCMC Convergence Assessment*, Volume 135 of *Lecture Notes in Statistics.* New York: Springer-Verlag.

Robert, C. P. and G. Casella (2010). *Monte Carlo Statistical Methods.* Springer-Verlag, New York.

Scarrott, A. and C. J. MacDonald (2010). Risk assessment for critical temperature exceedance in nuclear reactors. *Journal of Risk and Reliability 224*(4), 239–252.

Scarrott, C. J. and Y. Hu (2015). evmix: Extreme value mixture modelling, threshold estimation and boundary corrected kernel density estimation. R Package Version 0.2.5.

Scarrott, C. J. and A. MacDonald (2012). A review of extreme value threshold estimation and uncertainty quantification. *REVSTAT Statistical Journal 10*(1), 33–60.

Scollnik, D. P. M. (2007). On composite lognormal-Pareto models. *Scandinavian Actuarial Journal 2007*, 20–33.

So, M. K. P. and R. K. S. Chan (2014). Bayesian analysis of tail asymmetry based on a threshold extreme value model. *Computational Statistics and Data Analysis 71*, 568–587.

Solari, S. and M. A. Losada (2004). A unified statistical model for hydrological variables including the selection of threshold for the peak over threshold method. *Water Resources Research 48*, W10541.

Tancredi, A., C. W. Anderson, and A. O'Hagan (2006). Accounting for threshold uncertainty in extreme value estimation. *Extremes 9*(2), 87–106.

Teodorescu, S. (2010). On the truncated composite lognormal-Pareto models. *Romanian Academy Mathematical Reports 10*(1), 71–84.

Teodorescu, S. and E. Panaitescu (2009). On the truncated composite Weibull-Pareto model. *Mathematical Reports — Bucharest 11*(3), 259–273.

Teodorescu, S. and R. Vernicu (2006). A composite exponential-Pareto distribution. *Annals of the Ovidius University of Constanta 14*(1), 99–108.

Teodorescu, S. and R. Vernicu (2009). Some composite exponential-Pareto models for actuarial prediction. *Romanian Journal of Economic Forecasting 4*(1), 82–100.

Teodorescu, S. and R. Vernicu (2013). On composite Pareto models. *Romanian Academy Mathematical Reports 15*(1), 11–30.

Venturini, S., F. Dominici, and G. Parmigiani (2008). Gamma shape mixtures for heavy-tailed distributions. *The Annals of Applied Statistics 2*(2), 756–776.

Vrac, M. and P. Naveau (2007). Stochastic downscaling of precipitation: From dry events to heavy rainfalls. *Water Resources Research 43*, W07402.

Wang, Z., A. Rodriguez, and A. Kottas (2014). A nonparametric mixture modeling framework for extreme value analysis. Technical Report SOE 2012-02, University of California, Santa Cruz.

Wong, T. S. T. and W. K. Li (2010). A threshold approach for peaks-over-threshold modeling using maximum product of spacings. *Statistica Sinica 20*, 1257–1272.

Zhao, X., C. J. Scarrott, L. Oxley, and M. Reale (2014). "Let the tails speak for themselves": Bayesian extreme value mixture modelling for estimating VaR. Available from: http://www.math.canterbury.ac.nz/~c.scarrott.

Zhao, X., C. J. Scarrott, M. Reale, and L. Oxley (2010). Extreme value modelling for forecasting the market crisis. *Applied Financial Economics 20*(1), 63–72.

Zheng, L., K. Ismail, and X. Meng (2014). Shifted gamma-generalized Pareto distribution model to map the safety continuum and estimate crashes. *Safety Science 64*(1), 155–162.

4

Threshold Selection in Extreme Value Analysis

Frederico Caeiro

Universidade Nova de Lisboa (FCT and CMA), Portugal

M. Ivette Gomes

Universidade de Lisboa (DEIO and CEAUL), Portugal

Abstract

The main objective of *statistics of extremes* is the prediction of rare events, and its primary problem has been the estimation of the *extreme value index* (EVI). When we are interested in large values, estimation is usually performed on the basis of the largest $k + 1$ order statistics in the sample or on the excesses over a high level u. A question that has been often addressed in practical applications of *extreme value theory* is the choice of either k or u, and an adaptive EVI-estimation. The choice can be either heuristic or based on sample path stability or on the minimization of a mean square error estimate as a function of k. Some of these procedures will be reviewed. All these methods can be applied, with adequate modifications, not only to the adaptive EVI-estimation but also to the adaptive estimation of other relevant right tail parameters. We shall illustrate the methods essentially for a positive EVI.

4.1 Introduction

As usual, let us consider a sample $(X_1, \ldots, X_n)$ of independent, identically distributed or possibly weakly dependent and stationary random variables from an underlying population with unknown *cumulative distribution function* (CDF) $F(\cdot)$. Let us further use the notation $(X_{1:n} \leqslant \cdots \leqslant X_{n:n})$ for the associated sample of ascending order statistics.

The main objective of statistics of univariate extremes is the prediction of rare events, i.e., estimation of parameters such as *high quantiles*, *return periods*, the *extremal index*, and many other parameters or functionals related to extreme events. The primary question is the estimation of the parameter ξ in the extreme value CDF,

$$G_{\xi,\mu,\sigma}(x) = \begin{cases} \exp\left(-\left(1+\xi\frac{x-\mu}{\sigma}\right)\right)^{-1/\xi}, & \text{if } \xi \neq 0, \\ \exp\left(-\exp\left(-\frac{x-\mu}{\sigma}\right)\right), & \text{if } \xi = 0, \end{cases} \quad \mu \in \mathbb{R}, \ \sigma > 0. \quad (4.1)$$

The parameter ξ, in (4.1), is the *extreme value index* (EVI) for large values or for maxima. But since $\min(X_1,\ldots,X_n) = -\max(-X_1,\ldots,-X_n)$, results for small values can be easily derived from the analogous results for large events, and therefore we shall deal with the right tail or survival function,

$$\overline{F}(x) := 1 - F(x),$$

of the underlying CDF, F. One of the most recent and general approaches to statistics of univariate extremes (see Beirlant et al., 2004, and de Haan and Fereira, 2006, among others) is the semiparametric method. The estimation of ξ is then based on the largest $k+1$ order statistics in the sample or on the excesses over a high level u, requiring thus the choice of a threshold, either random or deterministic, at which the tail of the underlying distribution begins. The question that has been often addressed in practical applications of extreme value theory is thus the choice of either k or u and an adaptive estimation of the parameter of interest. We shall address the choice of k, and concomitant estimation of ξ, in (4.1). Indeed, we can always consider that u is chosen in the interval $[X_{n-k:n}, X_{n-k+1:n})$ or even that $u = X_{n-k:n}$. Then, our functionals of u can be transformed in functionals of an integer k in $[1, n)$, related to the number of top order statistics to be used in the estimation.

Ideally, the estimates of ξ, as a function of k, should not be highly sensitive to small changes in the threshold. Unfortunately, this is not always the case, and the most trivial example of this is the Hill (H) EVI-estimator, based on the log-excesses over an order statistic $X_{n-k:n}, 1 \leqslant k < n$,

$$\widehat{\xi}_{k,n}^{\mathrm{H}} := \frac{1}{k} \sum_{i=1}^{k} \left\{ \ln X_{n-i+1:n} - \ln X_{n-k:n} \right\}, \quad (4.2)$$

the classical EVI-estimator of a positive EVI (Hill, 1975). For the H EVI-estimator an inadequate choice of k can lead to large expected errors, essentially due to the fact that for small values of k, $\widehat{\xi}_{k,n}^{H}$ has a large variance, whereas large values of k usually induce a high bias in $\widehat{\xi}_{k,n}^{H}$. This leads to a very peaked *mean square error* (MSE), as a function of k. The problem of excessive oscillation of the H EVI-estimator has been discussed by Resnick and Stărică (1997), who recommend smoothing the H EVI-estimator by integrating it over a moving window, i.e., averaging $\widehat{\xi}_{k,n}^{H}$ over adequate k. Kernel EVI-estimators (Csörgő et al., 1985; Groeneboom et al., 2003) and semiparametric probability-weighted-moment EVI-estimators, like the ones in Caeiro and Gomes (2011) are also nice alternatives, usually providing much less volatile sample paths as a function of k. And nowadays, at the current state-of-the-art, and particularly for heavy tails, there are several reduced-bias alternatives to the

H EVI-estimator that are easy to use and much less sensitive to the choice of k. For recent overviews on reduced-bias estimation of parameters of extreme events, see Reiss and Thomas (2007), Beirlant et al. (2012), and Gomes and Guillou (2015). Practitioners using this type of semiparametric EVI-estimators can then get close to optimal EVI-estimates by using an estimator that is smooth in the tuning parameter k and that thus facilitates the threshold selection.

The problem of choosing the threshold k remains of high theoretical and practical interest. The existing methods rely either on heuristic and on graphical procedures based on sample paths' stability as a function of k or minimization of MSE's estimates, also as functions of k. Some of these procedures will be reviewed and applied to the estimation of a positive EVI.

The most common methods of adaptive choice of the threshold k are based on the minimization of an MSE estimate. We mention the pioneering papers by Hall (1982) and Hall and Welsh (1985), and discuss the role of the bootstrap methodology in the selection of $k_0(n) := \arg\min_k \mathrm{MSE}(\widehat{\xi}_{k,n})$ (Hall, 1990; Draisma et al., 1999; Danielsson et al., 2001; Gomes and Oliveira, 2001; Gomes et al., 2011, 2012). The regression methodologies developed in Beirlant et al. (2002, 1996a,b) have also the objective of minimization of the MSE. In Beirlant et al. (2002), the exponential regression model, introduced in Beirlant et al. (1999), is considered and applied to the selection of the optimal sample fraction, k/n, in extreme value estimation. A connection between such a choice strategy and the diagnostic proposed in Guillou and Hall (2001), essentially based on bias behaviour, is also provided. Drees and Kaufmann (1998) also propose a method partially based on bias properties. Coles (2001) outlines the common graphical diagnostics for threshold choice: mean residual life (or mean excess) plots, threshold stability plots, and all the usual distribution fit diagnostics, such as probability plots, quantile plots and return level plots (see also Beirlant at al., 2004, and references therein). For comparisons of different adaptive procedures on the basis of extensive small sample simulations, see Matthys and Beirlant (2000), Gomes and Oliveira (2001), and Beirlant et al. (2002). A recent overview of the topic can be found in Scarrott and MacDonald (2012). The optimal sample fraction for second-order reduced-bias estimators is under development. Possible heuristic choices are provided in Gomes and Pestana (2007), Gomes et al. (2013, 2011, 2008), Figueiredo et al. (2012), and Neves et al. (2015), among others.

In Section 4.2, we review graphical tools for threshold selection. In Section 4.3, details on classical EVI-estimation for heavy right tails, together with a direct threshold estimation, are provided. We further mention an algorithm for the estimation of second-order parameters, denoted by (β, ρ), the parameters in the tail quantile function,

$$U(t) := F^{\leftarrow}(1 - 1/t) = Ct^{\xi}\left(1 + \xi\beta t^{\rho}/\rho + o(t^{\rho})\right), \qquad (4.3)$$

as $t \to \infty$, with $C, \xi > 0$, $\beta \in \mathbb{R}$, $\rho < 0$ and $F^{\leftarrow}(x) := \inf\{y : F(y) \geq x\}$ denoting the generalized inverse function of F. Section 4.4 is dedicated to methods that deal with the minimization of MSE estimates. In Section 4.5, we review methods essentially related to sample paths' stability. In Section 4.6, we deal with a few other heuristic procedures devised for an adaptive EVI-estimation. Finally, in Section 4.7, we provide applications of the algorithms to a simulated sample.

4.2   Graphical Threshold Selection

The plots next described can be found in Beirlant et al. (2004), among others. Other common graphical diagnostics for threshold choice can be seen in Coles (2001), de Sousa and Michailidis (2004), and Wager (2014), also among others.

4.2.1   The Zipf, the Hill, and Sum Plots

In the Zipf plot, the quantity $\ln((n+1)/k)$ is plotted against $\ln X_{n-k+1:n}$, for $k = 1, \ldots, n$, and the estimate of ξ is given by the least squares estimate of the slope of the part of the plot that exhibits a linear behavior. If such a linear behavior is anchored at the point $\left(\ln x_{n-k_0+1:n}, \ln((n+1)/k_0)\right)$, we have an immediate choice of the threshold. A nice feature of the Zipf plot is that the lack of linearity at the left part of the graph suggests departure from a Pareto tail behaviour. Note that for a strict Pareto model, $F(x) = 1 - x^{-1/\xi}$, $x \geqslant 1$, the points $\left(1 - F(x_{n-k+1:n})\right) = x_{n-k+1:n}^{-1/\xi}$, $k/(n+1)$, $1 \leqslant k \leqslant n$, should be approximately linear.

Based on the Zipf plot, Kratz and Resnick (1996) considered the well-known QQ plot as an alternative to the Hill plot, a plot of $\widehat{\xi}_{k,n}^H$, in (4.2), versus k. In the Hill plot, the value of ξ is inferred by identifying a stable horizontal region in the graph. If data comes from a Pareto CDF, the ideal case, the Hill plot, as well as the Zipf plot, perform quite satisfactorily, almost independent of the threshold, allowing the data analyst to estimate correctly the EVI. But this is not always the case, and we can easily come to the so-called Hill "horror" plots (Resnick, 1997). As recommended by Drees et al. (2000), it is more informative to view the Hill plot on the log scale rather than on the linear scale. As illustrated in de Sousa and Michailidis (2004), and despite the fact that the QQ plot tends to be smoother than the Hill plot, problems are still easy to detect when departures from the Pareto CDF occur.

The sum plots were devised, among others, in de Sousa and Michailidis (2004) for the H EVI-estimators, in Henry III (2009) for the harmonic mean EVI-estimators and in Beirlant et al. (2011) for EVI-estimators that use a set of extreme order statistics in the estimation of a real EVI. They are based on the plot $(k, \widehat{S}_{k,n}^\bullet := k\widehat{\xi}_{k,n}^\bullet)$, which should be approximately linear for the k-values where $\widehat{\xi}_{k,n}^\bullet \approx \xi$. The slope of the linear part of the graph $(k, \widehat{S}_{k,n}^\bullet)$ can then be used as an estimator of ξ. The plots $(k, \widehat{S}_{k,n}^\bullet)$ and $(k, \widehat{\xi}_{k,n}^\bullet)$ are statistically equivalent. The sum plot naturally leads to the estimation of the slope and the plot $(k, \widehat{\xi}_{k,n}^\bullet)$, which should be horizontal, leads to the estimation of the intercept. Taking the H EVI-estimator as an example of an estimator of the slope in a Pareto quantile plot, the Hill sum plot can be viewed as a regression plot of $\widehat{S}_{k,n}^H$ on k. In de Sousa and Michailidis (2004) an algorithm is provided for identifying the threshold from the Hill sum plot. This algorithm is compared with two algorithms in Beirlant et al. (2002) and Matthys and Beirlant (2000), both based on MSE minimization of the H EVI-estimators. The misspecification $\rho = -1$ is considered, and the conclusion is that there does not exist a uniformly better procedure for all possible underlying parents.

4.2.2 The Mean Excess Plot

The mean excess plot is a common graphical tool in the field of insurance. The plot is related to the mean excess function (MEF)

$$e(t) := \mathbb{E}(X - t | X > t). \tag{4.4}$$

Given a sample $(x_1, \ldots, x_n)$, the MEF is often estimated at the values $t = x_{n-k:n}$, $1 \leqslant k < n$, through

$$\widehat{e}_{k,n} := \widehat{e}_n(x_{n-k:n}) = \frac{1}{k} \sum_{j=1}^{k} x_{n-j+1:n} - x_{n-k:n}. \tag{4.5}$$

Note that $\widehat{e}_{k,n}$ can also be interpreted as an estimate of the slope of the exponential QQ plot to the right of a reference point with coordinates $(-\ln((k_0 + 1)/(n + 1)), x_{n-k_0:n})$. The MEF is constant for an exponential model. When the underlying CDF has a right tail heavier than the exponential, the MEF ultimately increases. For lighter tails the MEF ultimately decreases. A mean excess plot is a plot of $\widehat{e}_{k,n}$, in (4.5), either versus k or versus $x_{n-k:n}$. The choice of the threshold is related to the region where the graph is approximately linear (see Beirlant et al., 2004).

4.3 Direct Estimation of the Threshold

Note first that for $\xi > 0$, a necessary and sufficient condition to have F in the max-domain of attraction of the standard model $G_{\xi,0,1}$, with $G_{\xi,\mu,\sigma}$ given in (4.1), is that $\overline{F} \in \mathcal{RV}_{-1/\xi}$ or equivalently that $U \in \mathcal{RV}_\xi$, where $\mathcal{RV}_a$ denotes the class of regularly varying functions at infinity, with an index of regular variation a, i.e., positive measurable functions g such that $g(tx)/g(t) \to x^a$, as $t \to \infty$. For such a max-domain of attraction we shall use the notation $\mathcal{D}_\mathcal{M}^+ \equiv \mathcal{D}_{\mathcal{M}|1}^+$. We shall thus assume the validity of the so-called first-order condition, working with models

$$F \in \mathcal{D}_\mathcal{M}^+ \equiv \mathcal{D}_{\mathcal{M}|1}^+ \iff \overline{F} \in \mathcal{RV}_{-1/\xi} \iff U \in \mathcal{RV}_\xi. \tag{4.6}$$

One of the pioneering classes of semiparametric estimators of a positive EVI, usually known as H EVI-estimators, was given in (4.2). Consistency is achieved in the whole $\mathcal{D}_{\mathcal{M}|1}^+$, provided that $X_{n-k:n}$ is an *intermediate* order statistic, i.e., $k = k_n \to \infty$ and $k/n \to 0$, as $n \to \infty$. In order to derive the normality of the EVI-estimators in (4.2), among others, it is convenient to slightly restrict the class $\mathcal{D}_{\mathcal{M}|1}^{+}$, assuming the validity of a second-order condition either on $\overline{F}$ or on U. We often assume the existence of a function $A(t)$, going to zero as $t \to \infty$, such that

$$\lim_{t \to \infty} \frac{\ln U(tx) - \ln U(t) - \xi \ln x}{A(t)} = \begin{cases} \frac{x^{\rho}-1}{\rho}, & \text{if } \rho < 0, \\ \ln x, & \text{if } \rho = 0, \end{cases} \qquad (4.7)$$

where $\rho \leqslant 0$ is a second-order parameter, which measures the rate of convergence in the first-order condition, in (4.6). For such a class of models, we use the notation $\mathcal{D}_{\mathcal{M}|2}^{+}$. If the limit in the left-hand side of (4.7) exists, it is necessarily of the above-mentioned type and $|A| \in \mathcal{RV}_{\rho}$. If we assume the validity of the second-order framework in (4.7), EVI-estimators are asymptotically normal, provided that $\sqrt{k}A(n/k) \to \lambda_A$, finite, as $n \to \infty$, with A given in (4.7). Indeed, if we denote $\widehat{\xi}_{k,n}^{C}$, either the H EVI-estimator in (4.2) or more generally any other classical (C) EVI-estimator, we have, with Z_k^C asymptotically standard normal and for adequate $(b_C, \sigma_C) \in (\mathbb{R}, \mathbb{R}^{+})$, the validity of the asymptotic distributional representation

$$\widehat{\xi}_{k,n}^{C} \overset{d}{=} \xi + \xi \frac{\sigma_C Z_k^C}{\sqrt{k}} + b_C\, A(n/k)(1 + o_p(1)), \quad \text{as} \quad n \to \infty. \qquad (4.8)$$

In this chapter, we assume that (4.3) holds with $\rho < 0$. Then the function $A(t)$ in (4.7) is given by $A(t) = \xi \beta t^{\rho}$. If b_C, in (4.8), depends only on (β, ρ), i.e., $b_C = b_C(\beta, \rho)$, as happens with the H EVI-estimators, we can build a reduced-bias classical (RBC) EVI-estimator associated with the C EVI-estimator, given by

$$\widehat{\xi}_{k,n}^{RBC} = \widehat{\xi}_{k,n}^{C} \left(1 - b_C(\widehat{\beta}, \widehat{\rho})\, \widehat{\beta}(n/k)^{\widehat{\rho}} \right), \qquad (4.9)$$

with $(\widehat{\beta}, \widehat{\rho})$ an adequate estimator of the vector of second-order parameters (β, ρ), in (4.3) (see Caeiro et al., 2005, for details on RBH EVI-estimation). Caeiro et al. (2009), among others, show how to estimate (β, ρ), externally and a bit more than consistently, in the sense that we need to have $\widehat{\rho} - \rho = o_p(1/\ln n)$, so that we decrease the bias keeping the variance. Under an adequate third-order condition (see Caeiro et al., 2009), we then get the validity of the asymptotic distributional representation

$$\widehat{\xi}_{k,n}^{RBC} \overset{d}{=} \xi + \xi \frac{\sigma_C Z_k^C}{\sqrt{k}} + b_{RBC}\, A^2(n/k)(1 + o_p(1)), \quad \text{as} \quad n \to \infty. \qquad (4.10)$$

For models in (4.3), a possible substitute for the MSE of any classical EVI-estimator $\widehat{\xi}_{k,n}^{C}$ is, according with the asymptotic distributional representation in (4.8), $\text{AMSE}(\widehat{\xi}_{k,n}^{C}) = \xi^2 \left(\sigma_C^2/k + b_C^2\, \beta^2\, (n/k)^{2\rho} \right)$, depending on n and k, and with AMSE standing for *asymptotic mean square error*. For the RBC EVI-estimator, in (4.9), and on the basis of (4.10), we obviously have

$$\text{AMSE}(\widehat{\xi}_{k,n}^{RBC}) = \xi^2 \left(\sigma_C^2/k + b_{RBC}^2\, \xi^2 \beta^4\, (n/k)^{4\rho} \right).$$

Then, with $\bullet$ denoting either C or RBC, with $\sigma = \sigma_C$, and considering

$$c_\bullet = \begin{cases} 2 & \text{if} \quad \bullet = \text{C} \\ 4 & \text{if} \quad \bullet = \text{RBC}, \end{cases} \tag{4.11}$$

we get

$$
\begin{aligned}
k_{0|\bullet}(n) &:= \arg\min_k \text{AMSE}\big(\widehat{\xi}^\bullet_{k,n}\big) = \big((-c_\bullet\rho)\, b_\bullet^2\, \xi^{c_\bullet - 2}\beta^{c_\bullet}\, n^{c_\bullet\rho}/\sigma^2\big)^{-1/(1-c_\bullet\rho)} \\
&= k_0^\bullet(n)(1 + o(1)),
\end{aligned}
$$

with $k_0^\bullet(n) := \arg\min_k \text{MSE}\big(\widehat{\xi}^\bullet_{k,n}\big)$. This intuitively suggests the estimator

$$\widehat{k}_0^\bullet := \left\lfloor \big(\widehat{\sigma}^2 n^{-c_\bullet\widehat{\rho}}/(-c_\bullet\widehat{\rho}\,\widehat{b}_\bullet^2\, \widehat{\xi}^{c_\bullet - 2}\widehat{\beta}^{c_\bullet})\big)^{1/(1-c_\bullet\widehat{\rho})} \right\rfloor, \tag{4.12}$$

with $\lfloor x \rfloor$ denoting the integer part of x. For the H EVI-estimator, we have, in (4.8), $\sigma_{\text{H}} = 1$ and $b_{\text{H}} = 1/(1 - \rho)$. Consequently, with $(\widehat{\beta}, \widehat{\rho})$ any consistent estimator of the vector (β, ρ) of second-order parameters, Equation (4.12) justifies asymptotically the estimator suggested in Hall (1982).

At the current state-of-the-art, the estimation of the second-order parameters β and ρ in (4.3) can be considered almost trivial. Despite the great variety of recent classes of estimators now available in the literature, we suggest the use of the ρ–estimators in Fraga Alves et al. (2003) and β–estimators in Gomes and Martins (2002). The second-order parameters estimates can be obtained through algorithms already used in several articles, like Gomes and Pestana (2007), among others. We can thus consider *Algorithm 4.1* for the estimation of the threshold, followed by the estimation of the EVI.

Note that if the second side of Equation (4.12) depends on ξ, let us write $\xi^{1-c_\bullet/2}\sigma/b_\bullet = \varphi_\bullet(\xi)$, we can also try solving this equation numerically through the fixed point method, replacing **Step 4.**, in *Algorithm 4.1*, by:

Step 4.1. Compute an initial estimate $\widehat{\xi}_0^\bullet = \widehat{\xi}^\bullet_{\widehat{k}_{0,0},n}$, $\widehat{k}_{0,0}^\bullet = \lfloor\sqrt{n}\rfloor$.

Step 4.2. For $i \geqslant 1$, and with $c_\bullet$ defined in Equation (4.11), consider the iterative procedure

$$\widehat{k}_{0,i}^\bullet := \left\lfloor \big(\varphi_\bullet^2(\widehat{\xi}_{i-1}^\bullet) n^{-c_\bullet\widehat{\rho}}/(-c_\bullet\widehat{\rho}\,\widehat{\beta}^{c_\bullet})\big)^{1/(1-c_\bullet\widehat{\rho})} \right\rfloor, \qquad \widehat{\xi}_i^\bullet = \widehat{\xi}^\bullet_{\widehat{k}_{0,i},n},$$

and stop the procedure if we get $\widehat{k}_{0,i}^\bullet = \widehat{k}_{0,i-1}^\bullet =: \widehat{k}_0^\bullet$.

Remark 4.3.1. *Note that this procedure works essentially for classical EVI-estimation. Indeed, since the estimation of the third-order parameters is still an almost open topic and $\sigma/b_\bullet$ depends usually on those parameters, we cannot yet directly estimate the optimal number of top order statistics for $\widehat{\xi}_{k,n}^{\text{RBC}}$.*

Algorithm 4.1

Step 1. Given the sample $(x_1, \ldots, x_n)$, compute for $\tau = 0$ and $\tau = 1$, the observed values of $\widehat{\rho}_\tau(k)$, the most simple class of estimators in Fraga Alves et al. (2003). Such estimators have the functional form

$$\widehat{\rho}_\tau(k) := - \left| \frac{3(W_{k,n}^{(\tau)} - 1)}{W_{k,n}^{(\tau)} - 3} \right|,$$

dependent on the statistics

$$W_{k,n}^{(\tau)} := \frac{\left(M_{k,n}^{(1)} \right)^\tau - \left(M_{k,n}^{(2)}/2 \right)^{\tau/2}}{\left(M_{k,n}^{(2)}/2 \right)^{\tau/2} - \left(M_{k,n}^{(3)}/6 \right)^{\tau/3}}$$

where $M_{k,n}^{(j)} := \frac{1}{k} \sum_{i=1}^{k} \left(\ln X_{n-i+1:n} - \ln X_{n-k:n} \right)^j$, $j = 1, 2, 3$, and $a^{b\tau} = b \ln a$ when $\tau = 0$.

Step 2. Consider $\mathcal{K} = \left(\lfloor n^{0.995} \rfloor, \lfloor n^{0.999} \rfloor \right)$. Compute the median of $\{\widehat{\rho}_\tau(k)\}_{k \in \mathcal{K}}$, denoted χ_τ, and compute $I_\tau := \sum_{k \in \mathcal{K}} \left(\widehat{\rho}_\tau(k) - \chi_\tau \right)^2$, $\tau = 0, 1$. Next choose $\tau^* = 0$ if $I_0 \leqslant I_1$; otherwise, choose $\tau^* = 1$.

Step 3. Work with $\widehat{\rho} \equiv \widehat{\rho}_{\tau^*} = \widehat{\rho}_{\tau^*}(k_{01})$ and $\widehat{\beta} \equiv \widehat{\beta}_{\tau^*} := \widehat{\beta}_{\widehat{\rho}_{\tau^*}}(k_{01})$, with $k_{01} = \lfloor n^{0.999} \rfloor$, being $\widehat{\beta}_{\widehat{\rho}}(k)$ the estimator in Gomes and Martins (2002), given by

$$\widehat{\beta}_{\widehat{\rho}}(k) := \left(\frac{k}{n} \right)^{\widehat{\rho}} \frac{d_k(\widehat{\rho}) \, D_k(0) - D_k(\widehat{\rho})}{d_k(\widehat{\rho}) \, D_k(\widehat{\rho}) - D_k(2\widehat{\rho})},$$

dependent on the estimator $\widehat{\rho} = \widehat{\rho}_{\tau^*}(k_{01})$, and where, for any $\alpha \leqslant 0$, $d_k(\alpha) := \frac{1}{k} \sum_{i=1}^{k} (i/k)^{-\alpha}$ and $D_k(\alpha) := \frac{1}{k} \sum_{i=1}^{k} (i/k)^{-\alpha} U_i$, with $U_i = i \left(\ln X_{n-i+1:n} - \ln X_{n-i:n} \right)$, $1 \leqslant i \leqslant k < n$, the *scaled log-spacings*.

Step 4. On the basis of $(\widehat{\beta}, \widehat{\rho})$, in **Step 3.**, compute $\widehat{k}_0^\bullet$, in (4.12).

Step 5. Finally obtain $\widehat{\xi}^\bullet := \widehat{\xi}_{\widehat{k}_0^\bullet, n}^\bullet$.

4.4 Threshold Selection Based on MSE

4.4.1 Hall and Welsh's Threshold Selection

Hall and Welsh (1985) provided a choice for the optimal threshold associated with H EVI-estimation, quite close to the one written in *Algorithm 4.1*, but with a different estimation of second-order parameters, dependent upon three tuning parameters, $\sigma < \tau_1 < \tau_2$. Again, the method is quite sensitive to changes in these parameters, but the choice suggested by Hall and Welsh seems to work well, as shown in the comparative study performed in Gomes and Oliveira (2001).

4.4.2 Regression Diagnostics' Selections

The algorithm related to Beirlant et al. (1996a, 1996b, 2002) regression diagnostics procedure has again been fully described and simulated in Gomes and Oliveira (2001). The method provides interesting by-products: estimates of the MSE and of squared bias, but there appear a few computational problems related to this method. The procedure is quite time-consuming and very sensitive to the two main tuning parameters. However, these parameters may be used to overcome some non-convergence problems, like negative values for the MSE and non-admissible estimates of the second-order parameter, ρ. And although time-consuming, the method provides nice results, particularly for small samples.

4.4.3 Hall's Bootstrap Methodology

Bootstrap methodology for threshold selection was first introduced in Hall (1990). For a recent comparison between the simple-bootstrap and the double-bootstrap methodology, see Caeiro and Gomes (2014), where an improved version of Hall's bootstrap methodology is introduced.

Given the sample $\underline{X}_n = (X_1, \ldots, X_n)$ from an unknown model Γ and the functional $\widehat{\xi}_{k,n}^H = \phi_k(\underline{X}_n)$, $1 \leqslant k < n$, consider the bootstrap sample $\underline{X}_{n_1}^* = (X_1^*, \ldots, X_{n_1}^*)$, $n_1 \leqslant n$, from the model $F_n^*(x) = \frac{1}{n} \sum_{i=1}^n I_{\{X_i \leqslant x\}}$, the empirical CDF associated with the original sample $\underline{X}_n$. Then associate with that bootstrap sample the corresponding bootstrap estimator $\widehat{\xi}_{k_1,n_1}^{H*} = \phi_{k_1}(\underline{X}_{n_1}^*)$, $1 \leqslant k_1 < n_1$.

Given an initial value of k, say k_{aux}, such that $\widehat{\xi}_{k_{aux},n}^H$ is a consistent estimator of ξ, (i.e., we need to have $k_{aux} \to \infty$, $k_{aux}/n \to 0$, as $n \to \infty$), Hall suggests the minimization of the bootstrap estimate of the MSE of $\widehat{\xi}_{k_1,n_1}^H$,

$$\text{MSE}^*(n_1, k_1) = \mathbb{E}\left[\left\{ \widehat{\xi}_{k_1,n_1}^{H*} - \widehat{\xi}_{k_{aux},n}^H \right\}^2 \Big| \underline{X}_n \right], \qquad (4.13)$$

for a sub-sample size, n_1. This is done taking into account the fact that for a special sub-class of Hall's class, for which $\rho = -1$, and for suitable $n_1 = o(n)$, as $n \to \infty$, the optimal k_1-choice is of the same type, $Dn_1^\gamma(1 + o(1))$, with $\gamma = 2/3$, both for the original estimator $\widehat{\xi}_{k_1,n_1}^H$ and for the bootstrap statistic $\widehat{\xi}_{k_1,n_1}^{H*} - \widehat{\xi}_{k_{aux},n}^H | \underline{X}_n$. The most common choices for k_{aux} and n_1 are $k_{aux} = \lfloor 2\sqrt{n} \rfloor$ and $n_1 = \lfloor n^{0.955} \rfloor$, respectively.

More generally, under the validity of (4.3) and with $n_1 = O(n^{1-\epsilon})$, for some $0 < \epsilon < 1$, and k_1 intermediate, we can say that there exists a function $D(\rho)$ such that the optimal performance of $\widehat{\xi}_{k_1,n_1}^{H*} | \underline{X}_n$ (level k_1 where $\text{MSE}[\widehat{\xi}_{k_1,n_1}^* | \underline{X}_n]$ is minimal), is achieved at

$$k_0^*(n_1) = D(\rho)n_1^{-2\rho/(1-2\rho)}(1 + o_p(1)). \qquad (4.14)$$

This follows in a way similar to the proof of Theorem 4 in Gomes and Oliveira (2001). Also from the results in Section 4.3, the AMSE of the H EVI-estimator is, for the same $D(\rho)$ as in (4.14), minimal for

$$k_0^H(n) = D(\rho)n^{-2\rho/(1-2\rho)}(1 + o(1)). \qquad (4.15)$$

Given an estimate of ρ, $\widehat{\rho}$, and an estimate of $k_0^*(n_1)$, $\widehat{k}_0^*(n_1)$, we can consider $\widehat{k}_0^H :=$ $\widehat{k}_0^*(n_1)(n/n_1)^{-2\widehat{\rho}/(1-2\widehat{\rho})}$. As noted in Gomes and Oliveira (2001), and despite an almost independence on n_1, there is a disturbing sensitivity of the method to the initial value of k_{aux}. These facts suggest the suitability of the alternative double-bootstrap methodology investigated in the next section.

4.4.4 An Alternative Bootstrap Method

Comparing (4.14) and (4.15), $k_0^*(n_1)/k_0^H(n) = (n_1/n)^{\frac{2\rho}{2\rho-1}}(1 + o_p(1))$, as $n \to \infty$, and thus for another sample size n_2 and for every $\alpha > 1$, we have

$$\frac{[k_0^*(n_1)]^\alpha}{k_0^*(n_2)}\left(\frac{n_1^\alpha}{n^\alpha}\frac{n}{n_2}\right)^{\frac{2\rho}{1-2\rho}} = \{k_0(n)\}^{\alpha-1}(1 + o_p(1)).$$

It is then sufficient to choose $n_2 = n(n_1/n)^\alpha$, in order to have independence of ρ. If we put $n_2 = n_1^2/n$, i.e., $\alpha = 2$, we have $[k_0^*(n_1)]^2/k_0^*(n_2) = k_0^H(n)(1 + o_p(1))$, as $n \to \infty$. This argument led to the so-called double-bootstrap method, first used in Draisma et al. (1999) for the general max-domain of attraction and in Danielsson et al. (2001) and Gomes and Oliveira (2001) for heavy tailed models. More recently, Gomes et al. (2012) modified the double bootstrap algorithm for an adaptive choice of the thresholds for second-order corrected-bias estimators. See also Gomes et al. (2011) and Caeiro and Gomes (2014). *Algorithm 4.2*, provided in this section, follows closely Gomes et al. (2012) and Caeiro and Gomes (2014). We consider again an auxiliary statistic of the type of the one considered in Gomes and Oliveira (2001), directly related to the EVI-estimator under consideration, but going to the known value zero, i.e., the statistic

$$T_{k,n}^\bullet := \widehat{\xi}_{\lfloor k/2\rfloor,n}^\bullet - \widehat{\xi}_{k,n}^\bullet, \quad k = 2,\ldots,n-1, \quad \bullet = \text{H, RBH}. \qquad (4.16)$$

Notice that if $\bullet = $ H this approach is equivalent to replacing the fixed threshold k_{aux} by $\lfloor k/2\rfloor$ in (4.13). On the basis of results similar to the ones in Gomes and Oliveira (2001), and with $c_\bullet$ given in (4.11), we can get for $T_{k,n}^\bullet$, in (4.16), the asymptotic distributional representation,

$$T_{k,n}^\bullet \stackrel{d}{=} \frac{\xi\, P_k^\bullet}{\sqrt{k}} + b_\bullet(2^{c_\bullet\rho/2} - 1)\, A^{c_\bullet/2}(n/k)(1 + o_p(1)),$$

with $P_k^\bullet$ asymptotically standard normal. Then, the AMSE of $T_{k,n}^\bullet$ is minimal at a level $k_{0|T}^\bullet(n)$, such that

$$k_0^\bullet(n) = k_{0|T}^\bullet(n)(2^{c_\bullet\rho/2} - 1)^{\frac{2}{1-c_\bullet\rho}}.$$

We can thus consider *Algorithm 4.2*.

Algorithm 4.2

Step 1. Given a sample $(x_1, \ldots, x_n)$, compute the estimates $\widehat{\rho}$ and $\widehat{\beta}$ of the second-order parameters ρ and β as described in *Algorithm 4.1*. Next, consider a sub-sample size $n_1 = o(n)$ and $n_2 = \lfloor n_1^2/n \rfloor + 1$.

Step 2. For l from 1 until B, generate independently B bootstrap samples $(x_1^*, \ldots, x_{n_2}^*)$ and $(x_1^*, \ldots, x_{n_2}^*, x_{n_2+1}^*, \ldots, x_{n_1}^*)$, of sizes n_2 and n_1, respectively, from the empirical distribution $F_n^*(x) = \frac{1}{n} \sum_{i=1}^n I_{\{X_i \leqslant x\}}$ associated with the observed sample $(x_1, \ldots, x_n)$.

Step 3. Denoting $T_{k,n_i}^{\bullet\bullet*}$ the bootstrap counterpart of $T_{k,n_i}^\bullet$, in (4.16), obtain $t_{k,n_i,l}^{\bullet\bullet*}$, $1 \leqslant l \leqslant B$, $i = 1, 2$, the observed values of $T_{k,n}^{\bullet\bullet*}$. For $k = 2, \ldots, n_i - 1$, and $i = 1, 2$, compute $\mathrm{MSE}^{\bullet\bullet*}(n_i, k) = \frac{1}{B} \sum_{l=1}^B \left(t_{k,n_1,l}^{\bullet\bullet*} \right)^2$, and obtain $\widehat{k}_{0|T}^{\bullet\bullet*}(n_i) :=$ $\arg \min_{1 < k < n_i} \mathrm{MSE}^{\bullet\bullet*}(n_i, k)$.

Step 4. Compute the threshold estimate

$$\widehat{k}_0^{\bullet\bullet*} \equiv \left\lfloor (1 - 2^{c \bullet \widehat{\rho}/2})^{\frac{2}{1-c \bullet \widehat{\rho}}} \left(\widehat{k}_{0|T}^{\bullet\bullet*}(n_1) \right)^2 / \widehat{k}_{0|T}^{\bullet\bullet*}(n_2) \right\rfloor + 1.$$

If $\widehat{k}_0^{\bullet\bullet*} \notin [1, n)$ go back to **Step 2**, generating different bootstrap samples.

Step 5. Obtain $\widehat{\xi}^{\bullet\bullet*} \equiv \widehat{\xi}_{\widehat{k}_0^{\bullet\bullet*}, n}^*$.

Remark 4.4.1. *The use of the sample $(x_1^*, \ldots, x_{n_2}^*)$, and of the extended sample $(x_1^*, \ldots, x_{n_2}^*, \ldots, x_{n_1}^*)$, $n_2 < n_1$, lead us to an increased precision of the result with the same number B of bootstrap samples generated in* **Step 2**. *This is quite similar to the use of the 'Common Random Numbers' simulation technique (see,* Hammersley and Handscomb, 1964, *among others).*

Remark 4.4.2. *Bootstrap confidence intervals are easily obtained through the replication of this algorithm r times. The replication can also provide us more precise estimates, if we consider the estimate given by the mean or the median of the r bootstrap estimates.*

4.4.5 Semiparametric Bootstrap

Caers et al. (1999) proposed a simple alternative bootstrap method for the selection of the threshold u which does not require the choice of the sub-sample size, as in the previous two methods. Here, a generalized Pareto (GP) distribution, with CDF

$$\mathrm{GP}_{\xi,\sigma}(x) = \begin{cases} 1 - \left(1 + \frac{\xi}{\sigma}x\right)^{-1/\xi}, & 0 < x < (\max(0, -\xi/\sigma))^{-1}, \quad \text{if } \xi \neq 0, \\ 1 - \exp\left(-\frac{x}{\sigma}\right), & x > 0, \quad \text{if } \xi = 0, \end{cases}$$

and $\sigma > 0$ a scale parameter, is fitted to the excesses above the threshold u and used to generate bootstrap tail observations. As mentioned at the very beginning of this chapter, we shall consider $u = x_{n-k:n}$ and a slight modification of the method in Caers et al. (1999), provided in the *Algorithm 4.3*, where $\widehat{\xi}_{k,n}$ and $\widehat{\sigma}_{k,n}$ denote any semiparametric EVI and scale estimators, respectively.

4.5   Threshold Selection Partially Based on Bias

4.5.1   Drees and Kaufmann's Estimator

Drees and Kaufmann (1998) proposed an algorithm partially based on bias and MSE minimization. Apart from the need to estimate ρ, which relies on the choice of a tuning parameter, there are two other tuning parameters, and the method is quite sensitive to the choice of these parameters. However, the suggestions given by Drees and Kaufmann seem appropriate, and the method has an overall positive performance, as can be seen in the comparative simulation study performed in Gomes and Oliveira (2001).

4.5.2   Guillou and Hall's Estimator

Guillou and Hall (2001) suggest an interesting approach for choosing the threshold when fitting the H EVI-estimator of a tail exponent to extreme value data. The argument is the following: for the H EVI-estimator, we have the validity of the distributional representation in (4.8), with $\sigma_{\mathrm{H}} = 1$ and $b_{\mathrm{H}} = 1/(1 - \rho)$, and the AMSE of $\sqrt{k}\,\widehat{\xi}^{\mathrm{H}}_{k,n}$ is equal to $\xi^2(1 + \mu_k^2)$ where $\mu_k = \sqrt{k}\,A(n/k)/(\xi(1 - \rho))$. Then, the value k_0 of k that is optimal in the sense of minimizing the AMSE of $\xi^{\mathrm{H}}_{k,n}$ can be taken as a solution of $\mu_{k_0}^2 = 1/\sqrt{-2\rho}$. Similar to the bootstrap methodology in Section 4.4.4, the idea underlying Guillou and Hall (2001) is to replace the random

---

Algorithm 4.3

Step 1. Given a sample $(x_1, \ldots, x_n)$, take a set of integer values $1 < k_1 < k_2 < \cdots < k_m < n$.

Step 2. For each threshold $x_{n-k_i:n}, 1 \leqslant i \leqslant m$:

- Compute the estimates $\widehat{\xi}_i \equiv \widehat{\xi}_{k_i,n}$ and $\widehat{\sigma}_i \equiv \widehat{\sigma}_{k_i,n}$.
- For l from 1 until B, generate independently B bootstrap samples $(x_1^*, \ldots, x_n^*)$, of size n, from the semiparametric model

$$\widehat{F}_{k_i,n}(x) = \begin{cases} 1 - \frac{k_i}{n} + \frac{k_i}{n}\mathrm{GP}_{\widehat{\xi}_i,\widehat{\sigma}_i}(x - x_{n-k_i:n}), & \text{if} \quad x > x_{n-k_i:n}, \\ \widehat{F}_n(x), & \text{if} \quad x \leq x_{n-k:n}, \end{cases}$$

where $\widehat{F}_n(x) = \frac{1}{n}\sum_{i=1}^{n} \mathrm{I}_{\{x_i \leqslant x\}}$ is the empirical CDF associated with the original sample, and compute the estimates $\widehat{\xi}^{*(l)}_{k_i,n}, l = 1, \ldots, B$.

Step 3. Compute $\widehat{\mathrm{MSE}}^*(k_i) = \frac{1}{B}\sum_{l=1}^{B}\left(\widehat{\xi}^{*(l)}_{k_i,n} - \widehat{\xi}_i\right)^2, 1 \leqslant i \leqslant m$, and obtain $\widehat{k}_0^* := \arg\min_{1 \leqslant i \leqslant m} \widehat{\mathrm{MSE}}^*(k_i)$.

Step 4. Obtain $\widehat{\xi}^{*|\mathrm{GP}} \equiv \widehat{\xi}_{\widehat{k}_0^*,n}$.

Algorithm 4.4

Step 1. Given the sample $(x_1, \ldots, x_n)$, compute the observed value of $Q_n(k)$, in (4.17), for $1 \leqslant k$ and $k + \lfloor k/2 \rfloor < n$.

Step 2. Given $c_{crit} = 1.25$, consider the choice

$$k_0^{GH} := \inf\{k : |Q_n(j)| \geqslant c_{crit}, \, \forall j \geqslant k\}.$$

Step 3. Finally obtain $\widehat{\xi}^{GH} := \widehat{\xi}^H_{k_0^{GH}, n}$.

---

variable $\sqrt{k}\{\widehat{\xi}^H_{k,n} - \xi\}/\xi$, by a statistic which also converges to 0 and has similar bias properties. Such a statistic is merely a linear combination of the log-spacings $U_i = i\{\ln X_{n-i+1:n} - X_{n-i:n}\}$, $1 \leqslant i \leqslant k < n$, with null mean value. More precisely, the main role in this diagnostic technique is played by the auxiliary statistics $T_n(k) := \sqrt{\frac{3}{k^3}} \sum_{i=1}^{k} (k - 2i + 1) U_i / \left(\frac{1}{k} \sum_{i=1}^{k} U_i\right)$, for $1 \leqslant k < n$, together with a moving average of its squares,

$$Q_n(k) := \left\{ \frac{1}{2\lfloor k/2 \rfloor + 1} \sum_{j=k-\lfloor k/2 \rfloor}^{k+\lfloor k/2 \rfloor} T_n^2(j) \right\}^{1/2}, \qquad (4.17)$$

which dampens stochastic fluctuations of $T_n(k)$. On the basis of an ad hoc choice of c_{crit}, like any value in $[1.25, 1.5]$, as suggested in Guillou and Hall (2001), together with the auxiliary statistic in (4.17), we can thus consider *Algorithm 4.4*.

4.6 Other Heuristic Methods

Let $\widehat{\xi}^\bullet_{k,n} =: \mathrm{T}(k)$ be any consistent EVI-estimator. A heuristic algorithm in the line of the ones in Gomes et al. (2011, 2013) and Neves et al. (2015), essentially based on sample path stability (PS), and known to work in a quite reliable way for reduced-bias EVI-estimators, but that can be used for the estimation of any parameter of extreme events, is *Algorithm 4.5*.

Other heuristic algorithms, essentially devised for reduced-bias estimators of parameters of extreme events, can be found in Gomes and Pestana (2007), Gomes et al. (2008), and Figueiredo et al. (2012). Other heuristic methods for a classical estimation of a positive EVI can be found in Csörgő and Viharos (1998), who provide a data-driven choice of k for the kernel class of estimators. Bayesian approaches for threshold selection can be found in Frigessi et al. (2002), among others.

Algorithm 4.5

Step 1. Given an observed sample $(x_1, \ldots, x_n)$, compute, for $k = 1, \ldots, n-1$, the observed values of $\mathrm{T}(k)$.

Step 2. Obtain j_0, the minimum value of j, a nonnegative integer, such that the rounded values, to j decimal places, of the estimates in **Step 1.** are distinct. Define $a_k^{(\mathrm{T})}(j) = \mathrm{round}(\mathrm{T}(k), j)$, $k = 1, 2, \ldots, n-1$, the rounded values of $\mathrm{T}(k)$ to j decimal places.

Step 3. Consider the sets of k values associated with equal consecutive values of $a_k^{(\mathrm{T})}(j_0)$, obtained in **Step 2.** Set $k_{min}^{(\mathrm{T})}$ and $k_{max}^{(\mathrm{T})}$ the minimum and maximum values, respectively, of the set with the largest range. The largest run size is then $l_{\mathrm{T}} := k_{max}^{(\mathrm{T})} - k_{min}^{(\mathrm{T})} + 1$.

Step 4. Consider all those estimates, $\mathrm{T}(k)$, $k_{min}^{(\mathrm{T})} \leqslant k \leqslant k_{max}^{(\mathrm{T})}$, now with two extra decimal places, i.e., compute $\mathrm{T}(k) = a_k^{(\mathrm{T})}(j_0 + 2)$. Obtain the mode of $\mathrm{T}(k)$ and denote $\mathcal{K}_{\mathrm{T}}$ the set of k-values associated with this mode.

Step 5. Take $\widehat{k}_{\mathrm{T}}$ as the maximum value of $\mathcal{K}_{\mathrm{T}}$.

Step 6. Compute $\widehat{\xi}^{\bullet|\mathrm{PS}} = \widehat{\xi}^{\bullet}_{\widehat{k}_{\mathrm{T}}, n}$.

4.7 Application to a Simulated Sample

We shall now consider an illustration of the performance of the aforementioned algorithms when applied to the analysis of a simulated sample with a size $n = 500$ from a Burr model with $\xi = 0.3$ and $\rho = -0.7$. The CDF of a $\mathrm{Burr}_{\xi,\rho}$ model is given by $F(x) = 1 - (1 + x^{-\rho/\xi})^{1/\rho}$, $x > 0$, $\xi > 0$, and $\rho < 0$. We are interested in the selection of the threshold and in the concomitant EVI-estimation provided by the H EVI-estimator in (4.2) and the corresponding RBH EVI-estimator, given in (4.9) for any C EVI-estimator. Figure 4.1, left, shows the estimates of the EVI, ξ, provided by the selected EVI-estimators, as a function of the threshold k. A plot of the observed values of the statistic $Q_n(k)$ in (4.17), related to *Algorithm 4.4*, is provided in Figure 4.1, right.

Regarding *Algorithm 4.1*, note that since the estimation of the third-order parameters is an open topic, we cannot yet directly estimate the optimal number of top order statistics for $\widehat{\xi}_{k,n}^{\mathrm{RBH}}$ through this technique.

In *Algorithm 4.2* we have used $n_1 = 366$ and $B = 5000$, and in *Algorithm 4.3*, we have considered $B = 400$ and the scale parameter σ estimated through maximum likelihood. *Algorithm 4.4* has so far been devised only for the H EVI-estimator. All obtained results are summarized in Table 4.1.

As a final remark, we can say that most of the methods provide accurate estimates of the true and known EVI-value. The closest estimate was provided by the RBH EVI-estimator and *Algorithm 4.2*, based on the double-bootstrap methodology, followed by the same EVI-estimator and *Algorithm 4.5*, based on sample path stability. Also notice that the three standard errors for the RBH EVI-estimator were very similar, and smaller than the ones we got for the H EVI-estimation. The largest and also the worst estimate was provided by the H EVI-estimator and *Algorithm 4.5*.

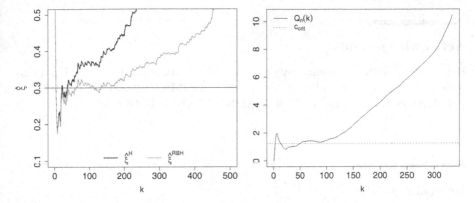

FIGURE 4.1
Left: H and RBH estimates of ξ, as a function of k, for a Burr$_{0.3,-0.7}$ sample. Right:
Values of the statistic $Q_n(k)$ in (4.17) for the simulated Burr$_{0.3,-0.7}$ sample.

TABLE 4.1
Estimates of the optimal threshold and the EVI, with standard errors in parentheses,
for the simulated Burr$_{0.3,-0.7}$ sample.

	EVI-estimator	$\widehat{k_0}$	$\widehat{\xi}$
Algorithm 4.1	H	57	0.333 (0.044)
Algorithm 4.2	H	74	0.362 (0.042)
	RBH	142	0.312 (0.026)
Algorithm 4.3	H	122	0.361 (0.033)
	RBH	128	0.283 (0.025)
Algorithm 4.4	H	51	0.323 (0.045)
Algorithm 4.5	H	291	0.579 (0.034)
	RBH	180	0.315 (0.023)

This is due to the fact that this algorithm is more suitable for reduced-bias estimators. The above results are from just one simulated sample and therefore we cannot
compare the performance of the different methods without an extended Monte-Carlo
comparison study.

Acknowledgements

Research partially supported by National Funds through FCT — Fundação para a Ciência e a Tecnologia, projects PEst-OE/MAT/UI0006/2014 (CEAUL), UID/MAT/00297/2013 and PEst-OE/MAT/UI0297/2014 (CMA/UNL).

References

Beirlant, J., E. Boniphace, and G. Dierckx (2011). Generalized sum plots. *REVSTAT Statistical Journal 9*(2), 181–198.

Beirlant, J., F. Caeiro, and M. I. Gomes (2012). An overview and open research topics in statistics of univariate extremes. *REVSTAT Statistical Journal 10*(1), 1–31.

Beirlant, J., G. Dierckx, Y. Goegebeur, and G. Matthys (1999). Tail index estimation and an exponential regression model. *Extremes 2*(2), 177–200.

Beirlant, J., G. Dierckx, A. Guillou, and C. Stărică (2002). On exponential representations of log-spacings of extreme order statistics. *Extremes 5*(2), 157–180.

Beirlant, J., Y. Goegebeur, J. Segers, and J. L. Teugels (2004). *Statistics of Extremes: Theory and Applications*. John Wiley & Sons.

Beirlant, J., P. Vynckier, and J. L. Teugels (1996a). Excess functions and estimation of the extreme-value index. *Bernoulli 2*(4), 293–318.

Beirlant, J., P. Vynckier, and J. L. Teugels (1996b). Tail index estimation, Pareto quantile plots regression diagnostics. *Journal of the American Statistical Association 91*(436), 1659–1667.

Caeiro, F. and M. I. Gomes (2011). Semi-parametric tail inference through probability-weighted moments. *Journal of Statistical Planning and Inference 141*(2), 937–950.

Caeiro, F. and M. I. Gomes (2014). On the bootstrap methodology for the estimation of the tail sample fraction. In M. Gilli, G. Gonzalez-Rodriguez, and A. Nieto-Reyes (Eds.), *Proceedings of COMPSTAT 2014*, pp. 545–552.

Caeiro, F., M. I. Gomes, and L. Henriques-Rodrigues (2009). Reduced-bias tail index estimators under a third order framework. *Communications in Statistics—Theory and Methods 38*(7), 1019–1040.

Caeiro, F., M. I. Gomes, and D. Pestana (2005). Direct reduction of bias of the classical Hill estimator. *REVSTAT Statistical Journal 3*(2), 113–136.

Caers, J., J. Beirlant, and M. A. Maes (1999). Statistics for modeling heavy tailed distributions in geology: Part I. methodology. *Mathematical Geology 31*(4), 391–410.

Coles, S. (2001). *An Introduction to Statistical Modeling of Extreme Values*. Springer.

Csörgő, S., P. Deheuvels, and D. Mason (1985). Kernel estimates of the tail index of a distribution. *The Annals of Statistics 13*(3), 1050–1077.

Csörgő, S. and L. Viharos (1998). Estimating the tail index. In B. Szyszkowicz (Ed.), *Asymptotic Methods in Probability and Statistics*, pp. 833–881. Elsevier.

Danielsson, J., L. de Haan, L. Peng, and C. G. de Vries (2001). Using a bootstrap method to choose the sample fraction in tail index estimation. *Journal of Multivariate Analysis 76*(2), 226–248.

de Haan, L. and A. Ferreira (2006). *Extreme Value Theory: An Introduction*. Springer.

de Sousa, B. and G. Michailidis (2004). A diagnostic plot for estimating the tail index of a distribution. *Journal of Computational and Graphical Statistics 13*(4), 1–22.

Draisma, G., L. de Haan, L. Peng, and T. T. Pereira (1999). A bootstrap-based method to achieve optimality in estimating the extreme-value index. *Extremes 2*(4), 367–404.

Drees, H., L. de Haan, and S. Resnick (2000). How to make a hill plot. *Annals of Statistics 28*, 254–274.

Drees, H. and E. Kaufmann (1998). Selecting the optimal sample fraction in univariate extreme value estimation. *Stochastic Processes and their Applications 75*(2), 149–172.

Figueiredo, F., M. I. Gomes, L. Henriques-Rodrigues, and M. C. Miranda (2012). A computational study of a quasi-port methodology for var based on second-order reduced-bias estimation. *Journal of Statistical Computation and Simulation 82*(4), 587–602.

Fraga Alves, M. I., M. I. Gomes, and L. de Haan (2003). A new class of semi-parametric estimators of the second order parameter. *Portugaliae Mathematica 60*(2), 193–214.

Frigessi, A., O. Haug, and H. Rue (2002). A dynamic mixture model for unsupervised tail estimation without threshold selection. *Extremes 5*, 219–235.

Gomes, M. I., F. Figueiredo, and M. M. Neves (2012). Adaptive estimation of heavy right tails: Resampling-based methods in action. *Extremes 15*(4), 463–489.

Gomes, M. I. and A. Guillou (2015). Extreme value theory and statistics of univariate extremes: A review. *International Statistical Review 83*(2), 263–292.

Gomes, M. I., L. Henriques-Rodrigues, M. I. Fraga Alves, and B. G. Manjunath (2013). Adaptive PORT–MVRB estimation: An empirical comparison of two heuristic algorithms. *Journal of Statistical Computation and Simulation 83*(6), 1129–1144.

Gomes, M. I., L. Henriques-Rodrigues, and M. C. Miranda (2011). Reduced-bias location-invariant extreme value index estimation: A simulation study. *Communications in Statistics—Simulation and Computation 40*(3), 424–447.

Gomes, M. I., L. Henriques-Rodrigues, B. Vandewalle, and C. Viseu (2008). A heuristic adaptive choice of the threshold for bias-corrected hill estimators. *Journal of Statistical Computation and Simulation 78*(2), 133–150.

Gomes, M. I. and M. J. Martins (2002). Asymptotically unbiased estimators of the tail index based on external estimation of the second order parameter. *Extremes 5*(1), 5–31.

Gomes, M. I., S. Mendonça, and D. Pestana (2011). Adaptive reduced-bias tail index and var estimation via the bootstrap methodology. *Communications in Statistics— Theory and Methods 40*(16), 2946–2968.

Gomes, M. I. and O. Oliveira (2001). The bootstrap methodology in statistics of extremes—choice of the optimal sample fraction. *Extremes 4*(4), 331–358.

Gomes, M. I. and D. Pestana (2007). A sturdy reduced-bias extreme quantile (var) estimator. *Journal of the American Statistical Association 102*(477), 280–292.

Groeneboom, P., H. Lopuhaä, and P. De Wolf (2003). Kernel-type estimators for the extreme value index. *The Annals of Statistics 31*(6), 1956–1995.

Guillou, A. and P. Hall (2001). A diagnostic for selecting the threshold in extreme value analysis. *Journal of the Royal Statistical Society: Series B (Statistical Methodology) 63*(2), 293–305.

Hall, P. (1982). On some simple estimates of an exponent of regular variation. *Journal of the Royal Statistical Society: Series B (Statistical Methodology) 44*, 37–42.

Hall, P. (1990). Using the bootstrap to estimate mean squared error and select smoothing parameter in nonparametric problems. *Journal of Multivariate Analysis 32*(2), 177–203.

Hall, P. and A. Welsh (1985). Adaptive estimates of parameters of regular variation. *The Annals of Statistics 13*(1), 331–341.

Hammersley, J. and D. Handscomb (1964). *Monte Carlo Methods*. Methuen, London.

Henry III, J. (2009). A harmonic moment tail index estimator. *Journal of Statistical Theory and Applications 8*(2), 141–162.

Hill, B. M. (1975). A simple general approach to inference about the tail of a distribution. *The Annals of Statistics 3*(5), 1163–1174.

Kratz, M. and S. I. Resnick (1996). The QQ-estimator and heavy tails. *Stochastic Models 12*(4), 699–724.

Matthys, G. and J. Beirlant (2000). Adaptive threshold selection in tail index estimation. In P. Embrechts (Ed.), *Extremes and Integrated Risk Management*, pp. 37–49. London, UK: Risk Books.

Neves, M. M., M. I. Gomes, F. Figueiredo, and D. P. Gomes (2015). Modeling extreme events: Sample fraction adaptive choice in parameter estimation. *Journal of Statistical Theory and Practice 9*(1), 184–199.

Reiss, R.-D. and M. Thomas (2007). *Statistical Analysis of Extreme Values with Applications to Insurance, Finance, Hydrology and Other Fields*. Birkhäuser Verlag.

Resnick, S. and C. Stărică (1997). Smoothing the hill estimator. *Advances in Applied Probability 29*(1), 271–293.

Resnick, S. I. (1997). Heavy tail modeling and teletraffic data. *The Annals of Statistics 25*(5), 1805–1869.

Scarrott, C. and A. MacDonald (2012). A review of extreme value threshold estimation and uncertainty quantification. *REVSTAT Statistical Journal 10*(1), 33–60.

Wager, S. (2014). Subsampling extremes: From block maxima to smooth tail estimation. *Journal of Multivariate Analysis 130*, 335–353.

5

Threshold Modeling of Nonstationary Extremes

Paul J. Northrop
University College London, United Kingdom

Philip Jonathan
Shell Research Ltd., United Kingdom

David Randell
Shell Research Ltd., United Kingdom

Abstract

It is common for extremes of a variable to be nonstationary, varying systematically with covariate values. We consider the incorporation of covariate effects into threshold-based extreme value models, using parametric and nonparametric regression functions. We use quantile regression to set a covariate-dependent threshold. As an example we model storm peak significant wave heights as a function of storm direction, season, and a climate index.

5.1 Introduction

In many applications a response variable exhibits clear nonstationary behaviour, perhaps varying systematically in time, space, or with the values of other *covariates*. Regression modeling is a natural way to describe such effects.

To illustrate the importance of modeling covariate effects we consider a situation in which a variable Y is strongly seasonal, although the same argument applies for other types of covariate. If one wishes to answer the question "How does the extremal behaviour of Y depend on season?" then clearly one must estimate behaviour conditional on season. However, it is essential to model seasonal effects even if only overall *omni-seasonal* inferences are required. A seasonal model can be expected to explain the variation in Y better than a non-seasonal model, producing more precise inferences about extreme quantiles. Moreover, the consequences of ignoring seasonality can be considerable. Coles (2003) used a two-season extreme value model to analyze Venezuelan daily rainfall totals, obtaining point estimates of the extreme value shape parameter ξ of 0.37 in winter and 0.14 in summer. Ignoring season gives an estimate of 0.26, close to the mean of the seasonal estimates. Similar "averaging" also occurs in the other model parameters, but it is the effect on ξ that is crucial: extreme value extrapolations are nonlinear in ξ so the non-seasonal model will not, in general, produce inferences that are equivalent to the seasonal model. In this example, where positive estimates of ξ are involved, the non-seasonal model gives considerably smaller estimated probabilities of extreme events than the seasonal model.

5.2   Extreme Value Regression Modeling

The general aim of regression modeling is to describe how the distribution of a response variable Y depends on a vector of covariates X. Once the value x of the covariate vector is observed it is treated as fixed, and the focus is on the conditional distribution of Y given x. From an extreme value perspective conditional tail characteristics are of interest, for example, how extreme conditional quantiles of Y depend on x. We suppose that Y is univariate and that distinct responses $Y_1, \ldots, Y_n$ are conditionally independent given their respective covariate values $x_1, \ldots, x_n$.

A simple way to incorporate covariate information into an extreme value analysis is to allow the parameters of a univariate extreme value model to be covariate-dependent (Beirlant et al., 2004, chapter 7). A *regression function* f specifies the functional relationship between the covariates and a parameter, for example a scale parameter σ of a model.

In a *parametric regression*, the form of f is specified in advance and involves a number of interpretable regression parameters (Smith, 1989; Davison and Smith, 1990). Estimating the regression function equates to estimating the regression parameters. In Section 5.3 we use a *linear model*, in which a regression function depends only on a linear combination of the covariates (a *linear predictor* η). However, subject matter considerations may suggest a nonlinear function and, even then, simple parametric regression may not be sufficiently flexible to describe complex relationships in real data.

In a *nonparametric regression*, f is left unspecified and the aim is to estimate f directly, rather than to estimate parameters. One approach is to represent f using a flexible family of smooth functions that can adapt to local behaviour in the data almost independently in several regions of covariate-space at once (Chavez-Demoulin and Davison, 2005; Jonathan et al., 2014). In Section 5.4 we base these smooth functions on marginal cubic B-splines for each covariate. B-splines are a particular form of piecewise cubic function, constrained to be continuous and smooth at the *knots* where separate cubic functions join. An alternative is local-likelihood regression (Butler et al., 2007; Ramesh and Davison, 2002), in which a relatively simple parametric model is fitted locally at each covariate value x_0 of interest and individual log-likelihood contributions are weighted more greatly the closer their covariate value is to x_0, according to some distance metric.

In either approach (parametric or nonparametric) a monotone *link function g*, satisfying $g(\sigma) = f(x)$, can be used to constrain a parameter to a specified range, or to improve the fit of the assumed functional relationship. For example, it is common to use a log-link, $\log \sigma = f(x)$, to ensure positivity of σ. An alternative is to transform the response variable. Even in the stationary case, transforming the response nonlinearly can lead to quite different extremal inferences (Wadsworth et al., 2010).

Eastoe and Tawn (2009) and Dupuis (2012) use a two-step *preprocessing approach*: a regression model is fitted to all the data, followed by a threshold-based extreme value analysis of residuals from this model. A constant threshold is used in the second step but the first step implies a covariate-dependent threshold for the raw data. An alternative (Kyselý et al., 2010; Northrop and Jonathan, 2011) sets directly a covariate-dependent threshold for the raw data. We discuss the latter approach in Section 5.2.3.

5.2.1 Storm Severity in the Northern North Sea

The quantity of interest in this example is significant wave height H_S, a measure of sea surface roughness defined as four times the standard deviation of the surface elevation of a *sea state*: the ocean surface observed for a certain period of time. Quantifying the extremal behaviour of H_S is crucial for the design of marine structures. The raw data consist of hindcast time series (Reistad et al., 2009) for H_S, (dominant) wave direction d (degrees from north), and seasonal degree s (day of the year, for a standardized year of 360 days) for three-hour sea states for the period September 1957 to December 2011. As H_S exhibits strong temporal dependence we decluster the raw data into (5107) *storm peak* H_S values, and their associated values of d and s, using a procedure described in Ewans and Jonathan (2008). We supplement each storm peak by the NOAA's daily North Atlantic Oscillation (NAO) index (from www.cpc.ncep.noaa.gov/products/precip/CWlink/pna/nao.shtml) for the day on which the storm peak occurred. The NAO is a climate index associated with weather patterns in Northern Europe. For notational convenience we refer to storm peak values by H_S, d, s and NAO, rather than the usual convention of adding a superscript of sp.

Figure 5.1 shows a time series plot of H_S and scatter plots of H_S against d, s and NAO. There is a clear seasonal effect, with a tendency for the largest values of H_S to occur in the winter. The effect of NAO is perhaps weaker but large H_S values seem to be positively associated with NAO. Storm peak direction also has a noticeable effect: for example, the intense storms that cross the Atlantic are evident in the directional sector (230,280) degrees. Thus, the characteristics of H_S are nonstationary with respect to multiple covariates. Failure to accommodate nonstationarity can lead to incorrect estimation of design values (Jonathan et al., 2008).

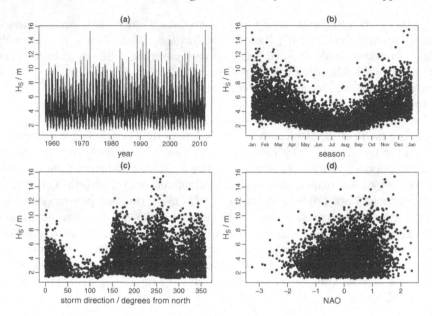

FIGURE 5.1
Northern North Sea hindcast data. (a) Time series plot of storm peak significant wave height (H_S). (b) H_S vs. seasonal degree. (c) H_S vs. storm peak direction. (d) H_S vs. NAO.

In Sections 5.3 and 5.4 we build extreme value regression models for these data, with H_S as the response variable and d, s and NAO as potential covariates. Ultimately, in Section 5.4.3, we concentrate on predicting future extreme observations, such as the largest value of H_S to be observed over a specified time horizon, given the data. These *predictive inferences* (Young and Smith, 2005, Chapter 10) take account of uncertainty in model parameters and are often more instructive than inferences about (functions of) model parameters, such as quantiles of the response conditioned on particular covariate values.

5.2.2 Extreme Value Regression Models

We accommodate nonstationarity in threshold-based extreme value models by allowing their parameters to be covariate-dependent. We consider two models: the nonhomogeneous Poisson process (NHPP) model (Smith, 1989) and the Poisson-GP model (Davison and Smith, 1990). In the absence of covariate effects these models are mathematically equivalent, but their differing parameterizations mean that this is not generally the case. We will see that there are reasons to prefer the NHPP model for parametric regression and the Poisson-GP model for nonparametric regression.

Suppose that over a period of T years independent responses Y_i are observed at times t_i, with associated vector-valued covariates $\boldsymbol{x}_i$, for $i = 1, \ldots, n$. For the dataset described in Section 5.2.1, Y_i is the ith storm peak significant wave height and x_i is based on seasonal degree, NAO, and storm direction.

The NHPP regression model

This model is parameterized in terms of the location $\mu_b(x)$, scale $\sigma_b(x)$, and shape $\xi(x)$ parameters of a GEV approximation to the distribution of the largest response over a (hypothetical) block of b years, during which the covariate remains constant at x. The choice of b is arbitrary: the value of b affects the interpretation of intercept parameters for $\mu_b(x)$ and $\sigma_b(x)$, but otherwise the model should be unaffected by b. It is common to choose $b = 1$, but the motivation for this weakened in cases where covariate values vary within a year. In the following, quantities with subscript t relate to a generic time t and depend on the covariate x_t, for example, u_t denotes $u(x_t)$, a covariate-dependent threshold applied when the covariate is x_t. To simplify the notation we often hide the dependence of μ and σ on b. So, μ_t denotes $\mu_b(x_t)$. We use subscript i to denote a quantity relating to observation i at time t_i, for example, $\mu_i = \mu_{t_i} = \mu(x_i)$.

Following Smith (2003, Section 1.2.5) we model the process of threshold exceedance times and values on $\{(t, y_t) : t \in [0, T], y_t \in [u_t, \infty)\}$ as a nonhomogeneous Poisson process with covariate-dependent intensity function

$$\lambda_b(t, y_t) = b^{-1}\sigma_t^{-1}\left(1 + \xi_t \frac{y_t - \mu_t}{\sigma_t}\right)^{-1/\xi_t - 1}, \qquad (5.1)$$

wherever $1 + \xi_t(y_t - \mu_t)/\sigma_t > 0$ (elsewhere $\lambda_b(t, y_t) = 0$), and we make the modification that u_t, which satisfies $1 + \xi_t(u_t - \mu_t)/\sigma_t > 0$ for all t, may be covariate-dependent. If, conditional on the covariate values, the responses are independent, then the likelihood under this model is

$$L_{NHPP}(\varpi) = \exp\left\{-\int_0^T \int_{u_t}^\infty \lambda_b(t, y_t) \, dy_t \, dt\right\} \prod_{i:y_i > u_i} \lambda_b(t_i, y_i), \qquad (5.2)$$

for a parameter vector ϖ. In practice, the integral in (5.2) is approximated by

$$\frac{T}{n}\sum_{i=1}^n \int_{u_i}^\infty \lambda_b(t_i, y_t) \, dy_t = n_b^{-1}\sum_{i=1}^n \left(1 + \xi_i \frac{u_i - \mu_i}{\sigma_i}\right)^{-1/\xi_i},$$

where $n_b = nb/T$ is the mean number of observations per block.

All parameters of the NHPP are invariant to threshold, so if particular parametric forms are used for $\mu(x)$ and/or $\sigma(x)$ then these forms apply for any threshold (Chavez-Demoulin et al., 2011). The Poisson-GP parameterization doesn't possess this desirable property, so it is usually preferable to specify parametric regression effects using the NHPP model. However, particular functional forms $\mu_{b_1}(x)$ and $\sigma_{b_1}(x)$ based on a block length b_1, may imply different forms for $\mu_{b_2}(x)$ and/or $\sigma_{b_2}(x)$, where $b_2 \neq b_1$. Consider a scalar covariate x, take $\xi(x) = \xi$ for all x and let $\delta = b_2/b_1$. The shape parameter is unaffected by a change of block length, but

$$\mu_{b_2}(x) = \mu_{b_1}(x) + \sigma_{b_1}(x)(\delta^\xi - 1)/\xi \quad \text{and} \quad \sigma_{b_2}(x) = \sigma_{b_1}(x)\delta^\xi.$$

Suppose that $\mu_{b_1}(x) = \alpha_0 + \alpha_1 x$ and $\sigma_{b_1}(x) = \exp(\gamma_0 + \gamma_1 x)$. Then $\mu_{b_2}(x) = \alpha_0 + \alpha_1 x + \exp(\gamma_0 + \gamma_1 x)(\delta^\xi - 1)/\xi$ and $\sigma_{b_2}(x) = \exp(\gamma_0' + \gamma_1 x)$, where $\gamma_0' = \gamma_0 + \xi \log \delta$. Therefore, in this commonly used model, the form of $\sigma_b(x)$ is invariant to changes in b but $\mu_b(x)$ is not, and an arbitrarily chosen b_1 is special because it is the only block length for which μ is linear in x. A remedy is $\mu_{b_1}(x) = \alpha_0 + \alpha_1 x + c \exp(\gamma_0 + \gamma_1 x)$ and $\sigma_{b_1}(x) = \exp(\gamma_0 + \gamma_1 x)$. Now the block length for which μ is linear in x is determined by the parameter c, rather than being fixed arbitrarily in advance. The practical impact of ignoring these considerations might not be great, but if this kind of invariance to b is desired then the following models also have this property:

(a) $\mu_b(x) = \alpha_0 + \alpha_1 x$ with $\sigma_b(x) = \sigma$ or $\sigma_b(x) = \gamma_0 + \gamma_1 x$;
(b) $\mu_b(x) = a + \exp(\alpha_0 + \alpha_1 x)$ with $\sigma_b(x) = \sigma$;
(c) $\sigma_b(x) \propto \mu_b(x)$, with $\mu_b(x) = \alpha_0 + \alpha_1 x$ or $\mu_b(x) = \exp(\alpha_0 + \alpha_1 x)$ or $\mu_b(x) = a + \exp(\alpha_0 + \alpha_1 x)$;

and this also holds if $\alpha_1 x$ and $\gamma_1 x$ are replaced by $\sum_{i=1}^p \alpha_i x_i$ and $\sum_{i=1}^p \gamma_i x_i$, respectively. This property is retained if $\xi(x)$ is piecewise constant, for example, when a different shape parameter is permitted within different regions of space, but otherwise we see no immediate way to ensure invariance to b when smooth functional forms are specified for $\xi(x)$.

The Poisson-GP regression model

In the NHPP likelihood (5.2) we make the parameter transformations $\rho_t = \int_{u_t}^\infty \lambda_b(t, y_t) \, dy_t = b^{-1}[1 + \xi_t(u_t - \mu_t)/\sigma_t]^{-1/\xi_t}$ and $\psi_t = \sigma_t + \xi_t(u_t - \mu_t)$. This gives (up to proportionality) the Poisson-GP likelihood

$$L_{PGP}(\varphi) = \exp\left\{-\int_0^T \rho_t \, dt\right\} \prod_{i:y_i > u_i} \rho_i \psi_i^{-1} \left(1 + \xi_i \frac{y_i - u_i}{\psi_i}\right)^{-1/\xi_i - 1}, \quad (5.3)$$

for $\rho_i > 0$ and $1 + \xi_i(y_i - u_i)/\psi_i > 0$ for all $i : y_i > u_i$ (elsewhere $L_{PGP}(\varphi) = 0$), where φ is a parameter vector.

The essential difference between the Poisson-GP and NHPP regression models is the parameterization of regression effects. The GP scale parameter ψ_t is not invariant to threshold so, apart from some simple special cases, the way in which covariates enter ψ_t depends on the threshold. Eastoe and Tawn (2009) discuss this issue in detail, including an example model: constant threshold u, constant ξ and (changing our notation for a moment so that $\psi_u(x)$ denotes the GP scale parameter at threshold u as a function of a scalar covariate x), $\psi_u(x) = \exp(\gamma_0 + \gamma_1 x)$. If the threshold is changed to v then $\psi_v(x) = \exp(\gamma_0 + \gamma_1 x) + (v - u)\xi$, which has a different functional form to $\psi_u(x)$. A remedy is to introduce a positive parameter a via $\psi_u(x) = a + \exp(\gamma_0 + \gamma_1 x)$. The situation may be complicated further if covariate-dependent thresholds, $u(x)$ and $v(x)$, are used. These issues mean that parametric regression effects are better incorporated using the NHPP parameterization.

Lack of functional stability to threshold is of little importance in a nonparametric regression approach because the functional form of the regression function, and estimates of the (many) parameters, are incidental and are not intended for interpretation. What matters are graphical representations of the fitted values that summarize the smooth approximation provided by the nonparametric regression function. We will see in Section 5.4.2 that when nonparametric regression is used the Poisson-GP parameterization has computational advantages over the NHPP parameterization. The rate of threshold exceedance and the GP parameters can be estimated separately and use of an orthogonal parameterization of the GP parameters can result in further computational stability and speed.

5.2.3   Covariate-Dependent Thresholds

Threshold-based extreme value models are motivated by asymptotic arguments, with the effect that the threshold must be suitably high. In practice a threshold is set empirically, to strike a balance between model mis-specification bias (threshold too low) and low precision of estimation (threshold too high). This is a difficult problem and many approaches have been suggested: see Scarrott and MacDonald (2012) for a review. A simple graphical diagnostic is a *threshold stability plot*: a plot of estimated model parameters against threshold, from which we aim to choose the lowest threshold above which the estimates are judged constant, taking into account sampling variability summarized by confidence intervals. Unlike most other threshold selection methods this approach extends easily to the regression situation. As no definitive threshold selection method exists, it is essential to examine the sensitivity of extreme value inferences to threshold.

In threshold-based extreme value modeling it seems natural to use a covariate-dependent threshold, because a threshold that is sufficiently high at one covariate value may be too low for another covariate value. Another argument is one of statistical design: to improve the precision of estimation of a covariate effect we should aim to have exceedances spread as far across the observed values of the covariate as possible. Setting a constant threshold will tend to narrow the range of covariates for which there are exceedances. Following Northrop and Jonathan (2011) we use quantile regression (Koenker, 2005) to set a threshold for which the probability $1 - \tau$ of threshold exceedance is approximately constant with respect to covariate.

Quantile regression estimates the $100\tau\%$ conditional quantile $y_\tau(\boldsymbol{x})$ of a response variable Y as a function $u(\boldsymbol{x}, \tau)$ of covariates $\boldsymbol{x}$. Specifically, $u(\boldsymbol{x}, \tau)$ is estimated by minimizing the fitting criterion

$$\ell_u = \tau \sum_{i:r_i \geqslant 0} |r_i| + (1 - \tau) \sum_{i:r_i < 0} |r_i|, \qquad (5.4)$$

for residuals $r_i = y_i - u(\boldsymbol{x}_i, \tau)$. If quantile regression functions are estimated independently for each τ, crossing of quantile curves for different values of τ can occur leading to inconsistency. A constrained version of quantile regression avoids this problem (see Bondell et al. (2010)). Jonathan et al. (2014) provide a simple algorithm for non-crossing penalized spline quantile regression, which we will use in Section 5.4.

Once a threshold $\widehat{u}(\boldsymbol{x}, \tau)$ is set it is treated as fixed when the extreme value model is fitted subsequently. However, in a parametric regression, the quantile regression model used to set the threshold is related to the parameterization of covariate effects in the NHPP model: each model should give the same functional form for $y_\tau(x)$. For example, suppose that $\xi(x) = \xi$ is constant. A constant probability of exceedance equates to $[1 + \xi(u(x) - \mu(x))/\sigma(x)]^{-1/\xi}$ being constant with respect to x, from which it follows that

$$u(x) = \mu(x) + k\,\sigma(x), \tag{5.5}$$

for some constant k. We consider some examples in Section 5.3. Northrop and Jonathan (2011) use (5.5) to assess whether $y_\tau(x)$ under the NHPP model differs significantly from $\widehat{u}(x, \tau)$.

In Figure 5.2 fitted quantile regression curves have been superimposed on the plots of H_S (and $\log H_S$) against season and against NAO, using separate models for season and for NAO. In Sections 5.3 and 5.4 we model the effects of different covariates together. Here we just wish to illustrate how quantile regression can be used to set a covariate-dependent threshold and to explore how regression effects depend on τ. In the seasonal plots, where $y_\tau(x) = A_\tau \cos(2\pi(s - \phi_\tau)/360)$, this simple function appears to represent well the seasonal effect for a given value of τ. However, when the response is H_S the fitted curves are not parallel (at least up to the 90% quantile), suggesting that the magnitude of the seasonal effect changes (increases) as τ increases. In contrast the curves for $\log H_S$ are approximately parallel. This is consistent with season having a multiplicative effect on H_S, rather than an additive effect. A similar phenomenon is seen in the NAO plots.

5.2.4   Statistical Inference

Typically parametric extreme value regression models are fitted using maximum likelihood (ML) estimation, because the general principle of maximizing the log-likelihood with respect to the parameters needs no special modification when covariate effects are introduced. A nonlinear optimization routine is used to maximize the log-likelihood, subject to constraints on the parameters. In the absence of covariates, for $\xi < -1$ there is a singularity on the edge of the parameter space at which the log-likelihood is unbounded above (see, for example, Hosking and Wallis (1987)) and this feature remains if covariates are included. However, provided that there is a local maximum away from the singularity, then this is used. In fact there may be multiple local maxima, so it is important to check whether results are sensitive to the initial estimates provided to the optimization routine. Smith (1994) shows that the asymptotic theory of ML estimation holds in regression problems of this type provided that $\xi_i > -1/2$ for all i. In that event standard errors and likelihood-based confidence intervals for parameters can be estimated in the usual way and nested models can be compared using a likelihood ratio test. However, the asymptotic results may provide poor approximations, especially for small sample sizes.

For a nonparametric model based on smooth functions, such as cubic B-splines, estimation based on the log-likelihood needs modification to prevent over-fitting. Suppose that q parameters, $\theta_i, i = 1, \ldots, q$, of a model have been modeled using

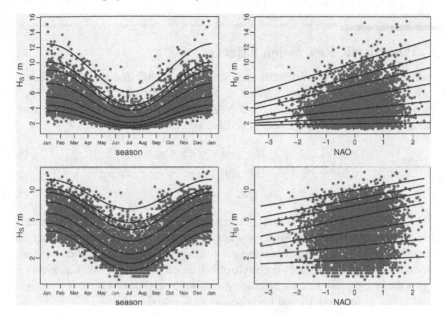

FIGURE 5.2
Parametric quantile regression for the northern North Sea data. Left: $y_\tau(x) = A_\tau \cos(s - \phi_\tau)$. Right: $y_\tau(x)$ linear in NAO. Top: response is H_S. Bottom: response is $\log H_S$. The lines are estimated $(10, 25, 50, 75, 90, 95, 99)\%$ conditional quantiles.

respective smooth functions $f_i(x)$ of x. These smooth functions are estimated by minimizing a penalized fitting criterion

$$\ell_\theta^*(\varrho) = \ell(\theta) + \sum_{i=1}^{q} \varrho_i R_i, \tag{5.6}$$

where $\ell(\theta)$ is a fitting criterion (the smaller the value the better the fit), $\theta' = (\theta_1, \ldots, \theta_q)$, $\varrho' = (\varrho_1, \ldots, \varrho_q)$ and R_i is a *roughness penalty*, which is large when $f_i(x)$ changes rapidly with x. For fitting extreme value models $l(\theta)$ is a negated log-likelihood and for estimating a threshold using quantile regression $l(\theta)$ is given by (5.4). The ϱ_is are *roughness coefficients* that control the trade-off between quality of fit to the data and roughness of the regression functions. The larger the values of ϱ_i the greater the weight given to the roughness penalties and the smoother is the fitted regression function. We will choose these constants, with the aim of providing good predictive performance, using leave-one-out cross-validation (CV), i.e., by those values for which $\sum_{i=1}^{n} \ell_{-i}(y_i, x_i)$ is minimized, where $\ell_{-i}(y_i, x_i)$ is the fitting criterion evaluated at (y_i, x_i) based on the penalized fit (5.6) using all the data except (y_i, x_i). In Section 5.4 we use bootstrapping to account for uncertainty in the roughness coefficients and in the model parameters.

5.3   Parametric Regression Effects

We build parametric regression effects of season and NAO into an NHPP model for H_S. In the notation of Section 5.2.2, Y_i is the value of (some function of) H_S for storm i, with associated season s_i and NAO value N_i. We consider three simple model structures, each with ξ constant and a predictor η_i for μ_i:

$$\text{Model 1}: \ Y_i = H_{Si} \text{ and } \mu_i = \eta_i, \ \sigma_i = \sigma, \qquad (5.7)$$

$$\text{Model 2}: \ Y_i = H_{Si} \text{ and } \log \mu_i = \eta_i, \ \sigma_i \propto \mu_i, \qquad (5.8)$$

$$\text{Model 3}: \ Y_i = \log H_{Si} \text{ and } \mu_i = \eta_i, \ \sigma_i = \sigma. \qquad (5.9)$$

Model 1 represents additive covariate effects on H_S. Models 2 and 3 both represent multiplicative effects, but in different ways: in model 2 the covariates affect multiplicatively location and scale parameters for H_S and in model 3 the response is log-transformed so that effects are multiplicative on the H_S-scale. Equation (5.5) suggests that we can use linear quantile regression to set thresholds for these models: with $u(\boldsymbol{x}_i) = \eta_i$ for model 1 and $\log u(\boldsymbol{x}_i) = \eta_i$ for models 2 and 3.

We aim to show that choice of model parameterization can have implications for threshold selection and for extreme value extrapolation, rather than to seek to build a definitive model for the data. In particular, we ignore the possibility of time trends or covariate effects in ξ. We take as a working model

$$\eta_i = \alpha_0 + A\cos(s_i - \phi) + (\alpha_N + A_N \cos(s_i - \phi_N))\,N_i, \qquad (5.10)$$

$$= \alpha_0 + \alpha_C \cos(s_i) + \alpha_S \sin(s_i) + \alpha_N N_i + \alpha_{CN} \cos(s)N_i + \alpha_{SN} \sin(s)N_i, \qquad (5.11)$$

where $\alpha_C = A\cos(\phi), \alpha_S = A\sin(\phi), \alpha_{CN} = A_N \cos(\phi_N)$ and $\alpha_{SN} = A_N \sin(\phi_N)$ and, here, season s has been scaled to $(0, 2\pi)$. In (5.10) seasonality is represented by a cosine wave with amplitude A and peak position ϕ, NAO has a linear main effect and a season-NAO interaction is incorporated by allowing the effect of NAO to vary seasonally according to a cosine wave. Equation (5.11) is useful for computational purposes because it is a *linear* predictor.

5.3.1   Threshold Estimation Using Quantile Regression

We use models based on (5.10) to illustrate the use of threshold stability plots to select an appropriate value of τ, i.e., the estimated conditional quantile at which the threshold is set. Some iteration is almost inevitable: the form of the covariate effects in the NHPP regression model determines the form of the threshold (Section 5.2.3) and the choice of model is informed by empirical analyses for a given threshold.

Figure 5.3 shows, for models 1, 2, and 3, plots of the maximum likelihood estimates (MLEs) of the shape parameter ξ and the amplitude A of the seasonal main effect against 100τ, for $\tau \in [0, 0.97]$. We include only these parameters because ξ is the most important parameter in the model and the variation in $\widehat{A}$ highlights the key

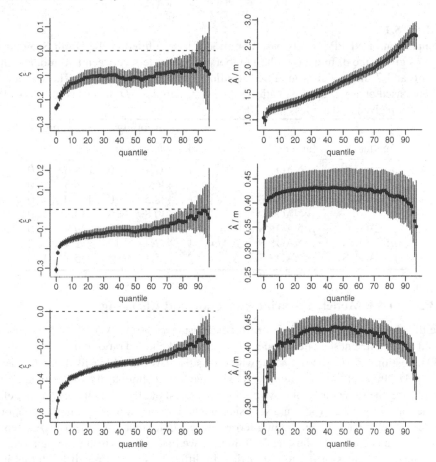

FIGURE 5.3
Threshold stability plots for $\widehat{\xi}$ and $\widehat{A}$ for the North Sea data. Top: model 1; middle: model 2; bottom: model 3. Parameter estimates and symmetric 95% confidence intervals against level of conditional quantile.

differences (and similarities) between the models. For model 1, $\widehat{A}$ increases substantially with τ, perhaps leveling off near $\tau = 0.9$. This behaviour is as expected based on plot (a) of Figure 5.2. For models 2 and 3 $\widehat{A}$ is quite stable across a wide range of τ before decreasing as smaller sample sizes of exceedances are reached. All three models have somewhat similar patterns for $\widehat{\xi}$ but model 3 has more strongly negative estimates than the other models and perhaps the estimates for model 1 stabilize at a lower threshold than the other models. This illustrates that choice of response scale and/or parameter scale can affect substantially the selection of threshold level and inferences about ξ.

TABLE 5.1
Comparison of NHPP models based on model 1 (5.7) fitted to the North Sea data, with df giving the difference in the numbers of parameters between two models. In rows 6 and 7 the model is modified to allow covariate effects in σ. The p-values (p) are based on a χ^2_{df} approximation to the distribution of D under the null model. Values of p given as 0 are less than 10^{-6}.

			dropping		adding	
smaller	larger	df	D	p	D	p
no covariates	μ:NAO	1	91.1	0	118.6	0
no covariates	μ:S	2	446.8	0	463.3	0
μ:NAO	μ:NAO+S	2	438.5	0	431.0	0
μ:S	μ:NAO+S	1	40.8	0	43.3	0
μ:NAO+S	μ:NAO$\times$S	2	28.1	0	29.9	0
μ:NAO$\times$S	μ:NAO$\times$S, σ:NAO	1	2.4	0.12	2.4	0.12
μ:NAO$\times$S	μ:NAO$\times$S, σ:S	2	5.1	0.08	5.1	0.08

5.3.2 NHPP Model: Covariate Selection and Checking

We illustrate covariate selection for Model 1, using $\tau = 0.9$. A model M_1 is compared to a smaller, nested, null model M_2 using the likelihood ratio statistic (Davison, 2003, Section 4.5) $D = 2(\widehat{l}_1 - \widehat{l}_2)$, where $\widehat{l}_i$ is the maximized log-likelihood under model M_i. In order to compare these maximized log-likelihoods the models must be fitted using the same threshold. We could use a threshold based on the larger model, of the form implied by (5.5) and set using quantile regression, (dropping terms), or based on the smaller model (adding terms). One would hope for little disagreement between the two comparisons and in Table 5.1 we find that this is the case, although the numerical values of the test statistics do differ. It is clear that all the terms in predictor (5.10) are required (this is also the case for models 2 and 3) and that there is some support for a seasonal effect on σ.

Table 5.2 gives MLEs and standard errors (SEs) for models 1, 2, and 3, based on predictor (5.10). The important differences relate to ξ: for model 3 the point estimate is rather more negative than for models 1 and 2 and the SE of $\widehat{\xi}$ is larger for model 1 due to use of a higher threshold level. As expected, the estimated regression effects in models 2 and 3 are similar.

If a model is a good fit to the data then, for $i : y_i > u_i$, i.e., for all threshold exceedances, $\widehat{\xi}^{-1} \log(1 + \widehat{\xi}(y_i - u_i)/\widehat{\psi}_i)$, where $\widehat{\psi}_i = \widehat{\sigma}_i + \widehat{\xi}_i(u_i - \widehat{\mu}_i)$, should look approximately like a sample from a standard exponential distribution (Coles, 2001, Section 6.2). Exponential QQ plots (not shown) do not reveal systematic lack-of-fit for any of these three models and are therefore of little help in choosing between them. However, we have not used important covariate information from storm peak direction, which we address in the next section.

TABLE 5.2
MLEs and SEs (in brackets) based on predictor (5.10), fitted to the North Sea data. For model 2 the values in the column labeled σ relate to the ratio σ_i/μ_i. Here $b = T/n$, so α_0 and σ relate to the distribution of individual storm peak H_S values. A $100\tau\%$ conditional threshold estimated using quantile regression is used, with $\tau = 0.9$ for model 1 and $\tau = 0.75$ for models 2 and 3.

model	α_0	α_N	A	ϕ	A_N	ϕ_N	σ	ξ
1	3.2	0.56	2.6	3.3	0.55	2.3	1.4	-0.05
	(0.34)	(0.07)	(0.08)	(1.8)	(0.10)	(10.0)	(0.22)	(0.05)
2	1.0	0.09	0.4	4.2	0.08	11.5	0.6	-0.06
	(0.04)	(0.02)	(0.02)	(2.6)	(0.03)	(17.9)	(0.04)	(0.02)
3	1.1	0.08	0.4	3.8	0.07	10.4	0.4	-0.23
	(0.02)	(0.01)	(0.01)	(1.4)	(0.01)	(10.7)	(0.02)	(0.02)

5.4 Nonparametric Regression Effects

The approach taken in this section follows Jonathan et al. (2014). Recall that Y_i is the value of H_S for storm i, with associated season s_i and now direction d_i. Physical considerations suggest that the threshold u and the Poisson-GP parameters ρ, ψ, ξ should vary smoothly with d and s. We achieve this by expressing each parameter in terms of an appropriate basis for the domain D of covariates, where $D = D_d \times D_s$ and $D_d = D_s = [0, 360)$ are the (marginal) domains of direction and season. We calculate a $(32 \times p_d)$ periodic marginal B-spline basis matrix B_d for an index set of 32 directional knots, and a $(24 \times p_s)$ periodic marginal B-spline basis matrix B_s for an index set of 24 seasonal bins, yielding a total of $m(= 32 \times 24)$ combinations of covariate values, and a corresponding partitioning of the response sample $Y_1, \ldots, Y_n$ into m covariate bins. Using periodic splines avoids discontinuities at the start of the year and at a storm direction of zero degrees. We define a basis matrix for D using Kronecker products of the marginal basis matrices. Thus $B = B_s \otimes B_d$ provides an $(m \times p)$, where $p = p_d p_s$, basis matrix for modeling each of u, ρ, ξ and ψ (ζ, say, for brevity) any of which can be expressed in the form $\zeta = B\beta$ for some $(p \times 1)$ vector β of basis coefficients. We allocate one B-spline function to each covariate bin, so that $p_d = 32$ and $p_s = 24$. Model estimation therefore reduces to estimating appropriate sets of basis coefficients for each of u, ρ, ψ, ξ.

Following Eilers and Marx (2010), we define roughness on the index set as

$$R = \beta' P \beta, \tag{5.12}$$

where the $p \times p$ matrix P is such that R penalises the difference between marginally neighbouring values of spline coefficients, thereby penalising lack of local smoothness (see Currie et al. (2006) for details).

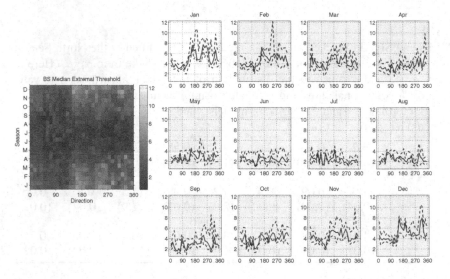

FIGURE 5.4
Directional-seasonal threshold u for non-exceedance probability 0.5. Left: bootstrap median threshold on storm peak direction, d, and storm peak season, s. Right: monthly directional thresholds with bootstrap median (solid lines) and 95% bootstrap uncertainty band (dashed lines).

Calculation of Kronecker products becomes challenging with increasing numbers of covariates. Fortunately, direct computation of B can be avoided using the generalized linear array model of Currie et al. (2006), resulting in exact low-memory, high-speed computations. Jonathan et al. (2014) describes strategies for calculating good initial estimates for model parameters.

In Sections 5.4.1 and 5.4.2 we describe the separate estimation of smooth cubic B-spline representations of u, of ρ and of (ψ, ξ) as functions of d and s. This is equivalent to direct estimation of a nonparametric NHPP regression model. We estimate each regression function using (5.6) with roughness penalty (5.12). In each case the roughness coefficient is chosen using CV. We account for uncertainty in model parameters, and in the CV-estimated roughness coefficients using bootstrap resampling. The entire analysis is repeated for 200 resamples of the original storm peak sample. We found that 200 bootstrap resamples were sufficient to ensure stability of the resulting inferences.

5.4.1   Threshold Estimation Using Quantile Regression

A covariate-dependent threshold is estimated by minimizing $\ell_u^* = \ell_u + \varrho_u R_u$, where ℓ_u is given by (5.4) with $r_i = y_i - u(d_i, s_i; \tau)$. We show results for $\tau = 0.5$, but findings are not critically sensitive to this choice. Figure 5.4 summarises how the bootstrap distribution of $u(d, s; 0.5)$ depends on d and s. Summer (Jun-Aug) is relatively calm, as are storm events from directions in $[0, 90)°$.

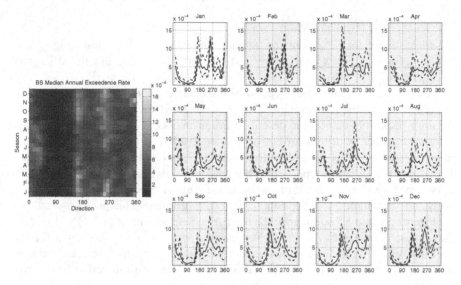

FIGURE 5.5
Directional-seasonal threshold exceedance rate ρ for non-exceedance probability 0.5.
Left: bootstrap median threshold on storm peak direction, d, and storm peak season,
s. Right: monthly directional thresholds with bootstrap median (solid lines) and 95%
bootstrap uncertainty band (dashed lines).

5.4.2 Poisson-GP Model: Fitting and Checking

Rate of threshold exceedance

A covariate-dependent rate ρ of threshold exceedance is estimated by minimizing
$\ell_\rho^* = \ell_\rho + \varrho_\rho R_\rho$, where, from (5.3), the Poisson negated log-likelihood is given by
$\ell_\rho = -\sum_{i:y_i > u_i} \log \rho(d_i, s_i) + \int \int \rho(d, s) \, \mathrm{d}d \, \mathrm{d}s$ and, following Chavez-Demoulin
and Davison (2005), is estimated by

$$\widehat{\ell}_\rho = -\sum_{j=1}^{m} c_j \log \rho(jA) + A \sum_{j=1}^{m} \rho(jA),$$

where $\{c_j\}_{j=1}^{m}$ are counts of numbers of threshold exceedances on an index set of
$m(\gg 1)$ bins on a regular grid that partitions the covariate domain into intervals of
constant area A.

Estimation of ρ is straightforward using a suitable iterative scheme. We use a
back-fitting algorithm (Davison, 2003), which converges quickly. Figure 5.5 summarises how the bootstrap distribution of $\rho(d, s)$ depends on d and s. The rate of
threshold exceedance is largest for winter storms from approximate directions $180°$
and $270°$.

Magnitudes of threshold excesses

Covariate-dependent GP parameters ψ and ξ are estimated by minimizing $\ell_{\xi,\psi}^* = \ell_{\xi,\psi} + \varrho_\xi R_\xi + \varrho_\psi R_\psi$, where, from (5.3), the GP negated log-likelihood is given by $\ell_{\xi,\psi} = \sum_{i:y_i > u_i} \{\log \psi_i + (1/\xi_i + 1) \log (1 + \xi_i(y_i - u_i)/\psi_i)\}$. We set $\varrho_\xi = \kappa \varrho_\psi$ for a constant κ, set by inspection of the relative smoothness of ξ and ψ with respect to covariates. Estimating the GP parameters is the most challenging step. We use a back-fitting algorithm described in Jonathan et al. (2014). MLEs of ψ and ξ are asymptotically dependent but reparameterization in terms of the orthogonal parameters $\nu = \psi(1 + \xi)$ and ξ (Davison, 2003, page 688) results in estimators that are asymptotically independent. This improves stability and speed of the fitting algorithm. Estimation of nonstationary ξ is particularly problematic; for this reason many authors assume ξ to be constant, often with insufficient justification.

Figures 5.6 and 5.7 summarise the bootstrap distributions of $\xi(d, s)$ and $\psi(d, s)$. The largest estimates of ξ occur in the months of October to March, but there is large directional variability for summer storms, with those emanating from directions close to $270°$ having relatively large estimates. Storms with direction in $(0,150)°$ show little seasonal variation in ψ. The largest estimates of ψ are for winter storms emanating from directions close to $270°$.

To check the fit of the directional-seasonal distributions of H_S, we simulate 5000 realizations of the same time span (54.3 years) as the data. Above the threshold, we simulate under the Poisson-GP model. Below the threshold, we resample with replacement the appropriate number of values from the subset of the data that fall below the threshold. We compare cumulative distribution functions (cdfs) of the data with simulated cdfs, overall, and partitioned by month or directional octant. Figure 5.8, where overall and monthly cdfs are compared, shows good agreement between model and data. Similar plots by direction suggest model fit of a similar quality.

5.4.3   Approximate Predictive Inference

Suppose that we are interested in the largest H_S value M_R observed over R years, conditional on direction and/or season and overall. We perform approximate predictive inference on M_R using bootstrapping (Fushiki et al., 2005). To each of 5000 bootstrap resamples of the original storm peak data we:
1. Fit the directional-seasonal Poisson-GP model to the resample.
2. For each directional-seasonal bin draw at random the number of storm peaks from a Poisson distribution with mean ρR, i.e., for a time period of R years.
    (a) Draw randomly a pair of values d^* and s^* for direction and season from the area corresponding to the covariate bin.
    (b) Draw randomly H_S^* corresponding to (d^*, s^*) from the GP model.
3. Accumulate maximum values of H_S^* per directional-seasonal bin.

This gives approximate samples of size 5000 from the predictive distribution of M_R within each bin, from which cdfs of M_R conditioned on any desired combination of covariate bins can be estimated. Since realizations based on models from different bootstrap resamples of the original sample are used, these cdfs incorporate

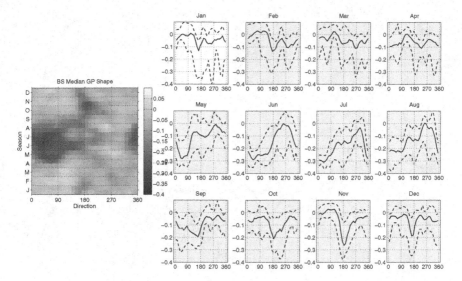

FIGURE 5.6
Directional-seasonal parameter plot for GP shape, ξ. The left-hand panel shows the bootstrap median shape on d and s. The right-hand panel shows 12 monthly directional shapes in terms of bootstrap median (solid) and 95% bootstrap uncertainty band (dashed).

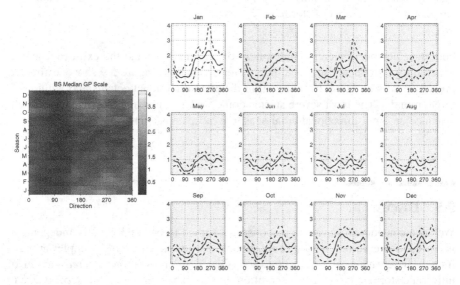

FIGURE 5.7
Directional-seasonal parameter plot for GP scale, ψ. The left-hand panel shows the bootstrap median scale on d and s. The right-hand panel shows 12 monthly directional scales in terms of bootstrap median (solid) and 95% bootstrap uncertainty band (dashed).

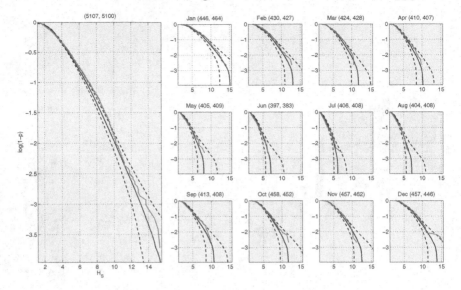

FIGURE 5.8
Validation of the directional-season Poisson-GP model. Left: omni-directional and omni-seasonal cdf p (plotted as $\log_{10}(1-p)$ to focus on the upper tail of H_S) for the data (gray lines), the simulated median (solid lines), and 2.5% and 97.5% quantiles (dashed lines). Right: monthly versions. The titles contain the actual number and simulated mean number of storm peaks, respectively.

both the (aleatory) inherent randomness of return values and the extra (epistemic) uncertainty introduced by model parameter estimation. Figure 5.9 shows estimated predictive cdfs of M_{100}, overall, within different directional octants and within different months. The most severe storms come from the north west in winter months. The median omni-directional omni-seasonal estimate is approximately 16.6m.

5.5 Discussion

We have considered univariate responses, assumed to be conditionally independent given covariate values. This is a reasonable assumption for our example: approximately independent storm events were isolated to create a *declustered* dataset of univariate storm maxima. However, often there will be non-negligible dependence in the tail behaviour of responses, for example over time and/or space, even after covariate effects have been modeled. In some cases it will be important to model the *joint* extremal behaviour of two or more responses. We conclude with brief comments on these and other issues.

In the stationary setting, Fawcett and Walshaw (2012) demonstrate that when ex-

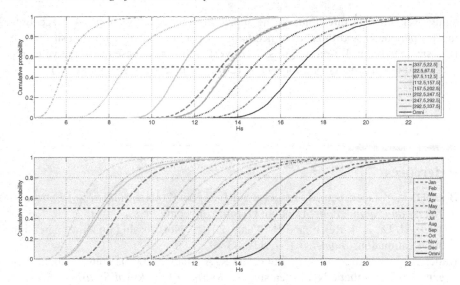

FIGURE 5.9
Estimated cdfs of 100-year maximum storm peak significant wave height from simulation under the directional-seasonal model, incorporating uncertainty in parameter estimation using bootstrap resampling. Top: by directional octants. Bottom: by
month. The common omni-directional omni-seasonal cdf is shown in black in both
plots.

treme values occur in clusters it is preferable to make inferences from *all* threshold
exceedances, rather than just from cluster maxima in a declustered series. Methods for the analysis of time series extremes are reviewed in Chavez-Demoulin and
Davison (2012). It is important to adjust estimates of parameter uncertainty for dependence in the data, and, in making extremal value inferences, to adjust for serial
dependence at extreme levels, as summarized by an estimate of the extremal index.
Similar considerations will apply if time trends are modeled, with the extremal index
playing a similar role (Hüsler, 1986). Dupuis (2012) estimates changepoints in the
extremal behaviour of daily maximum temperatures, while accounting for short-term
serial dependence. Northrop and Jonathan (2011) use the approach outlined in Section 5.3 to model spatial extremal effects, adjusting inferences for the strong spatial
dependence between series from nearby locations using the methodology of Chandler and Bate (2007).

Threshold-based modeling of spatial and spatial-temporal extremes is a subject
of current research. Examples are Huser and Davison (2014) and Turkman et al.
(2010). In Jonathan et al. (2014) the approach of Section 5.4 is extended to bivariate
responses based on the conditional extremes model of Heffernan and Tawn (2004).
There is scope to develop better methods for threshold selection for the regression
situation, and to account for uncertainty in the choice of response scale, tackled by
Wadsworth et al. (2010) for the IID case. Although one could also include an aver-

aging over different thresholds in the approach outlined in Section 5.4.3, accounting for parameter and threshold uncertainty is perhaps handled most easily in the Bayesian paradigm. Recent progress in Bayesian threshold-based extreme value regression modeling includes Cabras et al. (2011) and Reich and Smith (2013).

References

Beirlant, J., Y. Goegebeur, J. Teugels, and J. Segers (2004). *Statistics of Extremes: Theory and Applications*. London: Oxford University Press.

Bondell, H. D., B. J. Reich, and H. Wang (2010). Noncrossing quantile regression curve estimation. *Biometrika 97*, 825–838.

Butler, A., J. E. Heffernan, J. A. Tawn, and R. A. Flather (2007). Trend estimation in extremes of synthetic North Sea surges. *Journal of the Royal Statistical Society: Series C (Applied Statistics) 56*(4), 395–414.

Cabras, S., M. E. Castellanos, and D. Gamerman (2011). A default bayesian approach for regression on extremes. *Statistical Modelling 11*(6), 557–580.

Chandler, R. E. and S. B. Bate (2007). Inference for clustered data using the independence loglikelihood. *Biometrika 94*(1), 167–183.

Chavez-Demoulin, V. and A. C. Davison (2005). Generalized additive modelling of sample extremes. *Journal of the Royal Statistical Society: Series C (Applied Statistics) 54*(1), 207–222.

Chavez-Demoulin, V. and A. C. Davison (2012). Modelling time series extremes. *REVSTAT 10*, 109–133.

Chavez-Demoulin, V., A. C. Davison, and L. Frossard (2011). Discussion of 'threshold modelling of spatially dependent non-stationary extremes with application to hurricane-induced wave heights'. *Environmetrics 22*(7), 800–812.

Coles, S. G. (2001). *An Introduction to Statistical Modeling of Extreme Values*. Springer.

Coles, S. G. (2003). The use and misuse of extreme value models. In B. Finkenstädt and H. Rootzén (Eds.), *Extreme Values in Finance, Telecommunications and the Environment*, pp. 79–100. London: Chapman and Hall CRC.

Currie, I. D., M. Durban, and P. H. C. Eilers (2006). Generalized linear array models with applications to multidimensional smoothing. *Journal of the Royal Statistical Society: Series B (Statistical Methodology) 68*, 259–280.

Davison, A. C. (2003). *Statistical Models*. Cambridge University Press.

Davison, A. C. and R. L. Smith (1990). Models for exceedances over high thresholds (with discussion). *Journal of the Royal Statistical Society: Series B (Statistical Methodology) B52*, 393–442.

Dupuis, D. J. (2012). Modeling waves of extreme temperature: The changing tails of four cities. *Journal of the American Statistical Association 107*(497), 24–39.

Eastoe, E. F. and J. A. Tawn (2009). Modelling non-stationary extremes with application to surface level ozone. *Journal of the Royal Statistical Society: Series C (Applied Statistics) 58*, 25–45.

Eilers, P. H. C. and B. D. Marx (2010). Splines, knots and penalties. *Wiley Interscience Reviews: Computational Statistics 2*, 637–653.

Ewans, K. C. and P. Jonathan (2008). The effect of directionality on northern North Sea extreme wave design criteria. *Journal of Offshore Mechanics and Arctic Engineering 130*, 10.

Fawcett, L. and D. Walshaw (2012). Estimating return levels from serially dependent extremes. *Environmetrics 23*, 272–283.

Fushiki, T., F. Komaki, and K. Aihara (2005, 04). Nonparametric bootstrap prediction. *Bernoulli 11*(2), 293–307.

Heffernan, J. E. and J. A. Tawn (2004). A conditional approach for multivaiate extremes. *Journal of the Royal Statistical Society: Series B (Statistical Methodology) 66*(3), 497–546.

Hosking, J. R. M. and J. R. Wallis (1987). Parameter and quantile estimation for the generalized Pareto distribution. *Technometrics 29*(3), 339–349.

Huser, R. and A. C. Davison (2014). Space-time modelling of extreme events. *Journal of the Royal Statistical Society: Series B (Statistical Methodology) 76*, 439–461.

Hüsler, J. (1986). Extreme values of non-stationary random sequences. *Journal of Applied Probability 23*, 937–950.

Jonathan, P., K. C. Ewans, and G. Z. Forristall (2008). Statistical estimation of extreme ocean environments: The requirement for modelling directionality and other covariate effects. *Ocean Engineering 35*, 1211–1225.

Jonathan, P., K. C. Ewans, and D. Randell (2014). Non-stationary conditional extremes of northern North Sea storm characteristics. *Environmetrics 25*, 172–188.

Jonathan, P., D. Randell, Y. Wu, and K. Ewans (2014). Return level estimation from non-stationary spatial data exhibiting multidimensional covariate effects. *Ocean Engineering 88*(0), 520–532.

Koenker, R. (2005). *Quantile Regression*. Cambridge University Press.

Kyselý, J., J. Picek, and R. Beranová (2010). Estimating extremes in climate change simulations using the peaks-over-threshold method with a non-stationary threshold. *Global and Planetary Change 72*, 55–68.

Northrop, P. J. and P. Jonathan (2011). Threshold modelling of spatially dependent non-stationary extremes with application to hurricane-induced wave heights. *Environmetrics 22*(7), 799–809.

Ramesh, N. and A. Davison (2002). Local models for exploratory analysis of hydrological extremes. *Journal of Hydrology 256*(1), 106–119.

Reich, B. J. and L. B. Smith (2013). Bayesian quantile regression for censored data. *Biometrics 69*(3), 651–660.

Reistad, M., Ø. Breivik, H. Haakenstad, O. J. Aarnes, and B. R. Furevik (2009). A high-resolution hindcast of wind and waves for the North Sea, the Norwegian Sea and the Barents Sea. Technical report, Norwegian Meteorological Institute.

Scarrott, C. and A. MacDonald (2012). A review of extreme value threshold estimation and uncertainty quantification. *REVSTAT 10*(1), 33–60.

Smith, R. L. (1989). Extreme value analysis of environmental time series: An application to trend detection in ground-level ozone. *Statistical Science 4*, 367–377.

Smith, R. L. (1994). Nonregular regression. *Biometrika 81*, 173–183.

Smith, R. L. (2003). Statistics of extremes, with applications in environment, insurance and finance. In B. Finkenstädt and H. Rootzén (Eds.), *Extreme Values in Finance, Telecommunications and the Environment*, pp. 1–78. London: Chapman and Hall CRC.

Turkman, K. F., M. A. A. Turkman, and J. M. Pereira (2010). Asymptotic models and inferences for extremes of spatio-temporal data. *Extremes 13*, 375–397.

Wadsworth, J. L., J. A. Tawn, and P. Jonathan (2010). Accounting for choice of measurement scale in extreme value modelling. *The Annals of Applied Statistics 4*, 1558–1578.

Young, G. A. and R. L. Smith (2005). *Essentials of Statistical Inference*. Cambridge: Cambridge University Press.

6

Block-Maxima of Vines

Matthias Killiches
Technische Universität München, Germany

Claudia Czado
Technische Universität München, Germany

Abstract

This chapter examines the dependence structure of finite block-maxima of multivariate distributions. We provide a closed form expression for the copula density of the vector of the block-maxima. Further, we show how partial derivatives of three-dimensional vine copulas can be obtained by only one-dimensional integration. Combining these results allows the numerical treatment of the block-maxima of any three-dimensional vine copula for finite block-sizes. We look at certain vine copula specifications and examine how the density of the block-maxima behaves for different block-sizes. Additionally, a real data example from hydrology is considered. In extreme-value theory for multivariate normal distributions, a certain scaling of each variable and the correlation matrix is necessary to obtain a non-trivial limiting distribution when the block-size goes to infinity. This scaling is applied to different three-dimensional vine copula specifications.

6.1 Copula Density of the Distribution of Block-Maxima

Basically, block-maxima have been used in extreme-value theory as one approach to derive the family of generalized extreme-value (GEV) distributions (McNeil et al., 2010). In the recent past the block-maxima method has been studied more thoroughly and compared to the peaks-over-threshold (POT) method in Ferreira and de Haan (2014) and Jarušková and Hanek (2006). Dombry (2013) justifies the usage of a maximum-likelihood estimator for the extreme-value index within the block-maxima framework. The numerical convergence of the block-maxima approach to the GEV distribution is examined in Faranda et al. (2011). Moreover, the block-maxima method has found its way into many application areas: Marty and Blanchet (2012) investigate long-term changes in annual maximum snow depth and snowfall in Switzerland. Temperature, precipitation, wind extremes over Europe are analyzed in Nikulin et al. (2011). A spatial application can be found in Naveau et al. (2009). Rocco (2014) provides an overview over the concepts of extreme-value theory being used in finance. While many of the articles use univariate concepts, Bücher and Segers (2014) treat how to estimate extreme-value copulas based on block-maxima of a multivariate stationary time series. In contrast to the existing literature (known to the authors), in the following we will consider finite block-maxima of multivariate random variables focusing on the dependence structure.

Let $\mathbf{U} = (U_1, \ldots, U_d)'$ be a random vector with $\mathcal{U}[0,1]$-distributed margins, copula C and copula density c. We consider n i.i.d. copies $\mathbf{U}_i = (U_{i,1}, \ldots, U_{i,d})'$ of $\mathbf{U}$, $i = 1, \ldots, n$. We apply the inverse probability integral transform to each component of $\mathbf{U}_i$ to obtain marginally normalized data (called z-scale):

$$\mathbf{Z}_i = (Z_{i,1}, \ldots, Z_{i,d})' \text{ with } Z_{i,j} := \Phi^{-1}(U_{i,j}) \sim \mathcal{N}(0,1)$$

for $i = 1, \ldots, n$, $j = 1, \ldots, d$, where Φ^{-1} is the quantile function of the standard normal distribution $\mathcal{N}(0,1)$. We consider this normalized scale since later we want to compare this to the limiting approach used to derive the multivariate Hüsler–Reiss extreme-value copula. We are interested in the distribution $F^{(n)}$ of the vector of componentwise block-maxima

$$\mathbf{M}^{(n)} = (M_1^{(n)}, \ldots, M_d^{(n)})' \text{ with } M_j^{(n)} := \max_{i=1,\ldots,n} Z_{i,j}$$

for $j = 1, \ldots, d$. According to Sklar (1959) the dependency structure is determined by the corresponding copula $C_{\mathbf{M}^{(n)}}$. Since $Z_{i,j}, i = 1, \ldots, n$, are i.i.d., we know that the marginal distribution functions of $M_j^{(n)}$ are given by

$$F_j^{(n)}(m_j) = \mathbb{P}\left(Z_{1,j} \leqslant m_j, \ldots, Z_{n,j} \leqslant m_j\right) = \Phi(m_j)^n \qquad (6.1)$$

and hence the corresponding densities are

$$f_j^{(n)}(m_j) = n\Phi(m_j)^{n-1}\varphi(m_j) \qquad (6.2)$$

for $m_j \in \mathbb{R}$, $j = 1, \ldots, d$. Here Φ and φ denote the distribution function and the density of the standard normal distribution, respectively. Thus, the copula $C_{\mathbf{M}^{(n)}}$ is the distribution function of

$$\mathbf{V} = (V_1, \ldots, V_d)' \text{ with } V_j := \Phi\left(M_j^{(n)}\right)^n \sim \mathcal{U}[0,1].$$

For $n \in \mathbb{N}$ the copula of the componentwise maxima $C_{\mathbf{M}^{(n)}}$ can be expressed in terms of the underlying copula C as follows

$$C_{\mathbf{M}^{(n)}}(u_1, \ldots, u_d) = C\left(u_1^{1/n}, \ldots, u_d^{1/n}\right)^n, \qquad (6.3)$$

where $u_j \in [0,1]$, $j = 1, \ldots, d$. Since C is assumed to have a density c, Equation 6.3 yields that $C_{\mathbf{M}^{(n)}}$ also has a density, denoted by $c_{\mathbf{M}^{(n)}}$. Using Sklar's Theorem, Equations 6.1 and 6.3 imply that the joint distribution function of $\mathbf{M}^{(n)}$ is given by

$$F^{(n)}(m_1, \ldots, m_d) = C_{\mathbf{M}^{(n)}}\left(F_1^{(n)}(m_1), \ldots, F_d^{(n)}(m_d)\right) = C\left(\Phi(m_1), \ldots, \Phi(m_d)\right)^n .$$

Theorem 6.1.1. *The density of the copula of the vector of block-maxima satisfies for $u_j \in [0,1]$, $j = 1 \ldots, d$:*

$$c_{\mathbf{M}^{(n)}}(u_1, \ldots, u_d) = \frac{1}{n^d}\left(\prod_{j=1}^{d} u_j\right)^{\frac{1}{n}-1} \sum_{j=1}^{d \wedge n}\left\{\frac{n!}{(n-j)!}C\left(u_1^{1/n}, \ldots, u_d^{1/n}\right)^{n-j}\right.$$

$$\left. \cdot \sum_{\mathcal{P} \in \mathcal{S}_{d,j}} \prod_{M \in \mathcal{P}} \partial_M C\left(u_1^{1/n}, \ldots, u_d^{1/n}\right)\right\}, \quad (6.4)$$

where $d \wedge n := \min\{d, n\}$, $\mathcal{S}_{d,j} := \{\mathcal{P} | \mathcal{P} \text{ partition of } \{1, \ldots, d\} \text{ with } |\mathcal{P}| = j\}$ and

$$\partial_M C\left(u_1^{1/n}, \ldots, u_d^{1/n}\right) := \frac{\partial^p C(v_1, \ldots, v_d)}{\partial v_{m_1} \cdots \partial v_{m_p}}\bigg|_{v_1 = u_1^{1/n}, \ldots, v_d = u_d^{1/n}}$$

for $M = \{m_1, \ldots, m_p\} \subseteq \{1, \ldots, d\}$.

The proof of Theorem 6.1.1 as well as all other proofs can be found in the appendix at the end of this chapter (pages 124 ff.).

For the joint density $f^{(n)}$ of the block-maxima with marginally normalized data (on the z-scale) we also obtain an explicit expression.

Corollary 6.1.2. *For $m_j \in \mathbb{R}$, $j = 1, \ldots, d$, we have*

$$f^{(n)}(m_1, \ldots, m_d) = \left(\prod_{j=1}^{d} \varphi(m_j)\right) \cdot \sum_{j=1}^{d \wedge n}\left\{\frac{n!}{(n-j)!} \cdot C\left(\Phi(m_1), \ldots, \Phi(m_d)\right)^{n-j}\right.$$

$$\left. \cdot \sum_{\mathcal{P} \in \mathcal{S}_{d,j}} \prod_{M \in \mathcal{P}} \partial_M C\left(\Phi(m_1), \ldots, \Phi(m_d)\right)\right\}. \quad (6.5)$$

Example 6.1.3. *Let $d = 3$, $n \in \mathbb{N}$, i.e., $\mathbf{U}_i = (U_{i,1}, U_{i,2}, U_{i,3})'$, $\mathbf{Z}_i = (Z_{i,1}, Z_{i,2}, Z_{i,3})'$ and $\mathbf{M}^{(n)} = (M_1^{(n)}, M_2^{(n)}, M_3^{(n)})'$, $i = 1, \ldots, n$. If $n \geqslant 3$ the copula density of the vector of the block-maxima is given by*

$$c_{\mathbf{M}^{(n)}}(u_1, u_2, u_3)$$

$$= \frac{(u_1 u_2 u_3)^{\frac{1}{n}-1}}{n^3} \cdot \left\{nC\left(u_1^{1/n}, u_2^{1/n}, u_3^{1/n}\right)^{n-1} c\left(u_1^{1/n}, u_2^{1/n}, u_3^{1/n}\right)\right.$$

$$+ n(n-1)C\left(u_1^{1/n}, u_2^{1/n}, u_3^{1/n}\right)^{n-2}$$

$$\cdot \left[\partial_1 C\left(u_1^{1/n}, u_2^{1/n}, u_3^{1/n}\right) \partial_{23} C\left(u_1^{1/n}, u_2^{1/n}, u_3^{1/n}\right)\right.$$

$$+ \partial_2 C \left(u_1^{1/n}, u_2^{1/n}, u_3^{1/n} \right) \partial_{13} C \left(u_1^{1/n}, u_2^{1/n}, u_3^{1/n} \right)$$

$$+ \partial_3 C \left(u_1^{1/n}, u_2^{1/n}, u_3^{1/n} \right) \partial_{12} C \left(u_1^{1/n}, u_2^{1/n}, u_3^{1/n} \right) \Big]$$

$$+ n(n-1)(n-2) C \left(u_1^{1/n}, u_2^{1/n}, u_3^{1/n} \right)^{n-3} \partial_1 C \left(u_1^{1/n}, u_2^{1/n}, u_3^{1/n} \right)$$

$$\cdot \, \partial_2 C \left(u_1^{1/n}, u_2^{1/n}, u_3^{1/n} \right) \partial_3 C \left(u_1^{1/n}, u_2^{1/n}, u_3^{1/n} \right) \Big\}.$$

6.2 Vine Copulas

While the catalog of bivariate copula families (Joe, 1997, see, e.g.,) is large, this is not the case for multivariate copula families. They were initially dominated by Archimedean and elliptical copulas, however for complex dependency patterns such as asymmetric dependence in the tails these classes are insufficient. The class of vine copulas (Aas et al., 2009; Bedford and Cooke, 2002; Kurowicka and Cooke, 2006; Kurowicka and Joe, 2011) can accommodate such patterns. See Stöber and Czado (2012) for a tutorial introduction and Czado (2010) and Czado et al. (2013) for recent reviews. Basically vine copulas are constructed using bivariate copulas called pair copulas as building blocks which are combined to form multivariate copulas using conditioning. The pair copulas represent the copula associated with bivariate conditional distributions. The conditioning variables are determined with the help of a sequence of linked trees called the vine structure. Further it is commonly assumed that the conditioning value does not influence the copula and its parameter. See Stöber et al. (2013) for a discussion of this simplifying condition. Further they show that multivariate Clayton copula is the only Archimedean copula which can be represented as a vine copula, while the multivariate t-copula is the only scale elliptical one. For the multivariate Gaussian copula and the t-copula the needed pair copulas are bivariate Gaussian or t-copulas, respectively. The corresponding parameters are given by (partial) correlation parameters.

Vine copulas allow for product expressions of the density. We only consider three-dimensional vine copulas which can be expressed as

$$c(u_1, u_2, u_3) = c_{12}(u_1, u_2) c_{23}(u_2, u_3) c_{13;2}(C_{1|2}(u_1|u_2), C_{3|2}(u_3|u_2)). \quad (6.6)$$

Here $c_{ij;k}$ denotes the bivariate copula density corresponding to bivariate distribution (U_i, U_j) given $U_k = u_k$ and $C_{i|j}(u_i|u_j)$ denotes the conditional distribution function of U_i given $U_j = u_j$, which can be expressed as

$$C_{i|j}(u_i|u_j) = \frac{\partial}{\partial u_j} C_{ij}(u_i, u_j).$$

Further we write the bivariate copula densities in terms of their copula, i.e.,

$$c_{ij}(u_i, u_j) = \frac{\partial^2}{\partial u_i \partial u_j} C_{ij}(u_i, u_j).$$

For the pair copulas $C_{12}, C_{23}, C_{13;2}$ arbitrary bivariate copulas can be utilized. Many bivariate families including rotations are implemented in the R package VineCopula (see Schepsmeier et al. (2014)), which allows for parameter estimation and model selection of vine copulas in arbitrary dimensions.

Now we will consider the three-dimensional case and derive expressions for the partial derivatives needed in Theorem 6.1.1 for the expression of the copula density for the block-maxima.

Theorem 6.2.1. *For the vine copula density* (6.6) *we have:*

1. $C(u_1, u_2, u_3) = \int_0^{u_2} C_{13;2}\left(C_{1|2}(u_1|v_2), C_{3|2}(u_3|v_2)\right) dv_2,$
2. (a) $\partial_1 C(u_1, u_2, u_3) = \int_0^{u_2} \partial_1 C_{13|2}(C_{1|2}(u_1|v_2), C_{3|2}(u_3|v_2)) c_{12}(u_1, v_2) dv_2,$
   (b) $\partial_2 C(u_1, u_2, u_3) = C_{13;2}(C_{1|2}(u_1|u_2), C_{3|2}(u_3|u_2)),$
   (c) $\partial_3 C(u_1, u_2, u_3) = \int_0^{u_2} \partial_3 C_{13|2}(C_{1|2}(u_1|v_2), C_{3|2}(u_3|v_2)) c_{23}(v_2, u_3) dv_2,$
3. (a) $\partial_{12} C(u_1, u_2, u_3) = \partial_1 C_{13;2}(C_{1|2}(u_1|u_2), C_{3|2}(u_3|u_2)) c_{12}(u_1, u_2),$
   (b) $\partial_{13} C(u_1, u_2, u_3) = \int_0^{u_2} c_{13;2}(C_{1|2}(u_1|v_2), C_{3|2}(u_3|v_2)) c_{23}(v_2, u_3) c_{12}(u_1, v_2) dv_2,$
   (c) $\partial_{23} C(u_1, u_2, u_3) = \partial_3 C_{13;2}(C_{1|2}(u_1|u_2), C_{3|2}(u_3|u_2)) c_{23}(u_2, u_3),$
4. $c(u_1, u_2, u_3) = c_{12}(u_1, u_2) c_{23}(u_2, u_3) c_{13;2}(C_{1|2}(u_1|u_2), C_{3|2}(u_3|u_2)).$

Theorem 6.2.1 shows that the copula density corresponding to the three-dimensional vector of block-maxima based on an arbitrary vine copula is numerically tractable since only one-dimensional integration is needed. In particular this allows a numerical treatment for the block-size n in a finite setting. Additionally we can use the vine decomposition for a three-dimensional Gaussian or t-copula instead of requiring three-dimensional integration to calculate the corresponding density of the block-maxima. Two examples illustrate this way of proceeding.

Example 6.2.2. *As a first example we take a three-dimensional Clayton-vine, i.e., all three pair-copulas are bivariate Clayton copulas. As parameters we choose $\delta_{12} = 6$, $\delta_{23} = 7.09$ and $\delta_{13;2} = 4.67$ corresponding to Kendall's τ values of $\tau_{12} = 0.75$ and $\tau_{23} = 0.78$, $\tau_{13;2} = 0.70$. Figure 6.1 shows the copula density of the block-maxima of this vine with normalized margins (i.e., on the z-level) for block-sizes $n = 10, 50, 10^3$. Each row represents one block-size and contains three contour plots. Since it is difficult to plot three-dimensional objects in a simple way we decided not to show the isosurfaces but cut the three-dimensional object into three slices parallel to the z_1-z_2-plane. Each column presents the contourplot of one slice, where the z_3-value is fixed to $\Phi^{-1}(0.2)$, $\Phi^{-1}(0.5)$, or $\Phi^{-1}(0.8)$, respectively. Furthermore, we plotted the contours of the independence copula with normalized margins. One can see that already for $n = 10^3$ the contours of the copula density of the block-maxima with normalized margins practically coincide with the ones of the independence copula.*

Remark 6.2.3. *Even though it is not known whether all Clayton-vines lie in the domain of attraction of the independence copula, one can show that the Clayton-copula, which can be represented as a Clayton-vine with specific parameter restrictions (Stöber et al., 2013), lies in the domain of attraction of the independence copula. According to Gudendorf and Segers (2010) an Archimedean copula with gener-*

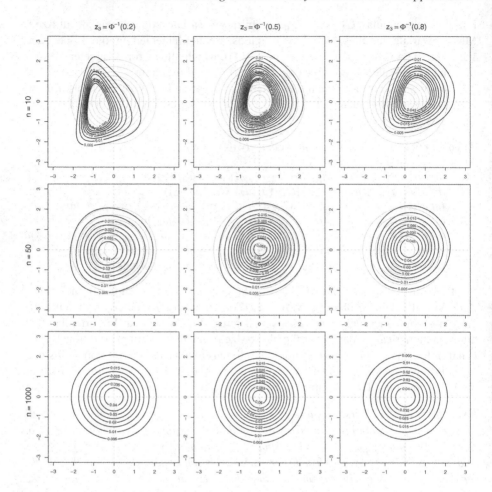

FIGURE 6.1
Two-dimensional slices of the copula density for block-maxima of a three-dimensional Clayton-vine with normalized margins ($\tau_{12} = 0.75$, $\tau_{23} = 0.78$, $\tau_{13;2} = 0.7$).

ator φ lies in the domain of attraction of the Gumbel-copula with parameter

$$\theta := -\lim_{s\downarrow 0} \frac{s\varphi'(1-s)}{\varphi(1-s)} \in [1, \infty)$$

if the limit exists. For the Clayton-copula this limit is equal to 1. Therefore, the copula of the block-maxima of a Clayton-copula converges to the Gumbel-copula with $\theta = 1$, which is the independence copula.

Example 6.2.4. *The second example we present is a three-dimensional Gaussian*

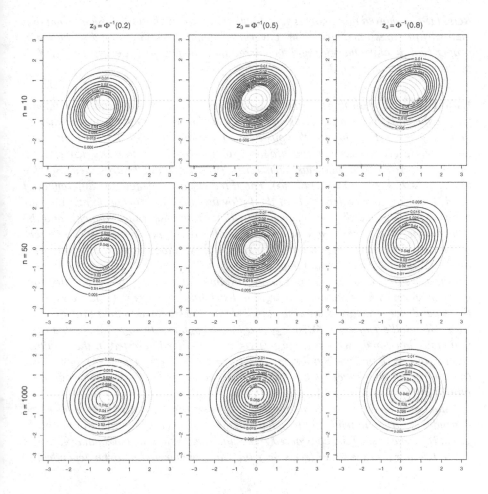

FIGURE 6.2

Two-dimensional slices of the copula density for block-maxima of a three-dimensional Gaussian vine with normalized margins ($\tau_{12} = 0.5$, $\tau_{23} = 0.57$, $\tau_{13;2} = 0.35$).

vine, i.e., all three pair-copulas are bivariate Gaussian copulas. As parameters we choose $\rho_{12} = 0.71$, $\rho_{23} = 0.78$ and $\rho_{13;2} = 0.52$ corresponding to Kendall's τ values of $\tau_{12} = 0.50$ and $\tau_{23} = 0.57$, $\tau_{13;2} = 0.35$.

Figure 6.2 shows the copula density of the block-maxima of this vine with normalized margins (i.e., on the z-level) for block-sizes $n = 10, 50, 10^3$. As above each row represents one block-size and contains three contour plots corresponding to z_3-values fixed to $\Phi^{-1}(0.2)$, $\Phi^{-1}(0.5)$, or $\Phi^{-1}(0.8)$, respectively. Again we detect convergence to the independence copula. This is also what one would expect: Hüsler and

Reiss (1989) showed that in order to achieve that the distribution of the maxima of a multivariate Gaussian distribution converges to a non-trivial limiting distribution, a proper scaling of the margins and the correlation coefficients is necessary. This will be discussed in Section 6.3.

Example 6.2.5. *Hydrology is one of the areas where block-maxima are important. Especially, the water levels of rivers can be interesting when it comes to analyzing the risk of floods. We consider a three-dimensional dataset containing the water levels of rivers in and around Munich, Germany, from August 1, 2007 to July 31, 2013. The data has been taken from Bavarian Hydrological Service (*http://www.gkd.bayern.de*). The three variables denote the differences of the 12-hour average water levels at the following three measuring points: the Isar measured in Munich, the Isar measured in Baierbrunn (south of Munich), and the Schwabinger Bach measured in Munich (a small stream entering the Isar in Garching, north of Munich). Since we only consider the hydrological winter (November 1 to April 30), we have 2176 data points.*

First, we transform the margins to the unit interval applying the probability integral transform with the empirical marginal distribution functions. Then, we estimate the dependence structure using vine copulas[1]: c_{12} is estimated to be a Frank copula with a Kendall's τ of $\tau_{12} = 0.76$, c_{23} is a Frank copula with $\tau_{23} = 0.23$, and $c_{13;2}$ is a Gaussian copula with $\tau_{13;2} = -0.18$. Now we are interested in the resulting copula density of the maxima for one day ($n = 2$), one week ($n = 14$), one month ($n = 60$), and one winter ($n = 362$). The respective contours (on the z-scale) are plotted in Figure 6.3.

Similar to the examples from above we see that with increasing n the observed dependence structure tends to the independence copula (gray contours in the background). In the case of the considered rivers this means that the maximal differences of the 12-hour average water levels over the entire winter are almost independent.

6.3   Copula Density of Scaled Block-Maxima

Examples 6.2.2 and 6.2.4 show that scaling of the block-maxima is necessary to achieve a limiting copula. These limiting copulas are called extreme value copulas and are characterized by max-stability. A recent introduction to extreme value copulas is given by Gudendorf and Segers (2010).

Since Hüsler and Reiss (1989) derived the scaling for the block-maxima of the multivariate normal distribution with standard normally distributed margins $X_1, \ldots, X_d$ to a non-trivial extreme value copula, we use the same marginal scaling for the block-maxima $\mathbf{M}^{(n)}$ on the z-scale given by

---

[1] In order to assure that the necessary integrals are numerically tractable we had to exclude some pair-copula families (e.g., the t-copula).

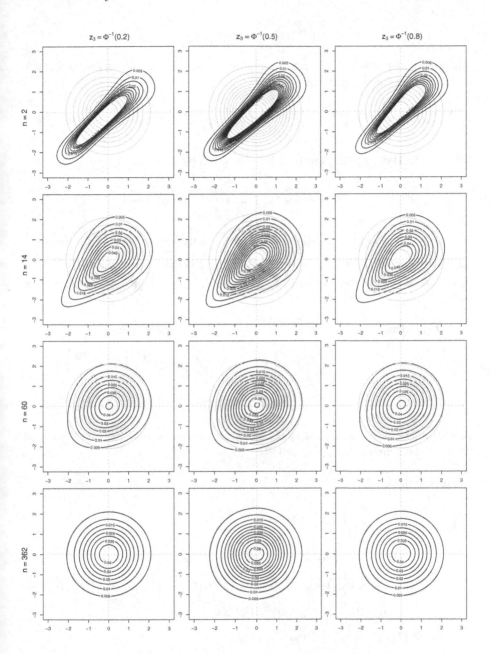

FIGURE 6.3
Two-dimensional slices of the copula density for block-maxima of the water level differences for one day, one week, one month, and one winter with normalized margins.

$$W_j^{(n)} := b_n \left(M_j^{(n)} - b_n \right),$$

where b_n satisfies $b_n = n \cdot \varphi(b_n)$ for φ the standard normal density. Univariate extreme value theory gives that

$$F_{W_j^{(n)}}(w_j) = P \left(W_j^{(n)} \leqslant w_j \right) = \Phi^n \left(b_n + \frac{w_j}{b_n} \right) \to \exp\{-e^{-w_j}\} \text{ as } n \to \infty$$

for $w_j \in \mathbb{R}$. The marginal density of $W_j^{(n)}$ is given by

$$f_{W_j^{(n)}}(w_j) = \frac{n}{b_n} \Phi^{n-1} \left(b_n + \frac{w_j}{b_n} \right) \varphi \left(b_n + \frac{w_j}{b_n} \right)$$

for $w_j \in \mathbb{R}$, $j = 1, \ldots, d$. Since $W_j^{(n)}$ is a strictly increasing transformation of $M_j^{(n)}$, the copula of $\mathbf{W}^{(n)} := \left(W_1^{(n)}, \ldots, W_d^{(n)} \right)$ is the same as the one of $\mathbf{M}^{(n)}$. Therefore, using (6.3) we obtain the following expression for the joint distribution of $\mathbf{W}^{(n)}$

$$\begin{aligned}
F_{\mathbf{W}^{(n)}}(w_1, \ldots, w_d) &= P \left(W_1^{(n)} \leqslant w_1, \ldots, W_d^{(n)} \leqslant w_d \right) \\
&= C_{\mathbf{M}^{(n)}} \left(\Phi^n \left(b_n + \frac{w_1}{b_n} \right), \ldots, \Phi^n \left(b_n + \frac{w_d}{b_n} \right) \right) \\
&= \left[C \left(\Phi \left(b_n + \frac{w_1}{b_n} \right), \ldots, \Phi \left(b_n + \frac{w_d}{b_n} \right) \right) \right]^n.
\end{aligned}$$

Similar arguments as in Corollary 6.1.2 can be used to express the joint density of $\mathbf{W}^{(n)}$ in three dimensions for $n \geqslant 3$ as

$$\begin{aligned}
&f_{\mathbf{W}^{(n)}}(w_1, w_2, w_3) \\
&= \frac{1}{b_n^3} \prod_{j=1}^d \varphi \left(b_n + \frac{w_j}{b_n} \right) \Big\{ nC(u_1, u_2, u_3)^{n-1} c(u_1, u_2, u_3) \\
&\quad + n(n-1)C(u_1, u_2, u_3)^{n-2} \Big[\partial_1 C(u_1, u_2, u_3) \partial_{23} C(u_1, u_2, u_3) \\
&\quad\quad + \partial_2 C(u_1, u_2, u_3) \partial_{13} C(u_1, u_2, u_3) + \partial_3 C(u_1, u_2, u_3) \partial_{12} C(u_1, u_2, u_3) \Big] \\
&\quad + n(n-1)(n-2)C(u_1, u_2, u_3)^{n-3} \partial_1 C(u_1, u_2, u_3) \\
&\quad\quad \cdot \partial_2 C(u_1, u_2, u_3) \partial_3 C(u_1, u_2, u_3) \Big\},
\end{aligned}$$

where $u_j := \Phi \left(b_n + \frac{w_j}{b_n} \right)$ for $j = 1, 2, 3$.

According to Hüsler and Reiss (1989), besides scaling $M_1^{(n)}, \ldots, M_d^{(n)}$, it is also necessary to change the correlation matrix $\Sigma(n) = (\rho_{i,j}(n))_{1 \leqslant i,j \leqslant d}$ of the underlying joint distribution of standard normal random variables $X_1, \ldots, X_n$, over whose i.i.d. copies $X_{i,1}, \ldots, X_{i,d}$, $i = 1, \ldots, n$, we take the maximum. The correlation matrices $\Sigma(n)$ have to satisfy the following condition

$$(1 - \rho_{i,j}(n)) \cdot \log(n) \to \lambda_{i,j}^2 \text{ as } n \to \infty, \tag{6.7}$$

where $\lambda_{i,j} \in (0, \infty)$ are some constants for $1 \leqslant i, j \leqslant d$, $i \neq j$ and $\lambda_{i,i} = 0$ for $i = 1, \ldots, d$. Since $\rho_{i,j}(n) = \rho_{j,i}(n)$, we also have $\lambda_{i,j} = \lambda_{j,i}$ for $1 \leqslant i, j \leqslant d$. Note that (6.7) implies that $\rho_{i,j}(n) \to 1$ as $n \to \infty$. The limiting distribution H_Λ of the scaled maxima depends on $\Lambda := (\lambda_{i,j})_{1 \leqslant i, j \leqslant d}$.

In the following we will examine the three-dimensional case and choose values for $\lambda_{12}, \lambda_{13}, \lambda_{23} \in (0, \infty)$. For simplicity, we set

$$\rho_{i,j}(n) := 1 - \frac{\lambda_{i,j}^2}{\log(n)} \tag{6.8}$$

for $1 \leqslant i, j \leqslant 3$, $n \in \mathbb{N}$, such that Equation 6.7 is always satisfied. However, for arbitrary $\lambda_{i,j}$ it is not trivial to decide whether we obtain a valid correlation matrix through this particular choice of $\rho_{i,j}(n)$ for any $n \in \mathbb{N}$. By construction the matrices

$$\Sigma(n) = \begin{pmatrix} 1 & \rho_{12}(n) & \rho_{13}(n) \\ \rho_{12}(n) & 1 & \rho_{23}(n) \\ \rho_{13}(n) & \rho_{23}(n) & 1 \end{pmatrix}$$

are symmetric and have ones on their diagonals. The only property we have to check is whether $\Sigma(n)$ is positive definite. For this we only need to check if the determinant of each leading principal minor is positive. Since $1 > 0$ and $1 - \rho_{12}(n)^2 > 0$ are trivially satisfied, the only real requirement is that

$$\det(\Sigma(n)) = 1 - \rho_{12}(n)\rho_{13}(n)\rho_{23}(n) - \rho_{12}(n)^2 - \rho_{13}(n)^2 - \rho_{23}(n)^2 > 0.$$

Using Equation 6.8 we obtain that $\det(\Sigma(n)) > 0$ if and only if

$$2(\lambda_{12}^2\lambda_{13}^2 + \lambda_{12}^2\lambda_{23}^2 + \lambda_{13}^2\lambda_{23}^2) - (\lambda_{12}^4 + \lambda_{13}^4 + \lambda_{23}^4) > \frac{2\lambda_{12}^2\lambda_{13}^2\lambda_{23}^2}{\log(n)}. \tag{6.9}$$

We denote the left-hand side of Equation 6.9 by $h(\lambda_{12}^2, \lambda_{13}^2, \lambda_{23}^2)$. Since the right-hand side of Equation 6.9 is always positive, it can only be satisfied if $h(\lambda_{12}^2, \lambda_{13}^2, \lambda_{23}^2) > 0$. If $h(\lambda_{12}^2, \lambda_{13}^2, \lambda_{23}^2) > 0$, then (6.9) is satisfied for all $n \in \mathbb{N}$ with

$$n \geqslant n^* := \left\lfloor \exp\left\{ \frac{2\lambda_{12}^2\lambda_{13}^2\lambda_{23}^2}{h(\lambda_{12}^2, \lambda_{13}^2, \lambda_{23}^2)} \right\} + 1 \right\rfloor,$$

where $\lfloor \cdot \rfloor$ denotes the floor function. Table 6.1 shows the values of h and n^* (if existing) for 10 different combinations of $\lambda_{12}, \lambda_{13}, \lambda_{23}$.

Hüsler and Reiss (1989) derived this scaling for multivariate normal distributions. Since we want to apply the scaling to vines, we need to transform the parameters of the normal distribution (correlations) to the parameters of the vine.

Considering the vine structure from (6.6) we further assume that the pair-copulas are one-parametric. Having fixed $\lambda_{12}, \lambda_{13}, \lambda_{23}$ such that $h(\lambda_{12}^2, \lambda_{13}^2, \lambda_{23}^2) > 0$, we can perform the following procedure for $n \geqslant n^*$:

TABLE 6.1
Different λ-combinations and the corresponding values of h and n^*.

#	λ_{12}^2	λ_{13}^2	λ_{23}^2	$h(\lambda_{12}^2, \lambda_{13}^2, \lambda_{23}^2)$	n^*
1	1	1	1	3	2
2	2	2	2	12	4
3	1	2	3	8	5
4	0.5	0.5	0.5	0.75	2
5	0.3	0.2	0.1	0.08	2
6	0.2	5	0.75	-15.8	–
7	15	20	15	800	76880
8	100	0.1	20	-6376.01	–
9	1.05	0.21	0.84	0.71	2
10	4	3	3	32	10

1. Calculate $\rho_{12}(n)$, $\rho_{13}(n)$ and $\rho_{23}(n)$ with the help of (6.8).
2. Determine the corresponding partial correlation via

$$\rho_{13;2}(n) = \frac{\rho_{13}(n) - \rho_{12}(n)\rho_{23}(n)}{\sqrt{1 - \rho_{12}(n)^2}\sqrt{1 - \rho_{23}(n)^2}}.$$

3. Translate the (partial) correlations $\rho_{12}(n)$, $\rho_{23}(n)$, and $\rho_{13;2}(n)$ into (partial) Kendall's τ values $\tau_{12}(n)$, $\tau_{23}(n)$, and $\tau_{13;2}(n)$ using the relation for elliptical distributions

$$\tau = \frac{2}{\pi}\arcsin(\rho).$$

4. Determine the parameters $\theta_{12}(n)$, $\theta_{13}(n)$, and $\theta_{23}(n)$ of the pair copulas from the corresponding τ values.[2]

Recall that $\rho_{12}(n) \to 1$, $\rho_{13}(n) \to 1$, and $\rho_{23}(n) \to 1$ as $n \to \infty$. Therefore, we also have $\tau_{12}(n) \to 1$, $\tau_{13}(n) \to 1$ and $\tau_{23}(n) \to 1$ as $n \to \infty$. However, the behavior of convergence of $\rho_{13;2}(n)$ and hence $\tau_{13;2}(n)$ is not trivial. We use (6.8) to obtain

$$\rho_{13;2}(n) = \frac{\left(1 - \frac{\lambda_{13}^2}{\log(n)}\right) - \left(1 - \frac{\lambda_{12}^2}{\log(n)}\right)\left(1 - \frac{\lambda_{23}^2}{\log(n)}\right)}{\sqrt{1 - \left(1 - \frac{\lambda_{12}^2}{\log(n)}\right)^2}\sqrt{1 - \left(1 - \frac{\lambda_{23}^2}{\log(n)}\right)^2}} \to \frac{\lambda_{12}^2 + \lambda_{23}^2 - \lambda_{13}^2}{2\lambda_{12}\lambda_{23}}$$

as $n \to \infty$. Thus,

$$\tau_{13;2}(n) \to \frac{2}{\pi}\arcsin\left(\frac{\lambda_{12}^2 + \lambda_{23}^2 - \lambda_{13}^2}{2\lambda_{12}\lambda_{23}}\right) \text{ as } n \to \infty.$$

For illustration, we will now take combinations 9 and 10 from Table 6.1. We show

---

[2]In the `VineCopula` package this transformation can be performed by the function `BiCopTau2Par`.

TABLE 6.2
Overview of the (partial) correlations and (partial) Kendall's τ values for different n for combinations 9 ($\lambda_{12}^2 = 1.05$, $\lambda_{13}^2 = 0.21$, $\lambda_{23}^2 = 0.84$) and 10 ($\lambda_{12}^2 = 4$, $\lambda_{13}^2 = 3$, $\lambda_{23}^2 = 3$).

	n	$\rho_{12}(n)$	$\rho_{23}(n)$	$\rho_{13;2}(n)$	$\tau_{12}(n)$	$\tau_{23}(n)$	$\tau_{13;2}(n)$
Combination 9	10	0.54	0.64	0.87	0.37	0.44	0.67
	50	0.73	0.79	0.88	0.52	0.57	0.69
	10^3	0.85	0.88	0.89	0.64	0.68	0.69
	∞	1	1	0.89	1	1	0.70
Combination 10	10	-0.74	-0.30	-0.82	-0.53	-0.20	-0.61
	50	-0.02	0.23	0.25	-0.01	0.15	0.16
	10^3	0.42	0.57	0.44	0.28	0.38	0.29
	∞	1	1	0.58	1	1	0.39

the (partial) correlations from Step 2 of the above procedure as well as the (partial) Kendall's τ values since they can be compared independently from the choice the respective pair-copulas.

If we compare the values from Table 6.2 for combinations 9 and 10, it is eye-catching that the choice of λ_{12}, λ_{13}, and λ_{23} has a crucial influence on the behavior of the (partial) correlations and the (partial) Kendall's τ values. In the first case the parameters are already relatively close to their limiting values for $n = 10^3$, whereas in the second case they are still rather far from their limits for $n = 10^3$. Further, we see that the limiting values of $\rho_{13;2}(n)$ and $\tau_{13;2}(n)$ can be very different depending on the choice λ_{12}, λ_{13}, and λ_{23}.

Now we examine the behavior of the three-dimensional density of the scaled block-maxima $f_{\mathbf{W}^{(n)}}$ for increasing values of n.

Example 6.3.1. *First we look at a Clayton-vine and choose $\lambda_{12}^2 = 1.05$, $\lambda_{13}^2 = 0.21$, and $\lambda_{23}^2 = 0.84$ (combination 9). The parameters and Kendall's τ values depend on the block-size n.*

Figure 6.4 shows the density of the scaled block-maxima of the Clayton-vine for block-sizes $n = 10, 50, 10^3$. Each row represents one block-size (and thus parameter set) and contains three contour plots corresponding to z_3-values fixed to $F^{-1}_{W_3^{(n)}}(0.2)$, $F^{-1}_{W_3^{(n)}}(0.5)$, or $F^{-1}_{W_3^{(n)}}(0.8)$, respectively.

Example 6.3.2. *As a second example we choose a Gaussian vine with $\lambda_{12}^2 = 4$, $\lambda_{13}^2 = 3$, and $\lambda_{23}^2 = 3$ (combination 10).*

Figure 6.5 shows the density of the scaled block-maxima of the Gaussian vine for block-sizes $n = 10, 50, 10^3$. Again, three contour plots corresponding to z_3-values fixed to $F^{-1}_{W_3^{(n)}}(0.2)$, $F^{-1}_{W_3^{(n)}}(0.5)$, or $F^{-1}_{W_3^{(n)}}(0.8)$, respectively, are displayed per row. The block-size and Kendall's τ values are denoted on the left for each row.

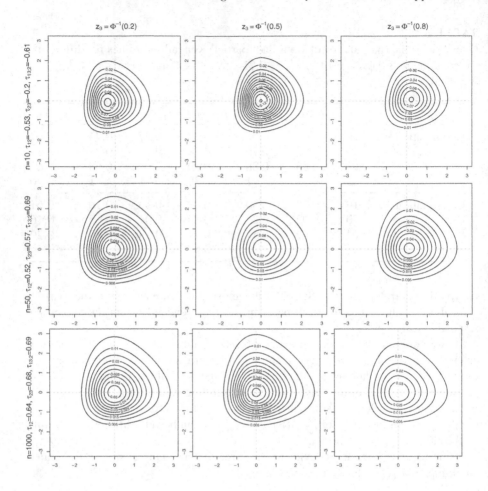

FIGURE 6.4

Two-dimensional slices of the density of the scaled block-maxima of a three-dimensional Clayton-vine ($\lambda_{12}^2 = 1.05$, $\lambda_{13}^2 = 0.21$, $\lambda_{23}^2 = 0.84$).

Conclusion

In this chapter we showed that the copula density of the block-maxima of multivariate distributions can be expressed explicitly. For three-dimensional vine copulas we made use of the fact that we can compute their partial derivatives by one-dimensional integration, which makes the evaluation of the copula density for block-maxima numerically tractable. The advantage of our method is that we can use the entire sample for estimation instead of reducing the sample size by taking the maximum over n

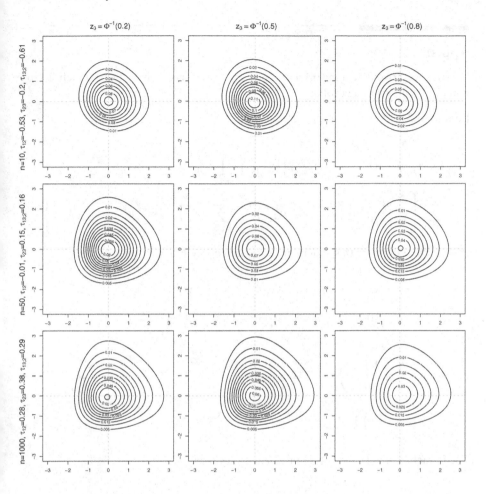

FIGURE 6.5
Two-dimensional slices of the density of the scaled block-maxima of a three-dimensional Gaussian vine ($\lambda_{12}^2 = 4$, $\lambda_{13}^2 = 3$, $\lambda_{23}^2 = 3$).

observations. Once we have estimated the underlying dependence structure we can derive the copula density of the block-maxima for any block-size (even larger than the original sample size). From the Clayton and the Gauss examples we have seen that without proper scaling the block-maxima do not approach a non-trivial limiting distribution for increasing block-size.

Appendix

In order to prove Theorem 6.1.1 we prove an auxiliary lemma from which Theorem 6.1.1 follows as a corollary.

Lemma 6.3.3. *For $k \in \{1, \ldots, d\}$ and $u_j \in [0,1]$, $j = 1 \ldots, d$, we have*

$$\frac{\partial^k}{\partial u_1 \cdots \partial u_k} \left[C(u_1^{1/n}, \ldots, u_d^{1/n})^n \right] = \frac{1}{n^k} \left(\prod_{j=1}^{k} u_j \right)^{\frac{1}{n}-1}$$

$$\cdot \sum_{j=1}^{k \wedge n} \left\{ \frac{n!}{(n-j)!} \cdot C\left(u_1^{1/n}, \ldots, u_d^{1/n}\right)^{n-j} \sum_{\mathcal{P} \in \mathcal{S}_{k,j}} \prod_{M \in \mathcal{P}} \partial_M C\left(u_1^{1/n}, \ldots, u_d^{1/n}\right) \right\}.$$

Proof. We will prove this statement using induction. For $k = 1$ we have

$$\frac{\partial}{\partial u_1} \left[C(u_1^{1/n}, \ldots, u_d^{1/n})^n \right]$$

$$= nC(u_1^{1/n}, \ldots, u_d^{1/n})^{n-1} \partial_1 C(u_1^{1/n}, \ldots, u_d^{1/n}) \frac{1}{n} u_1^{\frac{1}{n}-1}$$

$$= \frac{1}{n^1} \left(\prod_{j=1}^{1} u_j \right)^{\frac{1}{n}-1} \sum_{j=1}^{1 \wedge n} \left\{ \frac{n!}{(n-j)!} C(u_1^{1/n}, \ldots, u_d^{1/n})^{n-j} \sum_{\mathcal{P} \in \mathcal{S}_{1,j}} \right.$$

$$\left. \cdot \prod_{M \in \mathcal{P}} \partial_M C(u_1^{1/n}, \ldots, u_d^{1/n}) \right\}.$$

The inductive step $(k \to k+1)$ proceeds as follows

$$\frac{\partial^{k+1}}{\partial u_1 \cdots \partial u_{k+1}} \left[C(u_1^{1/n}, \ldots, u_d^{1/n})^n \right]$$

$$= \frac{\partial}{\partial u_{k+1}} \left\{ \frac{\partial^k}{\partial u_1 \cdots \partial u_k} \left[C(u_1^{1/n}, \ldots, u_d^{1/n})^n \right] \right\} =: (*)_1.$$

Applying the inductive assumption yields

$$(*)_1 = \frac{\partial}{\partial u_{k+1}} \left\{ \frac{1}{n^k} \left(\prod_{j=1}^{k} u_j \right)^{\frac{1}{n}-1} \sum_{j=1}^{k \wedge n} \left\{ \frac{n!}{(n-j)!} \cdot C\left(u_1^{1/n}, \ldots, u_d^{1/n}\right)^{n-j} \right. \right.$$

$$\left. \left. \cdot \sum_{\mathcal{P} \in \mathcal{S}_{k,j}} \prod_{M \in \mathcal{P}} \partial_M C\left(u_1^{1/n}, \ldots, u_d^{1/n}\right) \right\} \right\} =: (*)_2.$$

We will consider the cases $n > k$ and $n \leqslant k$ separately.

We begin with Case 1 ($n > k$). We have $k \wedge n = k$ and hence

$$(*)_2 = \frac{1}{n^k} \left(\prod_{j=1}^{k} u_j \right)^{\frac{1}{n}-1} \sum_{j=1}^{k} \left\{ \frac{n!}{(n-j)!} \cdot \left\{ \frac{\partial}{\partial u_{k+1}} \left[C\left(u_1^{1/n}, \ldots, u_d^{1/n}\right)^{n-j} \right] \right. \right.$$

$$\cdot \sum_{\mathcal{P} \in \mathcal{S}_{k,j}} \prod_{M \in \mathcal{P}} \partial_M C\left(u_1^{1/n}, \ldots, u_d^{1/n}\right) + C\left(u_1^{1/n}, \ldots, u_d^{1/n}\right)^{n-j}$$

$$\cdot \frac{\partial}{\partial u_{k+1}} \left[\sum_{\mathcal{P} \in \mathcal{S}_{k,j}} \prod_{M \in \mathcal{P}} \partial_M C\left(u_1^{1/n}, \ldots, u_d^{1/n}\right) \right] \right\} = (*)_3.$$

Now we use that fact that for $k \in \{1, \ldots, d-1\}$ and $j \in \{1, \ldots, k \wedge n\}$ we have

$$\frac{\partial}{\partial u_{k+1}} \left[\sum_{\mathcal{P} \in \mathcal{S}_{k,j}} \prod_{M \in \mathcal{P}} \partial_M C\left(u_1^{1/n}, \ldots, u_d^{1/n}\right) \right]$$

$$= \frac{u_{k+1}^{\frac{1}{n}-1}}{n} \sum_{\substack{\mathcal{P} \in \mathcal{S}_{k+1,j} \\ \{k+1\} \notin \mathcal{P}}} \prod_{M \in \mathcal{P}} \partial_M C\left(u_1^{1/n}, \ldots, u_d^{1/n}\right). \tag{6.10}$$

Applying Equation 6.10 yields

$$(*)_3 = \frac{1}{n^k} \left(\prod_{j=1}^{k} u_j \right)^{\frac{1}{n}-1} \left\{ \sum_{j=1}^{k} \frac{n!}{(n-j)!} \cdot (n-j) \cdot C\left(u_1^{1/n}, \ldots, u_d^{1/n}\right)^{n-j-1} \right.$$

$$\cdot \partial_{k+1} C\left(u_1^{1/n}, \ldots, u_d^{1/n}\right) \frac{1}{n} u_{k+1}^{\frac{1}{n}-1} \cdot \sum_{\mathcal{P} \in \mathcal{S}_{k,j}} \prod_{M \in \mathcal{P}} \partial_M C\left(u_1^{1/n}, \ldots, u_d^{1/n}\right)$$

$$+ \sum_{j=1}^{k} \frac{n!}{(n-j)!} \cdot C\left(u_1^{1/n}, \ldots, u_d^{1/n}\right)^{n-j} \cdot \frac{1}{n} u_{k+1}^{\frac{1}{n}-1} \sum_{\substack{\mathcal{P} \in \mathcal{S}_{k+1,j} \\ \{k+1\} \notin \mathcal{P}}} \prod_{M \in \mathcal{P}} \partial_M C\left(u_1^{1/n}, \ldots, u_d^{1/n}\right) \right\}$$

$$= \frac{1}{n^{k+1}} \left(\prod_{j=1}^{k+1} u_j \right)^{\frac{1}{n}-1} \left\{ \sum_{j=1}^{k} \frac{n!}{(n-(j+1))!} \cdot C\left(u_1^{1/n}, \ldots, u_d^{1/n}\right)^{n-(j+1)} \right.$$

$$\cdot \sum_{\substack{\mathcal{P} \in \mathcal{S}_{k+1,j+1} \\ \{k+1\} \in \mathcal{P}}} \prod_{M \in \mathcal{P}} \partial_M C\left(u_1^{1/n}, \ldots, u_d^{1/n}\right)$$

$$+ \sum_{j=1}^{k} \frac{n!}{(n-j)!} \cdot C\left(u_1^{1/n}, \ldots, u_d^{1/n}\right)^{n-j} \cdot \sum_{\substack{\mathcal{P} \in \mathcal{S}_{k+1,j} \\ \{k+1\} \notin \mathcal{P}}} \prod_{M \in \mathcal{P}} \partial_M C\left(u_1^{1/n}, \ldots, u_d^{1/n}\right) \right\} = (*)_4.$$

We perform an index shift in the first sum such that $j+1$ is replaced by j and make use of the following two properties:

(a) For all $\mathcal{P} \in \mathcal{S}_{l,1} = \{\{\{1, \ldots, l\}\}\}$ holds that $\{l\} \notin \mathcal{P}$.
(b) For all $\mathcal{P} \in \mathcal{S}_{l,l} = \{\{\{1\}, \ldots, \{l\}\}\}$ holds that $\{l\} \in \mathcal{P}$.

This results in

$$(*)_4 = \frac{1}{n^{k+1}} \left(\prod_{j=1}^{k+1} u_j \right)^{\frac{1}{n}-1} \left\{ \sum_{j=1}^{k+1} \frac{n!}{(n-j)!} \cdot C\left(u_1^{1/n}, \ldots, u_d^{1/n} \right)^{n-j} \right.$$

$$\cdot \sum_{\substack{\mathcal{P} \in \mathcal{S}_{k+1,j} \\ \{k+1\} \in \mathcal{P}}} \prod_{M \in \mathcal{P}} \partial_M C\left(u_1^{1/n}, \ldots, u_d^{1/n} \right)$$

$$\left. + \sum_{j=1}^{k+1} \frac{n!}{(n-j)!} \cdot C\left(u_1^{1/n}, \ldots, u_d^{1/n} \right)^{n-j} \cdot \sum_{\substack{\mathcal{P} \in \mathcal{S}_{k+1,j} \\ \{k+1\} \notin \mathcal{P}}} \prod_{M \in \mathcal{P}} \partial_M C\left(u_1^{1/n}, \ldots, u_d^{1/n} \right) \right\}$$

$$= \frac{1}{n^{k+1}} \left(\prod_{j=1}^{k+1} u_j \right)^{\frac{1}{n}-1} \left\{ \sum_{j=1}^{(k+1)\wedge n} \frac{n!}{(n-j)!} \cdot C\left(u_1^{1/n}, \ldots, u_d^{1/n} \right)^{n-j} \right.$$

$$\left. \cdot \sum_{\mathcal{P} \in \mathcal{S}_{k+1,j}} \prod_{M \in \mathcal{P}} \partial_M C\left(u_1^{1/n}, \ldots, u_d^{1/n} \right) \right\},$$

where we used the fact that $k + 1 = (k + 1) \wedge n$ since $n > k$. This concludes the first case.

Case 2 ($n \leqslant k$) is similar to the first one. The main difference is that $C(u_1^{1/n}, \ldots, u_d^{1/n})^{n-j} = 1$ for $j = n$, which was not possible before since $j \leqslant k < n$. Now $k \wedge n = n$ and therefore we obtain

$$(*)_2 = \frac{1}{n^k} \left(\prod_{j=1}^{k} u_j \right)^{\frac{1}{n}-1} \left\{ \sum_{j=1}^{n-1} \frac{n!}{(n-j)!} \cdot (n-j) \cdot C\left(u_1^{1/n}, \ldots, u_d^{1/n} \right)^{n-j-1} \right.$$

$$\cdot \partial_{k+1} C\left(u_1^{1/n}, \ldots, u_d^{1/n} \right) \frac{1}{n} u_{k+1}^{\frac{1}{n}-1} \cdot \sum_{\mathcal{P} \in \mathcal{S}_{k,j}} \prod_{M \in \mathcal{P}} \partial_M C\left(u_1^{1/n}, \ldots, u_d^{1/n} \right)$$

$$\left. + \sum_{j=1}^{n} \frac{n!}{(n-j)!} C\left(u_1^{1/n}, \ldots, u_d^{1/n} \right)^{n-j} \frac{\partial}{\partial u_{k+1}} \left[\sum_{\mathcal{P} \in \mathcal{S}_{k,j}} \prod_{M \in \mathcal{P}} \partial_M C\left(u_1^{1/n}, \ldots, u_d^{1/n} \right) \right] \right\}$$

$$= \frac{1}{n^{k+1}} \left(\prod_{j=1}^{k+1} u_j \right)^{\frac{1}{n}-1} \left\{ \sum_{j=1}^{n-1} \frac{n!}{(n-(j+1))!} \cdot C\left(u_1^{1/n}, \ldots, u_d^{1/n} \right)^{n-(j+1)} \right.$$

$$\cdot \sum_{\substack{\mathcal{P} \in \mathcal{S}_{k+1,j+1} \\ \{k+1\} \in \mathcal{P}}} \prod_{M \in \mathcal{P}} \partial_M C\left(u_1^{1/n}, \ldots, u_d^{1/n} \right)$$

$$\left. + \sum_{j=1}^{n} \frac{n!}{(n-j)!} \cdot C\left(u_1^{1/n}, \ldots, u_d^{1/n} \right)^{n-j} \cdot \sum_{\substack{\mathcal{P} \in \mathcal{S}_{k+1,j} \\ \{k+1\} \notin \mathcal{P}}} \prod_{M \in \mathcal{P}} \partial_M C\left(u_1^{1/n}, \ldots, u_d^{1/n} \right) \right\}$$

$$= \frac{1}{n^{k+1}} \left(\prod_{j=1}^{k+1} u_j \right)^{\frac{1}{n}-1} \left\{ \sum_{j=1}^{n} \frac{n!}{(n-j)!} \cdot C\left(u_1^{1/n}, \ldots, u_d^{1/n} \right)^{n-j} \right.$$

$$\cdot \sum_{\substack{\mathcal{P} \in \mathcal{S}_{k+1,j} \\ \{k+1\} \in \mathcal{P}}} \prod_{M \in \mathcal{P}} \partial_M C\left(u_1^{1/n}, \ldots, u_d^{1/n} \right) + \sum_{j=1}^{n} \frac{n!}{(n-j)!} \cdot C\left(u_1^{1/n}, \ldots, u_d^{1/n} \right)^{n-j}$$

$$\left. \cdot \sum_{\substack{\mathcal{P} \in \mathcal{S}_{k+1,j} \\ \{k+1\} \notin \mathcal{P}}} \prod_{M \in \mathcal{P}} \partial_M C\left(u_1^{1/n}, \ldots, u_d^{1/n} \right) \right\}$$

$$= \frac{1}{n^{k+1}} \left(\prod_{j=1}^{k+1} u_j \right)^{\frac{1}{n}-1} \left\{ \sum_{j=1}^{(k+1)\wedge n} \frac{n!}{(n-j)!} \cdot C\left(u_1^{1/n}, \ldots, u_d^{1/n} \right)^{n-j} \right.$$

$$\cdot \sum_{\mathcal{P} \in \mathcal{S}_{k+1,j}} \prod_{M \in \mathcal{P}} \partial_M C\left(u_1^{1/n}, \ldots, u_d^{1/n}\right)\Bigg\},$$

where we applied Equation 6.10 in the second equality. In the third equality we performed an index shift in the first sum and used property (a) above. Since $n \leqslant k$ we have $n = (k+1) \wedge n$. This concludes the second case and hence the proof of Lemma 6.3.3. □

Having proved the auxiliary lemma we can now easily prove the statement from Theorem 6.1.1.

Proof of Theorem 6.1.1. Using Equation 6.3 we obtain

$$c_{\mathbf{M}^{(n)}}(u_1, \ldots, u_d) = \frac{\partial^d}{\partial u_1 \cdots \partial u_d} C_{\mathbf{M}^{(n)}}(u_1, \ldots, u_d)$$

$$= \frac{\partial^d}{\partial u_1 \cdots \partial u_d}\left[C(u_1^{1/n}, \ldots, u_d^{1/n})^n\right].$$

Now Theorem 6.1.1 follows directly from Lemma 6.3.3 for $k = d$. □

Proof of Theorem 6.2.1. Expression 4) follows directly by the definition of $c(u_1, u_2, u_3)$ (see Equation (6.6)). For expression 3(a) we can write

$$\partial_{23} C(u_1, u_2, u_3) \overset{Eq.(6.6)}{=} c_{23}(u_2, u_3) \int_0^{u_1} c_{12}(v_1, u_2) c_{13;2}(C_{1|2}(v_1|u_2), C_{3|2}(u_3|u_2)) dv_1$$

$$= c_{23}(u_2, u_3) \int_0^{u_1} \partial_{12} C_{12}(v_1, u_2) \partial_{13} C_{13;2}(C_{1|2}(v_1|u_2), C_{3|2}(u_3|u_2)) dv_1$$

$$= c_{23}(u_2, u_3) \int_0^{u_1} \frac{\partial}{\partial v_1}\underbrace{\left[\partial_2 C_{12}(v_1, u_2)\right]}_{C_{1|2}(v_1|u_2)=w_1} \partial_{13} C_{13;2}(C_{1|2}(v_1|u_2), C_{3|2}(u_3|u_2)) dv_1$$

$$= c_{23}(u_2, u_3) \int_0^{u_1} \frac{\partial w_1}{\partial v_1} \frac{\partial}{\partial w_1} \partial_3 C_{13;2}(w_1, C_{3|2}(u_3|u_2))\Bigg|_{w_1=C_{1|2}(v_1|u_2)} dv_1$$

$$= c_{23}(u_2, u_3) \partial_3 C_{13;2}(C_{1|2}(u_1|u_2), C_{3|2}(u_3|u_2)).$$

Expression 3(a) follows similarly. For expression 2(c) we have

$$\partial_3 C(u_1, u_2, u_3) = \int_0^{u_1} \int_0^{u_2} c(v_1, v_2, u_3) dv_1 dv_2$$

$$= \int_0^{u_2} c_{23}(v_2, u_3) C_{1|23}(u_1|v_2, u_3) dv_2$$

$$= \int_0^{u_2} c_{23}(v_2, u_3) \partial_3 C_{13;2}(C_{1|2}(u_1|v_2), C_{3|2}(u_3|v_2)) dv_2.$$

Further, $\partial_3 C(u_1, u_2, u_3)$ can be written in another way which we use for the calculation of the copula $C(u_1, u_2, u_3)$. In particular we have

$$
\begin{aligned}
\partial_3 C(u_1, u_2, u_3) &= \int_0^{u_1} \int_0^{u_2} c(v_1, v_2, u_3) dv_1 dv_2 \\
&= \int_0^{u_2} \frac{\partial^2}{\partial v_2 \partial u_3} C_{23}(v_2, u_3) \frac{\partial}{\partial w_2} C_{13;2}(C_{1|2}(u_1|v_2), w_2) \bigg|_{w_2 = C_{3|2}(u_3|v_2)} dv_2 \\
&= \int_0^{u_2} \frac{\partial w_2}{\partial u_3} \frac{\partial}{\partial w_2} C_{13;2}(C_{1|2}(u_1|v_2), w_2) \bigg|_{w_2 = C_{3|2}(u_3|v_2)} dv_2 \quad\quad (6.11) \\
&= \int_0^{u_2} \frac{\partial}{\partial u_3} C_{13;2}(C_{1|2}(u_1|v_2), C_{3|2}(u_3|v_2)) dv_2 \\
&= \frac{\partial}{\partial u_3} \left[\int_0^{u_2} C_{13;2}(C_{1|2}(u_1|v_2), C_{3|2}(u_3|v_2)) dv_2 \right].
\end{aligned}
$$

Similarly one can derive expression 2(a). For the copula in expression 1 we have

$$
\begin{aligned}
C(u_1, u_2, u_3) &\overset{(6.11)}{=} \int_0^{u_3} \frac{\partial}{\partial v_3} \left[\int_0^{u_2} C_{13;2}(C_{1|2}(u_1|v_2), C_{3|2}(v_3|v_2)) dv_2 \right] dv_3 \\
&= \int_0^{u_2} \left[\int_0^{u_3} \frac{\partial}{\partial v_3} C_{13;2}(C_{1|2}(u_1|v_2), C_{3|2}(v_3|v_2)) dv_3 \right] dv_2 \\
&= \int_0^{u_2} C_{13;2} \left(C_{1|2}(u_1|v_2), C_{3|2}(u_3|v_2) \right) dv_2.
\end{aligned}
$$

Expression 2(b) follows by differentiating expression 1 with respect to v_2. It remains to show expression 3(b). For this we differentiate 2(c) with respect to v_1. Therefore we have

$$
\begin{aligned}
\partial_{13} C(u_1, u_2, u_3) &= \frac{\partial}{\partial u_1} \left[\int_0^{u_2} \partial_3 C_{13;2}(C_{1|2}(u_1|v_2), C_{3|2}(u_3|v_2)) c_{23}(v_2, u_3) dv_2 \right] \\
&= \int_0^{u_2} \partial_{13} C_{13;2}(C_{1|2}(u_1|v_2), C_{3|2}(u_3|v_2)) c_{12}(u_1, v_2) c_{23}(v_2, u_3) dv_2 \\
&= \int_0^{u_2} c_{13;2}(C_{1|2}(u_1|v_2), C_{3|2}(u_3|v_2)) c_{12}(u_1, v_2) c_{23}(v_2, u_3) dv_2.
\end{aligned}
$$

$\square$

Acknowledgment

The first author is thankful for the support from Allianz Deutschland AG. The authors are grateful to the referee for a number of helpful suggestions and comments.

References

Aas, K., C. Czado, A. Frigessi, and H. Bakken (2009). Pair-copula constructions of multiple dependence. *Insurance, Mathematics and Economics 44*, 182–198.

Bedford, T. and R. M. Cooke (2002). Vines — A new graphical model for dependent random variables. *Annals of Statistics 30*(4), 1031–1068.

Bücher, A. and J. Segers (2014). Extreme value copula estimation based on block maxima of a multivariate stationary time series. *Extremes 17*(3), 495–528.

Czado, C. (2010). Pair-copula constructions of multivariate copulas. In F. Durante, W. Härdle, P. Jaworki, and T. Rychlik (Eds.), *Workshop on Copula Theory and Its Applications*, pp. 93–109. Dordrecht: Springer.

Czado, C., E. Brechmann, and L. Gruber (2013). Selection of vine copulas. In P. Jaworski, F. Durante, and W. K. Härdle (Eds.), *Copulae in Mathematical and Quantitative Finance*, pp. 17–37. Springer.

Dombry, C. (2013). Maximum likelihood estimators for the extreme value index based on the block maxima method. arXiv preprint arXiv:1301.5611.

Faranda, D., V. Lucarini, G. Turchetti, and S. Vaienti (2011). Numerical convergence of the block-maxima approach to the generalized extreme value distribution. *Journal of Statistical Physics 145*(5), 1156–1180.

Ferreira, A. and L. de Haan (2014). On the block maxima method in extreme value theory: PWM estimators. *The Annals of Statistics 43*(1), 276–298.

Gudendorf, G. and J. Segers (2010). Extreme-value copulas. In F. Durante, W. Härdle, P. Jaworki, and T. Rychlik (Eds.), *Copula theory and its applications*, pp. 127–145. Springer.

Hüsler, J. and R.-D. Reiss (1989). Maxima of normal random vectors: between independence and complete dependence. *Statistics & Probability Letters 7*(4), 283–286.

Jarušková, D. and M. Hanek (2006). Peaks over threshold method in comparison with block-maxima method for estimating high return levels of several Northern Moravia precipitation and discharges series. *Journal of Hydrology and Hydromechanics 54*(4), 309–319.

Joe, H. (1997). *Multivariate Models and Dependence Concepts*. London: Chapman & Hall.

Kurowicka, D. and R. Cooke (2006). *Uncertainty Analysis with High Dimensional Dependence Modelling*. Chichester: Wiley.

Kurowicka, D. and H. Joe (2011). *Dependence Modeling — Handbook on Vine Copulae*. Singapore: World Scientific Publishing Co.

Marty, C. and J. Blanchet (2012). Long-term changes in annual maximum snow depth and snowfall in Switzerland based on extreme value statistics. *Climatic Change 111*(3-4), 705–721.

McNeil, A. J., R. Frey, and P. Embrechts (2010). *Quantitative Risk Management: Concepts, Techniques, and Tools*. Princeton University Press.

Naveau, P., A. Guillou, D. Cooley, and J. Diebolt (2009). Modelling pairwise dependence of maxima in space. *Biometrika 96*(1), 1–17.

Nikulin, G., E. Kjellström, U. Hansson, G. Strandberg, and A. Ullerstig (2011). Evaluation and future projections of temperature, precipitation and wind extremes over Europe in an ensemble of regional climate simulations. *Tellus A 63*(1), 41–55.

Rocco, M. (2014). Extreme value theory in finance: A survey. *Journal of Economic Surveys 28*(1), 82–108.

Schepsmeier, U., J. Stoeber, E. C. Brechmann, and B. Graeler (2014). *VineCopula: Statistical Inference of Vine Copulas*. R package version 1.3/r66.

Sklar, A. (1959). Fonctions dé repartition á n dimensions et leurs marges. *Publ. Inst. Stat. Univ. Paris 8*, 229–231.

Stöber, J. and C. Czado (2012). Pair copula constructions. In J. Mai and M. Scherer (Eds.), *Simulating Copulas: Stochastic Models, Sampling Algorithms, and Applications*. Imperial College Press.

Stöber, J., H. Joe, and C. Czado (2013). Simplified pair copula constructions — Limitations and extensions. *Journal of Multivariate Analysis 119*, 101–118.

7

Time Series of Extremes

Brian J. Reich
North Carolina State University, United States

Benjamin A. Shaby
Pennsylvania State University, United States

Abstract

Data collected for the purpose of an extreme value analysis very often exhibit temporal dependence. Examples include daily time series of stock prices, temperature, or water levels. In this chapter, we address two fundamental inferential objectives: estimating extreme marginal quantiles while accounting for serial dependence, and estimating the strength of serial dependence in extreme values. We review the literature for methods aimed at each of these objectives, including a survey of computational methods used for each approach. We illustrate the methods using an analysis of hourly wind gust speeds. R code to implement the methods is available online.

7.1 Motivating Data: Santa Ana Winds

Santa Ana winds are a weather phenomenon that occur in coastal regions of Southern California, predominantly in winter. Strong winds carry warm, dry air from the Mojave Desert west and up the San Gabriel and San Bernardino mountains. As this air descends the mountains and gains speed through canyons toward the coast, the increase in barometric pressure makes it contract quickly (adiabatically), further warming it. The induced increase in temperature in turn results in a decrease in relative humidity of the already dry air. The end result is hot, dry, and strong winds over the populated regions of coastal Southern California. In addition to folklore about strange psychological effects (Sergius et al., 1962), these winds present substantial hazards for fire ignition and spread.

The problem is twofold. First, strong gusts can knock down utility poles, some-

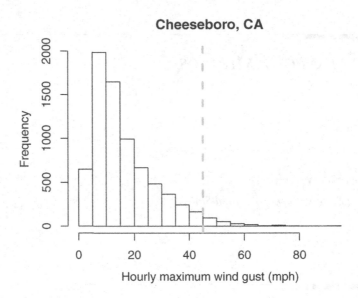

FIGURE 7.1

Histogram of the Cheeseboro January wind gust data. The dashed line shows a threshold of 45mph, chosen in Section 7.7.

times resulting in electrical arcing and fire ignition. Second, once fires start, strong winds, in particular hot, dry winds like the Santa Ana, cause them to grow and spread, greatly amplifying the resulting damage. The longer the winds persist, the greater their effect on fire spread. The fire damage impacts of Santa Ana winds thus depend on both the strength and persistence of the extreme events.

To begin to characterize extreme wind events in Southern California, we study hourly maximum gusts observed at the Cheeseboro weather station near Thousand Oaks, California (data can be downloaded from the Remote Automated Weather Stations network at http://www.raws.dri.edu/). This station was selected because it sits in a region associated with very high fire risk (Moritz et al., 2010). Hourly gusts during the month of January (winds are strongest there in the winter) for the 10 years from 2000 through 2009 were considered. The data are plotted in Figure 7.1; the data are clearly right skewed.

Figure 7.2 shows hourly gust speeds at Cheeseboro in January, 2003 (the year was chosen arbitrarily). The first feature of the data that is immediately evident is the presence of temporal dependence, which is easily seen by visual inspection. A second feature is that two wind events stand out as remarkably severe, with the event centered at around January 6 achieving hurricane speeds. This dataset will be used in Section 7.7 to illustrate the methods outlined in this chapter.

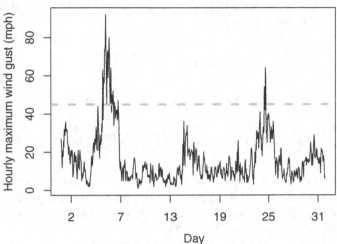

FIGURE 7.2
Hourly maximum wind gusts in January, 2003 recorded at the Cheeseboro weather station. The dashed line shows a threshold of 45mph, chosen in Section 7.7.

7.2 Time Series Data

Let Z_t be the response at time $t \in \{1, 2, ...\}$. In this chapter, we discuss primarily methods for modeling data above a predetermined threshold, u, which does not change with t. Methods for selecting the threshold are discussed in Chapter 4. The data are converted to the excesses

$$Y_t = \begin{cases} 0 & Z_t < u \\ Z_t - u & Z_t \geq u. \end{cases} \tag{7.1}$$

A common objective is to estimate the marginal distribution of the excess of u, Y_t, including the probability of exceeding the threshold, $P(Y_t > 0) = \lambda$, and the survival function, $S(y) = P(Y_t > y | Y_t > 0)$, while accounting for dependence between observations. Another objective may be to estimate the dependence between observations at different time lags and thus joint exceedance probabilities.

The theory summarized in Chapter 1 of this volume suggests the generalized Pareto distribution (GPD) as a model for the marginal distribution of the non-zero excesses of u. Therefore, the methods below assume that the excesses follow the GPD with

$$S(y) = \begin{cases} \left[1 + \frac{\xi}{\sigma}y\right]_+^{-1/\xi} & \xi \neq 0 \\ \exp\left(-\frac{y}{\sigma}\right) & \xi = 0, \end{cases} \tag{7.2}$$

where $[x]_+ = \max\{0, x\}$ is the positive part of x. The GPD has two parameters: σ is the scale and ξ is the shape. The support of the distribution is determined by ξ, with $y \in (0, \infty)$ for $\xi \geq 0$ and $y \in (0, -\sigma/\xi)$ for $\xi < 0$. The CDF of Y_t can be written concisely for $\xi \neq 0$ and $y \geq 0$ as

$$P(Y_t \leq y) = F_{\text{GPD}}(y) = 1 - \lambda \left[1 + \frac{\xi}{\sigma} y \right]_+^{-1/\xi}, \qquad (7.3)$$

so that $\text{Prob}(Y_t = 0) = F_{\text{GPD}}(0) = 1 - \lambda$. Similarly, the density function is

$$f_{\text{GPD}}(y) = (1 - \lambda) I(y = 0) + \frac{\lambda}{\sigma} \left(1 + \frac{\xi}{\sigma} y \right)_+^{-(1/\xi+1)} I(y > 0), \qquad (7.4)$$

where $I(y > 0) = 1$ if $y > 0$ and is zero otherwise. It is possible to allow λ, σ, and ξ to depend on t, perhaps through covariates, but for simplicity we assume the stationary model with GPD parameters held constant over time (methods to deal with non-stationarity are discussed in Chapter 5).

In addition to estimating parameters in the marginal distribution, we would also like to estimate the dependence function. Standard time series analysis focuses on the autocorrelation function. However, correlation is not conducive to extreme value analysis because heavy-tailed distributions often do not have moments, and correlation focuses on dependence in the center of the distribution and not the tails. A common measure of tail dependence between observations separated by time lag l is the conditional probability

$$\chi(l, c) = P(Y_{t+l} > c | Y_t > c). \qquad (7.5)$$

Asymptotic dependence at time lag l can be quantified using the limit

$$\chi(l) = \lim_{c \to \infty} \chi(l, c). \qquad (7.6)$$

If $\chi(l) = 0$, then we say the time series is asymptotically independent at lag l, and if $\chi(l) > 0$ then the time series has asymptotic dependence at lag l.

Another quantity that measures extremal dependence is the extremal index (Leadbetter et al., 1983). If $Y_1, \ldots, Y_n$ is a stationary sequence of random variables with common distribution function F, then for large enough n and u_n,

$$P(\max\{Y_1, \ldots, Y_n\} < u_n) = F^{n\theta}(u_n) \qquad (7.7)$$

under weak regularity conditions (Leadbetter et al., 1983). The quantity $\theta \in [0, 1]$ is called the extremal index, and it describes the strength of dependence of the sequence. A useful interpretation is that $1/\theta$ is the limiting mean size of clusters of exceedances of the threshold u.

The remainder of the chapter proceeds as follows. In Section 7.3, we describe approaches that aim to estimate marginal GPD parameters without specifying a model for extremal dependence. The advantage of this approach is that estimates of marginal parameters are free from parametric assumptions about the nature of extremal dependence. The rest of the chapter then discusses methods that estimate extremal dependence that can be used for prediction and simulation. We begin in Section 7.4 with Markov models, which are arguably the simplest model for extremal dependence. For more flexible models, we conclude with descriptions of copula (Section 7.5) and max-stable (Section 7.6) processes. These methods are then compared in Section 7.7 using the Santa Ana wind data described in Section 7.1.

7.3 Methods That Do Not Estimate Dependence

For the purpose of estimating the marginal GPD parameters $\Theta = (\sigma, \xi)$, it is tempting to ignore dependence between nearby observations. However this can lead to improper inference, particularly, underestimation of standard errors. A method of dealing with this issue is to remove dependence via declustering. In this approach, exceedances are placed into clusters, the maximum value in each cluster is computed, and these cluster maximums are modeled as independent GPD random variables. The key issue is then to define clusters to retain much of the information in the original time series, but remove temporal dependence. Davison and Smith (1990) and Coles et al. (1999) define a cluster as the time of the first exceedance until the first time r consecutive observations are below the threshold. The run length r should be chosen to be large enough to ensure clusters are independent. This results in two tuning parameters: the threshold u and the run length r. It is advisable to conduct the analysis for several values of u and r to illustrate sensitivity to these tuning parameters. Other popular approaches to declustering are Ferro and Segers (2003) and Süveges and Davison (2010).

An alternative to declustering is to ignore dependence for estimating Θ, but then account for dependence in standard error calculations as in Smith (1991) and Fawcett and Walshaw (2012). This avoids specifying a run length, r, and also allows all observations in each cluster to contribute to parameter estimation. However, this does require estimating the extremal index. Let $\widehat{\Theta}$ be the maximum likelihood estimate of Θ and H be the observed information matrix, both ignoring clustering and assuming all observations are independent. If the independence assumption where valid then H^{-1} would approximate the covariance of the sampling distribution of $\widehat{\Theta}$. However, if the observations are dependent, H^{-1} will underestimate variability of the sampling distribution. Smith (1991) and Fawcett and Walshaw (2007) propose an adjusted sandwich covariance estimate $H^{-1}VH^{-1}$, where the matrix V adjusts for dependence. Fawcett and Walshaw (2007) propose an empirical estimate of V and show that this yields proper statistical inference.

7.4   Markov Models

While the methods described in Section 7.3 provide computationally efficient estimates of the parameters in the marginal distribution, further modeling is required to estimate joint probabilities of exceedances on several days and to predict the next observation in the series. Perhaps the most natural model for temporally dependent data is a Markov model. A kth-order Markov model assumes that the response at time t is independent of the past conditioned on the previous k responses, $f(Y_t \mid Y_{t-l}, l = 1, 2, ...) = f(Y_t \mid Y_{t-l}, l = 1, 2, ..., k)$. We restrict attention to first-order models with $k = 1$ so that $f(Y_t \mid Y_{t-l}, l = 1, 2, ...) = f(Y_t \mid Y_{t-1})$. In this case, the joint distribution of n observations $\mathbf{Y} = (Y_1, ..., Y_n)^T$ can be written

$$f(\mathbf{Y}) = f(Y_1) \prod_{t=2}^{n} f(Y_t \mid Y_{t-1}). \tag{7.8}$$

Using a Markov assumption reduces the problem of modeling a time series to modeling a single conditional distribution, $f(Y_t \mid Y_{t-1})$. It remains to specify the conditional distribution so that the marginal distribution is GPD.

7.4.1   GPD Model for Data above a Threshold

Smith et al. (1997) note that the joint distribution defines the conditional distribution since

$$f(y_t \mid y_{t-1}) = f(y_t, y_{t-1}) / f_{\text{GPD}}(y_{t-1}). \tag{7.9}$$

Smith et al. (1997) propose a flexible class of models based on defining a bivariate extreme value distribution for successive pairs of observations, and then using the induced conditional distribution for the second day given the first to define a Markov model (also see Fawcett and Walshaw, 2006).

For the joint distribution, Smith et al. (1997) use a bivariate GPD distribution with the form

$$P(Y_t \leq y_1, Y_{t-1} \leq y_2) = F_{\text{GPD}}(y_1, y_2) = 1 - V\left[v(y_1), v(y_2)\right] \tag{7.10}$$

for $y_1, y_2 \geq 0$, dependence function V, and $v(y) = \lambda^{-1}\left(1 + \frac{\xi}{\sigma} y\right)_{+}^{1/\xi}$. The dependence function can be any function satisfying: (1) $V(tv_1, tv_2) = V(v_1, v_2)/t$ and (2) $V(\infty, v) = V(v, \infty) = 1/v$. As an example, they use the bivariate logistic function $V(v_1, v_2) = \left(v_1^{-1/\alpha} + v_2^{-1/\alpha}\right)^{\alpha}$. The parameter $\alpha \in [0, 1]$ controls dependence, with $\alpha = 1$ corresponding to independence and small α corresponding to strong dependence. Many other choices for bivariate parametric families are also available (Coles, 2001, ch. 8).

By construction, the marginal distribution is $f_{\text{GPD}}(y)$ in (7.4). Data below the threshold is treated as censored, so the joint density induced by (7.10) is

$$
f(y_1, y_2) = \begin{cases}
K_1 K_2 V_{12}[v(y_1), v(y_2)] & y_1 > 0, y_2 > 0 \\
K_2 V_2[1/\lambda, v(y_2)] & y_1 = 0, y_2 > 0 \\
K_1 V_1[v(y_1), 1/\lambda] & y_1 > 0, y_2 = 0 \\
1 - V(1/\lambda, 1/\lambda) & y_1 = 0, y_2 = 0
\end{cases} \tag{7.11}
$$

where $K_j = -\lambda^{-\xi} \sigma^{-1} v(y_j)^{1-\xi}$, V_j is the partial derivative of V with respect to its jth argument, and V_{12} is the mixed partial derivative of V. Together, these distributions specify the Markov model in (7.8) and (7.9)

Using this Markov model, the joint distribution of any pair of consecutive extreme observations is $F_{\text{GPD}}(y_1, y_2)$, and so the Markov chain inherits the dependence properties of the bivariate GPD. In particular, the lag-one extremal dependence is simply $\chi(1) = 2 - 2V(1, 1)$. For the logistic model, this reduces to $\chi(1) = 2 - 2^\alpha$. This illustrates the role of α. If $\alpha = 0$, then $\chi(1) = 1$ and extremal dependence is strong; in contrast, $\alpha = 1$ gives $\chi(1) = 0$ and corresponds to extremal independence.

7.4.2 Markov-Switching Models

Shaby et al. (2015) present an alternative to Smith et al. (1997). Rather than analyzing only data above a predetermined threshold, they build a model for the complete response distribution. The model separates observations into two categories, "bulk" and "extreme." Hard classification based on a threshold is avoided by classifying times using a Markov model for latent indictors of an extreme. This allows for more flexibility in modeling the threshold, and the ability to generate realistic simulations and predictions. The latent state variables provide a convenient construct for building separate models for the tail, which is the focus of the analysis, and the main part of the distribution into a well-defined likelihood.

To define the latent two-state model, let $S_t \in \{0, 1\}$ denote the state of the process at each time point. S_t takes a value of 1 if time t is in the extreme state, and a value of 0 if time t is in the bulk state. The state variables $\mathbf{S} = (S_1, ..., S_n)$ are dependent in time according to a Markov chain structure with transition matrix

$$
\mathbf{A} = \begin{pmatrix} 1 - a_0 & a_0 \\ 1 - a_1 & a_1 \end{pmatrix}.
$$

The parameter $a_0 = P(S_t = 1 \mid S_{t-1} = 0)$ determines the probability of entering the extreme state, and the parameter $a_1 = P(S_t = 1 \mid S_{t-1} = 1)$ determines the probability of remaining in the extreme state.

The distribution of the non-thresholded data Z_t depends on the states S_{t-1} and S_t. Furthermore, to account for dependence between observations in the same state, Shaby et al. (2015) specify a dependence structure for $\mathbf{Z}$, even conditional on $\mathbf{S}$. Since a Markov structure is assumed for $\mathbf{Z} \mid \mathbf{S}$, the likelihood may be written as the product of conditional densities

$$
L(\mathbf{z} \mid \mathbf{s}) = f(z_1 \mid s_1) \prod_{t=2}^{n} f(z_t \mid z_{t-1}, s_t, s_{t-1}).
$$

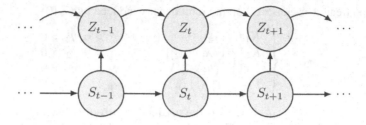

FIGURE 7.3
Graphical representation of the Markov-switching model. The state variables $S_1, \ldots, S_T$ are modeled as a two-state Markov chain. The distribution of each Z_t depends on its corresponding state variable S_t, for $t = 1, \ldots, T$. Finally, conditionally on $S_1, \ldots, S_T$, the observations $Z_1, \ldots, Z_T$ are also modeled as a Markov process.

Therefore, as in (7.11), the conditional likelihood of $\mathbf{Y} \mid \mathbf{S}$ may be completely specified by four families of conditional distributions $Z_t \mid Z_{t-1}, S_t = i, S_{t-1} = j$ for $i, j \in \{0, 1\}$. This type of model, depicted graphically in Figure 7.3, is sometimes referred to as a Markov-switching model (Frühwirth-Schnatter, 2006). Markov-switching models resemble hidden Markov models, the latter differing in that $\mathbf{Z}$ are conditionally independent given $\mathbf{S}$.

Conditioning on the state variables to separate the likelihood into two main components, one arising from the extreme state and one arising from the bulk state, endows each component with an immediate interpretation: a dedicated tail model for the extreme state and a model for the bulk of the distribution for the bulk state. The model is then determined by the conditional likelihoods $f(z_t \mid z_{t-1}, s_t, s_{t-1})$. In the extreme state, Z_t should follow a GPD and consecutive extreme observations with $S_{t-1} = 1$ and $S_t = 1$ should exhibit extremal dependence. To build the conditional likelihoods for the case where both Z_{t-1} and Z_t are in the extreme state, Shaby et al. (2015) follow Smith et al. (1997) and construct the conditional distribution using bivariate GPD as in the first case in (7.11).

The next case to be specified is when $S_{t-1} = S_t = 0$, indicating that times $t - 1$ and t are both members of the bulk state. This case is modeled simply as an AR(1) process with mean μ, variance σ_N^2, and autocorrelation parameter $\phi \in (0, 1)$

$$f(z_t \mid z_{t-1}, s_t = 0, s_{t-1} = 0) = N \left[\mu + \phi(z_{t-1} - \mu), (1 - \phi^2)\sigma_N^2 \right],$$

where $N(\mu, \sigma^2)$ is the Gaussian density with mean μ and variance σ^2. The Gaussian AR(1) process is often appropriate for the bulk distribution, but unlike the logistic Markov process specified for the extreme state, the AR(1) process is asymptotically independent and therefore inadequate as a model for the tail behavior. The Gaussian assumption for the marginal distribution of observations below the tail is convenient, but should be checked in applications. Examples of non-Gaussian models below the threshold include Behrens et al. (2004) and Reich et al. (2013).

The remaining heterogeneous cases $\{S_{t-1} = 0, S_t = 1\}$ and $\{S_{t-1} = 1, S_t = 0\}$

0} represent the transitions into and out of the extreme state. The approach taken in Shaby et al. (2015) is again similar to that of Smith et al. (1997). The conditional densities are defined through corresponding bivariate densities with logistic dependence. The difference from the $\{S_{t-1} = S_t = 1\}$ case is that here one of the marginal distributions is Gaussian rather than GPD. This necessitates two modifications. The first is that for $\{S_{t-1} = 0, S_t = 1\}$ (the transition into the extreme state), the density in the denominator of (7.9) is normal. The second is that in both heterogeneous cases, one of the v variables in (7.11) is the result of a transformation from normal to unit Fréchet, $v = -\log(\Phi[(z - \mu)/\sigma_N])^{-1}$, rather than both being the result of the transformation from GPD to unit Fréchet. A single dependence parameter α_{01} characterizes the temporal dependence during transitions between states.

7.5 Copula Methods

An alternative approach to time series analysis is copula modeling (Nelsen, 1999). Copulas offer a richer modeling framework than Markov models. Copula modeling fixes the marginal distribution of each response, and connects the marginal distributions using a latent process that accounts for temporal dependence. Denote the quantile function (inverse distribution function) of the response as $Q(\tau)$, so that $P[Y_t \leq Q(\tau)] = \tau \subset (0, 1)$ for all t. For example, the quantile function for the GPD with $\xi \neq 0$ in Section 7.2 is

$$Q(\tau) = \begin{cases} \frac{\sigma}{\xi}[(\frac{1-\tau}{\lambda})^{-\xi} - 1] & \tau > 1 - \lambda \\ 0 & \tau \leq 1 - \lambda. \end{cases} \tag{7.12}$$

These marginal distributions are connected using a latent temporal process. Let U_t be a temporal process with marginal distribution function $P(U_t < u) = F_U(u)$. Then the response is modelled using the probability integral transformation $Y_t = Q[F_U(U_t)]$. By construction, Y_t has the desired marginal distribution with quantile function Q, and inherits the temporal dependence structure of the latent U_t.

The most common copula is the Gaussian copula where U_t is a Gaussian process and $F_U(u) = \Phi(u)$ is the standard normal distribution function. This permits the use of the rich classes of autocorrelation functions that have been developed in the time series literature for Gaussian processes (Shumway and Stoffer, 2011). However, a drawback of the Gaussian copula is that despite accommodating autocorrelation, the process is asymptotically independent with $\chi(l) = 0$ for all $l > 0$. This is undesirable in extreme values analysis, and thus alternative copulas are needed. The max-stable process approach in Section 7.6.1 is one such alternative.

7.6   Max-Stable Processes

As described in Chapter 9, max-stable processes arise as the asymptotic distribution of block maximum data. However, because of their attractive representation of tail dependence, they are also used for points above a threshold. We first describe the approach of Reich et al. (2014) who use a copula model to preserve GPD margins and a latent max-stable process to account for dependence. We then discuss the methods of Huser and Davison (2014) and Raillard et al. (2014), who model data above a threshold directly using max-stable processes.

7.6.1   Max-Stable Copula

Let $\mathbf{U}_t = (U_1, ..., U_n)$ be a latent process with unit Fréchet marginal distribution function $F_U(u) = \exp(-1/u)$ and joint distribution function $F(u_1, ..., u_n) = \exp[-V(u_1, ..., u_n)]$ for some dependence function V (Beirlant et al., 2004). The process U_t is max-stable if and only if the dependence function satisfies: (1) $V(tu_1, ..., tu_n) = V(u_1, ..., u_n)/t$ and (2) $V(u_1, ..., u_n) = 1/u_j$ when $u_l = \infty$ for all $l \neq j$. As in Section 7.5, the response is modeled as $Y_t = Q[F_U(U_t)]$. Therefore, the marginal distribution is determined by the (GPD) quantile function Q, and dependence is determined by V.

Specifying a max-stable latent process gives a nice expression for the asymptotic dependence function. Let $\vartheta(i, j) \in (1, 2)$ be the extremal dependence function (Smith, 1990) defined by

$$P(U_i < c, U_j < c) = P(U_i < c)^{\vartheta(i,j)}. \tag{7.13}$$

If $\vartheta(i, j) = 2$, then the joint distribution function is the product of the marginal distribution functions and the responses are independent; if $\vartheta(i, j) = 1$, then the joint and marginal distributions are equal and the responses are completely dependent. For max-stable processes, $\vartheta(i, j) = V(u_1, ..., u_n)$ where $u_i = u_j = 1$ and all other $u_k = \infty$. If U_t is a stationary process and $\vartheta(i, j)$ depends on only $|i - j|$, and thus denoted $\vartheta(|i - j|)$, then the asymptotic dependence function is

$$\chi(l) = 2 - \vartheta(l). \tag{7.14}$$

This provides a convenient expression for specifying the dependence function V. Reich et al. (2014) use the asymmetric logistic (Tawn, 1990) dependence function

$$V(u_1, ..., u_n) = \sum_{l=1}^{L} \left[\sum_{j=1}^{J} u_j^{-1/\alpha} K_{jl}^{1/\alpha} \right]^{\alpha}, \tag{7.15}$$

where $\alpha \in [0, 1]$ and K_{jl} are weights that determine the dependence structure. The extremal coefficient is

$$\vartheta(i, j) = \sum_{l=1}^{L} \left[K_{il}^{1/\alpha} + K_{jl}^{1/\alpha} \right]^{\alpha}. \tag{7.16}$$

Reich et al. (2014) consider several choices of weighting function and explore the resulting asymptotic dependence properties. One example is

$$K_{jl} = \begin{cases} \gamma^{j-1} & l = 1 \\ I(j \geq l)(1-\gamma)\gamma^{j-l} & l > 1. \end{cases} \tag{7.17}$$

This choice of weights leads to a first-order Markov model with asymptotic dependence function $\chi(l) = 1+\gamma^l-(\gamma^{l/\alpha}+1)^\alpha$. In this Markov model, $\gamma \in (0,1)$ controls the range of dependence, with $\chi(l) \approx \gamma^l$ for $\alpha \approx 0$. Another possible weight function is the Gaussian kernel weights with $K_{jl} = \exp[-(j-l)^2/\rho^2]/c_j$, where ρ is the kernel bandwidth that controls the range of dependence, and the proportionality constant $c_j = \sum_l \exp[-(j-l)^2/\rho^2]$.

Computation is challenging for this model because evaluating the asymmetric logistic density function is slow in high-dimensions (large n). However, Reich et al. (2014) use a latent variable scheme (Reich and Shaby, 2013; Stephenson, 2009) to facilitate a Bayesian analysis using Markov chain Monte Carlo (MCMC). This permits analysis of the exact posterior distribution, which naturally provides measures of uncertainty and opens the door to model complex phenomena such as non-stationarity.

7.6.2 Censored Max-Stable Processes

Another approach is to conceive of the data as a censored max-stable process as in Huser and Davison (2014) for spatiotemporal data and Raillard et al. (2014) for temporally dependent data. That is, the uncensored data Z_t is modeled as a max-stable process over time. The marginal distribution of a max-stable process is $Z_t \sim$ GEV(μ, σ, ξ), the generalized extreme value (GEV) distribution with location μ, scale $\sigma > 0$, shape ξ, and distribution function (for $\xi \neq 0$)

$$F_{\text{GEV}}(z) = \exp\left[-\left(1 + \xi\frac{z-\mu}{\sigma}\right)_+^{1/\xi}\right].$$

Although the margins are GEV and not GPD, it has been shown that the GEV and GPD distributions have similar tail behavior (Drees et al., 2006).

Assuming Z_t is a max-stable process with GEV(μ, σ, ξ) margins, the joint distribution function is $F_Z(z_1, ..., z_n) = \exp[-V(v_1, ..., v_n)]$ where $v_j = -1/\log[F_{\text{GEV}}(z_j)]$ and V is the extremal dependence measure given in Section 7.6.1. This model for Z_t defines the joint distribution of the censored data Y_t. However, directly computing the joint likelihood of $Y_1, ..., Y_n$ is intractable. Instead, Huser and Davison (2014) and Raillard et al. (2014) use pairwise likelihood methods. The pairwise likelihood replaces the full n-dimensional likelihood with the product of the bivariate densities of pairs of observations. Therefore, all that is required is the bivariate density

$$f(y_i, y_j) = \begin{cases} F_Z(u, u) & y_i = y_j = 0 \\ \frac{\partial}{\partial y_i} F_Z(y_i - u, 0) & y_i > 0, y_j = 0 \\ \frac{\partial}{\partial y_j} F_Z(0, y_j - u) & y_i = 0, y_j > 0 \\ \frac{\partial^2}{\partial y_i \partial y_j} F_Z(y_i - u, y_j 0u) & y_i > 0, y_j > 0. \end{cases} \tag{7.18}$$

Temporal dependence is captured by the dependence measure V. While many dependence measures from spatial extremes (Chapter 9) could be used, Raillard et al. (2014) use the Gaussian extreme value model (Smith, 1990)

$$V(u_i, u_j)$$
$$= \frac{1}{u_1} \Phi \left[\frac{|i-j|}{2\rho} + \frac{\rho}{|i-j|} \log \left(\frac{u_2}{u_1} \right) \right] + \frac{1}{u_2} \Phi \left[\frac{|i-j|}{2\rho} + \frac{\rho}{|i-j|} \log \left(\frac{u_1}{u_2} \right) \right]$$

where $\rho > 0$ controls the range of temporal dependence. Under this model the tail dependence function is

$$\chi(l) = 2 \left[1 - \Phi \left(\frac{l}{2\rho} \right) \right], \tag{7.19}$$

which decays smoothly with lag l.

7.7   Hourly Wind Analysis

We now illustrate the methods described above on the hourly wind speed data from Cheeseboro, California. In the context of fire risk analysis, both the magnitude of the extreme wind gust speeds and the persistence of extreme events are of interest. We therefore compute and compare estimates of marginal GPD parameters and dependence parameters across methods. Since engineering guidelines state that the expected lifespan of a utility pole is 50 years, one quantity of particular interest is the 50-year return level for wind gusts (the value exceeded with probability 1/50 in any one year), computed as

$$\text{rl}_{50} = u + \frac{\sigma}{\xi} \left[\left(\lambda_u^{-1} \{ 1 - [1 - (50n_y)^{-1}]^{1/\theta} \} \right)^{-\xi} - 1 \right],$$

where $\lambda_u = P(Y > u)$, n_y is the number of hourly gust readings per observation year, and θ is the extremal index of Equation (7.7). The 50-year return level depends on marginal parameters λ_u, σ, and ξ as well as the dependence index θ.

This section is to a large degree an exercise in comparing apples to oranges. Methods that model dependence do so in different ways, so their estimated dependence parameters are not necessarily directly comparable. For Bayesian methods, reported point estimates are posterior medians, and intervals are 95% credible intervals built from MCMC samples. For maximum likelihood and maximum pairwise likelihood methods, 95% confidence intervals constructed using either profile likelihoods (Coles, 2001, Chapter 2.6.6) or the block bootstrap (Davison and Hinkley, 1997, Chapter 8). Throughout, we treat each January as an independent replicate of a time series of hourly maximum gust speeds.

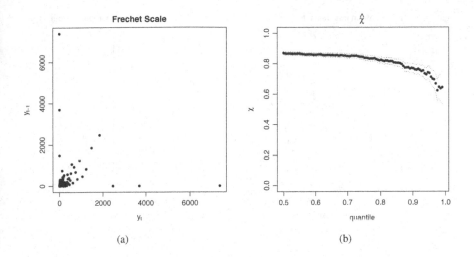

(a) (b)

FIGURE 7.4
Exploratory plots to check for extremal dependence. Panel (a) plots consecutive values on the Fréchet scale and panel (b) plots the sample χ statistic by quantile level.

7.7.1 Exploratory Analysis

To check whether extremal dependence is present in the data, we perform two diagnostics. The first is to transform the data to unit Fréchet using a rank transformation and plot the time series against itself with a lag-1 offset. The result, shown in Figure 7.4(a), depicts a number of points lying in the interior of the plot, rather than all crowded up against the axes. This is evidence of the presence of extremal dependence at lag 1.

The second diagnostic is to estimate the dependence measure $\chi(1, c)$, (see Equation (7.5)) for different values of c. Figure 7.4(b) shows an empirical estimate of $\chi(1, c)$, where the data has been transformed to Uniform(0,1) so that the x-axis is on the quantile scale, and the limit in Equation (7.6) becomes a limit as $c \to 1$. The plot in Figure 7.4(b) does not show $\chi(1, c)$ converging to 0 as $c \to 1$, so we again conclude that extremal dependence is present at lag 1.

Identifying an appropriate threshold is a notoriously difficult problem, and is discussed in details in Chapter 4 of this volume as well as Scarrott and MacDonald (2012) and Süveges and Davison (2010) specifically for time series data. To select a threshold for model fitting, we use the mean residual life plot (Coles, 2001, Chapter 4.3), shown in Figure 7.5(a), as a guide. These plots are difficult to interpret, but Figure 7.5(a) seems to suggest that any threshold above 40mph is a reasonable choice. We set the threshold at 45mph, which is the 0.975 empirical quantile of the wind speed data.

Since we consider only data from January, the pronounced seasonality in wind speeds is not an issue. One might also be concerned with diurnal cycles, where per-

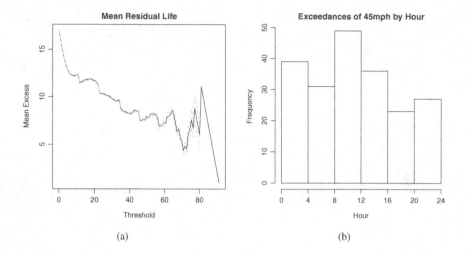

(a)   (b)

FIGURE 7.5
Exploratory plots to aid in specifying the model. The mean residual life plot in panel
(a) suggests that a threshold of 45mph is a reasonable choice. This corresponds to the
0.975 empirical quantile, and leaves 205 total exceedances. To check for a diurnal
cycle, a histogram of excesses by hour is plotted in panel (b).

haps winds are strong during the day and calm during the night. To investigate this,
we plot a histogram of all exceedances of 45mph by hour of the day (Figure 7.5(b)).
This histogram shows that the number of extreme wind observations is fairly uniformly spread throughout the day. There may be a hint of a diurnal cycle, but Figure
7.5(b) is plausibly uniform and so we make the assumption that the distribution of
extremes is stationary.

With the threshold set at 45mph and a stationary model chosen, we fit the various models described above. Results for marginal parameters are reported in Table
7.1, and results for dependence characteristics are reported in Table 7.2. The first approach identifies clusters of exceedances and fits a GPD to the cluster maxima (see
Section 7.3). Clusters were identified using the runs method, where the number of
clusters was estimated using the intervals method of Ferro and Segers (2003). Estimates of σ and ξ were then found by maximum likelihood. The reported estimate
of extremal index θ was computed using the intervals method. Confidence intervals
for σ, ξ, and rl$_{50}$ were constructed using the profile likelihood method, with critical
values set according to a χ_1^2 distribution. This "peaks over threshold" analysis was
performed using the R package extRemes (Gilleland and Katz, 2011).

The next analysis consisted of using all threshold exceedances, rather than just
cluster maxima, to estimate GPD parameters. This strategy has been shown to yield
preferable parameter estimates (Fawcett and Walshaw, 2012), but within-cluster dependence must be accounted for when constructing confidence intervals (referred

TABLE 7.1
Estimates and 95% intervals for the GPD parameter scale (σ) and shape (ξ) parameters and 50-year return level (rl_{50}) for various statistical models (rows).

	σ	ξ	rl_{50}
Cluster peaks	8.982	0.020	108.560
	(5.452, 14.296)	(-0.230, 0.494)	(84.449, 401.804)
Sandwich	9.27	-0.094	100.58
	(5.971, 13.799)	(-0.220, 0.100)	(86.012, 194.228)
Censored Markov	9.27	-0.094	98.370
	(5.971, 13.799)	(-0.220, 0.100)	(68.617, 138.975)
Markov switching	12.075	-0.074	97.902
	(8.596, 16.259)	(-0.190, 0.066)	(84.784, 124.050)
MS copula	10.657	0.118	219.609
	(8.394, 13.832)	(-0.089, 0.330)	(123.025, 614.568)
Censored MSP	9.27	-0.094	128.041
	(5.971, 13.799)	(-0.220, 0.100)	(72.540, 346.960)

to as "sandwich" in Tables 7.1 and 7.2). This is accomplished by applying the profile likelihood method and replacing critical values based on a χ_1^2 distribution for those based on an estimated sandwich matrix (Fawcett and Walshaw, 2012; Smith, 1991). The resulting confidence intervals (see Table 7.1) are indeed narrower than those obtained from estimates based only on cluster peaks. Code to estimate these sandwich-based profile likelihood confidence intervals may be found at http://www.personal.psu.edu/bas59/research.html.

We now turn to methods that explicitly model extremal dependence, rather than removing it or accounting for it only in interval estimation. To compare the strength of dependence captured by each fitted model, we estimate the extremal index θ, reported in Table 7.2. Since closed form expressions for θ as a function of model parameters are typically not available, we instead estimate θ by drawing long simulations (500 years of Januaries) from each fitted model and apply the intervals estimate of Ferro and Segers (2003). Since, for the processes we consider here, the intervals method is consistent, these long simulations result in stable estimates of θ.

The first model for extremal dependence we consider is the Markov model of Smith et al. (1997) with logistic dependence (Section 7.4.1). We fit this model using a two-step procedure, first estimating marginal parameters by maximum likelihood based on all threshold exceedances, and then using those fitted parameters to transform to unit Fréchet. The transformed data was then used to maximize the censored likelihood (7.11) to obtain an estimate of the logistic dependence parameter α. Confidence intervals for α were estimated using a block bootstrap, where, to capture variability from both the marginal estimation and the dependence estimation, the entire two-step procedure was repeated for each bootstrap sample. Maximization of (7.11) and simulation of the fitted model to estimate θ have been implemented in the R package POT (Ribatet, 2012).

We next apply the Markov-switching model of Shaby et al. (2015) and the max-

TABLE 7.2
Estimates and 95% intervals (when available) for the dependence parameters.

	θ	α	kernel bandwidth
Sandwich	0.18		
Censored Markov	0.31	0.475	
		(0.394, 0.623)	
Markov-Switching	0.19	0.621	
		(0.503, 0.759)	
MS copula	0.25		3.136
			(2.774, 3.688)
Censored MSP	0.24		1.601
			(0.963, 2.456)

stable copula model of Reich et al. (2014). These are both fully Bayesian models that concurrently estimate marginal and dependence parameters (see Reich et al. (2014); Shaby et al. (2015) for details on prior specification). Endpoints of the reported 95% credible intervals for all parameters are the 0.025 and 0.975 quantiles of the MCMC samples. Code to fit and draw from the Markov-switching and max-stable copula models are provided at http://www.personal.psu.edu/bas59/research.html and http://www4.stat.ncsu.edu/~reich/code/, respectively.

Finally, we fit the Gaussian extreme value max-stable process to the threshold exceedances using censored pairwise likelihoods, as described in Section 7.6.2. Just as with the Smith et al. (1997) Markov model, we employed a two-step procedure, first fitting marginal and then dependence parameters, and used a block bootstrap to construct confidence intervals. There is a choice to be made about which pairs of observations to include in the pairwise likelihood. To help capture possible dependence at different time scales, we included all pairs at lags of 1, 2, 5, 10, 20, 50, and 100 hours. Experimentation with different sets of lags indicated that results were quite insensitive to this choice.

Tables 7.1 and 7.2 give the estimates of the marginal and dependence parameters, respectively, for the various approaches fit to the Cheeseboro data. There are noteworthy differences between the models. For example, although all intervals include zero, the shape parameter ξ is estimated to be positive for the cluster peaks and MS copula model, and negative for all the exceedances and Markov switching methods. Similarly, the estimated 50-year return level varies from 98mph using the Markov switching model to 220mph using the MS copula model. This MS copula model gives a substantially higher estimate of the 50-year return level, perhaps due to the large estimate of the shape parameter, ξ. Also, some of the interval estimates of the 50-year return level are wide and may include infeasible values. Therefore, further analysis is required to evaluate the fit of each model to the data.

7.7.2 Assessing Model Fit

We randomly generated 500 datasets of the same dimension as the Cheeseboro hourly data from each fitted model to evaluate whether each adequately fit the data. One sample month from each model is plotted in Figure 7.6. Visually, all models seem to have similar operating characteristics. Each simulation has a few isolated clusters of observations above the threshold u (dashed gray line). The MS copula model does not generate data below the threshold, but the exceedances have similar shape and frequency as the observed data. The observations are unrealistically smooth for the max-stable process model, and thus a more flexible dependence model than the Gaussian extreme value process is required to simulate extreme wind.

To quantify goodness-of-fit, for each of 500 simulated datasets, we compute several summary measures, including the sample extremal index θ, the sample 0.99 and 0.999 quantiles (across all days and years), and the sample χ dependence measure for several thresholds and time lags. Table 7.3 gives the 95% intervals of these 500 draws for each summary statistic and model. Comparing the predictive distributions to the summaries of the observed data illustrates the fit of each model. Ideally, the 95% prediction intervals should contain the value from the dataset. This is similar in spirit to the posterior predictive checks advocated by Gelman et al. (2013). The procedures outlined in Gelman et al. (2013) average over the posterior distribution of the unknown parameters, which is easy for the Bayesian models we fit, but impossible otherwise. Therefore, instead of averaging, we fix the parameters at their estimated values, which results in goodness-of-fit assessments that are overly stringent relative to those of Gelman et al. (2013).

The predictive interval for the extremal measure θ does not capture the sample statistic for the censored Markov model, indicating misspecification of the extremal dependence. The Markov switching intervals exclude the data's value for both extreme quantile levels. The quantile intervals are more narrow for the Markov switching model, suggesting the model does not properly address uncertainty in the marginal GPD parameters. All predictive intervals contain the observed value for the MS copula model. However the 0.999 quantile interval is quite wide, perhaps due to the large estimate of the GPD shape parameter ξ. Finally, the censored MSP model has poor coverage for both extreme quantile and extremal dependence measures. As noted visually in the discussion of Figure 7.6, the max-stable process with Gaussian dependence is unrealistically smooth and provides a poor fit. Therefore, these diagnostics clearly demonstrate the need for a more flexible dependence model, such as the Brown–Resnick dependence model (Kabluchko et al., 2009).

7.8 Discussion

We have described and applied several methods for studying time series of extremes, but there are of course many approaches that we did not include in this review.

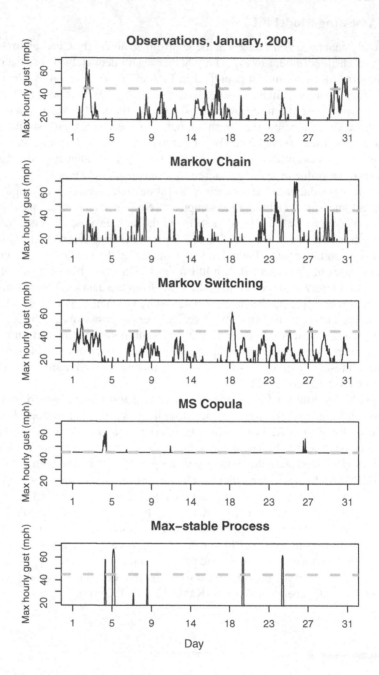

FIGURE 7.6
Observed wind speeds are plotted in the top panel. The remaining four panels show draws from the four fitted models.

TABLE 7.3
Observed summary statistics of the Cheeseboro data and 95% prediction intervals under different statistical models. Intervals that exclude the observed value are denoted with a *.

	θ	$0.99q$	$0.999q$	$\hat{\chi}(1, 60)$	$\hat{\chi}(5, 60)$
Data	0.18	53.6	72.6	0.47	0.38
Censored Markov	$(0.22, 0.43)^*$	$(49.9, 61.7)$	$(62.9, 89.5)$	$(0.34, 0.77)$	$(0, 0.50)$
Markov-Switching	$(0.15, 0.33)$	$(46.7, 50.5)^*$	$(52.2, 60.6)^*$	$(0, 0.79)$	$(0, 0.38)$
MS copula	$(0.14, 0.30)$	$(51.1, 73.4)$	$(71.4, 178)$	$(0.45, 0.82)$	$(0.18, 0.55)$
Censored MSP	$(0.18, 0.33)$	$(83.9, 93.6)^*$	$(98.6, 130.3)^*$	$(0.72, 0.79)^*$	$(0.064, 0.22)^*$

One interesting example is that of Bortot and Gaetan (2014), who construct dependent sequences of heavy-tailed GPDs using gamma mixtures of exponential random variables. Kottas et al. (2012) construct extreme value time series models by nonparametrically estimating the time-varying intensity measure of a Poisson process that scales parametric kernels. Another is Chavez-Demoulin and Davison (2012), who model high-frequency threshold exceedances in financial data using Hawkes processes. For further review, see Chavez-Demoulin and Davison (2012).

There are many needs for future work in this emerging field. For example, as far as we are aware there are no multivariate time series models to jointly analyze several responses. There is also additional work needed to address complex features often present in time series data such as non-stationarity, volatility, and cyclical dependence. In addition, although all code used to implement the methods discussed in this section are available online, more efficient software would be needed to scale to the size of many modern datasets.

Acknowledgments

Thanks to Raphael Huser for providing code for pairwise likelihood estimation.

References

Behrens, C., H. Lopes, and D. Gamerman (2004). Bayesian analysis of extreme events with threshold estimation. *Statistical Modelling 4*(3), 227.

Beirlant, J., Y. Goegebeur, J. Segers, J. Teugels, D. D. Waal, and C. Ferro (2004). *Statistics of Extremes: Theory and Applications*. New York: Wiley.

Bortot, P. and C. Gaetan (2014). A latent process model for temporal extremes. *Scandinavian Journal of Statistics 41*(3), 606–621.

Chavez-Demoulin, V. and A. C. Davison (2012). Modelling time series extremes. *REVSTAT Statistical Journal 10*(1), 109–133.

Coles, S. (2001). *An Introduction to Statistical Modeling of Extreme Values*. Springer Series in Statistics. London: Springer-Verlag London Ltd.

Coles, S. G., J. Heffernan, and J. Tawn (1999). Dependence measures for extreme value analysis. *Extremes 2*, 339–365.

Davison, A. and R. Smith (1990). Models for exceedances over high thresholds. *Journal of the Royal Statistical Society: Series B (Statistical Methodology) 52*(3), 393–442.

Davison, A. C. and D. V. Hinkley (1997). *Bootstrap Methods and Their Application*. Cambridge University Press, Cambridge.

Drees, H., R. De Haan, and D. Li (2006). Approximations to the tail empirical distribution function with application to testing extreme value conditions. *Journal of the Statistical Planning and Inference 136*(10), 3498–3538.

Fawcett, L. and D. Walshaw (2006). A hierarchical model for extreme wind speeds. *Journal of the Royal Statistical Society: Series C (Applied Statistics) 55*(5), 631–646.

Fawcett, L. and D. Walshaw (2007). Improved estimation for temporally clustered extremes. *Environmetrics 18*(2), 173–188.

Fawcett, L. and D. Walshaw (2012). Estimating return levels from serially dependent extremes. *Environmetrics 23*, 272–283.

Ferro, C. A. T. and J. Segers (2003). Inference for clusters of extreme values. *Journal of the Royal Statistical Society: Series B (Statistical Methodology) 65*(2), 545–556.

Frühwirth-Schnatter, S. (2006). *Finite Mixture and Markov Switching Models*. Springer Series in Statistics. New York: Springer.

Gelman, A., J. B. Carlin, H. S. Stern, D. B. Dunson, A. Vehtari, and D. B. Rubin (2013). *Bayesian Data Analysis*. New York: Chapman & Hall/CRC.

Gilleland, E. and R. W. Katz (2011). New software to analyze how extremes change over time. *Eos 92*(2), 13–14.

Huser, R. and A. Davison (2014). Space-time modelling of extreme events. *Journal of the Royal Statistical Society: Series B (Statistical Methodology) 76*(2), 439–461.

Kabluchko, Z., M. Schlather, and L. de Hann (2009). Stationary max-stable fields associated to negative definite function. *Annals of Probability 5*, 2042–2065.

Kottas, A., Z. Wang, and A. Rodrguez (2012). Spatial modeling for risk assessment of extreme values from environmental time series: A Bayesian nonparametric approach. *Environmetrics 23*(8), 649–662.

Leadbetter, M. R., G. Lindgren, and H. Rootzn (1983). *Extremes and Related Properties of Random Sequences and Processes*. Springer-Verlag, New York-Berlin.

Moritz, M. A., T. J. Moody, M. A. Krawchuk, M. Hughes, and A. Hall (2010). Spatial variation in extreme winds predicts large wildfire locations in chaparral ecosystems. *Geophysical Research Letters 37*(4), L04801.

Nelsen, R. B. (1999). *An Introduction to Copulas*. New York: Springer-Verlag.

Raillard, N., P. Ailliot, and J. Yao (2014). Modeling extreme values of processes observed at irregular time steps: Application to significant wave height. *The Annals of Applied Statistics 8*(1), 622–647.

Reich, B., D. Cooley, K. Foley, S. Napelenok, and B. A. Shaby (2013). Extreme value analysis for evaluating ozone control strategies. *Annals of Applied Statistics 7*, 739–762.

Reich, B. and B. Shaby (2013). A hierarchical max-stable spatial model from extreme precipitation. *The Annals of Applied Statistics 6*, 1430–1451.

Reich, B. J., B. A. Shaby, and D. Cooley (2014). A hierarchical model for serially-dependent extremes: A study of heat waves in the Western US. *Journal of Agricultural, Biological, and Environmental Statistics 19*(1), 119–135.

Ribatet, M. (2012). *POT: Generalized Pareto Distribution and Peaks over Threshold.* R package version 1.1-3.

Scarrott, C. and A. MacDonald (2012). A review of extreme value threshold estimation and uncertainty quantification. *REVSTAT Statistical Journal 10*, 33–60.

Sergius, L. A., G. R. Ellis, and R. M. Ogden (1962). The Santa Ana winds of Southern California. *Weatherwise 15*(3), 102–121.

Shaby, B. A., B. J. Reich, D. Cooley, and C. G. Kaufman (2015). A Markov-switching model for heat waves. ArXiv e-prints: http://arxiv.org/abs/1405.3904.

Shumway, R. H. and D. S. Stoffer (2011). *Time Series Analysis and Its Applications.* New York: Springer.

Smith, R. (1990). Max-stable processes and spatial extremes. Unpublished manuscript.

Smith, R. L. (1991). Regional estimation from spatially dependent data. Unpublished manuscript.

Smith, R. L., J. A. Tawn, and S. G. Coles (1997). Markov chain models for threshold exceedances. *Biometrika 84*(2), 249–268.

Stephenson, A. G. (2009). High-dimensional parametric modelling of multivariate extreme events. *Australian & New Zealand Journal of Statistics 51*, 77–88.

Süveges, M. and A. C. Davison (2010). Model misspecification in peaks over threshold analysis. *Annals of Applied Statistics 4*, 203–221.

Tawn, J. A. (1990). Modelling multivariate extreme value distributions. *Biometrika 77*, 245–253.

8

Max-Autoregressive and Moving Maxima Models for Extremes

Zhengjun Zhang

University of Wisconsin at Madison, United States

Liang Peng

Georgia State University, United States

Timothy Idowu

University of Wisconsin at Madison, United States

Abstract

Multivariate extreme value distributions are useful in modeling extreme dependence within a multivariate random vector. In order to model dynamic dependency of extreme values in spatial and time series data, some parametric models have been proposed in the literature, which share some important properties of a multivariate extreme value distribution and are employed to studies of weather extremes, extreme co-movements of financial markets, economic contagions, material reliability, and internet traffic analyses. This chapter provides an overview of the current development of these models.

8.1 Introduction

Since George Box and Gwilym Jenkins introduced autoregressive moving average (ARMA) (Box and Jenkins, 1970) models to model time series data, there has been tremendous development on proposing parametric time series models to capture dynamic dependence structures such as conditional means and conditional variances. Many of existing parametric time series models and studies typically focus on the central part of the data. On the other hand, extreme events or rare events (data outside the central part) have major impacts (bad or good) on the real world such as damages caused by a disaster hurricane. Modeling extreme events needs much more effort in characterizing variable dependency outside the central region of data, i.e., one has to deal with asymptotic (in)dependence in time series models.

In general, a multivariate distribution function characterizes the dependence structure within a random vector, and it does not portray dynamic dependence structures of a (univariate/vector) time series. As a result, models capable of characterizing asymptotically (in)dependent observations are practically useful. In the probability side of extreme value theory, it is known that max-stable processes are a generalization of multivariate extreme value distributions into an infinite dimensional one, and they are useful in modeling dynamic dependency of extreme values in spatial and time series data; see de Haan (1984) and Resnick (1987) for details. In the statistics side of extreme value theory, Smith and Weissman (1996) showed that extreme values of a multivariate stationary process may be characterized in terms of a limiting max-stable process under quite general conditions; see also Deheuvels (1983). As a result, it is very natural to model extreme processes by max-stable processes either parametrically or nonparametrically. Furthermore, if dynamic dependence structures are concerned, parametric models are preferred. In this note, we will review some parametric models, which can be used to capture dynamic extreme dependence structures. Before doing this, we first summarize some "stylized" properties in finite dimensional extreme value theory.

8.2   Finite Dimensional Extreme Value Theory

This section reviews some basic properties and measures in finite dimensional extreme value theory.

8.2.1   Extreme Value Theory for Dependent Random Variables

Let us first consider the simplest case of a sequence of independent random variables $\{X_i\}_{i=1}^n$ with a common distribution function $F(x)$. Let

$$M_n = \max(X_1, X_2, \ldots, X_n). \tag{8.1}$$

What we are interested in is the following equation

$$\lim_{n \to \infty} \Pr\left(\frac{M_n - b_n}{a_n} \le x\right) = \lim_{n \to \infty} F^n(a_n x + b_n) = H(x) \tag{8.2}$$

for some suitable normalizing constants $a_n > 0$, b_n and a function H. The limit distribution $H(x)$ can be expressed as the so-called generalized extreme value distribution

$$H(x; \mu, \sigma, \xi) = \exp\left\{ -\left[1 + \xi \frac{x - \mu}{\sigma}\right]^{-1/\xi}\right\}, \quad 1 + \xi\frac{x - \mu}{\sigma} > 0, \tag{8.3}$$

where μ, σ, ξ are location, scale, and shape parameter, respectively, and $\xi > 0$, $=$ 0, < 0 correspond to three types of extreme value distributions, i.e., Fréchet, Gumbel, and Weibull, respectively.

If (8.2) holds, we say F (or X_i) belongs to the (maximum) domain of attraction of an extreme value distribution H and write $F \in MDA(H)$ (or $X_i \in MDA(H)$). We say a nondegenerate distribution H is *max-stable*, if $H^n(a_n x + b_n) = H(x)$ holds for some constants $a_n > 0$ and b_n for each $n = 2, 3, \ldots$. Indeed, the extreme value distributions are *max-stable* distributions; see Theorem 1.4.1 in Leadbetter et al. (1983).

When extreme values from a stationary time series are concerned, (8.2) does not hold in general. Suppose now $\{X_i, i = 1, 2, \ldots, \}$ is a stationary sequence with a continuous marginal distribution function $F(x)$ and $\{\widehat{X}_i, i = 1, 2, \ldots, \}$ is the so-called associated sequence of independent random variables with the same marginal distribution function F. M_n stands for the maximum as usual, defined by (8.1), while $\widehat{M}_n$ denotes the corresponding maximum of $\{\widehat{X}_1, \ldots, \widehat{X}_n\}$. The limit distribution of M_n can be related to the limit distribution of $\widehat{M}_n$ via a quantity θ defined below.

Definition 8.2.1. *If for every $\tau > 0$, there exists a sequence of thresholds $\{u_n\}$ such that*

$$\Pr\{\widehat{M}_n \leq u_n\} \to e^{-\tau}, \tag{8.4}$$

then, under quite mild additional conditions (e.g., Leadbetter, 1983), one can get

$$\Pr\{M_n \leq u_n\} \to e^{-\theta\tau}. \tag{8.5}$$

Here θ is called the extremal index *of the sequence $\{X_n\}$.*

This concept originated in papers by Cartwright (1958), Newell (1964), Loynes (1965), and O'Brien (1974). A formal definition is given by Leadbetter (1983).

The index θ can take any values in [0,1] and $1/\theta$ is interpreted as the mean cluster size of exceedances over some high threshold. When $\theta = 0$, it corresponds to a strong dependence (infinite cluster sizes) but not so strong that all the values can be the same. While $\theta = 1$ is a form of asymptotic independence of extremes, it does not mean that the original sequence is independent. If (8.5) holds for some τ and corresponding $\{u_n\}$, then it holds for all τ' (equal or not equal to τ) and its corresponding $\{u_n'\}$.

Equation (8.5) shows that the extremal index is an important indicator of clustered extremes in time series data. Beirlant et al. (2004) calculated the extremal index for several examples of stationary time series. Estimators for the extremal index have been proposed by Leadbetter et al. (1989), Nandagopalan (1990), and Hsing (1993). Smith and Weissman (1994) gave a review of estimating the extremal index and proposed two estimation methods, i.e., blocks method and runs method. Other references can be found in Chapter 8 of Embrechts et al. (1997).

8.2.2 Limit Laws of Multivariate Extremes of Dependent Sequences

Suppose $\{\mathbf{X}_i = (X_{i1}, \ldots, X_{iD}), i = 1, 2, \ldots\}$ is a D-dimensional independent sequence with a common distribution $F(\mathbf{x}) = F(x_1, \ldots, x_D)$ and marginal distributions $F_d(x)$, $d = 1, \ldots, D$. Let $\mathbf{M}_n = (M_{n1}, \ldots, M_{nD})$ denote the vector of pointwise maxima, i.e., $M_{nd} = \max\{X_{id}, 1 \le i \le n\}$. If there exist normalizing constants $\mathbf{a}_n > 0$ and $\mathbf{b}_n$ such that as $n \to \infty$

$$\Pr\{\mathbf{M}_n \le \mathbf{a}_n\mathbf{x} + \mathbf{b}_n\}$$
$$= \Pr\{M_{nd} \le a_{nd}x_d + b_{nd}, d = 1, \dots, D\}$$
$$= F^n(a_{n1}x_1 + b_{n1}, a_{n2}x_2 + b_{n2}, \dots, a_{nD}x_D + b_{nD})$$
$$= F^n(\mathbf{a}_n\mathbf{x} + \mathbf{b}_n) \to H(\mathbf{x}) \tag{8.6}$$

for a distribution function H being nondegenerate and each H_d, $d = 1, \dots, D$, being nondegenerate, then the distribution H is called a D-dimensional multivariate extreme value (MEV) distribution and F is said to belong to the domain of attraction of H. We write $F \in D(H)$.

Multivariate extreme value distributions and their generalizations have long been studied in the literature; see de Haan and Resnick (1977), de Haan (1985), Pickands (1981), Resnick (1987), Joe (2014), and many others. In characterizing a multivariate extreme distribution, *max-stable (or min-stable)* distributions play a central role. Similar to the univariate case, we say a distribution $H(\mathbf{x})$ is *max-stable* if for every $t > 0$ there exist functions $\boldsymbol{\alpha}(t) > 0$ and $\boldsymbol{\beta}(t)$ such that

$$H^t(\mathbf{x}) = H(\boldsymbol{\alpha}(t)\mathbf{x} + \boldsymbol{\beta}(t)) = H(\alpha_1(t)x_1 + \beta_1(t), \dots, \alpha_D(t)x_D + \beta_D(t)). \tag{8.7}$$

The following theorem describes the equivalence between multivariate extreme value distributions and max-stable distributions.

Theorem 8.2.2. *The class of multivariate extreme value distributions is precisely the class of max-stable distribution functions with nondegenerate marginals.*

This is Proposition 5.9 in Resnick (1987). After a slight change of Pickands' representation of a min-stable multivariate exponential into a representation for a max-stable multivariate Fréchet distribution, we have

Theorem 8.2.3. *Suppose $H(\mathbf{x})$ is a limit distribution satisfying (8.6), then*

$$H(\mathbf{x}) = \exp\left\{ -\int_{S_D} \max_{1 \le i \le D} \left[\frac{w_i}{x_i}\right] dG(w) \right\}, \tag{8.8}$$

where G is a positive finite measure on the unit simplex

$$S_D = \left\{(w_1, \dots, w_D) : \sum_{i=1}^{D} w_i = 1, w_i \ge 0, i = 1, \dots, D\right\}$$

and G satisfies

$$\int_{S_D} w_i dG(w) = 1, \quad i = 1, \dots, D. \tag{8.9}$$

Note that

$$v(\mathbf{x}) = \int_{S_D} \max_{1 \le i \le D} \left(\frac{w_i}{x_i}\right) dG(w)$$

is called exponent measure by de Haan and Resnick (1977). So modeling a multivariate extreme value distribution function is equivalent to modeling the measure function G. A simple nonparametric estimation procedure for G is given by de Haan (1985). When one wants to simultaneously estimate the exponent measure and the dependence structure, Coles and Tawn (1991) argued that parametric models are preferable.

In Section 8.2.1, we look at the limit distribution of a dependent sequence with univariate random variables. Indeed some of those results can be extended to the multivariate case. Suppose now $\{\mathbf{X}_i = (X_{i1}, \dots, X_{iD}), i = 1, 2, \dots\}$ is a D-dimensional stationary stochastic process with a distribution function F and marginals F_d. Also let $\{\widehat{\mathbf{X}}_i\}$ be the associated sequence of i.i.d. random vectors having the same distribution function F. $\mathbf{M}_n$ and $\widehat{\mathbf{M}}_n$ are both pointwise maxima of $\{\mathbf{X}_i\}$ and $\{\widehat{\mathbf{X}}_i\}$, respectively. Suppose

$$\begin{cases} \lim_{n \to \infty} \Pr\{M_{n1} \leq u_{n1}, \dots, M_{nD} \leq u_{nD}\} = H(\tau), \\ \lim_{n \to \infty} \Pr\{\widehat{M}_{n1} \leq u_{n1}, \dots, \widehat{M}_{nD} \leq u_{nD}\} = \widehat{H}(\tau) \end{cases} \qquad (8.10)$$

hold for nonzero functions $H(\tau)$ and $\widehat{H}(\tau)$. Then a quantity, called *multivariate extremal index* (Nandagopalan, 1990, 1994), can relate the extreme value properties of a stationary process to those of i.i.d. sequence. More specifically, the *multivariate extremal index* is defined by

$$H(\tau) = \{\widehat{H}(\tau)\}^{\theta(\tau)} \qquad (8.11)$$

where $\theta(\tau)$ satisfies

(i) $0 \leq \theta(\tau) \leq 1$ for all τ;

(ii) $\theta(0, \dots, 0, \tau_d, 0, \dots, 0) = \theta_d$ for $\tau_d > 0$, where θ_d is the extremal index of the dth component process;

(iii) $\theta(c\tau) = \theta(\tau)$ for all $c > 0$ (Theorem 1.1 of Nandagopalan (1994).

However, Smith and Weissman (1996) pointed out that these properties are not sufficient to characterize the function $\theta(\tau)$. They argued why one needs to obtain a more precise characterization to cover a much broader range of processes and to correspond to real stochastic processes, for instance, multivariate maxima of moving maxima processes which we are going to address in this note. Two specific reasons are: i) the number of examples for which the multivariate extreme index has been calculated is currently very small and it is important to be able to extend this class to cover a much broader range of processes; ii) why we need a characterization is statistical: crude estimators of $\theta(\tau)$ are easy to construct, but would not correspond to multivariate extreme index of any real stochastic process.

8.2.3 Basic Properties of Multivariate Extreme Value Distributions

In this subsection, we review some basic properties of multivariate extreme value distribution functions. The following lemma (Galambos, 1987, Theorem 5.1.1) is very general, not restricted to MEV.

Lemma 8.2.4. *Let $F(\mathbf{x})$ be a D-dimensional distribution function with marginals $F_d(x)$, $1 \leq d \leq D$. Then, for all $x_1, x_2, \ldots, x_D$,*

$$\max\left(0, \sum_{d=1}^{D} F_d(x_d) - D + 1\right) \leq F(x_1, x_2, \ldots, x_D)$$

$$\leq \min\left(F_1(x_1), F_2(x_2), \ldots, F_D(x_D)\right).$$

This lemma can be used as a criterion for theoretically evaluating the flexibility of a joint distribution. For example, when one proposes a new class of multivariate extreme value distributions, or a new class of extreme value copulas, or a new class of time series models for dependent extremes, if both the lower bound and the upper bound can be reached, the proposed class can then be thought of as a good one.

The *copula*, or *dependence function*, is a very useful concept in the investigation of limit distributions for normalized extremes. A copula is a multivariate distribution with all marginals being a uniform distribution on $(0, 1)$.

Definition 8.2.5. *Let $F(\mathbf{x})$ be a D-dimensional distribution function, with continuous marginals $F_d's$. The copula associated with F, is a distribution function $C : [0, 1]^D \to [0, 1]$ that satisfies*

$$F(x_1, x_2, \ldots, x_D) = C\left(F_1(x_1), F_2(x_2), \ldots, F_D(x_D)\right).$$

The existence of a copula function is due to Sklar's theorem (Sklar, 1959). Based on the function $C(\mathbf{y})$, we now restate theorems which relate the univariate marginals and the multivariate or dependence structure of the limit distributions.

Theorem 8.2.6. *If (8.6) holds, then the copula function C_H of the limit $H(\mathbf{x})$ satisfies*

$$C_H^k(y_1^{1/k}, y_2^{1/k}, \ldots, y_D^{1/k}) = C_H(y_1, y_2, \ldots, y_D)$$

where $k \geq 1$ is an arbitrary integer (see Galambos, 1987, Theorem 5.2.1).

Theorem 8.2.7. *A D-dimensional distribution function $H(\mathbf{x})$ is a limit of (8.6) if and only if its univariate marginals are of the same type as one of three types of extreme value distributions in (8.3) and if its copula C_H satisfies the condition of Theorem 8.2.6 (see Galambos, 1987, theorem 5.2.4).*

Theorem 8.2.7 tells in principle that if we want to determine $\mathbf{a}_n$ and $\mathbf{b}_n$ we just need to determine the components from the marginal limit convergence forms, which can be illustrated by the following simple example.

Example 8.2.8. *Let (X, Y) have a bivariate exponential distribution function $F(x, y)$. If $(\mathbf{M}_n - \mathbf{a}_n)/\mathbf{b}_n$ converges weakly to a nondegenerate distribution function $H(x, y)$, we can choose*

$$\mathbf{a}_n = (\log n, \log n) \quad and \quad \mathbf{b}_n = (1, 1).$$

Although the usual construction of copula distributions is over [0,1], in the extreme value literature, however, either unit Fréchet marginal or Gumbel marginal is used. In this note, models in Section 8.4 are constructed using unit Fréchet marginal.

8.2.4 Asymptotic Dependence Index and Coefficient of Tail Dependence

There are two commonly used tail dependence measures in the literature. One is directly defined by the limit of a conditional probability, while another one is indirectly derived from a probabilistic model. The interpretation of the former is straightforward, while the interpretation of the latter has to rely on its probabilistic assumption.

Definition 8.2.9. *For a bivariate random variable* (X_1, X_2) *with an identically distributed distribution, denote* $x_F = \sup\{x \in \mathbb{R} : \Pr(X_1 \leq x) < 1\}$. *Then the asymptotic dependence index* λ *between the two random variables is defined by*

$$\lambda = \lim_{x \to x_F} \Pr(X_2 > x | X_1 > x) \tag{8.12}$$

as long as the limit exists.

The asymptotic dependence index λ quantifies the amount of dependence of the bivariate upper tails. If $\lambda > 0$, (X_1, X_2) is called asymptotically dependent; while (X_1, X_2) is called asymptotically independent when $\lambda = 0$.

This definition was due to Sibuya (1959). It was extended to the case of multivariate random variables by de Haan and Resnick (1977). In the literature, the asymptotic dependence indices between some paired random variables have been explicitly derived for many distributions. For example, when X and Y are normally distributed with correlation $\rho \subset [-1, 1)$, then $\lambda = 0$. When X_1 and X_2 have a bivariate standard t-distribution with ν degrees of freedom and correlation $\rho > -1$, then $\lambda = 2\bar{t}_{\nu+1}(\sqrt{\nu+1}\sqrt{1-\rho}/\sqrt{1+\rho})$, where $\bar{t}_{\nu+1}$ is the tail distribution function of the standard t distribution with $n + 1$ degrees of freedom. For additional cases, see Heffernan (2000), and Embrechts et al. (2002).

In practice, overly relying on multivariate normal distributions and Gaussian processes in which the asymptotic dependence index between any pair of random variables is zero can cause very serious inference errors. Indeed, identifying whether or not two random variables are asymptotically independent is extremely important in complex data analysis and modeling. Some tests and estimators have been proposed (e.g., Bacro et al., 2010; Falk and Michel, 2006; Hüsler and Li, 2009; Zhang, 2008).

Remark 8.2.10. *While the asymptotic dependence index is defined for identically distributed random variables and the same threshold, it can easily be extended to cases of nonidentically distributed random variables. See Lemma 14 of Heffernan et al. (2007) for a sufficient condition.*

Ledford and Tawn (2003), Zhang (2005), and Zhang and Huang (2006) extend the definition of asymptotic dependence between two random variables to lag-k asymptotic dependence of sequences of random variables with an identical marginal distribution. The definition of lag-k asymptotic dependence for sequences of random variables is given below.

Definition 8.2.11. *Suppose* $\{X_{1d}, X_{2d}, \ldots, X_{nd}, d = 1, \ldots, D\}$ *is a stationary*

D-dimensional multivariate time series with marginal distributions satisfying the conditions of Lemma 14 of Heffernan et al. (2007). Define

$$\lambda_{d_1 d'_{i+1}} = \lim_{x \to x_F} \Pr(X_{i+1,d'} > x | X_{1,d} > x),$$

where $x_F = \sup\{x \in \mathbb{R} : \Pr(X_{1d} \leq x) < 1\}$. *Then* $\lambda_{d_1 d'_{i+1}}$ *is called the lag-i asymptotic dependence index of the dth series on the d'th series. When* $d = d'$, $\lambda_{d_1 d'_{i+1}}$ *is the lag-i asymptotic dependence index within the dth series. When* $i = 0$, $\lambda_{d_1 d'_{i+1}}$ *is the asymptotic dependence index between* X_{1d} *and* $X_{1d'}$. *If*

$$\lambda_{d_1 d'_{k+1}} > 0 \quad and \quad \lambda_{d_1 d'_{k+j}} = 0 \quad for\ all \quad j > 1, \tag{8.13}$$

the dth series is said to be maximal lag-k asymptotically dependent on the d'th series.

One of the important ideas in studying asymptotic dependence, also called tail dependence or extreme dependence in the applied literature, between random variables is that as long as $\lambda_{d_1 d'_{i+1}}$ is zero, the construction of the joint limit extreme value distribution (or an extreme value type time series model) will become much easier since one only needs to deal with marginal distributions or the joint distributions of subsets of high-dimensional vectors or time series.

For a broad range of joint distributions, Ledford and Tawn (1996, 1997) consider the following model:

$$\Pr(X_1 > x,\ X_2 > x) \sim L\left(\frac{1}{\Pr(X_1 > x)}\right) \Pr(X_1 > x)^{1/\eta} \tag{8.14}$$

for some $\eta \in (0, 1]$ as $x \to x_F$, where L is a slowly varying function, i.e., $L(tx)/L(x) \to 1$ as $x \to \infty$. Using their terminology, the value of η effectively determines the decay rate of the joint bivariate survival function evaluated at the same large x, and η is termed as the coefficient of tail dependence. Two marginal variables are called positively associated when $1/2 < \eta \leq 1$; nearly independent when $\eta = 1/2$; and negatively associated when $0 < \eta < 1/2$.

Rewrite (8.14) as

$$\Pr(X_2 > x | X_1 > x) \sim L\left(\frac{1}{\Pr(X_1 > x)}\right) \Pr(X_1 > x)^{1/\eta - 1} \tag{8.15}$$

as $x \to x_F$. Then it is easy to see that the two variables X_1 and X_2 are asymptotically dependent ($\lambda > 0$) when $\eta = 1$ and $L(x) \nrightarrow 0$ as $x \to \infty$, and are asymptotically independent ($\lambda = 0$) when $0 < \eta < 1$. Based on these relations, one can see that λ and η function in different ways.

In the literature, there are quite a few published research papers involving η or $\bar{\chi} = 2\eta - 1$ (Coles, 2001), but not λ. Coefficient of tail dependence for different joint distributions has been calculated and presented in Heffernan (2000). For testing asymptotic dependence using η, in addition to Ledford and Tawn (1996, 1997), there are several continuing research papers such as Peng (1999) and Draisma et al. (2004).

In summary, the asymptotic dependence index and the extremal index are two important quantities in building multivariate extreme value distributions and extreme types of time series models. With the above-established notations and definitions, we shall review several parametric models proposed in the literature. These models share some important properties in multivariate extreme value theory, and can be used to capture the dynamic dependence structures of extremes in time series.

8.3 Max-Autoregressive Models

In the time series literature, autoregressive time series models have drawn much attention both theoretically and practically. In general, these models cannot characterize asymptotic dependency across time and across sections, which motivates the introduction of a new class of time series models called MARMA(p, q) processes.

8.3.1 MARMA(p, q) Processes

Davis and Resnick (1989) studied the so-called max-autoregressive moving average (MARMA(p, q)) process, which is a stationary process $\{X_n\}$ and satisfies the MARMA recursion,

$$X_n = \phi_1 X_{n-1} \vee \cdots \vee \phi_p X_{n-p} \vee Z_n \vee \theta_1 Z_{n-1} \vee \cdots \vee \theta_q Z_{n-q}$$

for all n where $\phi_i, \theta_j \geq 0, 1 \leq i \leq p, 1 \leq j \leq q$ and $\{Z_n\}$ is a sequence of independent random variables with a common distribution function $F(x) = \exp\{-\sigma x^{-1}\}$. For any given $\{\phi_i\}, \{\theta_j\}$, the corresponding process is a max-stable process. Davis and Resnick (1989) argued that it is unlikely that another subclass of the max-stable processes can be found which is as broad and tractable as the MARMA class. Some basic properties of the MARMA processes have been shown and the prediction of a max-stable process has been studied relatively completely. However, much less is known about estimation of a MARMA process. A naive estimation procedure for ϕ_i, θ_j's, when the order $q = 1$, is given in that paper. For prediction, see also Davis and Resnick (1993).

Certainly MARMA(p, q) processes are comparable to ARMA(p, q) time series models. Due to the estimation difficulty and the inherent signature patterns (ratios of observations equal constants; see Figure 8.1), applications and further development of MARMA(p, q) models are not seen in the literature. Instead, efforts have been made to introduce more realistic models. Next, we review a more workable extension of MARMA(p, q) models.

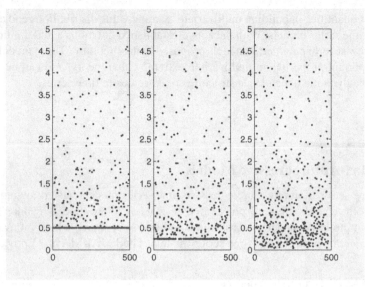

FIGURE 8.1
Illustration of the behavior of two processes. The left panel (X_{i+1}/X_i) and the middle panel (X_{i+2}/X_i) are generated from a MARMA(1,1) process with $\phi = 0.5$ and the right panel is generated from (8.16) with $p = 1, \alpha = 0.5, c = 1, a = 1$.

8.3.2   Max Positive Alpha Stable Processes and Applications

In an effort to overcome the parameter estimation difficulty and the abnormal signature patterns of MARMA processes, Naveau et al. (2011) considered model

$$Z_{t,\alpha} = c \max \left(\left[S_{t,\alpha} \max_{1 \le i \le p} \{ a_i Z_{t-i,\alpha} \} \right]^{\alpha}, \epsilon_t \right). \qquad (8.16)$$

where $c \in (0, 1]$ is a scale parameter, $a_i's \in (0, 1]$ are autoregressive parameters, and the sequence $\{ S_{t,\alpha} \}$ is independent of $\{ \epsilon_t \}$ and represents a sequence of independent and identically distributed positive α-stable variables defined by a Laplace transform

$$\mathbb{E}(\exp(-uS_{t,\alpha})) = \exp(-u^{\alpha}), \text{ for all } u \ge 0, \quad \alpha \in (0, 1]. \qquad (8.17)$$

A logarithm transformation of (8.16) with $p = 1$ yields the time series model

$$X_{t,\alpha} = \mu + \max \{ \gamma + \alpha \log(S_{t,\alpha}) + \alpha X_{t-1,\alpha}, \xi_t \}, \qquad (8.18)$$

where $\mu = \log(c)$, $\gamma = \alpha \log(a)$, $X_{t,\alpha} = \log(Z_{t,\alpha})$, $\xi_t = \log(\epsilon_t)$. In this model, μ is a location parameter, and both $X_{t,\alpha}$ and ξ_t are Gumbel distributed. This model can be regarded as a time series model with log of positive α stable noises and hidden max Gumbel shocks. Model (8.18) shares the spirit of the seminal threshold autoregression model introduced by Tong and Lim (1980). In fact, model (8.18) uses a

TABLE 8.1
Fitted values and Monte Carlo confidence intervals from simulated process.

	$a = 0.75$			$c = 0.75$			$d = 1.420$
size	95% MCI	mean	median	95% MCI	mean	median	mean
1000	(0.447,1.000)	0.743	0.730	(0.630,0.939)	0.762	0.757	1.422
2500	(0.543,1.000)	0.750	0.745	(0.647,0.867)	0.755	0.752	1.422
5000	(0.594,0.933)	0.749	0.746	(0.670,0.844)	0.753	0.751	1.421
10000	(0.597,0.873)	0.748	0.746	(0.694,0.831)	0.752	0.752	1.421

	$a = 0.25$			$c = 0.25$			$d = 0.283$
size	95% MCI	mean	median	95% MCI	mean	median	mean
1000	(0.000,0.591)	0.272	0.245	(0.210,0.290)	0.249	0.247	0.283
2500	(0.085,0.405)	0.283	0.251	(0.222,0.277)	0.247	0.246	0.283
5000	(0.130,0.403)	0.282	0.246	(0.225,0.269)	0.246	0.248	0.283
10000	(0.163,0.402)	0.288	0.249	(0.228,0.263)	0.246	0.249	0.283

	$a = 0.15$			$c = 0.85$			$d = 0.976$
size	95% MCI	mean	median	95% MCI	mean	median	mean
1000	(0.026,0.307)	0.152	0.147	(0.735,0.975)	0.852	0.850	0.977
2500	(0.067,0.244)	0.151	0.149	(0.777,0.928)	0.851	0.851	0.977
5000	(0.092,0.215)	0.151	0.150	(0.798,0.904)	0.850	0.850	0.976
10000	(0.107,0.194)	0.150	0.149	(0.813,0.888)	0.850	0.850	0.976

random threshold ξ_t and takes the value of the threshold itself when the threshold is larger than the computed autoregressive sum. This new model can also be regarded as a model with an infinite number of random change points.

With $p = 1$ and $\alpha = 0.5$, Naveau et al. (2011) showed that $\{Z_{t,\alpha}\}$ is an asymptotically independent process, estimated these three parameters by applying the method of moments approach, and presented Monte Carlo confidence intervals for both simulation examples and modeling real data of weekly maxima of river flow rates.

Even with the simple structure of $p = 1$, Naveau et al. (2011) demonstrated the usefulness and applicability of model (8.16) in modeling maxima of time series data. It is hoped that with $p > 1$ and varying α values model (8.16) can be more attractive in practice. However, further development in this direction has not yet been reported. In Section 8.3.3, we use numerical examples to illustrate how this class of models can be applied to real data modeling.

8.3.3 Numerical Examples

Following the simulation and estimation procedure proposed in Naveau et al. (2011), we consider $p = 1$ and $\alpha = 0.5$ in model (8.16) while the parameter values for a and

c vary. We simulate the data with different sample sizes. With the simulated data, we re-estimate the parameters $\theta = (d, a, c)$ using the generalized method of moments (GMM) approach, i.e.,

$$
\begin{aligned}
\widehat{\theta} \;=\; & \arg\min_{\theta \in \mathbb{R}^+ \times (0,1]^2} \left[\left(\frac{1}{n} \sum_{i=1}^{n} \exp\{-\frac{1}{Z_{t+i,\frac{1}{2}}}\} - \frac{d}{d+1} \right)^2 \right. \\
& \left. + \left(\frac{1}{n} \sum_{i=1}^{n} \exp\{-\frac{1}{Z_{t+i,\frac{1}{2}}} - \frac{1}{Z_{t-1+i,\frac{1}{2}}}\} - \frac{cd + acd\sqrt{d+1}}{(d+1)(c+1+ac\sqrt{d+1})} \right)^2 \right]
\end{aligned}
$$

with $ac\sqrt{d} + c = d$. In the simulation, we discard the initial 3000 values for achieving a stationary process, and keep different sizes of the remaining sequence. This procedure is done 5000 times for each set of parameter values. Each time we estimate the parameters and finally obtained the Monte Carlo mean, confidence interval, and median. The results presented in Table 8.1 indicate that the estimation procedure provides consistent estimators. Here we only report the confidence intervals for a and c at level 95%.

For the real data analysis, we apply model (8.16) to the deseasonalized weekly maxima of river flow rates of the Eagle River in Colorado, USA. The data was obtained from the US Geological Survey (USGS). The site number of the Eagle River is 0907000. The data is on Instantaneous Data Archive (IDA) starting from 1990-10-01 to 2014-08-15 with a reading at each 15-minute interval. Also, we collected the daily mean flow rates obtained from 1946-10-01 to 2014-08-15. Analysis is done on the deseasonalized flow rate in cubic feet per second. The weekly maxima and the deseasonalized weekly maxima are plotted in the top left panel and the top right panel in Figure 8.2. The plot in the bottom left panel of the present week and the following week shows a pattern of asymptotic independence, and hence the model (8.16) is applicable. We note that Naveau et al. (2011) analyzed the data with a shorter time window. What we have here contains the most recent data collection. The idea is to check whether or not there is any climate condition change indications by comparing the estimated parameter values in both studies.

To deseasonalize the data, the following procedure is used. Standardize the IDA data by using the mean and standard deviation of the daily mean data. Means and standard deviations are obtained from the daily mean data with a 13-day window centered on the day of the IDA values. After the standardization, the final deseasonalized data is obtained as the weekly maxima of the standardized values. Missing values were imputed using regression of daily maxima on daily mean. Upon fitting the model on the deseasonalized data the following estimates in Table 8.2 are obtained. Again we only report the confidence intervals with level 95% for a and c.

Although $\widehat{a} = 1$, the fact that $\widehat{ac} < 1$ implies the process is still stationary. The estimated values in Table 8.2 are similar to those presented in Naveau et al. (2011). Such similarity indicates that in terms of deseasonalized weekly maxima water flow rates in the Eagle River, the pattern has not been changed in recent years.

TABLE 8.2
Estimated parameter values with Monte Carlo confidence intervals for river flow rates.

Parameter	Estimate	CI
a	1.000	(0.564, 1.000)
c	0.584	(0.553, 0.775)
d	1.232	

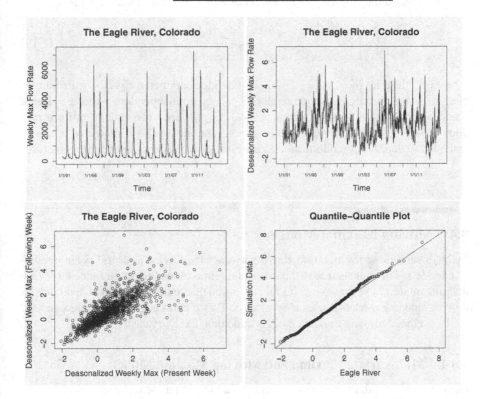

FIGURE 8.2
Illustration of river flow rates (cfs) of the Eagle River. Top left: Weekly maxima; top right: Deseasonalized weekly maxima; bottom left: Deseasonalized weekly maxima flow rates, present week vs the following week; bottom right: Quantile-quantile plot for deseasonalized weekly maxima and simulated data using the fitted model.

For diagnostics, a new process is simulated using the estimated values and compared with the real data using a QQ plot as shown in Figure 8.2. The QQ plot suggests that the model works well in approximating the weekly maxima even in the tail regions.

8.3.4   Extensions

Like the ARMA model, MARMA models, especially models with random coefficients, have flexibility and versatility in modeling long-range dependent time series. In addition to studying versatile random coefficient models, effort in developing efficient estimation procedures is needed. Moreover, flexible models for multivariate time series are also in demand.

Before closing this section we propose the following sparse random coefficient models for modeling multivariate maxima of multivariate stationary time series:

$$
Z_{t,\alpha_d,d} = \quad c_d \max\left(\left[S_{t,\alpha_d,d}\max_{1\le i\le p}\{a_{id}\max(U_{t,i,d}Z_{t,\alpha_d,d},\right.\right.
$$
$$
\left.\left.(1-U_{t,i,d})Z_{t-i,\alpha_d,d})\}\right]^{\alpha_d},\epsilon_t\right), \tag{8.19}
$$

where for dth series ($d = 1,\ldots,D$), α_d is the shape parameter, c_d is the scale parameter, a_{id} is the autoregressive parameter, $S_{t,\alpha_d,d}$'s are independent positive α-stable random variables with shape parameters α_d, $U_{t,i,d}$'s are independent uniform random variables on [0,1]. Here ϵ_t is a common random shock variable, and it can be extended to dependent random shock variables ϵ_{td} with some dependence structure.

8.4   Moving Maxima Models

In the linear time series literature, moving average models are as useful as autoregressive models. The same is true in the max-linear time series literature. Indeed, moving maxima models have drawn much more attention than max-autoregressive models as moving maxima models are more workable. This section is going to review several well-developed moving maxima models and some likely extensions.

8.4.1   Multivariate Maxima and Moving Maxima Models

Deheuvels (1983) defined the so-called moving minimum(MM) process as

$$
T_i = \min\{\delta_k Z_{i-k}, -\infty < k < \infty\}, -\infty < i < \infty,
$$

where $\delta_k > 0$, and $\{Z_k\}$ is a sequence of independent random variables with the standard exponential distribution. The following theorem of Deheuvels (1983) states that every joint multivariate distribution for minima can be approximated by a moving minimum process. As a result, moving minimum processes can be used in many applications as long as the joint dependencies for minima are concerned. This phenomenon may be compared with the wide use of multivariate normal distributions, implied by the central limit theorem, in many applications as long as means are concerned.

Theorem 8.4.1. *If* $(T_0,\ldots,T_m)$ *follows a joint multivariate extreme value distribution for minima with each margin being a standard exponential distribution, then* *there exist $m + 1$ sequences $\{a_k^i(n), -\infty < k < \infty\}$ depending on $n = 1, 2, \ldots,$ of positive numbers, such that, if $T_i(n) = \min\{a_k^i(n)Z_{-k}, -\infty < k < \infty\},$ $i = 0, \ldots, m$, then $(T_0(n), \ldots, T_m(n))$ converges in distribution to $(T_0, \ldots, T_m)$ as $n \to \infty$.*

The results of Deheuvels (1983) are very strong, but the model itself is still not tractable when estimating the parameters is concerned. Notice that the reciprocal of $1/T_i$ gives the moving maximum processes as

$$\frac{1}{T_i} = \max\{\frac{1}{\delta_k}Z'_{i-k}, -\infty < k < \infty\}, \quad -\infty < i < \infty$$

where $\{Z'_k\}$ is a sequence of independent random variables with the unit Fréchet distribution. Smith and Weissman (1996) extended this definition to a more general framework which is more realistic and is called multivariate maxima of moving maxima (henceforth M4) process. The definition is

$$Y_{id} = \max_l \max_k a_{l,k,d}Z_{l,i-k}, \quad d = 1, \ldots, D, \tag{8.20}$$

where $\{Z_{li}, l \geq 1, -\infty < i < \infty\}$ is an array of independent unit Fréchet random variables. The constants $\{a_{l,k,d}, l \geq 1, -\infty < k < \infty, 1 \leq d \leq D\}$ are nonnegative and satisfy

$$\sum_{l-1}^{\infty} \sum_{k=-\infty}^{\infty} a_{l,k,d} = 1 \text{ for } d = 1, \ldots, D. \tag{8.21}$$

It is straightforward (though tedious) to derive the joint distribution of $\{Y_{id}\}$. Particularly, we have

$$\begin{aligned}
&\Pr\{Y_{id} \leq y_{id}, 1 \leq i \leq r, 1 \leq d \leq D\} \\
&= \Pr\{Z_{l,i-k} \leq \tfrac{y_{id}}{a_{l,k,d}} \text{ for } l \geq 1, -\infty < k < \infty, 1 \leq i \leq r, 1 \leq d \leq D\} \\
&= \Pr\{Z_{l,m} \leq \min_{1-m \leq k \leq r-m} \min_{1 \leq d \leq D} \tfrac{y_{m+k,d}}{a_{l,k,d}}, l \geq 1, -\infty < m < \infty\} \\
&= \exp[-\textstyle\sum_{l=1}^{\infty} \sum_{m=-\infty}^{\infty} \max_{1-m \leq k \leq r-m} \max_{1 \leq d \leq D} \tfrac{a_{l,k,d}}{y_{m+k,d}}]
\end{aligned} \tag{8.22}$$

(Smith and Weissman, 1996, (2.5)), and

$$\Pr^n\{Y_{id} \leq ny_{id}, 1 \leq i \leq r, 1 \leq d \leq D\} = \Pr\{Y_{id} \leq y_{id}, 1 \leq i \leq r, 1 \leq d \leq D\},$$

which, following de Haan (1984), implies that $\{\mathbf{Y}_i\}$ is a max-stable process. Smith and Weissman (1996) argued that the extreme values of a multivariate stationary process may be characterized in terms of a limiting max-stable process under quite general conditions. They also showed that a very large class of max-stable processes may be approximated by the M4 processes mainly because those processes have the same multivariate extremal index (Smith and Weissman, 1996, Theorem 2.3). The theorem and conditions appear below.

Now fix $\tau = \{\tau_1, \ldots, \tau_D\}$ with $0 \leq \tau_d < \infty, d = 1, \ldots, D$. Let $\{u_{nd}, n \geq 1\}$ be a sequence of thresholds such that $n\{1 - F_d(u_{nd})\} \to \tau_d$ under the model assumption. Since Z_{lk} is unit Fréchet, we can take $u_{nd} = \frac{n}{\tau_d}$. Put $\mathbf{u}_n = (u_{n1}, \ldots, u_{nd})$

and let $\mathcal{B}_j^k(\mathbf{u}_n)$ denote the σ-field generated by the events $\{X_{id} \leq u_{nd}, j \leq i \leq k, 1 \leq d \leq D\}$ for $1 \leq j \leq k \leq n$. Define

$$\alpha_{nt} = \sup\{|P(A \cap B) - P(A)P(B)| : A \in \mathcal{B}_1^k(\mathbf{u}_n), B \in \mathcal{B}_{k+t}^n(\mathbf{u}_n)\}, \quad (8.23)$$

where the supremum is taken over $1 \leq k \leq n - t$ and the two respective σ-fields. If there exists a sequence $\{t_n, n \geq 1\}$ such that

$$t_n \to \infty, \ t_n/n \to 0, \ \alpha_{n,t_n} \to 0 \text{ as } n \to \infty, \quad (8.24)$$

the mixing condition $\triangle(\mathbf{u}_n)$ is said to hold (Nandagopalan, 1994; Smith and Weissman, 1996). Furthermore, suppose there exists a sequence $\{k_n, n \geq 1\}$ such that

$$k_n \to \infty, \ k_n t_n/n \to 0, \ k_n \alpha_{n,t_n} \to 0 \text{ as } n \to \infty. \quad (8.25)$$

Let $r_n = [n/k_n]$ be the integer part of n/k_n. Then Lemma 2.2 and Theorem 2.3 of Smith and Weissman (1996) are stated as follows.

Lemma 8.4.2. *Suppose (8.23)–(8.25) hold. Then the multivariate extremal index in (8.20) is given by*

$$\theta(\boldsymbol{\tau}) = \lim_{n \to \infty} \Pr\{Y_{id} \leq u_{nd}, 2 \leq i \leq r_n, 1 \leq d \leq D | \max_d(\frac{Y_{1d}}{u_{nd}}) > 1\}. \quad (8.26)$$

Alternatively, if we assume

$$\lim_{r \to \infty} \lim_{n \to \infty} \sum_{i=r}^{r_n} \sum_{d=1}^{D} \Pr\{Y_{id} > u_{nd} | \max_d(\frac{Y_{1d}}{u_{nd}}) > 1\} = 0, \quad (8.27)$$

then (8.26) is equivalent to

$$\theta(\boldsymbol{\tau}) = \lim_{r \to \infty} \lim_{n \to \infty} \Pr\{Y_{id} \leq u_{nd}, 2 \leq i \leq r, 1 \leq d \leq D | \max_d(\frac{Y_{1d}}{u_{nd}}) > 1\}. \quad (8.28)$$

We remark that this lemma is also a result in O'Brien (1987).

Theorem 8.4.3. *Suppose $\triangle(\mathbf{u}_n)$ and (8.27) hold for $\{\mathbf{Y}_i\}$, so that the multivariate extremal index $\theta^{\mathbf{Y}}(\boldsymbol{\tau})$ is given by (8.28). Suppose also the same assumptions hold for $\{\mathbf{X}_i\}$ (with the same t_n, k_n sequences). So the multivariate extremal index $\theta^{\mathbf{X}}(\boldsymbol{\tau})$ is also given by (8.28) with X_{id} replacing Y_{id} everywhere. Then $\theta^{\mathbf{Y}}(\boldsymbol{\tau}) = \theta^{\mathbf{X}}(\boldsymbol{\tau})$.*

The extension of MM processes to M4 processes results in an explicit expression of the extremal index of the process defined by (8.20), which is

$$\theta(\boldsymbol{\tau}) = \frac{\sum_l \max_k \max_d a_{l,k,d}\tau_d}{\sum_l \sum_k \max_d a_{l,k,d}\tau_d}. \quad (8.29)$$

Notice that an M4 model (8.20) contains an infinite number of parameters, hence parameter estimation will be quite challenging. Efforts have been put on studying M4 models with finite moving ranges, which led to a more workable M4 model defined as:

$$Y_{id} = \max_{1 \leq l \leq L} \max_{0 \leq k \leq K} a_{l,k,d} Z_{l,i-k}, \quad -\infty < i < \infty, \quad d = 1, \ldots, D, \qquad (8.30)$$

where L and K are finite and the coefficients satisfy $\sum_{l=1}^{L} \sum_{k=0}^{K} a_{l,k,d} = 1$ for each d. We note that in (8.30), some $a'_{l,k,d}$s may be zero. For example, $a_{l,k,d} = 0$ when $k < K_{1,l,d}$ or $k > K_{2,l,d}$, and $a_{l,k,d} > 0$ when $K_{1,l,d} \leq k \leq K_{2,l,d}$, for some parameters L_d, $K_{1,l,d}$ and $K_{2,l,d}$, $l = 1, \ldots, L_d$, $d = 1, \ldots, D$. Without loss of generality and for notation convenience, we simply use L and K in (8.30).

Under model (8.30), a direct calculation of the tail dependence index $\lambda_{dd'}$ between Y_{id} and $Y_{id'}$ gives

$$\lambda_{dd'} = 2 - \sum_{l=1}^{L} \sum_{m=1-K}^{1} \max\left\{a_{l,1-m,d}, a_{l,1-m,d'}\right\}.$$

Moreover, for fixed d, $Y_{i_1 d}$ and $Y_{i_2 d}$ have positive tail dependence whenever $|i_1 - i_2| \leq k (\leq K)$, and are therefore said to be lag-K tail dependent. The tail dependence index between $Y_{i_1 d}$ and $Y_{i_2 d}$, when $|i_2 - i_1| = k$, is

$$\lambda_{d(k)} = 2 - \sum_{l=1}^{L} \sum_{m=1-K}^{1+k} \max\left\{a_{l,1-m,d}, a_{l,1+k-m,d}\right\}.$$

Probabilistic properties of the model (8.30) have been studied in Zhang and Smith (2004), Martins and Ferreira (2005), Heffernan et al. (2007), Zhang (2009), Zhang and Smith (2010), among others. More specifically Heffernan et al. (2007) and Zhang (2009) demonstrated that M4 processes can model various extreme dependence structures. Zhang (2009) also demonstrated that the simplest M4 process performs as well as or better than the widely used Gumbel copula in modeling dependencies between bivariate random variables or among multivariate random variables, respectively.

8.4.2 Estimation

Hall et al. (2002) discussed moving maximum models

$$Y_i = \sup\{a_{j-i} Z_i, \ -\infty < i < \infty\}$$

where the distribution of Z_i is assumed to be either $F(z|\theta) = \exp(-z^{-\theta})$ or the generalized Pareto distribution $F(z|\theta) = 1 - (1 + z)^{-\theta}$. Then for a finite number of parameters, they chose $(\theta, a_{(m)})$ to minimize

$$D_m(\theta, a_{(m)}) = \int (\widehat{G}(y) - \prod_{i=2-m}^{k} F[\min\{a_{j-i}^{-1} y_j, \\ \max(i,1) \leq j \leq \min(i+m,k)\}|\theta])^2 w(y) dy, \qquad (8.31)$$

where the integral is over $y = (y_1, \ldots, y_k) \in \mathbb{R}_+^k$ and

$$\widehat{G}(y) = (n-k)^{-1} \sum_{i=1}^{n-k} I_{(Y_{i+j-1} \le y_j \text{ for } 1 \le j \le k)}, \qquad (8.32)$$

and w is a nonnegative weight function. We state their main theorem as follows.

Theorem 8.4.4. *Assume*
- *F has support on the positive half-line, and is in the domain of attraction of a Type II extreme value distribution,*
- *each a_i is nonnegative and, for some $\epsilon \in (0, r)$, $0 < \sum_i a_i^{r-\epsilon} < \infty$.*

Then

$$\sup_{-\infty < y_1, \ldots, y_k < \infty} |\Pr(Y_1^* \le y_1, \ldots, Y_k^* \le y_k | Y_1, \ldots, Y_n) - \Pr(Y_1 \le y_1, \ldots, Y_k \le y_k)| \to 0,$$

$$(8.33)$$

where Y_j^* is defined by

$$Y_j^* = \sup\{\widehat{a}_{j-i} Z_i^*, \ -\infty < i < \infty\},$$

$\widehat{a}_{j-i}$ and $\widehat{\theta}$ are solutions of (8.31) and Z_i^* has distribution function $F(.|\widehat{\theta})$. Moreover, if $m \ge C_4 (\log n)^2$ for C_4 sufficiently large, the rate of convergence in (8.33) is $O_p(n^{-(1/2)+\delta})$ for any given $\delta > 0$.

For inferences of the model (8.30), Zhang and Smith (2010) followed the idea of Hall et al. (2002) by defining a class of estimators based on empirical processes. In contrast to the bootstrapped processes which Hall et al. (2002) used to construct confidence intervals and prediction intervals for moving maxima models, they directly constructed parameter estimators and proved their asymptotic properties for the M4 processes. Zhang and Smith (2010) applied model (8.30) to a set of financial return data after transforming marginals to the unit Fréchet distribution.

We note that although M4 models are capable of modeling asymptotically dependent random processes with clustered extreme observations, they cannot simultaneously model asymptotically dependent and asymptotically independent random processes. There exist signature patterns in simulation examples (Zhang and Smith, 2004). Also, there are still many parameters in the model, which makes some signature patterns not estimable for a finite sample of observations. One remedy is to reduce the number of signature patterns. For example, Zhang (2008) studied M4 models with geometric signature patterns and applied them to sea wave data. Next, we review a class of M4 models with sparse random coefficients.

8.4.3 Sparse Parametric Moving Maxima Models with Random Coefficients

In studying large financial returns, Zhang (2005) noticed that large negative returns have extreme impacts on the observations in several lagged days, but it is rare to observe very large negative returns appearing in more than two consecutive days.

Bearing this in mind, Zhang (2005) proposed a sparse moving maxima process model and fitted it to transformed financial return data.

Tang et al. (2013) further generalized the sparse idea to allow for random coefficients. The model for univariate time series is defined as:

$$X_t = \max\left[\mathbf{B}^{(t)} \cdot \mathbf{Z}^{(t)}\right], \quad -\infty < t < \infty, \tag{8.34}$$

where

$$\mathbf{B}^{(t)} = \begin{pmatrix} \beta_{00}^{(t)} & 0 & 0 & \cdots & 0 \\ \beta_{10}^{(t)} & \beta_{11}^{(t)} & 0 & \cdots & 0 \\ \beta_{20}^{(t)} & 0 & \beta_{22}^{(t)} & \cdots & 0 \\ \vdots & \vdots & \vdots & \ddots & \vdots \\ \beta_{L0}^{(t)} & 0 & 0 & \cdots & \beta_{LL}^{(t)} \end{pmatrix},$$

$$\mathbf{Z}^{(t)} = \begin{pmatrix} Z_{0t} & Z_{0,t-1} & \cdots & Z_{0,t-L} \\ Z_{1t} & Z_{1,t-1} & \cdots & Z_{1,t-L} \\ \vdots & \vdots & \ddots & \vdots \\ Z_{Lt} & Z_{L,t-1} & \cdots & Z_{L,t-L} \end{pmatrix},$$

with $\beta_{00}^{(t)} = b_0$, $\beta_{l0}^{(t)} = a_{lt}b_l$ and $\beta_{ll}^{(t)} = (1 - a_{lt})b_l$, $l = 1, ..., L$, $\{a_{lt}\}$ are independent and identically distributed random variables on interval $[0, 1]$ and are assumed to be independent of $\{Z_{lt}\}$, and $\{b_l\}$ are positive constants with $\sum_{l=0}^{L} b_l = 1$. Here $\{b_l\}$ are scale factors, while $\{a_{lt}\}$ are understood as random effects which change the magnitude of shock variables Z_{lt}.

For any set of indices of $t = 1, ..., r$ and positive constants $\mathbf{x} = \{x_t, 1 \leqslant t \leqslant r\}$, conditioning on the generic random vector $\mathbf{a}$ representing all the a_{lk}'s involved, one can have

$$\Pr\left(X_t \leqslant x_t, 1 \leqslant t \leqslant r \mid \mathbf{a}\right) = \exp\left\{-V\left(\mathbf{x}, \mathbf{a}; \mathbf{b}\right)\right\} \tag{8.35}$$

with $V\left(\mathbf{x}, \mathbf{a}; \mathbf{b}\right)$ defined as

$$\sum_{t=1}^{r} \frac{b_0}{x_t} + \sum_{l=1}^{L} b_l \left[\sum_{j=1}^{\min(l,r)} \frac{1 - a_{lj}}{x_j} + \sum_{j=1}^{(r-l)_+} \frac{a_{lj}}{x_j} \bigvee \frac{1 - a_{l,j+l}}{x_{j+l}} + \sum_{j=(r-l)_+ +1}^{r} \frac{a_{lj}}{x_j} \right],$$

where $y_+ = \max(y, 0)$ and $\mathbf{y} \bigvee \mathbf{z} = (\max(y_1, z_1), \ldots, \max(y_n, z_n))$ for any two vectors $\mathbf{y}$ and $\mathbf{z}$ with the same dimension n. By the dominated convergence theorem, one gets

$$\begin{aligned} H\left(\mathbf{x}\right) &= \lim_{T \to \infty} \left[\Pr\left(X_t \leqslant Tx_t, 1 \leqslant t \leqslant r\right)\right]^T \\ &= \lim_{T \to \infty} \left\{1 + \frac{1}{T} E\left[T\left(e^{-\frac{1}{T}V(\mathbf{x},\mathbf{a};\mathbf{b})} - 1\right)\right]\right\}^T \\ &= \exp\left\{-E\left[V\left(\mathbf{x}, \mathbf{a}; \mathbf{b}\right)\right]\right\}. \end{aligned}$$

Hence, $\{X_t\}_{t=1}^{r}$ is in the maximum domain of attraction of max-stable distribution $H\left(\cdot\right)$ defined before (e.g., de Haan, 1984).

By introducing random coefficients in an M4 model, the number of signature patterns and the number of parameters have been greatly reduced, and hence the difficulty of estimating parameters is reduced. This model is called a sparse M3 model with random coefficients (SM3R). The following theoretical results are directly taken from Tang et al. (2013).

Proposition 8.4.5. *The extremal index of the SM3R process $\{X_t\}$ defined by (8.34) is*

$$\theta = b_0 + \sum_{l=1}^{L} b_l \phi_l, \tag{8.36}$$

where

$$\phi_l = \mathrm{E}\left[\max\left(a_l, 1 - a_l^*\right)\right], \tag{8.37}$$

a_l *and* a_l^* *are independent random variables having the same distribution as* a_{lt}.

Proposition 8.4.6. *The SM3R process $\{X_t\}$ defined by (8.34) is serially maximal lag-L tail-dependent with*

$$\lambda_l = \begin{cases} \lim_{u \to \infty} \Pr\left(X_{t+l} > u \mid X_t > u\right) = (1 - \phi_l) b_l, \ 1 \leqslant l \leqslant L, \\ \lim_{u \to \infty} \Pr\left(X_{t+l} > u \mid X_t > u\right) = 0, \ l > L, \end{cases} \tag{8.38}$$

where ϕ_l is defined by (8.37).

The above SM3R model can easily be extended to multivariate time series. Motivated by empirical findings that the cross-sectional tail dependence is in general restricted to the "same time" dependence (e.g., Morales, 2005; Poon et al., 2004; Zhang and Smith, 2010), Tang et al. (2013) introduced a new class of sparse M4 model with random coefficients (SM4R) model:

$$X_{td} = \max\left[\mathbf{B}_d^{(t)} \cdot \mathbf{Z}_d^{(t)}\right], \quad d = 1, \ldots, D, \ -\infty < t < \infty, \tag{8.39}$$

where

$$\mathbf{B}_d^{(t)} = \begin{pmatrix} \beta_{00d}^{(t)} & 0 & 0 & \cdots & 0 \\ \beta_{10d}^{(t)} & \beta_{11d}^{(t)} & 0 & \cdots & 0 \\ \beta_{20d}^{(t)} & 0 & \beta_{22d}^{(t)} & \cdots & 0 \\ \vdots & \vdots & \vdots & \ddots & \vdots \\ \beta_{L_d 0 d}^{(t)} & 0 & 0 & \cdots & \beta_{L_d L_d d}^{(t)} \end{pmatrix},$$

$$\mathbf{Z}_d^{(t)} = \begin{pmatrix} Z_{0td} & Z_{0,t-1,d} & \cdots & Z_{0,t-L_d,d} \\ Z_{1td} & Z_{1,t-1,d} & \cdots & Z_{1,t-L_d,d} \\ \vdots & \vdots & \ddots & \vdots \\ Z_{L_d td} & Z_{L_d,t-1,d} & \cdots & Z_{L_d,t-L_d,d} \end{pmatrix},$$

$\{\mathbf{Z}_{0t} = (Z_{0t1}, \ldots, Z_{0tD})\}$ is a sequence of independent and identically distributed D-dimensional random vectors (across t) having a multivariate extreme value distribution function with unit Fréchet margins, $\{Z_{ltd}\}$ are independent and identically distributed unit Fréchet random variables for $l \geqslant 1$, $\beta_{00d}^{(t)} = b_{0d}$, $\beta_{l0d}^{(t)} =$

$a_{ltd}b_{ld}$ and $\beta_{lld}^{(t)} = (1 - a_{ltd})b_{ld}$, $l = 1, ..., L_d$, $d = 1, ..., D$, $\{a_{ltd}\}$ are independent and identically distributed random variables on interval $[0, 1]$, $\mathbf{b} = \{b_{ld}, l = 1, ..., L_d, d = 1, ..., D\}$ are positive constants with $\sum_{l=0}^{L_d} b_{ld} = 1$ for any d, and $\{Z_{0t}\}$, $\{a_{ltd}\}$, and $\{Z_{ltd} : l \geqslant 1\}$ are independent with each other.

For any $t = 1, ..., r$ and positive constants $\mathbf{x} = \{x_{td}, t = 1, ..., r, d = 1, ..., D\}$, the joint distribution function of $\{X_{td}, t = 1, ..., r, d = 1, ..., D\}$ conditional on the generic random vector $\mathbf{a}$ representing all the a_{lkd}'s involved is

$$\Pr\left(X_{td} \leqslant x_{td}, 1 \leqslant t \leqslant r, 1 \leqslant d \leqslant D \mid \mathbf{a}\right) = \exp\left\{-V\left(\mathbf{x}, \mathbf{a}; \mathbf{b}\right)\right\}, \qquad (8.40)$$

where $V\left(\mathbf{x}, \mathbf{a}; \mathbf{b}\right)$ is defined as

$$\sum_{t=1}^{r} V_*\left(\frac{x_{t1}}{b_{01}}, ..., \frac{x_{tD}}{b_{0D}}\right) + \sum_{d=1}^{D} \sum_{l=1}^{L} b_{ld} \left[\sum_{j=1}^{\min(l,r)} \frac{1 - a_{ljd}}{x_{jd}} + \right.$$
$$\left. \sum_{j'=1}^{\max(r-l,0)} \max\left(\frac{a_{lj'd}}{x_{j'd}}, \frac{1 - a_{l,j'+l,d}}{x_{j'+l,d}}\right) + \sum_{j''=\max(r-l,0)+1}^{r} \frac{a_{lj''d}}{x_{j''d}}\right],$$

$\exp\left[-V_*\left(\cdot\right)\right] = \exp\left[-V_*\left(\cdot|\theta_{Z_0}\right)\right]$ is the multivariate extreme value distribution of Z_{0t}, and $V_*\left(\cdot\right)$ is called the exponent measure of Z_{0t} (e.g., Resnick, 1987).

To make model (8.39) more flexible, Z_{0t} is assumed to follow Gumbel–Hougaard copula with unit Fréchet margins, i.e.,

$$G_{\log}\left(\mathbf{x}; \alpha\right) = \exp\left\{-\left(\sum_{d=1}^{D} x_d^{-\frac{1}{\alpha}}\right)^{\alpha}\right\}, \qquad (8.41)$$

where $\alpha \in (0, 1]$. Then the joint distribution function becomes

$$\Pr\left(X_{td} \leqslant x_{td}, 1 \leqslant d \leqslant D\right)$$
$$= \exp\left\{-\sum_{d=1}^{D} \frac{1}{x_{td}}\left(1 - b_{0d}\right) - \left[\sum_{d=1}^{D} \left(\frac{b_{0d}}{x_{td}}\right)^{\frac{1}{\alpha}}\right]^{\alpha}\right\}. \qquad (8.42)$$

The following propositions (Tang et al., 2013) show that SM4R models can model asymptotically dependent random variables with extreme clusters.

Proposition 8.4.7. *The multivariate extremal index $\theta\left(\tau\right)$ of the SM4R process (8.39) is given by*

$$\theta\left(\tau\right) = \frac{V_*\left(\frac{1}{\tau_1 b_{01}}, ..., \frac{1}{\tau_D b_{0D}}\right) + \sum_{d=1}^{D} \tau_d \sum_{l=1}^{L} b_{ld}\phi_{ld}}{V_*\left(\frac{1}{\tau_1 b_{01}}, ..., \frac{1}{\tau_D b_{0D}}\right) + \sum_{d=1}^{D} \tau_d \left(1 - b_{0d}\right)}, \qquad (8.43)$$

where $\tau \geqslant 0$ and

$$\phi_{ld} = E\left[\max\left(a_{ld}, 1 - a_{ld}^*\right)\right]. \qquad (8.44)$$

Proposition 8.4.8. *The SM4R process $\{X_{td}\}$ defined by (8.39) is serially maximal lag-L tail-dependent and cross-sectionally maximal lag-0 tail-dependent.*

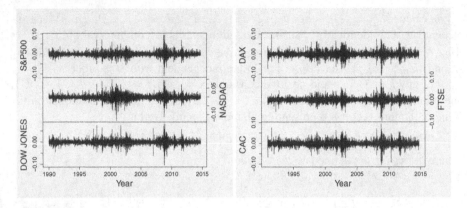

FIGURE 8.3
Left panel: Negative log-return plots of the S&P500 (GSPC), NASDAQ (IXIC), and Dow Jones Industrial Average (DJI) ranging from 1990-01-02 to 2014-09-29. Right panel: Negative log-return plots of the DAX (GDAXI), FTSE 100 (FTSE), and CAC 40 (FCHI) ranging from 1991-01-02 to 2014-09-29.

8.4.3.1 Estimation of SM4R Model

Tang et al. (2013) applied the generalized method of moments (GMM) to estimate all parameters in the model and employed the SM4R model to evaluate financial risk in a portfolio, which showed that the fitted SM4R model outperforms other models in terms of back-testing. In Section 8.4.3.2, we present numerical examples to illustrate its financial applications.

8.4.3.2 Real Data Modeling

Following the estimation procedure in Tang et al. (2013), we present two model fitting examples. The first example contains three US stock index returns, S&P500 (GSPC), NASDAQ (IXIC), and Dow Jones Industrial Average (DJI), from 1990-01-02 to 2014-09-29. The second example contains three Europe index returns, DAX (GDAXI), FTSE 100 (FTSE), and CAC 40 (FCHI), from 1991-01-02 to 2014-09-29. The data are plotted in Figure 8.3 showing the negative log returns for each financial time series. Define $r_t = -\log(V_t/V_{t-1})$, then r_t is the negative log return for day t where V_t is the closing value for day t. One can see that those time series show some similar patterns.

We follow the estimation procedure in Tang et al. (2013): 1) Fit an ARMA-GARCH type model to each univariate financial return time series and extract the devolatized returns also called standardized returns from the fitted models; 2) Model the marginal distributions of the devolatized returns using GPD based on POT with the threshold set at the 95th sample percentile; 3) Transform the devolatized returns to unit Fréchet distribution using the fitted marginal distributions; 4) Fit the proposed (time series) models to the transformed data with the unit Fréchet distribution; 5) Es-

TABLE 8.3

Estimated parameter values and standard errors (in parentheses) from S&P500 (GSPC), NASDAQ (IXIC), and Dow Jones Industrial Average (DJI), DAX (GDAXI), FTSE 100 (FTSE), and CAC 40 (FCHI).

α	US STOCK 0.3541(0.0182)			EUROPEAN STOCK 0.4429(0.0222)		
	S&P500	NASDAQ	DJI	DAX	FTSE	CAC
b_1	0.059(0.026)	0.083(0.026)	0.114(0.029)	0.126(0.023)	0.147(0.067)	0.148(0.033)
b_2	0.257(0.032)	0.287(0.036)	0.201(0.031)	0.231(0.035)	0.151(0.058)	0.199(0.037)
b_3	0.081(0.022)	0.064(0.030)	0.101(0.023)	0.169(0.028)	0.143(0.062)	0.144(0.035)
b_4	0.196(0.028)	0.030(0.033)	0.191(0.028)	0.178(0.029)	0.124(0.061)	0.106(0.031)

timate parameters by the generalized method of moments (GMM). Table 8.3 reports the final estimation results.

Note that Tang et al. (2013) analyzed three US stock index returns ranged from 1972 to 2010. From Table 8.3, we find that the estimated α parameter values for the two markets are similar, but the lagged time effect parameter patterns are different, which is an indication of different market dependencies between US stock markets and Europe stock markets.

8.4.4 Extensions

One of the models in Heffernan et al. (2007), which is an extension of M4 processes and called AIM4 model, is defined as

$$Y_{td} = \max(W_{td}^{1/\beta}, X_{td}^*), \quad d = 1, \ldots, D, \quad -\infty < t < \infty, \tag{8.45}$$

where X_{td}^* follows a M4 process defined in (8.30), $\beta > 0$, and $\{W_{td}, -\infty < t < \infty, d = 1, \ldots, D\}$ is an array of independent unit Fréchet random variables. In this model W_{td}'s are termed as random shocks which are independent of Z_{lt}'s.

Probabilistic properties of model (8.45) were presented in Heffernan et al. (2007). The model (8.45) allows variables Y_{td}'s to be either asymptotically independent (when $\beta < 1$) or asymptotically dependent (when $\beta \geq 1$). However, these two kinds of dependencies cannot coexist in the model. There is also an infinite number of parameters in an AIM4 model, or more accurately in the M4 part of X_{td}^*.

The proposed AIM4 models are more flexible and may deserve more attention in statistical inference and applications. One of the potential extensions of AIM4 models is to have a similar structure as SM4R models, and to let β vary with different marginals. The latter case has been studied by Ferreira and Ferreira (2014). Martins and Ferreira (2014) also extended AIM4 models to a more flexible setting with one set of variables being asymptotically independent and another set of variables being asymptotically dependent.

References

Bacro, J.-N., L. Bel, and C. Lantuéjoul (2010). Testing the independence of maxima: From bivariate vectors to spatial extreme fields. *Extremes 13*(2), 155–175.

Beirlant, J., Y. Goegebeur, J. Segers, and J. Teugels (2004). *Statistics of Extremes: Theory and Applications*. Wiley.

Box, G. and G. Jenkins (1970). *Time Series Analysis: Forecasting and Control*. San Francisco: Holden-Day.

Cartwright, D. E. (1958). On estimating the mean energy of sea waves from the highest waves in a record. *Proceedings of the Royal Society of London. Series A, Mathematical and Physical Sciences 247*(1248), 22–48.

Coles, S. (2001). *An Introduction to Statistical Modeling of Extreme Values*. Springer.

Coles, S. G. and J. A. Tawn (1991). Modelling extreme multivariate events. *Journal of the Royal Statistical Society: Series B (Statistical Methodology) 53*(2), 377–392.

Davis, R. A. and S. I. Resnick (1989). Basic properties and prediction of max-ARMA processes. *Advances in Applied Probability 21*(4), 781–803.

Davis, R. A. and S. I. Resnick (1993). Prediction of stationary max-stable processes. *The Annals of Applied Probability 3*(2), 497–525.

de Haan, L. (1984). A spectral representation for max-stable processes. *The Annals of Probability 12*(4), 1194–1204.

de Haan, L. and S. I. Resnick (1977). Limit theory for multivariate sample extremes. *Zeitschrift für Wahrscheinlichkeitstheorie und verwandte Gebiete 40*(4), 317–337.

de Haan, L. F. M. (1985). Extremes in higher dimensions: The model and some statistics. In *Proceedings 45th Session International Statistical Institute*, Volume 4, pp. 185–192. Amsterdam: International Statistical Institute.

Deheuvels, P. (1983). Point processes and multivariate extreme values. *Journal of Multivariate Analysis 13*(2), 257–272.

Draisma, G., H. Dress, A. Ferreira, and L. de Haan (2004). Bivariate tail estimation: Dependence in asymptotic independence. *Bernoulli 10*(2), 251–280.

Embrechts, P., C. Klüppelberg, and T. Mikosch (1997). *Modelling Extremal Events*. Springer Science & Business Media.

Embrechts, P., A. McNeil, and D. Straumann (2002). Correlation and dependence in risk management: Properties and pitfalls. In M. A. H. Dempster (Ed.), *Risk Management: Value at Risk and Beyond*, pp. 176–223. Cambridge: Cambridge University Press.

Falk, M. and R. Michel (2006). Testing for tail independence in extreme value models. *Annals of the Institute of Statistical Mathematics 58*(2), 261–290.

Ferreira, H. and M. Ferreira (2014). Extremal behavior of pMAX processes. *Statistics & Probability Letters 93*, 46–57.

Galambos, J. (1987). *Asymptotic Theory of Extreme Order Statistics* (2 ed.). Malabar, Florida: Krieger.

Hall, P., L. Peng, and Q. Yao (2002). Moving-maximum models for extrema of time series. *Journal of Statistical Planning and Inference 103*, 51–63.

Heffernan, J. E. (2000). A directory of coefficients of tail dependence. *Extremes 3*, 279–290.

Heffernan, J. E., J. A. Tawn, and Z. Zhang (2007). Asymptotically (in)dependent multivariate maxima of moving maxima processes. *Extremes 10*, 57–82.

Hsing, T. (1993). Extremal index estimation for a weakly dependent stationary sequence. *The Annals of Statistics 21*(4), 2043–2071.

Hüsler, J. and D. Li (2009). Testing asymptotic independence in bivariate extremes. *Journal of Statistical Planning and Inference 139*, 990–998.

Joe, H. (2014). *Dependence Modeling with Copulas*. Chapman and Hall/CRC.

Leadbetter, M. R. (1983). Extremes and local dependence in stationary sequences. *Probability Theory and Related Fields 65*(2), 291–306.

Leadbetter, M. R., G. Lindgren, and H. Rootzén (1983). *Extremes and Related Properties of Random Sequences and Processes*. Berlin: Springer.

Leadbetter, M. R., I. Weissman, L. de Haan, and H. Rootzén (1989). On clustering of high values in statistically stationary series. In *Proc. 4th Int. Meet. Statistical Climatology*, pp. 217–222. Wellington: New Zealand Meteorological Service.

Ledford, A. W. and J. A. Tawn (1996). Statistics for near independence in multivariate extreme values. *Biometrika 83*, 169–187.

Ledford, A. W. and J. A. Tawn (1997). Modelling dependence within joint tail regions. *Journal of the Royal Statistical Society: Series B (Statistical Methodology) 59*(2), 475–499.

Ledford, A. W. and J. A. Tawn (2003). Diagnostics for dependence within time series extremes. *Journal of the Royal Statistical Society: Series B (Statistical Methodology) 65*(2), 521–543.

Loynes, R. M. (1965). Extreme values in uniformly mixing stationary stochastic processes. *The Annals of Mathematical Statistics 36*(3), 993–999.

Martins, A. P. and H. Ferreira (2005). The multivariate extremal index and the dependence structure of a multivariate extreme value distribution. *Test 14*, 433–448.

Martins, A. P. and H. Ferreira (2014). Extremal properties of M4 processes. *Test 23*, 388–408.

Morales, F. C. (2005). *Estimation of max-stable processes using Monte Carlo methods with applications to financial risk assessment*. Ph. D. thesis, University of North Carolina at Chapel Hill.

Nandagopalan, S. (1990). *Multivariate extremes and the estimation of the extremal index*. Ph. D. thesis, University of North Carolina at Chapel Hill.

Nandagopalan, S. (1994). On the multivariate extremal index. *Journal of Research of the National Institute of Standards and Technology 99*(4), 543–550.

Naveau, P., Z. Zhang, and B. Zhu (2011). An extension of max autoregressive models. *Statistics and Its Interface 4*, 253–266.

Newell, G. F. (1964). Asymptotic extremes for m-dependent random variables. *The Annals of Mathematical Statistics 35*(3), 1322–1325.

O'Brien, G. (1974). Limit theorems for the maximum term of a stationary process. *The Annals of Probability 2*(3), 540–545.

O'Brien, G. L. (1987). Extreme values for stationary and markov sequences. *The Annals of Probability 15*(1), 281–291.

Peng, L. (1999). Estimation of the coefficient of tail depdendence in bivariate extremes. *Statistics & Probability Letters 43*, 399–409.

Pickands, III, J. (1981). Multivariate extreme value distributions. In *Proceedings 43rd Session International Statistical Institute*, Volume 2, pp. 859–878. Amsterdam: International Statistical Institute.

Poon, S. H., M. Rockinger, and J. Tawn (2004). Extreme value dependence in financial markets: diagnostics, models, and financial implications. *The Review of Financial Studies 17*, 581–610.

Resnick, S. I. (1987). *Extreme Values, Regular Variation, and Point Processes*. Springer, New York.

Sibuya, M. (1959). Bivariate extreme statistics, i. *Annals of the Institute of Statistical Mathematics 11*(2), 195–210.

Sklar, M. (1959). Fonctions de répartition à n dimensions et leurs marges. *Publ. Inst. Statist. Univ. Paris 8*, 229–231.

Smith, R. L. and I. Weissman (1994). Estimating the extremal index. *Journal of the Royal Statistical Society: Series B (Statistical Methodology) 56*, 515–528.

Smith, R. L. and I. Weissman (1996). Characterization and estimation of the multivariate extremal index. Unpublished manuscript, University of North Carolina. http://www.stat.unc.edu/postscript/rs/extremal.pdf.

Tang, R., J. Shao, and Z. Zhang (2013). Sparse moving maxima models for tail dependence in multivariate financial time series. *Journal of the Royal Statistical Society: Series B (Statistical Methodology) 143*, 882–895.

Tong, H. and K. S. Lim (1980). Threshold autoregression, limit cycles and cyclical data (with discussion). *Journal of the Royal Statistical Society: Series B (Statistical Methodology) 42*, 245–292.

Zhang, Z. (2005). A new class of tail-dependent time series models and its applications in financial time series. *Advances in Econometrics 20(B)*, 323–358.

Zhang, Z. (2008). Quotient correlation: a sample based alternative to Pearson's correlation. *The Annals of Statistics 36*, 1007–1030.

Zhang, Z. (2009). On approximating max-stable processes and constructing extremal copula functions. *Statistical Inference for Stochastic Processes 12*, 89–114.

Zhang, Z. and J. Huang (2006). Extremal financial risk model and portfolio evaluation. *Computational Statistics and Data Analysis 51*, 2313–2338.

Zhang, Z. and R. L. Smith (2004). The behavior of multivariate maxima of moving maxima processes. *Journal of Applied Probability 41*, 1113–1123.

Zhang, Z. and R. L. Smith (2010). On the estimation and application of max-stable processes. *Journal of Statistical Planning and Inference 140*, 1135–1153.

9

Spatial Extremes and Max-Stable Processes

Mathieu Ribatet

University of Montpellier 2 / University of Lyon 1, France

Clément Dombry

Université de Franche-Comté, France

Marco Oesting

INRA / AgroParisTech, France

Abstract

This chapter aims at being a crash course on max-stable processes with an emphasis on their use for modeling spatial extremes. We will see how max-stable processes are defined through a simple spectral representation and how it is possible to derive the finite dimensional distributions from it. Because the goal of this crash course is also to be of practical interest, existing parametric max-stable models will be introduced and discussed. A useful measure of spatial dependence, the extremal coefficient function, will be introduced as well as several approaches to fit max-stable processes to spatial data. Finally we open the discussion with other alternatives to max-stable processes to model spatial extremes.

9.1 Introduction

In Chapter 2, the extreme value theory was introduced in a finite dimensional setting, i.e., extremes of random variables or vectors. In this chapter we go a bit further by investigating infinite dimensional extremes, i.e., extremes of stochastic processes. Although it is possible to work with weaker assumption, we will work with sample path continuous stochastic processes to ensure that random variables such as $\sup\{Y(x) : x \in \mathcal{X}\}$ where $\mathcal{X}$ is a compact subset of $\mathbb{R}^d$, $d \geqslant 1$, are well defined.

Similarly to univariate extreme value analysis, the aim of modelling spatial extremes is typically related to risk assessments which in a spatial context can take several forms. For example if one observes a precipitation field $Y(x)$ over a given catchment $\mathcal{X} \subset \mathbb{R}^2$, one could be interested in evaluating the probability that the total rainfall amount over this catchment exceeds a given critical quantity $z_{\text{crit}} > 0$, i.e.,

$$\Pr\left\{ \int_{\mathcal{X}} Y(x)\mathrm{d}x > z_{\text{crit}} \right\}.$$

One could also be interested in characterizing the distribution of the largest "pointwise" rainfall amount in this catchment, i.e., evaluating probabilities of the form

$$\Pr\left\{ \sup_{x \in \mathcal{X}} Y(x) > z \right\}, \qquad z > 0.$$

Clearly evaluating the above probabilities is even more challenging than what we usually do in a univariate or multivariate setting since, for instance, it requires the knowledge of the distribution of the random variable $Y(x)$ for all $x \in \mathcal{X}$ as well as capturing the spatial dependence of the process $\{Y(x) \colon x \in \mathcal{X}\}$.

9.2   Max-Stable Processes

9.2.1   Spectral Representation

Before introducing the spectral representation of max-stable processes, it seems necessary to motivate their use for modelling spatial extremes. Similarly to the asymptotic arguments justifying the use of the generalized extreme value and the generalized Pareto distributions (see Chapter 1), max-stable processes appear to be a sensible choice for modelling pointwise maxima.

Definition 9.2.1. *Let $Z_1, Z_2, \ldots,$ be a sequence of independent copies of a stochastic process $\{Z(x) \colon x \in \mathcal{X}\}$. If for each $n \geqslant 1$ there exist normalizing functions $a_n > 0$ and $b_n \in \mathbb{R}$ such that*

$$\frac{\max_{i=1,\ldots,n} Z_i - b_n}{a_n} \stackrel{\mathrm{d}}{=} Z, \tag{9.1}$$

then $\{Z(x) \colon x \in \mathcal{X}\}$ is said to be max-stable. Recall that equality in distribution for continuous sample path stochastic processes on a compact set means that all finite dimensional distributions are identical.

Based on (9.1) it is not clear why max-stable processes are especially relevant as far as spatial extremes are of concern. Similarly to the univariate case, their use is based on asymptotic arguments.

Theorem 9.2.2. *(de Haan, 1984) Let $Y_1, Y_2, \ldots$ be a sequence of independent copies of a stochastic process $\{Y(x) \colon x \in \mathcal{X}\}$ with continuous sample paths. If there exist* *continuous functions $c_n > 0$ and $d_n \in \mathbb{R}$ such that the limiting process $\{Z(x)\colon x \in \mathcal{X}\}$ defined by*

$$\frac{\max_{i=1,\ldots,n} Y_i(x) - d_n(x)}{c_n(x)} \longrightarrow Z(x), \qquad x \in \mathcal{X}, \quad n \to \infty, \qquad (9.2)$$

is non-degenerate, then the process $\{Z(x)\colon x \in \mathcal{X}\}$ has to be a max-stable process. Note that the convergence in (9.2) refers to weak convergence in the space of continuous functions on $\mathcal{X}$.

Remark 9.2.3. *To be consistent with the univariate extreme value theory, Theorem 9.2.2 implies that the marginal distribution of $\{Z(x)\colon x \in \mathcal{X}\}$ have to be generalized extreme value distributed.*

The statistical motivation for using max-stable processes for modelling spatial extremes is the following. Based on n independent replicates, we will assume that the limiting process $\{Z(x)\colon x \in \mathcal{X}\}$ is likely to be a good candidate for modeling the partial maxima process $\{\max_{i=1,\ldots,n} Y_i(x)\colon x \in \mathcal{X}\}$, as far as n is large enough. The logic beyond this is exactly the same as for univariate extreme value analysis where one prefers to work directly with the asymptotic distribution of block maxima, i.e., the generalized extreme value distribution, rather than estimating the distribution of $\{Y(x)\colon x \in \mathcal{X}\}$ and raising it to the power n to estimate the distribution of the partial maxima.

So far, neither Definition 9.2.1 nor Theorem 9.2.2 give a precise description of max-stable processes, and it would be nice to get a better picture of them as for the univariate and multivariate cases. This description is known as the spectral representation of max-stable processes. Since we know that the marginal distributions of $\{Z(x)\colon x \in \mathcal{X}\}$ have to be generalized extreme value distributed, it is more convenient to set the margins to a given distribution. A widely used choice is to use unit Fréchet margins, i.e., $\Pr\{Z(x) \leqslant z\} = \exp(-1/z)$ for all $x \in \mathcal{X}$ and $z > 0$, and with these specific margins $\{Z(x)\colon x \in \mathcal{X}\}$ is said to be a *simple max-stable process*.

Theorem 9.2.4. *(de Haan, 1984; Penrose, 1992) Any non-degenerate simple max-stable process $\{Z(x)\colon x \in \mathcal{X}\}$ defined on a compact set $\mathcal{X} \subset \mathbb{R}^d$, $d \geqslant 1$, with continuous sample paths satisfies*

$$Z(x) \stackrel{\mathrm{d}}{=} \max_{i \geqslant 1} \zeta_i f_i(x), \qquad x \in \mathcal{X}, \qquad (9.3)$$

where $\{(\zeta_i, f_i)\colon i \geqslant 1\}$ are the points of a Poisson process on $(0, \infty) \times \mathcal{C}$ with intensity $\zeta^{-2} d\zeta \nu(df)$ for some locally finite measure ν defined on the space $\mathcal{C}$ of non-negative continuous functions on $\mathcal{X}$ such that

$$\int f(x)\nu(df) = 1, \qquad x \in \mathcal{X}.$$

Before giving more details on Theorem 9.2.4, some comments are worth mentioning. First the spectral characterization (9.3) is not unique in the sense that different measures ν can lead to the same max-stable process $\{Z(x)\colon x \in \mathcal{X}\}$. Second the

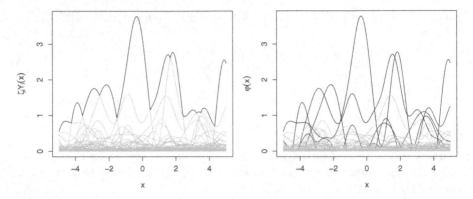

FIGURE 9.1
Simulation of a max-stable process on $\mathcal{X} = [-5, 5]$ from its spectral characterization. For this example we take $Y(x) = \sqrt{2\pi} \max\{0, W(x)\}$ where W is a standard Gaussian process with a Gaussian correlation function, i.e., $\rho(h) = \exp(-h^2)$. Left: The gray curves corresponds to $\{\zeta_i Y_i(x): i = 1, \ldots, 100\}$ and the black one to the pointwise maxima. Right: Decomposition of the spectral functions as extremal (black curves) and non extremal (gray curves) functions.

restriction on non-negative function is only required for convenience and, since the spectral characterization consists in taking pointwise maxima over an infinite number of such functions, one can consider real functions as long as $\nu\{f(x) > 0\} > 0$ for all $x \in \mathcal{X}$. Lastly an important special case of (9.3) is when ν is a probability measure since in that case the spectral representation may be rewritten as

$$Z(x) \overset{\mathrm{d}}{=} \max_{i \geqslant 1} \zeta_i Y_i(x), \qquad x \in \mathcal{X},$$

where $\{\zeta_i: i \geqslant 1\}$ are the points of a Poisson process on $(0, \infty)$, $Y_1, Y_2, \ldots$ a sequence of independent copies of a non-negative stochastic process $\{Y(x): x \in \mathcal{X}\}$ with continuous sample paths and such that $\mathbb{E}\{Y(x)\} = 1$ for all $x \in \mathcal{X}$.

Similarly to the radial / (pseudo) angular decomposition for multivariate extremes, the points $\{\zeta_i: i \geqslant 1\}$ in (9.3) play the role of the radius while the stochastic processes $\{Y_i(x): i \geqslant 1\}$ the role of the angle. The spectral representation suggests a rainfall storm based interpretation due to Smith (1990) which, even though it has no theoretical justification, has the merits of clarifying things to readers not familiar with point processes and stochastic processes. Think about a rainfall storm impacting a region $\mathcal{X}$ which has an overall intensity ζ and spatial extent driven by $\{Y(x): x \in \mathcal{X}\}$, i.e., $\zeta Y(x)$ corresponds to the amount of rain for this storm at location $x \in \mathcal{X}$. With this conceptualization, max-stable processes appear as the pointwise maxima over an infinite number of storms $\{(\zeta_i, \{Y_i(x): x \in \mathcal{X}\}): i \geqslant 1\}$.

The left panel of Figure 9.1 is an illustration of this rainfall storm interpretation where the simulated max-stable process was obtained from 100 storms. We can see that some storms did not contribute to the pointwise maxima at any location $x \in \mathcal{X}$. This is illustrated by the right panel of Figure 9.1 which decomposed the Poisson process $\Phi = \{\varphi_i = \zeta_i Y_i \colon i \geqslant 1\}$ into two sub-point processes

$$\Phi^+ = \{\varphi \in \Phi \colon \exists x \in \mathcal{X}, \ \varphi(x) = Z(x)\},$$
$$\Phi^- = \{\varphi \in \Phi \colon \varphi(x) < Z(x), \ x \in \mathcal{X}\}.$$

According to Dombry et al. (2013) and Dombry and Éyi-Minko (2013), the atoms of the sub-point processes Φ^+ and Φ^- are called the extremal and sub-extremal functions, respectively. We will see later in Section 9.4 that this decomposition of Φ into extremal and sub-extremal functions will be especially convenient for likelihood based inference as well as for deriving the regular conditional distributions of max-stable processes in the related chapter.

9.2.2 Parametric Max-Stable Process Families

In this section we introduce existing parametric max-stable process families and will do so in an historical order. The first model that appeared in the literature is what is called the *Smith process* (Smith, 1990) also sometimes referred to as the *Gaussian extreme value process* (Schlather, 2002). This process is based on a particular mixed moving maxima representation (see Section 10.2.2 for more details), i.e.,

$$Z(x) = \max_{i \geqslant 1} \zeta_i \varphi(x - U_i; 0, \Sigma), \qquad x \in \mathcal{X}, \tag{9.4}$$

where $\{(\zeta_i, U_i) \colon i \geqslant 1\}$ are the points of a Poisson process on $(0, \infty) \times \mathbb{R}^d$ with intensity measure $\zeta^{-2} d\zeta du$ and $\varphi(\cdot; 0, \Sigma)$ denotes the d-variable Gaussian density with mean 0 and covariance matrix Σ. Although this process is interesting for historical reasons, it is rarely useful because of a lack of flexibility — the shape of multivariate Gaussian densities being too restrictive.

The second model was introduced about 10 years after in the seminal paper of Schlather (2002) who introduced what is called the *Schlather process*, also sometimes referred to as the *extremal Gaussian process*. This model is defined by

$$Z(x) = \sqrt{2\pi} \max_{i \geqslant 1} \zeta_i \max\{0, W_i(x)\}, \qquad x \in \mathcal{X}, \tag{9.5}$$

where $\{W_i(x) \colon x \in \mathcal{X}\}$ are independent copies of a stationary Gaussian process with correlation function ρ. Note that the scaling factor $\sqrt{2\pi}$ is required to have $\sqrt{2\pi}\mathbb{E}[\max\{0, W(x)\}] = 1$ for all $x \in \mathcal{X}$.

The third model historically introduced is the *Brown–Resnick process* (Brown and Resnick, 1977) and, having a look at the date of publication, should have been introduced first. But this model was well known to be difficult to work with and, apart from Hüsler and Reiss (1989), no further work was done until that of Kabluchko et al. (2009). This process is defined by

$$Z(x) = \max_{i \geqslant 1} \zeta_i \exp\{W_i(x) - \gamma(x)\}, \qquad x \in \mathcal{X}, \tag{9.6}$$

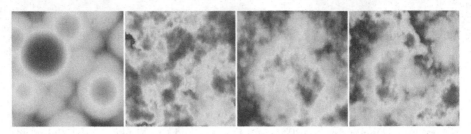

FIGURE 9.2
One realization on a 250×250 grid from each of the max-stable models introduced in
Section 9.2.2. From left to right: Smith where Σ is the identity matrix; Schlather with
$\rho(h) = \exp\{-(h/2)^{1.5}\}$; Brown–Resnick with $\gamma(h) = (h/2)^{1.5}$; and Extremal-t
with $\nu = 5$ and $\rho(h) = \exp\{-(h/4)^{1.5}\}$. Note that for visualization purposes the
margins were transformed to standard Gumbel margins.

where $\{W_i(x) \colon x \in \mathcal{X}\}$ are independent copies of a zero mean Gaussian process
with stationary increments and semi-variogram $\gamma(h) = \mathrm{Var}\{W(x+h) - W(x)\}/2$.
It is interesting to note that the Smith model is a special case of (9.6) with

$$W(x) = x^T \Sigma^{-1} X, \qquad X \sim N(0, \Sigma),$$

and whose semi-variogram therefore satisfies $2\gamma(x) = x^T \Sigma^{-1} \mathrm{Var}(X) \Sigma^{-1} x = x^T \Sigma^{-1} x$.

Finally the last model introduced so far is the extremal–t process. It was first
introduced in a multivariate setting by Nikoloulopoulos et al. (2009), in a spatial
context by Davison et al. (2012) and Ribatet and Sedki (2013). Finally Opitz (2013)
derived a spectral characterization for this process

$$Z(x) = c_\nu \max_{i \geqslant 1} \zeta_i \max\{0, W_i(x)\}^\nu, \qquad x \in \mathcal{X}, \tag{9.7}$$

where $\nu \geqslant 1$, $\{W_i(x) \colon x \in \mathcal{X}\}$ are independent copies of a stationary Gaussian
process with correlation function ρ and

$$c_\nu = \sqrt{\pi} 2^{-(\nu-2)/2} \Gamma\left(\frac{\nu+1}{2}\right)^{-1},$$

where Γ is the Gamma function. As is seen clearly in (9.7), the Schlather process is
a special case of the extremal-t model with $\nu = 1$.

Figure 9.2 plots one realization from each of the max-stable processes introduced
above. As expected we can see that the Smith process produces artificial surfaces,
thus confirming its lack of flexibility already mentioned earlier. The other processes
produce sample surfaces that are much more wiggly — though we set the dependence parameters such that the sample surfaces are smooth enough. Compared to the
Brown–Resnick and the extremal-t processes, we can see that the Schlather process
tends to give larger areas where the largest values occur. We will see later that is a
consequence of this model that does not allow for spatial independence.

9.2.3 Finite Dimensional Distributions

From (9.3), it is possible to derive the finite dimensional distribution of $\{Z(x)\colon x \in \mathcal{X}\}$. More precisely for any $\mathbf{x} = (x_1, \ldots, x_k) \in \mathcal{X}^k$, $k \geqslant 1$, and $\mathbf{z} = (z_1, \ldots, z_k) \in (0, \infty)^k$, we have

$$\Pr\{Z(\mathbf{z}) \leqslant \mathbf{z}\} = \Pr\left[\text{no atom } (\zeta, Y) \in \Phi\colon \zeta Y(x_j) > z_j \text{ for some } j \in \{1, \ldots, k\}\right]$$

$$= \exp\left[-\int_0^\infty \Pr\left\{\zeta > \min_{j=1,\ldots,k} \frac{z_j}{Y(x_j)}\right\} \zeta^{-2} \mathrm{d}\zeta\right]$$

$$= \exp\left\{-V_{\mathbf{x}}(z_1, \ldots, z_k)\right\},$$

where the function

$$V_{\mathbf{x}}(z_1, \ldots, z_k) = \mathbb{E}\left\{\max_{j=1,\ldots,k} \frac{Y(x_j)}{z_j}\right\}, \qquad (9.8)$$

called the *exponent function*, fully characterizes the joint distribution of $Z(\mathbf{x})$.

Working a bit with (9.8), it is straightforward to see that

$$V_{\mathbf{x}}(z, \ldots, z) = \frac{\theta(\mathbf{x})}{z}, \qquad \theta(\mathbf{x}) = \mathbb{E}\left\{\max_{j=1,\ldots,k} Y(x_j)\right\}. \qquad (9.9)$$

In the above equation $\theta(\mathbf{x})$ is called the (k-dimensional) *extremal coefficient* and is a summary measure of dependence across the element of the random vector $Z(\mathbf{x})$. Because of the independence property between the radial and angular components of multivariate extremes, the extremal coefficient is as expected independent of the radius, i.e., the level z appearing in $V_{\mathbf{x}}(z, \ldots, z)$, and focus only on the dependence.

In a spatial context, it is more convenient to restrict our attention to the bivariate case and define what is known as the *extremal coefficient function*, i.e.,

$$\theta\colon h \longmapsto \mathbb{E}\left[\max\left\{Y(x), Y(x+h)\right\}\right]. \qquad (9.10)$$

The extremal coefficient function takes values in $[1, 2]$ where the lower bound corresponds to perfect dependence and the upper one to independence. Indeed for these two limiting cases we have, respectively,

$$\Pr\left\{Z(x+h) \leqslant z \mid Z(x) \leqslant z\right\} = \Pr\{Z(x+h) \leqslant z\}^{\theta(h)-1}$$

$$= \begin{cases} 1, & \text{perfect dependence,} \\ \Pr\{Z(x+h) \leqslant z\}, & \text{independence.} \end{cases}$$

It has to be noted that (9.8) can be difficult to evaluate explicitly when $k > 2$ and this is why the finite dimensional distribution of max-stable processes were first restricted to the bivariate case only. These bivariate distributions are given in Table 9.1. However working a bit with (9.8), one can derive an expression for $V_{\mathbf{x}}$ much easier to work with when $k > 2$.

It is straightforward to see that the exponent function is homogeneous of order -1, i.e.,

$$V_{\mathbf{x}}(cz_1, \ldots, cz_k) = c^{-1} V_{\mathbf{x}}(z_1, \ldots, z_k), \qquad c > 0.$$

TABLE 9.1

Bivariate distributions for each parametric max-stable model introduced in Section 9.2.2. Smith model: $a^2(h) = h^T \Sigma^{-1} h$. Schlather: ρ is the correlation function of $\{W(x): x \in \mathcal{X}\}$ in (9.5). Brown–Resnick: γ is the semi-variogram of $\{W(x): x \in \mathcal{X}\}$ in (9.6). Extremal-t: ρ is the correlation function of $\{W(x): x \in \mathcal{X}\}$ and $b^2 = \{1 - \rho(h)^2\}/(\nu+1)$ in (9.7).

Model	$-\log \Pr\{Z(x) \leqslant z_1, Z(x+h) \leqslant z_2\}$
Smith	$\frac{1}{z_1}\Phi\left\{\frac{a(h)}{2} + \frac{1}{a(h)}\log\frac{z_2}{z_1}\right\} + \frac{1}{z_2}\Phi\left\{\frac{a(h)}{2} + \frac{1}{a(h)}\log\frac{z_1}{z_2}\right\}$
Schlather	$\frac{1}{2}\left(\frac{1}{z_1}+\frac{1}{z_2}\right)\left\{1 + \sqrt{1 - \frac{2\{1+\rho(h)\}z_1 z_2}{(z_1+z_2)^2}}\right\}$
Brown–Resnick	$\frac{1}{z_1}\Phi\left\{\sqrt{\frac{\gamma(h)}{2}} + \sqrt{\frac{1}{2\gamma(h)}}\log\frac{z_2}{z_1}\right\} +$
	$\frac{1}{z_2}\Phi\left\{\sqrt{\frac{\gamma(h)}{2}} + \sqrt{\frac{1}{2\gamma(h)}}\log\frac{z_1}{z_2}\right\}$
Extremal-t	$\frac{1}{z_1}T_{\nu+1}\left\{\frac{1}{b}\left(\frac{z_2}{z_1}\right)^{1/\nu} - \frac{\rho(h)}{b}\right\} + \frac{1}{z_2}T_{\nu+1}\left\{\frac{1}{b}\left(\frac{z_1}{z_2}\right)^{1/\nu} - \frac{\rho(h)}{b}\right\}$

In particular, since $V_{\mathbf{z}}$ is positive and provided it is continuously differentiable, Euler's homogeneous theorem implies that

$$V_{\mathbf{x}} = \sum_{j=1}^{k} z_j^{-1} \Pr\left\{\frac{Y(x_\ell)}{z_\ell} \leqslant \frac{Y(x_j)}{z_j}, \ \ell \neq j\right\}. \tag{9.11}$$

As an example, (9.11) is more tractable than (9.8) for Brown–Resnick processes since in that case we have (Huser and Davison, 2013)

$$\Pr\left\{\frac{Y(x_\ell)}{z_\ell} \leqslant \frac{Y(x_j)}{z_j}, \ \ell \neq j\right\}$$

$$= \Pr\left\{W(x_\ell) - W(x_j) \leqslant \gamma(x_\ell) - \gamma(x_j) + \log\frac{z_\ell}{z_j}, \ \ell \neq j\right\}$$

$$= \Pr\left\{W(x_\ell - x_j) \leqslant \gamma(x_\ell) - \gamma(x_j) + \log\frac{z_\ell}{z_j}, \ \ell \neq j\right\}$$

$$= \Phi\left(\log\frac{z_\ell}{z_j}; \ \ell \neq j; \mu_j, \Sigma_j\right),$$

where the second equality used the fact that W has stationary increments and where $\mu_j = \{\gamma(x_\ell) - \gamma(x_j), \ \ell \neq j\}$, $\Sigma_j = \{\gamma(x_m - x_j) + \gamma(x_n - x_j) - \gamma(x_m - x_n)\}_{m,n \neq j}$ and $\Phi(\cdot; \mu_j, \Sigma_j)$ denotes the $(k-1)$-variate cumulative normal distribution with mean vector μ_j and covariance matrix Σ_j.

Closed forms for the k-variate distribution for the extremal-t can be found using the same lines as for the Brown–Resnick process but by using the conditional distribution instead of the stationary increment property and is left as an exercise.

9.3 Dependence Measure for Spatial Extremes

In this section we investigate how the dependence between extreme events evolves across space. In what we shall call *conventional geostatistics*, to be opposed to what we aim for, i.e., a *geostatistics of extremes*, the underlying statistical model is usually based on a Gaussian process W whose semi variogram

$$\gamma(h) = \frac{1}{2}\mathbb{E}\left[\{W(x) - W(x+h)\}^2\right], \qquad x, h \in \mathcal{X}, \tag{9.12}$$

plays a fundamental role in driving how dependence evolves in space.

Unfortunately the semi-variogram γ is not suitable for spatial extremes that are of interest. For instance since for simple max-stable processes the margins are unit Fréchet, $\mathbb{E}\{Z(x)\} = \infty$ for all $x \in \mathcal{X}$ and the semi-variogram function does not exist. Therefore there is a pressing need to define a summary dependence measure similar to the semi-variogram but devoted to extreme values. Among the various tools introduced, one of the most relevant ones is the F-madogram (Cooley et al., 2006)

$$\nu_F(h) = \frac{1}{2}\mathbb{E}\left[|F\{Z(x+h)\} - F\{Z(x)\}|\right], \qquad x, h \in \mathcal{X}, \tag{9.13}$$

where F denotes the cumulative distribution function of $Z(x)$, $x \in \mathcal{X}$. Contrary to the semi-variogram, the F-madogram is well defined since $F\{Z(x)\} \sim U(0,1)$ and therefore has expectation $1/2$. The F-madogram is particularly convenient for spatial extremes since it has strong connections with the extremal coefficient function. Indeed using the fact that $|a - b| = 2\max(a, b) - a - b$, it is not difficult to show that

$$\theta(h) = \frac{1 + 2\nu_F(h)}{1 - 2\nu_F(h)}, \qquad h \in \mathcal{X}. \tag{9.14}$$

Clearly the F-madogram is easily estimated by its empirical counterpart, i.e.,

$$\widehat{\nu}_F(h) = \frac{1}{2n(n+1)}\sum_{i=1}^{n}|R_i(x) - R_i(x+h)|, \qquad R_i(x) = \sum_{\ell=1}^{n}1_{\{Z_\ell(x)\leqslant Z_i(x)\}}, \tag{9.15}$$

where $Z_1, \ldots, Z_n$ are independent replicates of a max-stable process (not necessarily simple) Z. To reduce sample variability, it is often a good idea to use a binned version of (9.15), i.e., to average $\widehat{\nu}_F$ over suitable classes of distance, i.e.,

$$\tilde{\nu}_F(h) = \frac{1}{2n(n+1)|\mathcal{N}_h|}\sum_{x,y\in\mathcal{N}_h}\sum_{i=1}^{n}|R_i(x) - R_i(y)|,$$

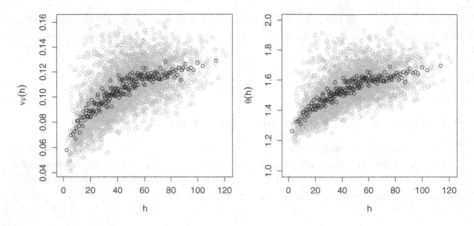

FIGURE 9.3
Empirical F-madogram (left) and corresponding extremal coefficients (right) for the Swiss precipitation data of the `SpatialExtremes` package (Ribatet et al., 2013). The gray points are pairwise estimates while the black ones are binned estimates obtained from 200 bins.

where
$$\mathcal{N}_h = \{x, y \in \{x_1, \dots, x_k\} \colon \big|\|x - y\| - h\big| < \delta\},$$
and some suitable binning radius $\delta > 0$.

Figure 9.3 plots the empirical estimates for the F-madogram and the related pairwise extremal coefficients for the Swiss precipitation data that are freely available from the `SpatialExtremes` package (Ribatet et al., 2013). As expected we can see that the spatial dependence decreases as the distance lag increases but appears to be still persistent beyond 100km.

9.4   Inference

Although theory for max-stable processes was well developed, several decades passed before a suitable framework was found to use max-stable processes for modeling spatial extremes. The main reason for this was the lack of closed form for the likelihood of such processes. To see this recall that the k-variate distribution function is given by

$$\Pr\{Z(x_1) \leqslant z_1, \dots, Z(x_k) \leqslant z_k\} = \exp\{-V_{\mathbf{x}}(z_1, \dots, z_k)\},$$

and hence the related probability density function is

$$f_{\mathbf{x}}(z_1, \ldots, z_k) = \exp\{-V_{\mathbf{x}}(z_1, \ldots, z_k)\} \sum_{\tau \in \mathscr{P}_k} w(\tau), \qquad (9.16)$$

with

$$w(\tau) = (-1)^{|\tau|} \prod_{j=1}^{|\tau|} \frac{\partial^{|\tau_j|}}{\partial z_{\tau_j}} V(z_1, \ldots, z_k),$$

and where $\mathscr{P}_k$ is the set of all possible partitions of the set $\{x_1, \ldots, x_k\}$, $\tau = (\tau_1, \ldots, \tau_\ell)$, $|\tau| = \ell$ is the size of the partition τ and $\partial^{|\tau_j|}/\partial z_{\tau_j}$ denotes the mixed partial derivatives with respect to the elements of the j-th element of the partition τ. As emphasized by Dombry et al. (2013) and Ribatet (2013), the number of possible partitions of $\{x_1, \ldots, x_k\}$ corresponds to Bell numbers and hence yield a combinatorial explosion even for moderate values of k — when $k = 10$ there is around 115,000 partitions.

9.4.1 Pairwise Likelihood

Due to this computational burden, the maximum likelihood estimator cannot be used and other approaches have been proposed. A first attempt was to use least squares to estimate the dependence parameters from the empirical extremal coefficient estimates (Smith, 1990) — using the F-madogram for instance. This approach is not completely satisfactory since it focuses only on the dependence parameters and hence prediction at unobserved locations is not possible. In some sense a likelihood based approach was missing and this is exactly what was proposed with the work of Padoan et al. (2010). Padoan et al. (2010) propose to use the maximum pairwise likelihood estimator, i.e., maximizing

$$\ell_{\mathrm{p}}\{\psi; z_1(\mathbf{x}), \ldots, z_n(\mathbf{x})\} = \sum_{\ell=1}^{n} \sum_{i=1}^{k-1} \sum_{j=i+1}^{k} w_{i,j} \log f\{z_\ell(x_i), z_\ell(x_j); \psi\}, \qquad (9.17)$$

where ψ are the parameter to be estimated and $\{w_{i,j}, 1 \leqslant i < j \leqslant k\}$ are suitable weights that can be used either to improve efficiency or to reduce the computational cost.

Composite likelihoods belong to *mis-specified models* or more accurately *under-specified models* as we typically assume that the bivariate densities appearing in (9.17) are correct.

Remark 9.4.1. *It is important to remember that composite likelihoods are not genuine likelihoods nor do they give an approximation of the likelihood. Composite likelihoods provide a framework to define estimators based on unbiased estimating equations.*

As any maximum composite likelihood estimator, and under mild regularity conditions, the maximum pairwise likelihood estimator shares similar properties to that of the maximum likelihood estimator. More precisely, it is consistent, asymptotically normal (e.g., Padoan et al., 2010).

The advantage of a likelihood based approach is that it is straightforward to extend it to more complex statistical models. For instance so far we restricted our attention to simple max-stable processes, i.e., with unit Fréchet margins, and this assumption is clearly unrealistic for concrete applications where it is expected that the intensity of extreme events may vary spatially. Up to slight modification of (9.17), one can allow for varying marginal parameters. More precisely let $\{\tilde{Z}(x),\ x \in \mathcal{X}\}$ be a max-stable process such that $\tilde{Z}(x) \sim \text{GEV}\{\mu(x), \sigma(x), \xi(x)\}$ for all $x \in \mathcal{X}$ and define the mapping

$$t_x : \tilde{z} \longmapsto \left\{1 + \xi(x)\frac{\tilde{z} - \mu(x)}{\sigma(x)}\right\}^{1/\xi(x)}.$$

Then the pointwise transformed stochastic process $\{t_x\{\tilde{Z}(x)\},\ x \in \mathcal{X}\}$ is a simple max-stable process and the log-pairwise likelihood for $\tilde{Z}$ is therefore

$$\ell_p\{\psi; \tilde{z}_1(\mathbf{x}), \dots, \tilde{z}_n(\mathbf{x})\} = \ell_p\{\psi; t_{\mathbf{x}}\{\tilde{z}_1(\mathbf{x})\}, \dots, t_{\mathbf{x}}\{\tilde{z}_n(\mathbf{x})\}\}+$$

$$\sum_{\ell=1}^{n}\sum_{i=1}^{k-1}\sum_{j=i+1}^{k} \log |J\{\tilde{z}_\ell(x_i)\}J\{\tilde{z}_\ell(x_i)\}|,$$

where the additional term corresponds to the logarithm of the Jacobian coming from the mapping t_x.

To gain efficiency some authors try to use the triplet-wise likelihood in place of the pairwise likelihood (Genton et al., 2011; Huser and Davison, 2013) but simulation studies indicate that the gain in efficiency is small compared to the large computational increase unless the spatial process is really smooth — which is unfortunately not the case for many environmental processes.

9.4.2   Full Likelihood

Recently some authors tried to perform full likelihood inferences for max-stable processes (Wadsworth and Tawn, 2014) whose framework relies on the derivation of the conditional distributions of max-stable processes given by Dombry and Éyi-Minko (2013) and Dombry et al. (2013) — see Chapter 11 for more details. Although (9.17) appears rather complicated it is possible to get a better understanding of this expression in terms of the extremal and non-extremal point processes Φ^+ and Φ^- introduced in Section 9.2.1.

Having observed one single realization of $\{Z(x) : x \in \mathcal{X}\}$ at locations $x_1, \dots, x_k \in \mathcal{X}$, we have $|\Phi^+| \in \{1, \dots, k\}$ since a single extremal function might contribute to $\{Z(x_1), \dots, Z(x_k)\}$ at more than one location. It thus defines a random partition θ of the set $\{x_1, \dots, x_k\}$ and using the terminology of Wang and Stoev (2011) and Dombry and Éyi-Minko (2013), each of these partitions defines a *hitting scenario*. Using the conditional independence property of the point processes Φ^-

and Φ^+ and that of the extremal functions (Dombry and Éyi-Minko, 2013), the contribution to the full likelihood of a single realization from any simple max-stable processes with spectral characterization (9.3) is

$$\exp\left\{-\int_{\{z(\mathbf{x}),\infty\}} \lambda_{\mathbf{x}}(\mathbf{u})d\mathbf{u}\right\} \tag{9.18}$$

$$\cdot \sum_{\tau\in\mathscr{P}_k} \prod_{j=1}^{|\tau|} \lambda_{\mathbf{x}_{\tau_j}}(\mathbf{x}_{\tau_j}) \int_{\{0,z(\mathbf{x}_{\tau_j^c})\}} \lambda_{\mathbf{x}_{-\tau_j}|z(\mathbf{x}_{\tau_j}),\mathbf{x}_{\tau_j}}(\mathbf{u})d\mathbf{u},$$

where $\mathbf{x}_{\tau_j} = \{x \in \{x_1,\ldots,x_k\}: x \in \tau_j\}$, $\mathbf{x}_{\tau_j^c} = \{x \in \{x_1,\ldots,x_k\}: x \notin \tau_j\}$, $\lambda_{\mathbf{x}}$ is the intensity function of the Poisson point process $\Phi_{\mathbf{x}} = \{\varphi(\mathbf{x}) \in (0,\infty)^k: \varphi \in \Phi\}$, i.e., for any Borel set $A \subset (0,\infty)^k$,

$$\Lambda_{\mathbf{x}}(A) = \int_0^\infty \Pr\{\zeta Y(\mathbf{x}) \in A\}\zeta^{-2}d\zeta = \int_A \lambda_{\mathbf{x}}(\mathbf{u})d\mathbf{u},$$

and where $\lambda_{x_1|\mathbf{z},x_2}(\mathbf{u})$, $x_1, x_2 \in \mathcal{X}$, $\mathbf{z}, \mathbf{u} > 0$, is the conditional intensity function, i.e.,

$$\lambda_{x_1|\mathbf{z},x_2}(\mathbf{u}) = \frac{\lambda_{(x_1,x_2)}\{(\mathbf{u},\mathbf{z})\}}{\lambda_{x_2}(\mathbf{z})}.$$

Dombry et al. (2013) and Ribatet (2013) give closed forms for the intensity and conditional intensity functions for the max-stable processes introduced in Section 9.2.2.

As expected (9.16) and (9.18) share the same structure and we now understand from the point process theory that

$$f_{\mathbf{x}}(z_1,\ldots,z_k) = \underbrace{\exp\{-V_{\mathbf{x}}(z_1,\ldots,z_k)\}}_{\substack{\text{contribution of the} \\ \text{sub-extremal functions}}} \times \underbrace{\sum_{\tau\in\mathscr{P}_k} w(\tau).}_{\substack{\text{contribution of each} \\ \text{hitting scenario}}}$$

Apart from getting a better understanding of (9.16), (9.18) is still not tractable and maximizing the log-likelihood appears to be numerically impossible. However if the hitting scenario is known to be $\tau = (\tau_1,\ldots,\tau_\ell)$, then the contribution to the full likelihood is much more tractable and becomes

$$\exp\left\{-\int_{\{z(\mathbf{x}),\infty\}} \lambda_{\mathbf{x}}(\mathbf{u})d\mathbf{u}\right\} \prod_{j=1}^{\ell} \lambda_{\mathbf{x}_{\tau_j}}(\mathbf{x}_{\tau_j}) \int_{\{0,z(\mathbf{x}_{\tau_j^c})\}} \lambda_{\mathbf{x}_{\tau_j^c}|z(\mathbf{x}_{\tau_j}),\mathbf{x}_{\tau_j}}(\mathbf{u})d\mathbf{u}.$$

This is what Wadsworth and Tawn (2014) suggest since when daily observations are available, one knows which partition yields to the annual maxima.

9.5   Discussion

The aim of this chapter was to introduce the basic foundations for using max-stable processes for modelling spatial extremes. Although max-stable processes are asymptotically justified models, other modelling strategies are possible and we will cover them briefly in this discussion.

Probably one of the most famous competitors to max-stable processes are latent variable models. These models rely on a univariate extreme value argument only and assume that

$$\tilde{Z}(x) \mid \mu(x), \sigma(x), \xi(x) \sim \text{GEV}\{\mu(x), \sigma(x), \xi(x)\},$$

i.e., for each location $x \in \mathcal{X}$, the random variable $Y(x)$ has a generalized extreme value distribution with location, scale, and shape parameters equal to $\{\mu(x), \sigma(x), \xi(x)\}$. Typically a *conditional independence* assumption is supposed, i.e., conditionally on the marginal parameters, $\tilde{Z}(x_1)$ is independent of $\tilde{Z}(x_2)$ for any $x_1, x_2 \in \mathcal{X}^2$. As $\{\mu(x), \sigma(x), \xi(x)\}$ are allowed to vary in space, it is typically assumed that $[\{\mu(x), \sigma(x), \xi(x)\}\colon x \in \mathcal{X}\}]$ is a Gaussian process whose mean function depends on some relevant covariates and whose covariance function is chosen among available parametric covariance function families, e.g., Whittle–Matern. Since the likelihood of this model involves intractable integrals, to bypass this hurdle one often has to resort to the Bayesian paradigm combined with Markov chain Monte Carlo algorithms — see Davison et al. (2012) for more details. The main drawback of this model is that, because of the conditional independence assumption, the spatial dependence is completely ignored and that, for any $x \in \mathcal{X}$, $\tilde{Z}(x)$ is not extreme value distributed anymore. However in practice if the aim of the study is to characterize the pointwise distribution of extremes then it is so far probably one of the best approaches.

To take into account the spatial dependence, one could think of using copulas (Sang and Gelfand, 2009). For instance one could use the Gaussian copula, i.e., having observed $\{\tilde{Z}(x)\colon x \in \mathcal{X}\}$ at locations $x_1, \ldots, x_k \in \mathcal{X}$, we have

$$\Pr\{\tilde{Z}(x_1) \leqslant \tilde{z}_1, \ldots, \tilde{Z}(x_k) \leqslant \tilde{z}_k\} = \Phi_\rho \left\{ \Phi^{-1}(u_1), \ldots, \Phi^{-1}(u_k) \right\},$$

where Φ_ρ is the multivariate normal distribution with zero mean and correlation matrix $\{\rho(x_i - x_j)\colon i, j = 1, \ldots, k\}$, ρ a parametric correlation function, Φ^{-1} the quantile function of a standard normal distribution and

$$u_j = \exp\left\{ -\left(1 + \xi(x_j)\frac{\tilde{z}_j - \mu(x_j)}{\sigma(x_j)}\right)^{-1/\xi(x_j)} \right\}, \qquad 1 + \xi(x_j)\frac{y_j - \mu(x_j)}{\sigma(x_j)} > 0,$$

for $j = 1, \ldots, k$. As for the latent variable model, one typically assumes that the marginal parameters $[\{\mu(x), \sigma(x), \xi(x)\}\colon x \in \mathcal{X}]$ are Gaussian processes.

Although this type of model might seem relevant, they have the same weaknesses as the use of copula for multivariate extremes: most often this modelling strategy is not able to capture the spatial dependence of extremes (Davison et al., 2012). One exception needs to be mentioned though: the use of extreme value copulas but in that case it is equivalent to the use of max-stable processes.

Finally one could be tempted to use asymptotic independent models for modelling spatial extremes. If from a strict asymptotic point of view such models might seem irrelevant as, in the tails, spatial extremes will behave as a pure noise process with generalized extreme valued margins, they possess interesting properties that might be true for environmental processes and that max-stable processes cannot have (Davison et al., 2013; Wadsworth and Tawn, 2012). More precisely as shown by (9.9) the dependence structure of max-stable process is independent of the level z. Asymptotically independent models allow that the spatial dependence structure becomes increasingly weaker as we go far in the tail.

References

Brown, B. M. and S. I. Resnick (1977). Extreme values of independent stochastic processes. *Journal of Applied Probability 14*, 732–739.

Cooley, D., P. Naveau, and P. Poncet (2006). Variograms for spatial max-stable random fields. In P. Bertail, P. Soulier, P. Doukhan, P. Bickel, P. Diggle, S. Fienberg, U. Gather, I. Olkin, and S. Zeger (Eds.), *Dependence in Probability and Statistics*, Volume 187 of *Lecture Notes in Statistics*, pp. 373–390. Springer New York.

Davison, A., R. Huser, and E. Thibaud (2013). Geostatistics of dependent and asymptotically independent extremes. *Mathematical Geosciences 45*(5), 511–529.

Davison, A., S. Padoan, and M. Ribatet (2012). Statistical modelling of spatial extremes. *Statistical Science 7*(2), 161–186.

de Haan, L. (1984). A spectral representation for max-stable processes. *The Annals of Probability 12*(4), 1194–1204.

Dombry, C. and F. Éyi-Minko (2013). Regular conditional distributions of max infinitely divisible random fields. *Electronic Journal of Probability 18*(7), 1–21.

Dombry, C., F. Éyi-Minko, and M. Ribatet (2013). Conditional simulations of max-stable processes. *Biometrika 100*(1), 111–124.

Genton, M. G., Y. Ma, and H. Sang (2011). On the likelihood function of Gaussian max-stable processes. *Biometrika 98*(2), 481–488.

Huser, R. and A. C. Davison (2013). Composite likelihood estimation for the Brown–Resnick process. *Biometrika 100*(2), 511–518.

Hüsler, J. and R.-D. Reiss (1989, February). Maxima of normal random vectors: Between independence and complete dependence. *Statistics & Probability Letters 7*(4), 283–286.

Kabluchko, Z., M. Schlather, and L. de Haan (2009). Stationary max-stable fields associated to negative definite functions. *Annals of Probability 37*(5), 2042–2065.

Nikoloulopoulos, A., H. Joe, and H. Li (2009). Extreme values properties of multivariate t copulas. *Extremes 12*, 129–148.

Opitz, T. (2013). Extremal-t process: Elliptical domain of attraction and a spectral representation. *Journal of Multivariate Analysis 122*, 409–413.

Padoan, S., M. Ribatet, and S. Sisson (2010). Likelihood-based inference for max-stable processes. *Journal of the American Statistical Association (Theory & Methods) 105*(489), 263–277.

Penrose, M. D. (1992). Semi-min-stable processes. *Annals of Probability 20*(3), 1450–1463.

Ribatet, M. (2013). Spatial extremes: Max-stable processes at work. *Journal de la Société Française de Statistique 154*(2), 156–177.

Ribatet, M. and M. Sedki (2013). Extreme value copulas and max-stable processes. *Journal de la Société Française de Statistique 154*(1), 138–150.

Ribatet, M., R. Singleton, and R. C. Team (2013). *SpatialExtremes: Modelling Spatial Extremes*. R package version 2.0-1.

Sang, H. and A. Gelfand (2009). Hierarchical modeling for extreme values observed over space and time. *Environmental and Ecological Statistics 16*(3), 407–426.

Schlather, M. (2002, March). Models for stationary max-stable random fields. *Extremes 5*(1), 33–44.

Smith, R. L. (1990). Max-stable processes and spatial extreme. Unpublished manuscript.

Wadsworth, J. L. and J. A. Tawn (2012). Dependence modelling for spatial extremes. *Biometrika 99*(2), 253–272.

Wadsworth, J. L. and J. A. Tawn (2014). Efficient inference for spatial extreme value processes associated to log-gaussian random functions. *Biometrika 101*(1), 1–15.

Wang, Y. and S. A. Stoev (2011). Conditional sampling for spectrally discrete max-stable random fields. *Advances in Applied Probability 443*, 461–483.

10

Simulation of Max-Stable Processes

Marco Oesting
INRA / AgroParisTech, France

Mathieu Ribatet
Université Montpellier 2 / University of Lyon 1, France

Clément Dombry
Université de Franche-Comté, France

Abstract

Simulation of stochastic phenomena has become an important tool to assess characteristics which are analytically intractable. In this chapter, we deal with the problem of sampling from max-stable processes, which, by the spectral representation (de Haan, 1984), require taking the pointwise maxima over an infinite number of spectral functions belonging to a Poisson point process. In practice, spectral functions are simulated in an appropriate order until some stopping rule takes effect. For mixed moving maxima processes with bounded shape functions with joint compact support, Schlather (2002) provides such a stopping criterion yielding an exact simulation. Oesting et al. (2013) consider stopping rules for a family of equivalent spectral representations in a very general setting, particularly focusing on the normalized spectral representation which allows for an exact simulation, as well. Although this representation exists under mild conditions, the distribution of the corresponding spectral functions might be inappropriate for sampling. In this case, approximative procedures are proposed. Here, the choice of the stopping rule and the spectral representation is crucial for the quality of approximation. In this context, we discuss measures of simulation efficiency and quality. Several examples of max-stable processes are analyzed. We particularly focus on the challenging case of Brown–Resnick processes which are stationary although originally constructed via nonstationary spectral functions. Finally, we review existing R packages on the simulation of max-stable processes.

10.1   Introduction

Max-stable processes have become frequently used models for spatial extremes. However, due to their rather sophisticated structure, in many cases, analytical expressions are available for the bivariate (or sometimes also trivariate) distributions of the process only. Thus, in order to assess some more advanced characteristics of the process which are of interest in practical applications, stochastic simulation is used. Further, the problem of simulation of max-stable processes is closely related to the questions of conditional simulation and prediction of max-stable processes given observations at some locations (see Chapter 11 for details).

In the following, we will deal with the problem of simulation of a sample-continuous max-stable process Z on some compact domain $\mathcal{X} \subset \mathbb{R}^d$. Without loss of generality, we may assume that Z has unit Fréchet margins. Realizations of a process with arbitrary generalized extreme value marginal distributions can be obtained by marginal transformations.

Recall that process Z allows for a spectral representation (cf. de Haan, 1984; Giné et al., 1990; Penrose, 1992) of the form

$$Z(x) = \max_{i \geqslant 1} \zeta_i \psi_i(x), \quad x \in \mathcal{X}, \tag{10.1}$$

where $\{(\zeta_i, \psi_i)\}_{i \geqslant 1}$ is a Poisson point process on $(0, \infty) \times \mathcal{C}$ with intensity measure $\zeta^{-2} \, d\zeta \times \nu(d\psi)$ for some locally finite measure ν on the space $\mathcal{C} = C(\mathcal{X}, [0, \infty))$ of continuous non-negative functions on $\mathcal{X}$ such that

$$\int \psi(x) \, \nu(d\psi) = 1, \quad x \in \mathcal{X}. \tag{10.2}$$

The aim of this chapter is the simulation of the process Z on $\mathcal{X}$. According to representation (10.1), Z is constructed as the pointwise maximum of an infinite number of functions. Thus, exact simulation of Z is in general not straightforward. In many cases, the Poisson points $\{(\zeta_i, \psi_i)\}_{i \geqslant 1}$ can be subsequently simulated such that $\zeta_{i+1} \leqslant \zeta_i$ for all $i \geqslant 1$ (see Section 10.2). This motivates the strategy of approximating Z by the maximum over the "first" N points where the finite number N is given by some stopping criterion.

Although this procedure might be quite efficient in some cases, there are examples where this straightforward approach also encounters severe problems. Such examples contain the important class of Brown–Resnick processes (Kabluchko et al., 2009) where the spectral measure ν in representation (10.1) is the probability measure of the stochastic process $Y(t) = \exp(W(t) - \sigma^2(t)/2)$. Here, $W(\cdot)$ is a centered Gaussian process with stationary increments, semi-variogram γ and variance $\sigma(\cdot)$. Thus, the Brown–Resnick process Z associated to the semi-variogram γ can be written as

$$Z(t) = \max_{i \geqslant 1} \zeta_i \exp(W_i(t) - \sigma^2(t)/2), \quad t \in \mathbb{R}^d, \tag{10.3}$$

where $\{\zeta_i\}$ are the points of a Poisson point process with intensity $\zeta^{-2} \, d\zeta$ and W_i, $i \geqslant 1$, are independent copies of W.

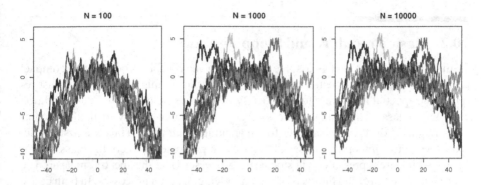

FIGURE 10.1
Ten approximations of $\log Z$ on the interval $[-50, 50]$ by taking the maximum over the first N spectral functions with $N = 100, 1000, 10000$ (from left to right). Here, Z is a Brown–Resnick process associated to the semi-variogram $\gamma(h) = \frac{1}{2}|h|$.

Then, the max-stable process Z is stationary although the spectral process Y is not. Thus, a challenge for an accurate simulation of Z is the visualization of stationarity. However, as Figure 10.1 shows for the example of a Brown–Resnick process associated to the semi-variogram $\gamma(h) = \frac{1}{2}|h|$, considered first by Brown and Resnick (1977), finite approximations obtained in the above way may depict clear trends even for a large number N of considered spectral functions. Thus, in this case, a more sophisticated procedure is needed for an accurate approximation. This motivates a careful analysis of different cases.

The chapter is organized as follows. In Section 10.2, we present the basic procedure of simulation. We focus on two important classes of spectral representations — the representation via stochastic processes and the mixed moving maxima representation — and consider subclasses which allow for an exact simulation by Schlather's (2002) algorithm. Section 10.3 deals with transformations that allow to switch between different, but equivalent spectral representations and the according simulation procedures. Here, we place particular emphasis on the so-called normalized spectral representation which enables exact simulation. Further, we provide a measure for efficiency to compare different exact simulation procedures. Due to the complexity of exact simulation, it may be worthwhile to consider approximative algorithms. Error estimates are given in Section 10.4. In Section 10.5, we discuss accurate simulation for several examples including some of the most commonly used classes of max-stable processes. Finally, we present an overview over R packages on the simulation of max-stable processes in Section 10.6.

10.2   Basic Procedure and Stopping Rules

In the following, we always assume that Z is a max-stable process on a compact domain $\mathcal{X} \subset \mathbb{R}^d$ with spectral representation (10.1). As this representation just requires the spectral measure ν to be a locally finite measure on $\mathcal{C}$ satisfying condition (10.2), the class of max-stable processes contains a large variety of models potentially requiring different procedures for an accurate simulation. Thus, we first restrict ourselves to two important classes of max-stable processes, namely the class of processes which are represented via stochastic processes (i.e., the spectral measure is a probability measure) and the class of mixed moving maxima processes. In both cases, we consider subclasses that allow for an exact simulation in finite time by algorithms devised by Schlather (2002). In Section 10.3, we will see how transformations of the spectral measure allow us to reduce simulation problems to these cases. Thus, the procedures presented in this section also prove useful in a more general setting.

10.2.1   Representation via Stochastic Processes

First, we consider the case that ν is a probability measure. Then, the corresponding process Z possesses a Penrose (1992)-type representation:

$$Z(x) = \max_{i \geqslant 1} \zeta_i Y_i(x), \qquad x \in \mathcal{X}, \tag{10.4}$$

where $\{\zeta_i\}_{i \geqslant 1}$ is a Poisson point process on $(0, \infty)$ with intensity measure $\zeta^{-2} d\zeta$ and Y_i, $i \geqslant 1$, are independently distributed according to the probability measure ν. According to condition (10.2), the processes Y_i necessarily satisfy $\mathbb{E} Y_i(x) = 1$ for all $x \in \mathcal{X}$. As the spectral functions Y_i allow for an interpretation as independent stochastic processes, we call the spectral representation (10.4) a *representation via stochastic processes*. Examples of max-stable processes with such a representation are Brown–Resnick processes (Kabluchko et al., 2009) where Y_i is a log-Gaussian process or extremal t processes (Opitz, 2013) where Y_i is the positive part of a Gaussian process to some positive power α, including the particular case of an extremal Gaussian process (Schlather, 2002) with $\alpha = 1$.

For simulation of the Poisson point process $\{\zeta_i\}_{i \geqslant 1}$, we note that, by the mapping theorem, $\{\zeta_i^{-1}\}_{i \geqslant 1}$ is a standard Poisson point process on $(0, \infty)$, i.e., $\zeta_1^{-1}, \zeta_2^{-1} - \zeta_1^{-1}, \zeta_3^{-1} - \zeta_2^{-1}, \ldots$ are independent and standard exponentially distributed. Thus, for Z as defined in (10.4), we obtain the equality

$$Z(x) =_d \max_{i \geqslant 1} \frac{Y_i(x)}{\sum_{j=1}^i E_j}, \qquad x \in \mathcal{X}, \tag{10.5}$$

for $E_j \sim_{i.i.d.} \mathrm{Exp}(1)$, $j \geqslant 1$, where $\mathrm{Exp}(\lambda)$ denotes the exponential distribution with parameter $\lambda > 0$ and $=_d$ denotes equality in distribution.

In the case that the spectral processes Y_i, $i \geqslant 1$, are a.s. bounded, that is,

$$\sup_{x \in \mathcal{X}} Y_i(x) \leqslant C \quad a.s. \tag{10.6}$$

Algorithm 10.1 Simulation of a max-stable process (10.4) with spectral function bounded by C.

Set $Z(x) = 0$, $x \in \mathcal{X}$, and simulate $\zeta^{-1} \sim \text{Exp}(1)$.
While $((\zeta C > \inf_{x \in \mathcal{X}} Z(x))$ {
    Simulate $Y \sim \nu$.
    Update $Z(x)$ by $\max\{Z(x), \zeta Y(x)\}$ for all $x \in \mathcal{X}$.
    Simulate $E \sim \text{Exp}(1)$ and update ζ^{-1} by $\zeta^{-1} + E$.
}
Return Z.

for some constant $C > 0$, the product $(\sum_{j=1}^{k+1} E_j)^{-1} Y_{k+1}$ cannot contribute to the pointwise maximum in (10.5) if

$$\frac{C}{\sum_{j=1}^{k+1} E_j} \leq \inf_{x \in \mathcal{X}} \left(\max_{i \geq 1} \frac{Y_i(x)}{\sum_{j=1}^{i} E_j} \right). \tag{10.7}$$

(cf. Oesting et al., 2013; Schlather, 2002). As the points $\{(\sum_{j=1}^{i} E_j)^{-1}\}_{i \geq 1}$ are in a descending order, this fact allows us to write the process Z as the pointwise maximum over a finite number of functions.

Proposition 10.2.1. *(analogous to Schlather, 2002, Thm. 4) Let $E_i \sim_{i.i.d.} \text{Exp}(1)$, $i \geq 1$, and, independently of $\{E_i\}_{i \geq 1}$, let Y_i, $i \geq 1$, be independent copies of a stochastic process Y satisfying (10.6) for some $C > 0$. Then,*

$$\max_{i=1}^{N} \left(\left(\sum_{j=1}^{i} E_j \right)^{-1} Y_i(x) \right) = \max_{i \geq 1} \left(\left(\sum_{j=1}^{i} E_j \right)^{-1} Y_i(x) \right), \qquad x \in \mathcal{X},$$

where N is an a.s. finite random number N, defined by

$$N = \min \left\{ k \geq 1 : \frac{C}{\sum_{j=1}^{k+1} E_j} \leq \inf_{x \in \mathcal{X}} \left(\max_{i=1}^{k} \frac{Y_i(x)}{\sum_{j=1}^{i} E_j} \right) \right\}.$$

In particular, the process Z given by (10.4) satisfies $Z(\cdot) =_d \max_{i=1}^{N} \frac{Y_i(\cdot)}{\sum_{j=1}^{i} E_j}$.

Thus, the process Z can be simulated exactly in finite time by Algorithm 10.1.

For many examples of processes represented via stochastic processes, condition (10.6) is not satisfied. In this case, Schlather (2002) proposes to approximate the process by choosing some constant C^* and applying Algorithm 10.1 with $C = C^*$. For an accurate approximation, C^* should be chosen large enough such that $\mathbb{P}(\sup_{x \in \mathcal{X}} Y(x) > C^*)$ is small. However, the larger C^* is, the more iterations of the algorithm are needed until the stopping criterion is met. Thus, the choice of C^* is a trade-off between simulation accuracy and running time. We will deal with this question in more detail in Section 10.4.

One way to handle the difficulties associated with the choice of an upper bound C^* might be the choice of a different but stochastically equivalent spectral representation, that is, a spectral representation that yields a max-stable process with the same distribution. Note that, by Corollary 9.4.5 in de Haan and Ferreira (2006), any sample-continuous max-stable process possesses a spectral representation satisfying $\sup_{x \in \mathcal{X}} Y(x) = c$ a.s. for some positive constant c. In particular, condition (10.6) is met and, thus, the process can be simulated exactly using the above algorithm. We will discuss the construction of such a representation and its effects on simulation in Section 10.3.

10.2.2   Mixed Moving Maxima Processes

As a second class of max-stable processes, we consider the class of mixed moving maxima processes on $\mathbb{R}^d$ (see Schlather, 2002; Stoev and Taqqu, 2005, for example). Here, let $\{(\zeta_i, S_i)\}_{i \geqslant 1}$ be the points of a Poisson point process on $(0, \infty) \times \mathbb{R}^d$ with intensity measure $\zeta^{-2} \mathrm{d}\zeta \times \mathrm{d}s$. Independently, let $F_i \sim_{i.i.d.} F$, $i \geqslant 1$, for some nonnegative random function F on $\mathbb{R}^d$ satisfying $\mathbb{E}\left(\int_{\mathbb{R}^d} F(x)\, \mathrm{d}x\right) = 1$. Then, the mixed moving maxima process

$$Z(x) = \max_{i \geqslant 1} \zeta_i F_i(x - S_i), \quad x \in \mathcal{X}, \tag{10.8}$$

is a stationary max-stable process with standard Fréchet margins. Note that the class of mixed moving maxima processes contains the processes considered by Smith (1990) with F being a deterministic function, for instance, a Gaussian density function. The work of Smith (1990) also provides an interpretation of mixed moving maxima as models for storms, where ζ_i is perceived as the strength, S_i as the center and F_i as the shape of a storm.

In contrast to the representation via stochastic processes considered in Subsection 10.2.1, here, the associated spectral measure ν, given by

$$\nu(A) = \int_{\mathbb{R}^d} \mathbb{P}(F(\cdot - s) \in A)\, \mathrm{d}s, \quad A \subset \mathcal{C} \text{ Borel},$$

is infinite. Due to this fact, the collection $\{\zeta_i\}_{i \geqslant 1}$ of first components of the points $\{(\zeta_i, S_i)\}_{i \geqslant 1}$ cannot be simulated in a descending order as the number of ζ_i, $i \geqslant 1$, with $a \leqslant \zeta_i \leqslant b$ is infinite for all $0 < a < b$. However, this number becomes finite if we restrict ourselves to those points whose second component is in some compact set.

Note that such a restriction can be made without affecting the corresponding max-stable process Z if the shape functions are a.s. supported in some compact domain only, i.e., if we have

$$\mathbb{P}(F(x) = 0 \text{ for all } x \in \mathbb{R}^d \setminus b(o, R)) = 1 \tag{10.9}$$

for some $R > 0$ where $b(o, R)$ denotes a d-dimensional ball around the origin with radius R. In this case, the process $\zeta_i F_i(\cdot - S_i)$ cannot contribute to the maximum $\{Z(X) : x \in \mathcal{X}\}$ if $S_i \notin K \oplus b(o, R)$ where $A \oplus B = \{a + b : a \in A, b \in B\}$ for $A, B \subset \mathbb{R}^d$. Thus, instead of considering all points $\{(\zeta_i, S_i)\}_{i \geqslant 1}$, the points $\{(\zeta_i, S_i) : S_i \in \mathcal{X} \oplus b(o, R)\}_{i \geqslant 1}$, i.e., the points of the Poisson point process restricted to $(0, \infty) \times (\mathcal{X} \oplus b(o, R))$, are sufficient to simulate the process

---

Algorithm 10.2 Simulation of a max-stable process (10.8) with shape function supported within $b(0, R)$ and bounded by C.

Set $Z(x) = 0$, $x \in \mathcal{X}$, and simulate $\zeta^{-1} \sim \text{Exp}(1)$.
While $(\zeta C > \inf_{x \in \mathcal{X}} Z(x)/|\mathcal{X} \oplus b(o, R)|)$ {
　　Simulate $F \sim \mathbb{P}(F \in \cdot)$ and $S \sim \text{Unif}(\mathcal{X} \oplus b(o, R))$.
　　Update $Z(x)$ by $\max\{Z(x), |\mathcal{X} \oplus b(o, R)| \cdot \zeta \cdot F(x - S)\}$ for all $x \in \mathcal{X}$.
　　Simulate $E \sim \text{Exp}(1)$ and update ζ^{-1} by $\zeta^{-1} + E$.
}
Return Z.

$\{Z(x), \ x \in \mathcal{X}\}$. These points can be simulated in the following way (cf. Schlather, 2002, Lemma 3): Let $E_j \sim_{i.i.d.} \text{Exp}(1)$, $j \geqslant 1$, and, independently of $\{E_j\}_{j \geqslant 1}$, let S_i, $i \geqslant 1$, be independent random variables uniformly distributed on $\mathcal{X} \oplus b(o, R)$, denoted by $S_i \sim_{i.i.d.} \text{Unif}(\mathcal{X} \oplus b(o, R))$. Then, the points

$$\left\{ \left(|\mathcal{X} \oplus b(o, R)| \left(\textstyle\sum_{j=1}^{i} E_i \right)^{-1}, S_i \right) \right\}_{i \geqslant 1}$$

form a Poisson point process on $(0, \infty) \times (\mathcal{X} \oplus b(o, R))$ with the desired intensity measure, where $|\cdot|$ denotes the Lebesgue measure. Thus, the mixed moving maxima process (10.8) can be written as the maximum over a finite number of functions provided that the shape function F is bounded and satisfies (10.9).

Proposition 10.2.2. *(Schlather, 2002, Thm. 4) Let F_i, $i \geqslant 1$, be independent copies of some stochastic process F that satisfies (10.9) for some $R > 0$ and*

$$\sup_{x \in \mathbb{R}^d} F(x) \leqslant C \quad a.s. \tag{10.10}$$

for some $C > 0$. Independently of $\{F_i\}_{i \geqslant 1}$, let $E_i \sim_{i.i.d.} \text{Exp}(1)$, $i \geqslant 1$, and $S_i \sim_{i.i.d.} \text{Unif}(\mathcal{X} \oplus b(o, R))$ be independent sequences. Then, the mixed moving maxima process $\{Z(x), \ x \in \mathcal{X}\}$ defined in (10.8) satisfies

$$Z(\cdot) =_d \max_{i=1}^{N} \frac{|\mathcal{X} \oplus b(o, R)|}{\sum_{j=1}^{i} E_j} F_i(\cdot - S_i)$$

for an a.s. finite number N defined by

$$N = \min \left\{ k \geqslant 1 : \frac{C}{\sum_{j=1}^{k+1} E_j} \leqslant \inf_{x \in \mathcal{X}} \left(\max_{i=1}^{k} \frac{F_i(x - S_i)}{\sum_{j=1}^{i} E_j} \right) \right\}.$$

　　This result allows for the implementation of an exact simulation procedure in Algorithm 10.2.

　　If not both condition (10.9) and (10.10) are met, similarly to the case of a representation via stochastic processes, Schlather (2002) proposes an approximation. We choose $R^* > 0$ and $C^* > 0$ such that $\mathbb{P}(\sup_{x \in \mathbb{R}^d} F(x) > C^*)$ is small and the shape function F is negligible outside $b(o, R^*)$ and apply Algorithm 10.2 with $R = R^*$ and $C = C^*$. Again, the choice of R^* and C^* is a tradeoff between accuracy and running time of the simulation algorithm. We will further address this issue in Section 10.4 providing some error bounds.

10.3   The Choice of the Spectral Representation

In this section, we will discuss the choice of the (non-unique) spectral representation of a max-stable process and its effects on simulation. We will consider several transformations that allow for switching from one spectral to another.

At first, we review a class of transformations presented in Oesting et al. (2013) leading to several equivalent representations via stochastic processes. We start with a max-stable process Z with general representation (10.1) and spectral measure ν. Let g be a probability density function with respect to ν, i.e., $g \geqslant 0$ and $\int_{\mathcal{C}} g(f) \nu(\mathrm{d}f) = 1$, such that

$$\nu\left(\{f \in \mathcal{C} : g(f) = 0, \, \sup_{x \in \mathcal{X}} f(x) > 0\}\right) = 0.$$

Then, we obtain that

$$Z(x) =_d \max_{i \geqslant 1} \frac{1}{\sum_{j=1}^{i} E_j} \frac{Y_i^*(x)}{g(Y_i^*)}, \quad x \in \mathcal{X}, \tag{10.11}$$

where Y_i^*, $i \geqslant 1$, are independent and identically distributed according to the probability measure $g\nu$, i.e., $\mathbb{P}(Y_i^* \in A) = \int_{\mathcal{C}} \mathbf{1}_A(f) g(f) \nu(\mathrm{d}f)$ for all Borel sets $A \subset \mathcal{C}$, (Oesting et al., 2013, Prop. 2.1). In particular, the transformed spectral functions $Y_i^*/g(Y_i^*)$ can be seen as independent and identically distributed processes. Thus, (10.11) is a representation of Z via stochastic processes.

Similarly to Subsection 10.2.1, the representation via stochastic processes allows us to give some stopping rule for the simulation. Following Oesting et al. (2013), this criterion will in general be sharper than the one presented in Subsection 10.2.1 as it is based on a pointwise bound instead of an overall bound: A function $(\sum_{j=1}^{k} E_j)^{-1} Y_k^*(x)/g(Y_k^*)$ cannot contribute to the pointwise maximum in (10.11) if

$$\frac{1}{\sum_{j=1}^{k+1} E_j} \operatorname*{ess\,sup}_{f \in \mathcal{C}} \frac{f(x)}{g(f)} \leqslant \max_{i \geqslant 1} \frac{1}{\sum_{j=1}^{i} E_j} \frac{Y_i^*(x)}{g(Y_i^*)} \quad \text{for all } x \in \mathcal{X},$$

or, equivalently,

$$\sum_{j=1}^{k+1} E_j \geqslant \sup_{x \in \mathcal{X}} \operatorname*{ess\,sup}_{f \in \mathcal{C}} \frac{f(x)}{g(f) Z^{(k)}(x)} \tag{10.12}$$

where

$$Z^{(k)}(x) = \max_{i=1}^{k} \frac{1}{\sum_{j=1}^{i} E_j} \frac{Y_i^*(x)}{g(Y_i^*)}, \quad x \in \mathcal{X},$$

and ess sup denotes the essential supremum with respect to the transformed measure $g\nu$. Thus, as $\inf_{x \in \mathcal{X}} Z(x) > 0$ a.s., the representation (10.11) allows for an exact simulation of Z in a.s. finite time if and only if g satisfies

$$\sup_{x \in \mathcal{X}} \operatorname*{ess\,sup}_{f \in \mathcal{C}} \frac{f(x)}{g(f)} < \infty. \tag{10.13}$$

Oesting et al. (2013) focus on a specific choice of g that satisfies (10.13) — the so-called *normalized spectral representation*. This representation — whose existence is shown by de Haan and Ferreira (2006), Cor. 9.4.5 — is a representation via stochastic processes (10.11) and is characterized by the fact that the stochastic processes Y_i, $i \geqslant 1$, (which correspond to $Y_i^*/g(Y_i^*)$ in representation (10.11)) satisfy

$$\sup_{x \in \mathcal{X}} Y_i(x) = c \quad a.s. \tag{10.14}$$

for some $c > 0$. In representation (10.11), this corresponds to the choice $g(f) = c^{-1} \sup_{x \in \mathcal{X}} f(x)$ where

$$c = \int_{\mathcal{C}} \sup_{x \in \mathcal{X}} f(x)\, \nu(\mathrm{d}f) = -\log \mathbb{P}\left(\sup_{x \in \mathcal{X}} Z(x) \leqslant 1\right)$$

which ensures that g is a probability density with respect to ν as well as (10.14) by which it is determined uniquely. Note that $c < \infty$ if and only if $\sup_{x \in \mathcal{X}} Z(x) < \infty$ a.s. (cf. Resnick and Roy, 1991) which holds as Z is sample-continuous. Further, the distribution of the spectral functions $Y_i^*/g(Y_i^*)$ is uniquely defined by (10.14) (cf. Oesting et al., 2013, Prop. 2.5). In particular, the distribution does not depend on the initial choice of ν.

Due to (10.14), the normalized spectral representation allows for an exact simulation in a.s. finite time. Although in general being slightly less precise than the stopping rule (10.12) with $g(f) = c^{-1} \sup_{x \in \mathcal{X}} f(x)$, in this case, in practice, the simulation is stopped if

$$\sum_{j=1}^{k+1} E_j \geqslant \frac{c}{\inf_{x \in \mathcal{X}} Z^{(k)}(x)}, \tag{10.15}$$

that is, simulation is performed according to Algorithm 10.1 with $C = c$.

Although simulation via the normalized spectral representation is exact and, as we will see in Section 10.4, seems to be quite efficient, for some models, the implementation may be difficult in practice due to the transformed measure $g\nu$. While the original spectral representation is usually chosen such that the spectral function can be simulated at rather little cost, this is not necessarily the case any more for the functions $Y_i^* \sim g\nu$ from representation (10.11). Thus, in view of the simulation efficiency, it may be useful to use a different representation. Engelke et al. (2014) present transformations that allow to switch between a representation via stochastic processes (10.4) and a mixed moving maxima representation (10.8). While a mixed moving maxima representation can always be transformed into a representation via stochastic processes via a transformation of the same type as in (10.11), the existence of a mixed moving maxima representation is linked to further assumptions on the spectral process $Y \sim \nu$ defined on $\mathbb{R}^d$. According to Kabluchko et al. (2009) and Engelke et al. (2014) sufficient conditions are:

(i) $\mathbb{E}(Y(x)) = 1$ for all $x \in \mathbb{R}^d$

(ii) $\lim_{\|x\| \to \infty} Y(x) = 0$ a.s.

(iii) $\mathbb{E}(\sup_{x \in K} Y(x)) < \infty$ for all compact sets $K \subset \mathbb{R}^d$

If all these conditions are satisfied, the max-stable process (10.4) can be written as a mixed moving maxima process (10.8) with shape function $F(\cdot) = $

$(\mathbb{E}(\int_{\mathbb{R}^d} \tilde{F}(x)\,dx))^{-1}\tilde{F}(\cdot)$ where the distribution of $\tilde{F}$ is given by

$$\mathbb{P}(\tilde{F} \in A) = \frac{\int_0^\infty m\mathbb{P}(Y(\cdot + \tau)/m \in A, \ \tau \in K \mid M = m)\,\mathbb{P}(M \in dm)}{\int_0^\infty m\mathbb{P}(\tau \in K \mid M = m)\,\mathbb{P}(M \in dm)} \qquad (10.16)$$

for any Borel set $A \subset C(\mathbb{R}^d)$ (Engelke et al., 2014, Thm. 4.3). Here, $K \subset \mathbb{R}^d$ is an arbitrary compact set and M and τ are random variables defined by $M = \max_{x \in \mathbb{R}^d} Y(x)$ and $\tau = \inf\{\arg\max_{x \in \mathbb{R}^d} Y(x)\}$, respectively. Note that, with probability one, the maximum of the shape function F has the value $\mathbb{E}(\int_{\mathbb{R}^d} \tilde{F}(x)\,dx))^{-1}$ and is attained at the origin. Thus, it is suitable to simulate the process via Algorithm 10.2 even though the support of F may be unbounded. In general, the distribution of the shape function given by (10.16) is difficult to handle. However, in some cases, it allows for simulation (see Subsection 10.5.3 for some examples).

10.4   Measures of Simulation Efficiency and Quality

In this section, we will investigate the efficiency and quality of the simulation procedures introduced above. Following Dombry and yi Minko (2012) and Dombry and yi Minko (2013), we separate the set of Poisson points $\{\zeta_i, \psi_i\}_{i \geqslant 1}$ from (10.1) into those processes that contribute to the maximum Z and those that do not. To this end, we define the set of *extremal functions*

$$\Phi^+ = \{(\zeta_k, \psi_k) : \zeta_k \psi_k(x) = \max_{i \geqslant 1} \zeta_i \psi_i(x) \text{ for some } x \in \mathcal{X}\}$$

and the set of *subextremal functions*

$$\Phi^- = \{(\zeta_k, \psi_k) : \zeta_k \psi_k(x) < \max_{i \geqslant 1} \zeta_i \psi_i(x) \text{ for all } x \in \mathcal{X}\}.$$

Note that both Φ^+ and Φ^- form point processes (cf. Dombry and yi Minko, 2012). In view of these definitions, a realization that is obtained by some simulation procedure can be seen as correct if all extremal functions are taken into consideration, and a simulation algorithm is an exact algorithm if the resulting realizations are correct with probability one. Thus, an appropriate measure for the quality of an approximative procedure is the probability of producing incorrect realizations. Besides providing accurate realizations, a simulation algorithm should also be efficient, that is the number of subextremal functions taken into consideration should be minimal.

10.4.1   Simulation Efficiency

We first investigate the efficiency of an exact algorithm which we measure in terms of the expected number of spectral functions that need to be simulated until a stopping criterion is satisfied.

Oesting et al. (2013) consider this value as criterion for a good choice of a spectral representation, that is, the choice of the density g in representation (10.11). In this case, according to the stopping rule (10.12), on average

$$Q_g = \mathbb{E} \min \left\{ m \in \mathbb{N} : \operatorname{ess\,sup}_{f \in \mathcal{C}} \sup_{x \in \mathcal{X}} \frac{f(x)}{g(f) Z^{(m)}(x)} \leqslant \sum_{j=1}^{m+1} E_j \right\}$$

spectral functions have to be considered. Decomposing the considered functions into extremal and subextremal functions, this number can be computed

$$Q_g = \mathbb{E} \left(\operatorname{ess\,sup}_{f \in \mathcal{C}} \sup_{x \in \mathcal{X}} \frac{f(x)}{g(f) Z(y)} \right). \tag{10.17}$$

Thus, a transformation that yields the most efficient simulation algorithm, can be found by minimizing (10.17) with respect to g. However, as this optimization problem is very difficult, Oesting et al. (2013) present a closely related class of replacement problems, which they show to be solved uniquely by the normalized spectral representation, i.e., by $g^*(f) = c^{-1} \sup_{x \in \mathcal{X}} f(x)$. For this choice, the expression (10.17) can be bounded by $c\mathbb{E}[(\inf_{x \in \mathcal{X}} Z(x))^{-1}]$ which corresponds to the number of spectral functions considered according to Algorithm 10.1 with $C = c$.

Adapting the stopping rules, we can modify the proof of Prop. 4.8 in Oesting et al. (2013) to obtain the expected number of functions considered in the exact procedures given by Algorithms 10.1 and 10.2, respectively.

Proposition 10.4.1. *Let $\{E_j\}_{j \geqslant 1}$ be independent standard exponentially distributed random variables.*

1. *Independently of the E_j's, let Y_j, $j \geqslant 1$, be independent copies of a nonnegative stochastic process Y satisfying $\mathbb{E}Y(x) = 1$ for all $x \in \mathcal{X}$ and $\sup_{x \in \mathcal{X}} Y(x) \leqslant C$ a.s. for some $C > 0$. Further, let Z be defined by (10.4). Then, we have*

$$\mathbb{E} \min \left\{ k \geqslant 1 : \frac{C}{\sum_{j=1}^{k+1} E_j} \leqslant \inf_{x \in \mathcal{X}} \max_{i=1}^{k} \frac{Y_i(x)}{\sum_{j=1}^{i} E_j} \right\} = C\mathbb{E} \left[(\inf_{x \in \mathcal{X}} Z(x))^{-1} \right].$$

2. *Independently of the E_j's, let F_j, $j \geqslant 1$, be independent copies of a nonnegative stochastic process F satisfying $\mathbb{E} \int_{\mathcal{X}} F(x) \, \mathrm{d}x = 1$, $\sup_{x \in \mathcal{X}} F(x) \leqslant C$ a.s. for some $C > 0$ and (10.9) for some $R > 0$. Further, let $S_j \sim_{i.i.d.} \text{Unif}(\mathcal{X} \oplus b(o, R))$, $j \geqslant 1$, and Z be defined by (10.8). Then, we have*

$$\mathbb{E} \min \left\{ k \geqslant 1 : \frac{C}{\sum_{j=1}^{k+1} E_j} \leqslant \inf_{x \in \mathcal{X}} \max_{i=1}^{k} \frac{F_i(x - S_i)}{\sum_{j=1}^{i} E_j} \right\}$$
$$= C \cdot |\mathcal{X} \oplus b(o, R)| \cdot \mathbb{E} \left[(\inf_{x \in \mathcal{X}} Z(x))^{-1} \right].$$

Remark 10.4.1. *By Thm. 2.2 in Dombry and yi Minko (2012), the quantity $\mathbb{E} \left[(\inf_{x \in \mathcal{X}} Z(x))^{-1} \right]$ is finite for any compact set $\mathcal{X} \subset \mathbb{R}^d$ and any sample-continuous max-stable process Z with unit Fréchet margins.*

10.4.2   Simulation Quality

In this subsection, we consider measures for the quality of simulation algorithms that are not exact. In particular, we provide error estimates for Algorithms 10.1 and 10.2 in the case that the underlying assumptions are not met such that an approximative stopping rule is applied. Here, a realization produced by this algorithm is not correct if there exists at least one extremal function that is not taken into consideration by the algorithm, i.e., some $(\zeta_i, \psi_i) \in \Phi$ which already satisfies the stopping criterion. Using Slivnyak's formula from Poisson point process theory (cf. Stoyan et al., 1995), we obtain the following assessments for the probability of obtaining an incorrect realization when applying Algorithms 10.1 and 10.2 with $C = C^*$ and $R = R^*$, respectively.

Proposition 10.4.2. *Let Z be a max-stable process with spectral representation* (10.4). *Then, for any $C^* > 0$, we have*

$$\mathbb{P}\left(\exists(\zeta_i, Y_i) \in \Phi^+ : \zeta_i C^* < \inf_{x \in \mathcal{X}} \max_{j=1}^{i-1} \zeta_j Y_j(x)\right) \tag{10.18}$$

$$\leqslant \sqrt{\mathbb{E}\left(\sup_{x \in \mathcal{X}} (Y(x)/Z(x))^2\right)} \cdot \sqrt{\mathbb{P}\left(\sup_{x \in \mathcal{X}} Y(x) > C^*\right)}.$$

Proof. First, we note that a point $(\zeta_i, Y_i) \in \Phi^+$ that is not considered due to the stopping criterion necessarily satisfies $\sup_{x \in \mathcal{X}} Y(x) > C^*$. Thus, we have

$$\mathbb{P}\left(\exists(\zeta_i, Y_i) \in \Phi^+ : \zeta_i C^* < \inf_{x \in \mathcal{X}} \max_{j=1}^{i-1} \zeta_j Y_j(x)\right)$$

$$\leqslant \mathbb{E}\left(\sum_{(\zeta,Y) \in \Phi} \mathbf{1}_{\{\zeta Y(x) \geqslant \max_{(\tilde\zeta, \tilde Y) \in \Phi \setminus \{(\zeta,Y)\}} \tilde\zeta \tilde Y(x) \text{ for some } x \in \mathcal{X}\}} \mathbf{1}_{\{\sup_{x \in \mathcal{X}} Y(x) > C^*\}}\right)$$

$$= \int_0^\infty \zeta^{-2} \mathbb{E}\left(\mathbf{1}_{\{\zeta Y(x) > Z(x) \text{ for some } x \in \mathcal{X}\}} \mathbf{1}_{\{\sup_{x \in \mathcal{X}} Y(x) > C^*\}}\right) \, \mathrm{d}\zeta$$

$$= \mathbb{E}\left(\sup_{x \in \mathcal{X}} \frac{Y(x)}{Z(x)} \mathbf{1}_{\{\sup_{x \in \mathcal{X}} Y(x) > C^*\}}\right),$$

where we use Slivnyak's formula (Stoyan et al., 1995) to obtain the equality of the second and the third lines. The assertion of the proposition is obtained by applying the Cauchy–Schwarz inequality.   □

For the error estimate for Algorithm 10.2, we assume that the shape function F is uniformly bounded by some constant $C > 0$, which is the case for all the examples considered in this chapter. In the case that F is unbounded, the additional error can be assessed in a similar way as in Proposition 10.4.2.

Proposition 10.4.3. *Let Z be a max-stable process with spectral representation* (10.8). *Further assume that the corresponding shape function F satisfies* (10.10). *Then, for any $R^* > 0$, we have*

$$\mathbb{P}\left(\exists(\zeta_i, S_i, F_i) \in \Phi^+ : S_i \notin \mathcal{X} \oplus b(o, R^*)\right) \tag{10.19}$$

$$\leqslant \mathbb{E}\left[(\inf_{x\in\mathcal{X}} Z(x))^{-1}\right] \cdot \int_{\mathbb{R}^d\setminus(\mathcal{X}\oplus b(o,R^*))} \mathbb{E}\left(\sup_{x\in\mathcal{X}} F(x-s)\right) \,\mathrm{d}s.$$

Proof. Using Slivnyak's Theorem (Stoyan et al., 1995), we obtain that

$$\mathbb{P}\left(\exists(\zeta_i, S_i, F_i) \in \Phi^+ : S_i \notin \mathcal{X} \oplus b(o, R^*)\right)$$

$$\leqslant \mathbb{E}\left(\sum_{(\zeta,F(\cdot-S))\in\Phi} \mathbf{1}_{\zeta F(x-S)\geqslant\max_{(\tilde{\zeta},\tilde{F}(\cdot-\tilde{S}))\in\Phi\setminus\{(\zeta,F(\cdot-S))\}} \tilde{\zeta}\tilde{F}(x-\tilde{S}) \text{ for some } x\in\mathcal{X}}\right)$$

$$= \int_{\mathbb{R}^d\setminus(\mathcal{X}\oplus b(o,R^*))} \int_0^\infty \zeta^{-2}\mathbb{E}\left(\mathbf{1}_{\zeta F(x-S)\geqslant Z(x) \text{ for some } x\in\mathcal{X}}\right) \,\mathrm{d}\zeta\,\mathrm{d}s$$

$$= \int_{\mathbb{R}^d\setminus(\mathcal{X}\oplus b(o,R^*))} \mathbb{E}\left(\sup_{x\in\mathcal{X}} \frac{F(x-s)}{Z(x)}\right) \,\mathrm{d}s.$$

The assertion of the proposition follows from the fact that $\sup_{x\in\mathcal{X}} \frac{F(x-s)}{Z(x)} \leqslant (\sup_{x\in\mathcal{X}} F(x-s)) \cdot (\inf_{x\in\mathcal{X}} Z(x))^{-1}$. □

Remark 10.4.2. *Note that the integral $\int_{\mathbb{R}^d} \mathbb{E}(\sup_{x\in\mathcal{X}} F(x-s)) \,\mathrm{d}s$ is finite if Z has continuous sample paths (Resnick and Roy, 1991). Thus, indeed, the error bound tends to zero as $R^* \to \infty$.*

Besides using these error estimates, there are also other diagnostic tools to assess the quality of a simulation algorithm. In particular, it can be checked whether the marginal distributions are reproduced correctly. As an example one can calculate the Kolmogorov–Smirnov distance between the simulated and the theoretical marginal distributions as Oesting et al. (2012) did in a simulation study to compare different simulation algorithms for Brown–Resnick processes. Also graphical tools, like Q-Q plots for the marginal distribution, or a plot of the extremal coefficient function θ which, for a stationary max-stable process Z, is defined by

$$\mathbb{P}(Z(t+h) \leqslant x, \, Z(t) \leqslant x) = \mathbb{P}(Z(t) \leqslant x)^{\theta(h)}, \quad h \in \mathbb{R}^d,$$

(Schlather and Tawn, 2003), can be used. See Subsection 10.5.2, for an example.

10.5 Examples

In this section, we discuss the choice of an accurate simulation procedure for several examples, including some of the most popular models of max-stable processes. For more details on these models, see Chapter 11.

10.5.1 Moving Maxima Processes

As the first example, we consider moving maxima processes, i.e., mixed moving maxima processes (10.8) with a deterministic shape function F. Of particular interest are the processes devised by Smith (1990) who uses probability densities as shape functions. If such a density function f is radially decreasing, the transformed Lebesgue density $g(f) = c^{-1} \sup_{x \in \mathcal{X}} f(x)$ from the normalized spectral representation allows for convenient sampling (cf. Oesting et al., 2013, for explicit formulae). Thus, the normalized spectral representation can be used for an exact simulation. Simulation studies in Oesting et al. (2013) for the case of f being a 1- or 2-dimensional Gaussian density show that the usage of the normalized spectral representation also yields remarkable improvements with respect to the number of considered spectral functions in comparison to Schlather's (2002) algorithm even though the latter one provides an approximation only. Thus, simulation via the normalized spectral representation is preferable both with respect to simulation accuracy and costs.

10.5.2 Extremal Gaussian and Extremal t Processes

Secondly, we consider extremal Gaussian processes (Schlather, 2002), i.e., processes of the form

$$Z(x) = \max_{i \geqslant 1} \zeta_i \sqrt{2\pi} \max\{0, W_i(x)\}, \quad x \in \mathcal{X},$$

where W_i, $i \geqslant 1$, are independent copies of some stationary standard Gaussian process W with correlation function $\rho(h)$, independently from the Poisson point process $\{\zeta_i\}_{i \geqslant 1}$ with intensity $\zeta^{-2}d\zeta$.

Here, for simulation based on the normalized spectral representation, one has to sample from the modified law $c^{-1} \max\{0, \sup_{x \in \mathcal{X}} w(x)\} \mathbb{P}(W \in dw)$. Although this can be done by MCMC methods, such a sampling procedure is very complex and time-consuming compared to sampling from W. Thus, we consider approximations via Algorithm 10.1. According to the approximation procedure described in Subsection 10.2.1, we choose some value C^* such that $\mathbb{P}(\sup_{x \in \mathcal{X}} \sqrt{2\pi} W(x) > B)$ is small and run Algorithm 10.1 with $C = C^*$.

To show the performance of the approximation procedure, we choose the bound $C^* = 5$ and simulate 1000 realizations of an extremal Gaussian process on $\{0, 0.5, \ldots, 4.5, 5\}$ based on Gaussian processes W_i with correlation function $\rho(h) = \exp(-|h|)$. Figure 10.2 shows a Q-Q plot of the empirical distribution of $\log Z(0)$ against the standard Gumbel distribution and the empirical extremal coefficients in comparison to the theoretical extremal coefficient function $\theta(h) = 1 + \sqrt{(1 - \rho(h))/2}$. As both the Q-Q plot and the extremal coefficient indicate, the approximation is quite accurate even though the bound $C = 5$ is rather small (this corresponds to a bound $B \approx 2$ for W).

The class of extremal Gaussian processes is generalized by extremal-t processes (Opitz, 2013)

$$Z(x) = \max_{i \geqslant 1} \zeta_i c_\alpha \max\{0, W_i(x)\}^\alpha, \quad x \in \mathcal{X},$$

where $c_\alpha = \sqrt{\pi} 2^{-(\alpha-2)/2} \Gamma((\alpha+1)/2)^{-1}$ for $\alpha > 0$. For small or moderate values of α, similar techniques as in the case of the extremal Gaussian process are suitable for simulation. If α gets large, the extremal t-process resembles a Brown–Resnick process which we will discuss in the following example.

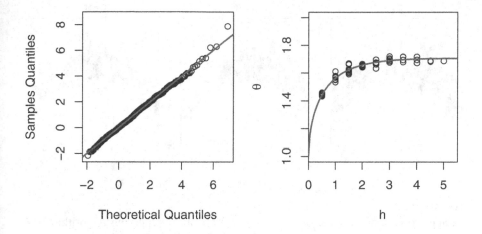

FIGURE 10.2

Diagnostic plots based on 1000 realizations of an extremal Gaussian process simulated via Algorithm 10.1 with $C = 5$. Left: Q-Q plot of the empirical distribution of $\log Z(0)$ against the standard Gumbel distribution. Right: Empirical extremal coefficients in comparison to the theoretical extremal coefficient function (solid line).

10.5.3 Brown–Resnick Processes

As the last example, we consider the challenging case of Brown–Resnick processes as defined in Equation (10.3). Kabluchko et al. (2009) show that the Brown–Resnick process is stationary and that its law — and, thus, the choice of an appropriate simulation procedure — depends on γ only. Therefore, Z is called Brown–Resnick process associated to the semi-variogram γ.

First, we note that, as for extremal Gaussian processes, the simulation of normalized spectral functions is very complex, but could be done by MCMC methods. Thus, in the following, we will focus on alternative approaches.

In the case that W is a stationary process, that is, Z is a so-called geometric Gaussian process (Davison et al., 2012), the semi-variogram is bounded and the variance function is constant. Thus, the process can be approximated by Algorithm 10.1 with an appropriate value C which, however, may be much larger than in the Gaussian case, in particular, if the variance σ^2 is large. As W is stationary, so are all finite approximations.

If, however, γ (and, thus, also the variance function) is unbounded, the single spectral functions $Y_i(x) = \exp(W_i(x) - \sigma^2(x)/2)$ and the finite approximations exhibit clear nonstationary trends. Similarly to the example in Section 10.1 (Figure 10.1) where a fixed number of spectral functions is considered, these problems also occur in case of Algorithm 10.1. Figure 10.3 shows realizations of the original Brown–Resnick process (Brown and Resnick, 1977) with W being a standard Brownian motion, i.e., a Brown–Resnick process associated to semi-variogram

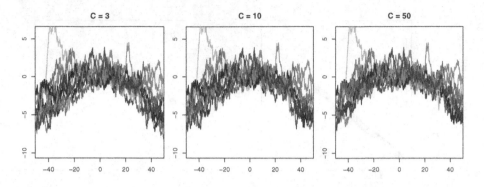

FIGURE 10.3
Ten approximations of $\log Z$ on the interval $[-50, 50]$ by applying Algorithm 10.1 with $C = 3, 10, 25$ (from left to right). Here, Z is a Brown–Resnick process associated to the semi-variogram $\gamma(h) = \frac{1}{2}|h|$.

$\gamma(h) = \frac{1}{2}|h|$, for several values of C. Note that the realizations look nonstationary even for large values of C.

Oesting et al. (2012) introduce several equivalent representations in order to generate approximations that indicate stationarity. For example, they consider random shifts of the spectral functions. Let $\zeta_i, W_i, i \geq 1$, be as above, and, independently from ζ_i, W_i, let $X_i, i \geq 1$, be independent random vectors in $\mathbb{R}^d$ distributed according to some measure μ. Then, for the Brown–Resnick process Z associated to the semi-variogram γ, we have

$$Z(x) =_d \max_{i \geq 1} \zeta_i \exp\left(W_i(x - X_i) - \sigma^2(x - X_i)/2\right), \quad x \in \mathcal{X}. \quad (10.20)$$

By these shifts, intuitively, the trend should vanish even in finite approximations. Thus, it is worthwhile to apply Algorithm 10.1 to this representation with shifted spectral functions. Results based on a uniform shift in $[-50, 50]$ are illustrated in Figure 10.4. Indeed, the approximations look stationary. However, even for large C, the values of Z are too small in general.

Further, one can also use Algorithm 10.2 to simulate Brown–Resnick processes provided that they allow for a mixed moving maxima representation. By construction — apart on some effects close to the boundary as a result of neglecting the spectral functions with $S_i \notin \mathcal{X} \oplus b(o, R)$ — this algorithm leads to stationary finite approximations. As already mentioned at the end of Section 10.3, a mixed moving maxima representation exists if

$$\lim_{\|x\| \to \infty} (W(x) - \sigma^2(x)) \quad a.s. \quad (10.21)$$

By Kabluchko et al. (2009), in the case $d = 1$, condition (10.21) is satisfied if $\lim_{x \to \infty} \gamma(x)/\log x > 8$.

In general, the simulation of shape functions according to the law given by 10.16

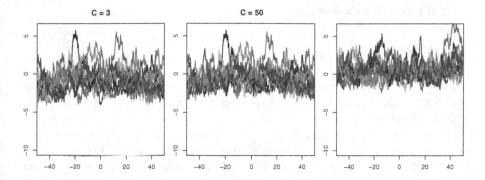

FIGURE 10.4

Ten approximations of $\log Z$ on the interval $[-50, 50]$ obtained by different procedures where Z is a Brown–Resnick process associated $\gamma(h) = \frac{1}{2}|h|$. Left/Middle: Approximations by Algorithm 10.1 with $C = 3$ and $C = 10$, respectively, applied to representation (10.20) with randomly shifted spectral functions. Here, the spectral functions are shifted according to a uniform distribution on $[-50, 50]$. Right: Approximations by Algorithm 10.2.

is rather sophisticated. However, some cases allow for simplifications. For the case of the original Brown–Resnick process where W is a standard Brownian motion, Engelke et al. (2011) make use of the Markovian structure of the Brownian motion to write F as a diffusion that allows for convenient simulation. Oesting et al. (2012) consider the case of a general Brown–Resnick process on some grid $p\mathbb{Z}^d$ for some $p > 0$ with a mixed moving maxima representation where the shifts S_i are restricted to the same grid. Using the fact that, without loss of generality, the process W can be assumed to be 0 at the origin o almost surely, they show that F has the same distribution as $\lambda \exp(W(\cdot) - \gamma(\cdot)) \mid \tau = o$ where $\tau = \inf\{\arg\max_{x \in p\mathbb{Z}^d} W(\cdot) - \gamma(\cdot)\}$ and $\lambda = p^{-d}\mathbb{P}(\tau = o)$. Thus, F can be simulated via an acceptance-rejection algorithm accepting only those functions $W - \gamma$ that attain their maximum at the origin. As the function F is bounded by λ, Algorithm 10.2 can be used for simulation neglecting the spectral functions with $S_i \notin \mathcal{X} \oplus b(o, R)$. Figure 10.4 shows realizations of the original Brown–Resnick process simulated by this method. Indeed, the realizations look stationary and also the marginal distributions fit well. However, in the case of a dense grid, the rejection rate might be very high. Thus, Oesting et al. (2012) propose a generalization for processes on $\mathbb{R}^d$, restricting to those spectral functions whose maximum is in some specific set. By an appropriate choice of this set, the simulation error can be controlled. See Oesting et al. (2012) for more details.

In a recent publication, Dieker and Mikosch (2015) propose an alternative representation that allows for an exact simulation of Brown–Resnick processes. For an arbitrary probability measure μ on $\mathbb{R}^d$, it holds that

$$Z(x) =_d \max_{i \geqslant 1} \zeta_i \frac{\exp(W_i(x - X_i) - \gamma(x - X_i))}{\int_{\mathbb{R}^d} \exp(W_i(y - X_i) - \gamma(y - X_i)) \, \mu(\mathrm{d}y)}, \quad x \in \mathcal{X},$$

where $\{(\zeta_i, X_i)\}_{i \geqslant 1}$ are the points of a Poisson point process on $(0, \infty) \times \mathbb{R}^d$ with intensity $\zeta^{-2} \, \mathrm{d}\zeta \times \mu(\mathrm{d}x)$, independently from the Gaussian processes W_i, $i \geqslant 1$, with $W(o) = 0$ a.s. and semi-variogram γ.

For simulating Z on some set $\{x_1, \ldots, x_n\}$, Dieker and Mikosch (2015) propose to choose $\mu = n^{-1} \sum_{k=1}^m \delta_{x_k}$. Noting further that $W(x - y) =_d W(x) - W(y)$ because of $W(o) = 0$ and the fact that W has stationary increments, we obtain $Z(x) =_d \max_{i \geqslant 1} \zeta_i Y_i(x)$ with

$$Y_i(x) = n \frac{\exp(W_i(x) - \gamma(x - T_i))}{\sum_{k=1}^n \exp(W_i(x_k) - \gamma(x_k - T_i))}, \quad x \in \{x_1, \ldots, x_n\}. \quad (10.22)$$

As Y_i is bounded by n, this representation allows for exact simulation on $\{x_1, \ldots, x_n\}$ via Algorithm 10.1 with $C = n$. For a comparison of this method to an exact simulation via the normalized spectral representation, it should be mentioned that the spectral functions Y_i in (10.22) can be simulated straightforwardly while the simulation of the normalized spectral functions requires MCMC techniques. However, by Proposition 10.4.1, the use of the bound $C = n$ for the stopping rule implies that the expected number of spectral functions increases at least linearly. This makes it very difficult to simulate the process on a dense grid in order to approximate the continuous sample paths. Contrarily, in the case that all the simulation locations are in some bounded area K, the bound for the normalized spectral representation is always below $\mathbb{E}(\sup_{x \in K} W(x))$ and thus, the expected number of considered spectral functions is smaller than $\mathbb{E}(\sup_{x \in K} W(x)) \mathbb{E}((\inf_{x \in K} W(x))^{-1})$.

As the above discussion shows, in general, it is quite difficult to decide which spectral representation to choose for the simulation of a Brown–Resnick process. This choice depends both on the corresponding semi-variogram and on the simulation domain. Oesting et al. (2012) conduct a simulation study to compare the performance of the simulation procedures based on the original spectral representation, random shifts, and the mixed moving maxima representation. We refer to their paper for more details on the results and indications for an appropriate choice of the simulation procedure.

10.6   Software Packages

Finally, we review two R packages that provide functions for the simulation of max-stable processes. The package `RandomFields` (Schlather et al., 2014) contains the max-stable models `RPsmith` for (mixed) moving maxima processes, `RPschlather` and `RPopitz` for extremal Gaussian and extremal t processes and `RPbrownresnick` for Brown–Resnick processes (including geometric Gaussian processes). The corresponding shape function, correlation function and semivariogram, respectively can be chosen from a large variety of basic models and combinations of them. All max-stable models can be simulated using the function `RFsimulate`. For mixed moving maxima processes, the exact simulation procedure via the normalized spectral representation (Oesting et al., 2013) is implemented. The simulation of Brown–Resnick processes is based on methods discussed in Oesting et al. (2012), namely simulation via the original representation, via random shifts and via the mixed moving maxima representation. If the user does not specify the simulation method (`RPbrorig`, `RPbrshifted` and `RPbrmixed`, respectively), an appropriate procedure is chosen automatically.

Further, the package `SpatialExtremes` (Ribatet et al., 2013) allows for the simulation of max-stable processes via the function `rmaxstab`. By specifying the parameter `cov.mod`, the user can choose between different models: for example, `gauss` is used for a moving maxima process with Gaussian density function as shape function and `brown` for Brown–Resnick processes associated to a semi-variogram of the type $\gamma(h) = a\|h\|^\alpha$, $a > 0$, $0 < \alpha \leqslant 2$. Further, extremal Gaussian, extremal t, and geometric Gaussian processes can be simulated for some classes of correlation functions. Here, the simulation procedures used are mainly based on Schlather's (2002) algorithms. For Brown–Resnick processes, random shifts are used as well.

Acknowledgements

The work of M. Oesting and M. Ribatet was supported by the ANR project McSim. The authors thank Liliane Bel, Christian Lantuéjoul, and Martin Schlather for numerous fruitful discussions.

References

Brown, B. M. and S. I. Resnick (1977). Extreme values of independent stochastic processes. *Journal of Applied Probability 14*(4), 732–739.

Davison, A. C., S. A. Padoan, and M. Ribatet (2012). Statistical modeling of spatial extremes. *Statistical Science 27*(2), 161–186.

de Haan, L. (1984). A spectral representation for max-stable processes. *Annals of Probability 12*(4), 1194–1204.

de Haan, L. and A. Ferreira (2006). *Extreme Value Theory: An Introduction*. Berlin: Springer.

Dieker, A. B. and T. Mikosch (2015). Exact simulation of Brown–Resnick random fields. *Extremes 18*(2), 301–314.

Dombry, C. and F. yi Minko (2012). Strong mixing properties of max-infinitely divisible random fields. *Stochastic Processes and Their Applications 122*(11), 3790–3811.

Dombry, C. and F. yi Minko (2013). Regular conditional distributions of continuous max-infinitely divisible random fields. *Electronic Journal Probability 18*(7), 1–21.

Engelke, S., Z. Kabluchko, and M. Schlather (2011). An equivalent representation of the Brown-Resnick process. *Statistics & Probability Letters 81*(8), 1150–1154.

Engelke, S., A. Malinowski, M. Oesting, and M. Schlather (2014). Statistical inference for max-stable processes by conditioning on extreme events. *Advances in Applied Probability 46*(2), 478–495.

Giné, E., M. G. Hahn, and P. Vatan (1990). Max-infinitely divisible and max-stable sample continuous processes. *Probability Theory and Related Fields 87*, 139–165.

Kabluchko, Z., M. Schlather, and L. de Haan (2009). Stationary max-stable fields associated to negative definite functions. *Annals of Probability 37*(5), 2042–2065.

Oesting, M., Z. Kabluchko, and M. Schlather (2012). Simulation of Brown-Resnick processes. *Extremes 15*, 89–107.

Oesting, M., M. Schlather, and C. Zhou (2013). On the normalized spectral representation of max-stable processes on a compact set. Available from http://arxiv.org/abs/1310.1813v1.

Opitz, T. (2013). Extremal t processes: Elliptical domain of attraction and a spectral representation. *Journal of Multivariate Analysis 122*, 409–413.

Penrose, M. D. (1992). Semi-min-stable processes. *Annals of Probability 20*(3), 1450–1463.

Resnick, S. I. and R. Roy (1991). Random usc functions, max-stable processes and continuous choice. *The Annals of Applied Probability 1*(2), 267–292.

Ribatet, M., R. Singleton, and R Core team (2013). *SpatialExtremes: Modelling Spatial Extremes*. R package version 2.0-0.

Schlather, M. (2002). Models for stationary max-stable random fields. *Extremes 5*(1), 33–44.

Schlather, M., A. Malinowski, M. Oesting, D. Boecker, K. Strokorb, S. Engelke, J. Martini, F. Ballani, P. J. Menck, S. Gross, U. Ober, K. Burmeister, J. Manitz, P. Ribeiro, R. Singleton, B. Pfaff, and R Core Team (2014). *RandomFields: Simulation and Analysis of Random Fields*. R package version 3.0.44.

Schlather, M. and J. A. Tawn (2003). A dependence measure for multivariate and spatial extreme values: Properties and inference. *Biometrika 90*(1), 139–156.

Smith, R. L. (1990). Max-stable processes and spatial extremes. Unpublished manuscript.

Stoev, S. A. and M. S. Taqqu (2005). Extremal stochastic integrals: A parallel between max-stable processes and α-stable processes. *Extremes 8*(4), 237–266.

Stoyan, D., W. Kendall, and J. Mecke (1995). *Stochastic Geometry and Its Applications* (second ed.). Chichester: John Wiley & Sons.

11

Conditional Simulation of Max-Stable Processes

Clément Dombry
Université de Franche-Comté, France

Marco Oesting
INRA / AgroParisTech, France

Mathieu Ribatet
Université Montpellier 2 / Université Lyon 1, France

Abstract

Max-stable processes are well-established models for spatial extremes. In this chapter, we address the prediction problem: assuming that a max-stable process is observed at some locations only, how can we use these observations to predict the behavior of the process at other unobserved locations? Mathematically, the prediction problem is related to the conditional distribution of the process given the observations. Recently, Dombry and Eyi-Minko (2013) provided an explicit theoretical formula for the conditional distributions of max-stable processes. The result relies on the spectral representation of the max-stable process as the pointwise maxima over an infinite number of spectral functions belonging to a Poisson point process. The effect of conditioning on the Poisson point process is analyzed, resulting in the notions of hitting scenario and extremal or subextremal functions. Due to the complexity of the structure of the conditional distributions, conditional simulation appears at the same time challenging and important to assess characteristics that are analytically intractable such as the conditional median or quantiles. The issue of conditional simulation was considered by Dombry et al. (2013) who proposed a three-step procedure for conditional sampling. As the conditional simulation of the hitting scenario becomes computationally very demanding even for a moderate number of conditioning points, a Gibbs sampler approach was proposed for this step. The results are illustrated on some simulation studies and we propose several diagnostics to check the performance.

11.1   Introduction: The Prediction Problem and Conditional Distribution

In classical geostatistics, Gaussian random fields play a central role in the statistical theory based on the central limit theorem. In a similar manner, max-stable random fields turn out to be fundamental models for spatial extremes since they extend the well-known univariate and multivariate extreme value theory to the infinite dimensional setting. Max-stable random fields arise naturally when considering the component-wise maxima of a large number of independent and identically random fields and seeking for a limit under a suitable affine normalization.

In this multivariate or functional setting, the notion of dependence is crucial: how does an extreme event occurring in some region affect the behavior of the random field at other locations? This is related to the *prediction problem* which is an important and long-standing challenge in extreme value theory. Suppose that a max-stable random field $Z = (Z(x))_{x \in \mathcal{X}}$ is observed at some stations $x_1, \ldots, x_k \in \mathcal{X}$ only, yielding

$$Z(x_1) = z_1, \ldots, Z(x_k) = z_k. \tag{11.1}$$

How can we take benefit from these observations and predict the random field Z at other places? We are naturally led to consider the *conditional distribution* of $(Z(x))_{x \in \mathcal{X}}$ given the observations (11.1).

In the classical Gaussian framework, i.e., if Z is a centered Gaussian random field, it is well known that the corresponding conditional distribution remains Gaussian and simple formulas give the conditional mean and covariance structure. This theory is strongly linked with the theory of Hilbert spaces: for example, the conditional expectation of $Z(x)$ can be obtained as the L^2-projection of the random field η onto the Gaussian subspace generated by the variables $\{Z(x_i),\ 1 \leqslant i \leqslant k\}$, resulting in the linear combination

$$\mathbb{E}\left[Z(x) \mid Z(x_1), \ldots, Z(x_k)\right] = \alpha_1(x)Z(x_1) + \cdots + \alpha_k(x)Z(x_k), \quad x \in \mathcal{X},$$

for some weight functions $\alpha_i(x)$, $i = 1, \ldots, n$, that are obtained via the kriging theory (e.g., Chils and Delfiner, 1999). A conditional simulation of Z given the observations (11.1) is then easily performed by setting

$$Z(x) = \tilde{Z}(x) + \alpha_1(x)(z_1 - \tilde{Z}(x_1)) + \cdots + \alpha_k(x)(z_k - \tilde{Z}(x_k)), \quad x \in \mathcal{X},$$

where $\tilde{Z}$ denotes the realization of an unconditional simulation of the random field Z.

In extreme value theory, the prediction problem turns out to be more difficult. A similar kriging theory for max-stable processes is very appealing and a first approach in that direction was done by Davis and Resnick (1989, 1993). They introduced a L^1-metric between max-stable variables and proposed a kind of projection onto max-stable spaces. To some extent, this work mimics the corresponding L^2-theory for Gaussian spaces. However, unlike the exceptional Gaussian case, there is no clear relationship between the predictor obtained by projection onto the max-stable space generated by the variables $\{Z(x_i), \ 1 \leqslant i \leqslant k\}$ and the conditional distributions of η with respect to these variables. Conditional distributions have been considered first by Weintraub (1991) in the case of a bivariate vector. A major contribution is the work by Wang and Stoev (2011) where the authors consider max-linear random fields, a special class of max-stable random fields with discrete spectral measure, and give an exact expression of the conditional distributions as well as efficient algorithms. The max-linear structure plays an essential role in their work and provides major simplifications since in this case Z admits the simple representation

$$Z(x) = \bigvee_{j=1}^{q} F_j f_j(x), \quad x \in \mathcal{X},$$

where the symbol $\bigvee$ denotes the maximum, $f_1, \ldots, f_q$ are deterministic functions, and $F_1, \ldots, F_q$ are i.i.d. random variables with unit Fréchet distribution. The authors determine the conditional distributions of $(F_j)_{1 \leqslant j \leqslant q}$ given the observations (11.1) and deduce the conditional distribution of Z. Another approach by Oesting and Schlather (2014) deals with max-stable random fields with a mixed moving maxima representation.

We present and discuss here the recent result from Dombry and Eyi-Minko (2013) and Dombry et al. (2013) providing formulas for the conditional distributions of max-stable random fields as well as efficient algorithm for conditional sampling. Note that the results can be stated in the more general framework of max-infinitely divisible processes, but for the sake of simplicity, we stick here to the max-stable case. Section 11.2 reviews the theoretical results on conditional distribution: we introduce the notion of extremal functions and hitting scenarios and explain the exact distribution of the max-stable process given the observations. Here, we focus on the framework of regular models that yields tractable formulas. In Section 11.3, we put the emphasis on efficient simulation and discuss a three-step procedure for conditional sampling. A difficult step is the conditional sampling for the hitting scenario for which a Gibbs sampler approach is proposed. Section 11.4 is devoted to simulation studies and we present several diagnostics to check the performance of the methods suggested.

11.2 Conditional Distribution of Max-Stable Processes

In the sequel, we consider Z a sample continuous max-stable random field on $\mathcal{X} \subset \mathbb{R}^d$. We can assume without loss of generality that Z has unit Fréchet margins. Thus, the process Z possesses a spectral representation (e.g., de Haan, 1984; de Haan and Ferreira, 2006; Penrose, 1992; Schlather, 2002)

$$Z(x) = \max_{i \geqslant 1} \zeta_i Y_i(x), \quad x \in \mathcal{X}, \tag{11.2}$$

where $\{\zeta_i\}_{i \geqslant 1}$ are the points of a Poisson process on $(0, \infty)$ with intensity $\zeta^{-2} d\zeta$, $(Y_i)_{i \geqslant 1}$ are independent replicates of a non-negative continuous sample path stochastic process Y such that $\mathbb{E}[Y(x)] = 1$ for all $x \in \mathcal{X}$, $(\zeta_i)_{i \geqslant 1}$ and $(Y_i)_{i \geqslant 1}$ are independent.

We now introduce a fundamental object in our analysis which is a function-valued Poisson point process associated with the representation (11.2). Let $\mathcal{C} = \mathcal{C}(\mathcal{X}, [0, +\infty))$ be the space of continuous non-negative functions on $\mathcal{X}$. We consider the $\mathcal{C}$-valued point process $\Phi = \{\phi_i\}_{i \geqslant 1}$ where $\phi_i(x) = \zeta_i Y_i(x)$ with ζ_i and Y_i as in (11.2). It is well known (de Haan and Ferreira, 2006) that Φ is a Poisson point process with intensity measure Λ given by

$$\Lambda(A) = \int_0^\infty \mathbb{P}[\zeta Y \in A] \zeta^{-2} d\zeta, \quad A \subset \mathcal{C} \text{ Borel.}$$

Our strategy is to work on the level of the Poisson point process Φ rather than on the level of the max-stable process Z and to derive the conditional distribution of Φ given the observations (11.1). The conditional distribution of Z is then deduced easily.

11.2.1 Extremal Functions, Subextremal Functions, and Hitting Scenarios

The observations (11.1) together with the spectral representation (11.2) yield the constraint

$$\max_{i \geqslant 1} \phi_i(x_1) = z_1, \dots, \max_{i \geqslant 1} \phi_i(x_k) = z_k.$$

A first step in our strategy is the analysis of how these constraints are met, i.e., how the maxima are attained. An important preliminary remark is that for each $j = 1, \cdots, k$, the maximum $Z(x_j) = \max_{i \geqslant 1} \phi_i(x_j)$ is almost surely attained by a unique function ϕ_i, which can be easily seen by the fact that the point process $\{\varphi_i(x_j), i \geqslant 1\}$ on $(0, \infty)$ has intensity $\zeta^{-2} d\zeta$. This leads to the following definition.

Definition 1. The (almost surely) unique function $\phi \in \Phi$ such that $\phi(x_j) = Z(x_j)$ is called the extremal function at x_j and denoted by $\varphi_{x_j}^+$. Furthermore, we define the extremal point process $\Phi^+ = \{\varphi_{x_j}^+\}_{1 \leqslant j \leqslant k}$ as the set of extremal functions with respect to the conditioning points $\{x_j\}_{1 \leqslant j \leqslant k}$.

Note that the notation $\Phi^+ = \{\varphi_{x_j}^+\}_{1 \leqslant j \leqslant k}$ may include the multiple occurrence of some functions, for example if $\varphi_{x_1}^+ = \varphi_{x_2}^+$. This is not taken into account in the extremal point process Φ^+ where each point has multiplicity one. The possible redundancies are captured by the so-called hitting scenario. We first introduce this notion with simple examples. Assume first that there are $k = 2$ conditioning points x_1, x_2 and hence two extremal functions $\varphi_{x_1}^+$ and $\varphi_{x_2}^+$. Two cases can occur:

- Either $\varphi_{x_1}^+ = \varphi_{x_2}^+$, i.e., the maxima at points x_1 and x_2 are reached by the same extremal function, in this case the hitting scenario is $\theta = \{x_1, x_2\}$;

- Or $\varphi_{x_1}^+ \neq \varphi_{x_2}^+$, i.e., the maxima at points x_1 and x_2 are reached by different extremal functions, in this case the hitting scenario is $\theta = (\{x_1\}, \{x_2\})$.

In the case of $k = 3$ conditioning points, there are 5 different possibilities for the hitting scenario:

$$(\{x_1, x_2, x_3\}), \quad (\{x_1, x_2\}, \{x_3\}), \quad (\{x_1, x_3\}, \{x_2\}),$$

$$(\{x_1\}, \{x_2, x_3\}) \quad \text{and} \quad (\{x_1\}, \{x_2\}, \{x_3\}).$$

The interpretation is straightforward: for instance, the hitting scenario $\theta = (\{x_1, x_3\}, \{x_2\})$ corresponds to the case when the maxima at x_1 and x_3 are attained by the same extremal function but the maximum at x_2 corresponds to a different extremal event, i.e., $\varphi_{x_1}^+ = \varphi_{x_3}^+ \neq \varphi_{x_2}^+$. The general definition is as follows.

Definition 2. The hitting scenario θ is a random partition $(\theta_1, \ldots, \theta_\ell)$ of the conditioning points $\{x_1, \cdots, x_k\}$ such that for any $j_1 \neq j_2$, x_{j_1} and x_{j_2} are in the same component of θ if and only if $\varphi_{j_1}^+ = \varphi_{j_2}^+$.

The hitting scenario θ takes into account the redundancies in $\Phi^+ = \{\varphi_{x_1}^+, \cdots, \varphi_{x_k}^+\}$ and it is straightforward that the number of blocks ℓ of θ is exactly the number of distinct extremal functions and hence the cardinality of Φ^+. Hence we can rewrite $\Phi^+ = \{\varphi_1^+, \cdots, \varphi_\ell^+\}$ where $\varphi_j^+ = \varphi_x^+$ for all $x \in \theta_j$.

So far, we have considered only those functions ϕ_i that hit the maximum Z at some conditioning points $\{x_j\}_{1 \leqslant j \leqslant k}$. The remaining functions are called subextremal, as in the following definition.

Definition 3. A function $\phi \in \Phi$ satisfying $\phi(x_j) < Z(x_j)$ for all $1 \leqslant j \leqslant k$ is called subextremal. The set of subextremal functions is called the subextremal point process Φ^-.

This yields a disjoint decomposition of the point process $\Phi = \Phi^+ \cup \Phi^-$ into its extremal and subextremal parts. This is illustrated in Figure 11.1 where the extremal functions (black) and the subextremal functions (gray) are depicted as well as the corresponding hitting scenarios.

A key result in the study of the conditional distribution is the following theorem providing the joint distribution of (θ, Φ^+, Φ^-). We denote by $\mathcal{P}_k$ the set of partitions of $\{x_1, \ldots, x_k\}$ and by $\mathcal{M}_p(\mathcal{C})$ the set of $\mathcal{C}$-valued point measures. We introduce some vectorial notation. Let $\mathbf{x} = (x_1, \ldots, x_k)$ and for $\tau = (\tau_1, \ldots, \tau_\ell) \in \mathcal{P}_k$, we introduce $\mathbf{x}_{\tau_j} = (x)_{x \in \tau_j}$. For any vector $\mathbf{s} = (s_1, \ldots, s_m) \in \mathcal{X}^m$ and any function $f : \mathcal{X} \to \mathbb{R}$, we note $f(\mathbf{s}) = (f(s_1), \ldots, f(s_m))$. If $f_1, f_2 : \mathcal{X} \to \mathbb{R}$ are two functions, the notation $f_1(\mathbf{s}) < f_2(\mathbf{s})$ means that $f_1(s_j) < f_2(s_j), j = 1, \ldots, m$.

Theorem 11.2.1. *For any partition* $\tau = (\tau_1, \ldots, \tau_\ell)$ *in* $\mathcal{P}_k$ *and any Borel sets* $A \subset \mathcal{C}^\ell$ *and* $B \subset \mathcal{M}_p(\mathcal{C})$, *we have*

$$\mathbb{P}[\theta = \tau, (\varphi_1^+, \ldots, \varphi_\ell^+) \in A, \Phi^- \in B]$$

$$= \int_{\mathcal{C}^\ell} 1_{\{\max_{j' \neq j} f_{j'}(\mathbf{x}_{\tau_j}) < f_j(\mathbf{x}_{\tau_j}), \, j=1,\ldots,\ell\}} 1_{\{(f_1,\ldots,f_\ell) \in A\}} \cdots$$

$$\cdots \mathbb{P}\left[\Phi \in B \text{ and } \forall \phi \in \Phi, \, \phi(\mathbf{x}) < \max_{1 \leqslant j \leqslant \ell} f_j(\mathbf{x})\right] \Lambda(df_1) \cdots \Lambda(df_\ell).$$

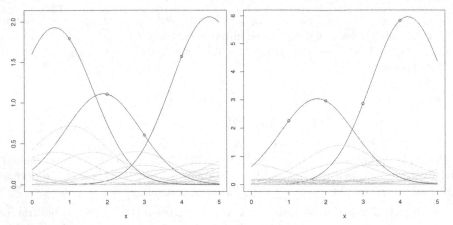

FIGURE 11.1
Two realizations of the Poisson point process Φ and of the corresponding hitting scenario θ with conditioning points $x_i = i$, $i = 1, \cdots, 4$ represented by the circles. Left: the hitting scenario is $\theta = (\{x_1\}, \{x_2, x_3\}, \{x_4\})$. Right: the hitting scenario is $\theta = (\{x_1, x_2\}, \{x_3, x_4\})$.

The main technical tool for this proof is the Slivniack–Mecke formula from stochastic geometry, see e.g., Stoyan et al. (1987). We sketch here the main lines of the proof.

Proof. First note that the event $\{\theta = \tau, (\varphi_1^+, \ldots, \varphi_\ell^+) \in A, \Phi^- \in B\}$ is realized if and only if there exists a ℓ-tuple $(\phi_1, \cdots, \phi_\ell) \in \Phi^\ell$ satisfying the following conditions:

i) $\Phi^+ = \{\phi_1, \ldots, \phi_\ell\}$
ii) $\Phi^- = \Phi \setminus \{\phi_1, \ldots, \phi_\ell\}$;
iii) $\theta = \tau$;
iv) $(\phi_1, \ldots, \phi_\ell) \in A$;
v) $\Phi^- \in B$.

Clearly, if such a ℓ-tuple does exist, it is necessarily unique and equal to $(\varphi_1^+, \ldots, \varphi_\ell^+)$. We deduce that the probability of interest can be written in the form

$$\mathbb{P}[\theta = \tau, (\varphi_1^+, \ldots, \varphi_\ell^+) \in A, \Phi^- \in B]$$

$$= \mathbb{E}\left[\sum_{(\phi_1, \cdots, \phi_\ell) \in \Phi^\ell} F(\phi_1, \ldots, \phi_\ell, \Phi \setminus \{\phi_1, \ldots, \phi_\ell\})\right] \qquad (11.3)$$

with $F : \mathcal{C}^\ell \times \mathcal{M}_p(\mathcal{C}) \to \{0, 1\}$ defined by

$$F(\phi_1, \ldots, \phi_\ell, \Phi \setminus \{\phi_1, \ldots, \phi_\ell\}) = \begin{cases} 1 & \text{if conditions i)-v) are satisfied,} \\ 0 & \text{otherwise.} \end{cases}$$

By the Slyvniak–Mecke formula (Stoyan et al., 1987), we deduce that the expectation

(11.3) is equal to

$$\int_{C^\ell} \mathbb{E}[F(f_1,\dots,f_\ell,\Phi)\Lambda(\mathrm{d}f_1)\cdots\Lambda(\mathrm{d}f_\ell).$$

The theorem follows after a more explicit description of the functional F. Condition i), ii), and iii) together are equivalent to

$$\max_{j'\neq j}\phi_{j'}(\mathbf{x}_{\tau_j}) < \phi_j(\mathbf{x}_{\tau_j}),\ j=1,\dots,\ell,$$

and

$$\forall\phi\in\Phi\setminus\{\phi_1,\dots,\phi_\ell\},\ \phi(\mathbf{x}) < \max_{1\leqslant j\leqslant\ell}\phi_j(\mathbf{x}).$$

Condition iv) is clear and condition v) is equivalent to $\Phi\setminus\{\phi_1,\dots,\phi_\ell\}\in B$. $\qquad\square$

11.2.2 The General Structure of Conditional Distributions

In the previous section, we have seen that we are able to compute the joint distribution of the hitting scenario θ, the extremal functions Φ^+ and the subextremal functions Φ^-. We now consider the conditional joint distribution given the observations (11.1). It can be computed on the basis of Theorem 11.2.1 only, because the observations $(Z(x_1),\dots,Z(x_k))$ can be expressed as a function of θ and Φ^+, i.e.,

$$Z(x_i) - \varphi_j^+(x_i)\quad\text{with } j \text{ such that } x_i \subset \theta_j.$$

Nevertheless, this computation is rather tedious and we give here the result without proof and we consider only the so-called regular case. We say that the intensity measure Λ of the point process Φ is regular at some point $\mathbf{s} = (s_1,\dots,s_p) \in \mathcal{X}^p$ if the marginal spectral measure Λ is absolutely regular with respect to the Lebesgue measure on $(0,+\infty)^k$. More precisely, we define the marginal spectral measure

$$\Lambda_{\mathbf{s}}(\mathrm{d}z_1,\dots,\mathrm{d}z_p) = \Lambda(f(s_1)\subset \mathrm{d}z_1,\dots,f(s_p)\in \mathrm{d}z_p)$$

and assume that on $(0,+\infty)^k$

$$\Lambda_{\mathbf{s}}(\mathrm{d}z_1,\dots,\mathrm{d}z_p) = \lambda_{\mathbf{s}}(z_1,\dots,z_p)\mathrm{d}z_1\cdots\mathrm{d}z_p.$$

The proof of the following result and more details are to be found in Dombry and Eyi-Minko (2013). Note that, in the non-regular case, some further analysis of the different hitting scenarios is needed, as some hitting scenarios have to be excluded due to the conditioning data. For examples of non-regular cases and conditional sampling procedures in these cases, see Wang and Stoev (2011) and Oesting and Schlather (2014) who consider max-linear and mixed moving maxima processes, respectively.

Theorem 11.2.2. *Assume that Λ is regular at* $\mathbf{x} = (x_1,\dots,x_k)$. *For* $\tau = (\tau_1,\dots,\tau_\ell) \in \mathcal{P}_k$ *and* $j = 1,\dots,\ell$, *define* $I_j = \{i\colon x_i \in \tau_j\}$, $\mathbf{x}_{\tau_j} = (x_i)_{i\in I_j}$, $\mathbf{z}_{\tau_j} = (z_i)_{i\in I_j}$, $\mathbf{x}_{\tau_j^c} = (x_i)_{i\notin I_j}$ *and* $\mathbf{z}_{\tau_j^c} = (z_i)_{i\notin I_j}$. *The conditional distribution of* (θ,Φ^+,Φ^-) *with respect to the observations (11.1) is obtained as follows:*

1. *For $\tau = (\tau_1, \ldots, \tau_\ell) \in \mathcal{P}_k$,*

$$\mathbb{P}\left[\theta = \tau \mid Z(\mathbf{x}) = \mathbf{z}\right]$$

$$= \frac{1}{C_{\mathbf{x},\mathbf{z}}} \prod_{j=1}^{\ell} \int_{\{u_j < \mathbf{z}_{\tau_j^c}\}} \lambda_{(\mathbf{x}_{\tau_j}, \mathbf{x}_{\tau_j^c})}(\mathbf{z}_{\tau_j}, \mathbf{u}_j) d\mathbf{u}_j, \qquad (11.4)$$

where the normalization constant $C_{\mathbf{x},\mathbf{z}}$ is such that

$$\sum_{\tau \in \mathcal{P}_k} \mathbb{P}\left[\theta = \tau \mid Z(\mathbf{x}) = \mathbf{z}\right] = 1.$$

2. *Conditionally on the observations (11.1) and on the hitting scenario $\tau = (\tau_1, \ldots, \tau_\ell) \in \mathcal{P}_k$, the extremal functions $\varphi_1^+, \ldots, \varphi_\ell^+$ are independent with distribution*

$$\mathbb{P}[\varphi_j^+ \in df \mid Z(\mathbf{x}) = \mathbf{z}, \theta = \tau]$$
$$= \Lambda[df \mid f(\mathbf{x}_{\tau_j}) = \mathbf{z}_{\tau_j}, f(\mathbf{x}_{\tau_j^c}) < \mathbf{z}_{\tau_j^c}], \quad j = 1, \ldots, \ell. \quad (11.5)$$

3. *Conditionally on the observations (11.1), Φ^- is independent of $\Phi^+ = \{\varphi_1^+, \ldots, \varphi_\ell^+\}$ and has the same distribution as a Poisson point process on $\mathcal{C}$ with intensity $1_{\{f(\mathbf{x}) < \mathbf{z}\}} \Lambda(df)$.*

This theorem allows us to reconstruct the conditional distribution of the point process Φ given the observations $Z(\mathbf{x}) = \mathbf{z}$ via a three-step procedure: construct first the conditional hitting scenario θ and the extremal functions $\varphi_1^+, \ldots, \varphi_\ell^+$ (step 1 and 2), this yields Φ^+; independently construct the conditional subextremal point process Φ^- (step 3); finally set $\Phi = \Phi^+ \cup \Phi^-$ to obtain the conditional point process Φ.

The conditional probability appearing in steps 2 and 3 are quite natural. Given the observations $Z(\mathbf{x}) = \mathbf{z}$ and the hitting scenario $\theta = \tau$, the extremal function φ_j^+ must satisfy the constraints $\varphi_j^+(\mathbf{x}_{\tau_j}) = \mathbf{z}_{\tau_j}$ and $\varphi_j^+(\mathbf{x}_{\tau_j^c}) < \mathbf{z}_{\tau_j^c}$ so that it seems natural to obtain the conditional intensity Λ given these constraints. Similarly, given $Z(\mathbf{x}) = \mathbf{z}$, the subextremal functions $\varphi \in \Phi^-$ must satisfy the constraint $\varphi(\mathbf{x}) < \mathbf{z}$ which naturally leads to the point process intensity Λ restricted to the set of functions $\{f \in \mathcal{C}, f(\mathbf{x}) < \mathbf{z}\}$.

11.2.3   Examples of Regular Models

Theorem 11.2.2 requires the assumption of regularity of the intensity measure Λ at point $\mathbf{x} = (x_1, \ldots, x_k)$. We now recall some popular models for max-stable process that are regular and give the corresponding intensity function $\lambda_{\mathbf{x}}$. These models are mainly based on Gaussian or related distributions and provide a nice framework for explicit computations.

Example 1: Brown–Resnick process

The Brown–Resnick process introduced by Kabluchko et al. (2009) is a stationary max-stable random field on $\mathcal{X} = \mathbb{R}^d$ corresponding to the case where $Y(x) = \exp\{W(x) - \gamma(x)\}$ in (11.2) with W a centered Gaussian process with stationary increments, semi-variogram γ and such that $W(o) = 0$ almost surely. One can show that the intensity measure Λ is regular at $\mathbf{x} \in \mathcal{X}^k$ as long as the covariance matrix $\Sigma_{\mathbf{x}}$ of the random vector $W(\mathbf{x})$ is positive definite. We denote by $g_{\mathbf{x}}$ the Gaussian density

$$g_{\mathbf{x}}(\mathbf{u}) = (2\pi)^{-k/2}\det(\Sigma_{\mathbf{x}})^{-1/2}\exp\left\{-\frac{1}{2}\mathbf{u}^T\Sigma_{\mathbf{x}}^{-1}\mathbf{u}\right\}.$$

The marginal intensity measure is computed as follows: for all Borel set $A \subset (0, +\infty)^k$

$$
\begin{aligned}
\Lambda_{\mathbf{x}}(A) &= \int_0^\infty \mathbb{P}[\zeta Y(\mathbf{x}) \in A]\zeta^{-2}\mathrm{d}\zeta \\
&= \int_0^\infty \mathbb{P}[\zeta\exp\{W(\mathbf{x}) - \gamma(\mathbf{x})\} \in A]\zeta^{-2}\mathrm{d}\zeta \\
&= \int_0^\infty \int_A g_{\mathbf{x}}(\log\mathbf{z} - \log\zeta + \gamma(\mathbf{x}))\prod_{i=1}^k z_i^{-1}\zeta^{-2}\mathrm{d}\zeta\mathrm{d}\mathbf{z} \\
&= \int_A \lambda_{\mathbf{x}}(\mathbf{z})\mathrm{d}\mathbf{z}
\end{aligned}
$$

with

$$\lambda_{\mathbf{x}}(\mathbf{z}) = \int_0^\infty g_{\mathbf{x}}(\log\mathbf{z} - \log\zeta + \gamma(\mathbf{x}))\prod_{i=1}^k z_i^{-1}\zeta^{-2}\mathrm{d}\zeta.$$

Some standard but tedious computations for Gaussian integrals reveal that

$$\lambda_{\mathbf{x}}(\mathbf{z}) = C_{\mathbf{x}}\exp\left(-\frac{1}{2}\log\mathbf{z}^T Q_{\mathbf{x}}\log\mathbf{z} + L_{\mathbf{x}}\log\mathbf{z}\right)\prod_{i=1}^k z_i^{-1}, \quad \mathbf{z} \in (0, \infty)^k,$$

with $1_k = (1)_{i=1,\dots,k}$, $\gamma_{\mathbf{x}} = \{\gamma(x_i)\}_{i=1,\dots,k}$,

$$Q_{\mathbf{x}} = \Sigma_{\mathbf{x}}^{-1} - \frac{\Sigma_{\mathbf{x}}^{-1}1_k 1_k^T\Sigma_{\mathbf{x}}^{-1}}{1_k^T\Sigma_{\mathbf{x}}^{-1}1_k}, \qquad L_{\mathbf{x}} = \left(\frac{1_k^T\Sigma_{\mathbf{x}}^{-1}\gamma_{\mathbf{x}} - 1}{1_k^T\Sigma_{\mathbf{x}}^{-1}1_k}1_k - \gamma_{\mathbf{x}}\right)^T\Sigma_{\mathbf{x}}^{-1},$$

$$C_{\mathbf{x}} = (2\pi)^{(1-k)/2}\det(\Sigma_{\mathbf{x}})^{-1/2}(1_k^T\Sigma_{\mathbf{x}}^{-1}1_k)^{-1/2}\dots$$

$$\dots\exp\left\{\frac{1}{2}\frac{(1_k^T\Sigma_{\mathbf{x}}^{-1}\gamma_{\mathbf{x}} - 1)^2}{1_k^T\Sigma_{\mathbf{x}}^{-1}1_k} - \frac{1}{2}\gamma_{\mathbf{x}}^T\Sigma_{\mathbf{x}}^{-1}\gamma_{\mathbf{x}}\right\}.$$

See Dombry et al. (2013) for more details.

Example 2: Schlather process
The Schlather process (Schlather, 2002), also called the extremal Gaussian process, is a max-stable random field on $\mathcal{X} = \mathbb{R}^d$ corresponding to the case where

$Y(x) = (2\pi)^{1/2} \max\{0, W(x)\}$ in (11.2) with W a standard Gaussian process with correlation function ρ. For $\mathbf{x} \in \mathcal{X}^k$ and provided the covariance matrix Σ_x of the random vector $W(\mathbf{x})$ is positive definite, the intensity function is

$$\lambda_{\mathbf{x}}(\mathbf{z}) = \pi^{-(k-1)/2} \det(\Sigma_{\mathbf{x}})^{-1/2} a_{\mathbf{x}}(\mathbf{z})^{-(k+1)/2} \Gamma\left(\frac{k+1}{2}\right), \qquad \mathbf{z} \in (0, +\infty)^k,$$

where $a_{\mathbf{x}}(\mathbf{z}) = \mathbf{z}^T \Sigma_{\mathbf{x}}^{-1} \mathbf{z}$. See Dombry et al. (2013) for more details on the computation.

Example 3: Extremal t process

The extremal t process is a generalization of the Schlather process above obtained with an extra parameter $\nu > 0$ that yields more flexibility in the model. It is a max-stable random field on $\mathcal{X} = \mathbb{R}^d$ obtained with $Y(x) = c_\nu \max\{0, W(x)^\nu\}$ in (11.2) with W a standard Gaussian process with correlation function ρ and $c_\nu = \sqrt{\pi} 2^{-(\nu-2)/2} \Gamma((\nu+1)/2)^{-1}$. Provided the covariance matrix Σ_x of the random vector $W(\mathbf{x})$ is positive definite, the intensity function is

$$\lambda_{\mathbf{x}}(\mathbf{z}) = c_\nu \nu^{-k+1} 2^{(\nu-2)/2} \pi^{-k/2} \det(\Sigma_{\mathbf{x}})^{-1/2} a_{\mathbf{x}}(\mathbf{z}, \nu)^{-(k+\nu)/2} \cdots$$

$$\cdots \Gamma\left(\frac{k+\nu}{2}\right) \prod_{j=1}^{k} z_j^{(1-\nu)/\nu}$$

for $\mathbf{z} \in (0, +\infty)^k$ and with $a_{\mathbf{x}}(\mathbf{z}, \nu) = (\mathbf{z}^{1/\nu})^T \Sigma_{\mathbf{x}}^{-1} \mathbf{z}^{1/\nu}$. See Ribatet (2013).

11.2.4 Distribution of the Extremal Functions

The distribution of the extremal function appearing in Equation (11.5) is still quite theoretical at this stage and it is unclear how one can sample from this distribution. There are also theoretical questions related to its definition: the intensity measure Λ has an infinite total mass and the conditioning event $\{f(\mathbf{x}_{T_j}) = \mathbf{z}_{T_j}, f(\mathbf{x}_{T_j^c}) < \mathbf{z}_{T_j^c}\}$ has zero measure.

A way to bypass these difficulties is to consider only the finite dimensional margins of the extremal functions φ_j^+ and not the full random process. In practice, this is enough for simulation purposes since one simulates the random field on a finite grid only and not on the whole state space $\mathcal{X}$. Let $\mathbf{s} = (s_1, \ldots, s_m) \in \mathcal{X}^m$ be the set of new locations for the conditional sampling. We focus on the conditional distribution of $Z(\mathbf{s}) \mid Z(\mathbf{x}) = \mathbf{z}$. Equation (11.5) can be simplified if we assume the regularity of the intensity measure Λ at $(\mathbf{x}, \mathbf{s}) \in \mathcal{X}^{k+m}$. We can then introduce the conditional intensity function

$$\lambda_{\mathbf{s}|\mathbf{x},\mathbf{z}}(\mathbf{u}) = \frac{\lambda_{(\mathbf{s},\mathbf{x})}(\mathbf{u}, \mathbf{z})}{\lambda_{\mathbf{x}}(\mathbf{z})}, \qquad \mathbf{u} \in (0, \infty)^m.$$

Equation (11.5) can be rewritten as

$$\mathbb{P}\left[\varphi_j^+(\mathbf{s}) \in d\mathbf{v} \mid Z(\mathbf{x}) = \mathbf{z}, \theta = \tau\right]$$

$$= \frac{1}{C_j} \left(\int 1_{\{\mathbf{u_j} < \mathbf{z}_{\tau_j^c}\}} \lambda_{(\mathbf{s}, \mathbf{x}_{\tau_j^c}) | \mathbf{x}_{\tau_j}, \mathbf{z}_{\tau_j}} (\mathbf{v}, \mathbf{u_j}) d\mathbf{u_j} \right) d\mathbf{v} \qquad (11.6)$$

with C_j the normalization constant

$$C_j = \int 1_{\{\mathbf{u_j} < \mathbf{z}_{\tau_j^c}\}} \lambda_{(\mathbf{s}, \mathbf{x}_{\tau_j^c}) | \mathbf{x}_{\tau_j}, \mathbf{z}_{\tau_j}} (\mathbf{v}, \mathbf{u_j}) d\mathbf{u_j} d\mathbf{v}.$$

In words, the conditional law of $\varphi_j^+(\mathbf{s})$ is equal to the distribution of the random variable $\mathbf{V}$ obtained as the first component of $(\mathbf{V}, \mathbf{U_j})$ with density $\lambda_{(\mathbf{s}, \mathbf{x}_{\tau_j^c}) | \mathbf{x}_{\tau_j}, \mathbf{z}_{\tau_j}} (\mathbf{v}, \mathbf{u})$ conditioned to the event $\mathbf{U_j} < \mathbf{z}_{\tau_j^c}$.

Example 1 continued: Brown–Resnick process

In the case of Brown–Resnick process, for all $(\mathbf{s}, \mathbf{x}) \in \mathcal{X}^{m+k}$, $(\mathbf{u}, \mathbf{z}) \in (0, \infty)^{m+k}$ and provided the covariance matrix $\Sigma_{(\mathbf{s}, \mathbf{x})}$ is positive definite, the conditional intensity function corresponds to a multivariate log-normal probability density function

$$\lambda_{\mathbf{s}|\mathbf{x}, \mathbf{z}}(\mathbf{u}) = (2\pi)^{-m/2} \det(\Sigma_{\mathbf{s}|\mathbf{x}})^{-1/2} \cdots$$

$$\cdots \exp \left\{ -\frac{1}{2} (\log \mathbf{u} - \mu_{\mathbf{s}|\mathbf{x}, \mathbf{z}})^T \Sigma_{\mathbf{s}|\mathbf{x}}^{-1} (\log \mathbf{u} - \mu_{\mathbf{s}|\mathbf{x}, \mathbf{z}}) \right\} \prod_{i=1}^{m} u_i^{-1},$$

where $\mu_{\mathbf{s}|\mathbf{x}, \mathbf{z}} \in \mathbb{R}^m$ and $\Sigma_{\mathbf{s}|\mathbf{x}}$ are the mean and covariance matrix of the underlying normal distribution and are given by

$$\Sigma_{\mathbf{s}|\mathbf{x}}^{-1} = J_{m,k}^T Q_{(\mathbf{s}, \mathbf{x})} J_{m,k}$$

$$\mu_{\mathbf{s}|\mathbf{x}, \mathbf{z}} = \left\{ L_{(\mathbf{s}, \mathbf{x})} J_{m,k} - \log \mathbf{z}^T \tilde{J}_{m,k}^T Q_{(\mathbf{s}, \mathbf{x})} J_{m,k} \right\} \Sigma_{\mathbf{s}|\mathbf{x}},$$

with

$$J_{m,k} = \begin{bmatrix} \mathrm{Id}_m \\ 0_{k,m} \end{bmatrix}, \qquad \tilde{J}_{m,k} = \begin{bmatrix} 0_{m,k} \\ \mathrm{Id}_k \end{bmatrix},$$

where Id_k denotes the $k \times k$ identity matrix and $0_{m,k}$ the $m \times k$ null matrix.

Example 2 continued: Schlather process

For $(\mathbf{s}, \mathbf{x}) \in \mathcal{X}^{m+k}$, $(\mathbf{u}, \mathbf{z}) \in \mathbb{R}^{m+k}$ and provided that the covariance matrix $\Sigma_{(\mathbf{s}, \mathbf{x})}$ is positive definite, the conditional intensity function $\lambda_{\mathbf{s}|\mathbf{x}, \mathbf{z}}(\mathbf{u})$ corresponds to a multivariate Student distribution

$$\lambda_{\mathbf{s}|\mathbf{x}, \mathbf{z}}(\mathbf{u}) = \pi^{-m/2} (k+1)^{-m/2} \det(\tilde{\Sigma})^{-1/2} \cdots$$

$$\cdots \left\{ 1 + \frac{(\mathbf{u} - \mu)^T \tilde{\Sigma}^{-1} (\mathbf{u} - \mu)}{k+1} \right\}^{-(m+k+1)/2} \frac{\Gamma\left(\frac{m+k+1}{2}\right)}{\Gamma\left(\frac{k+1}{2}\right)},$$

with $k+1$ degrees of freedom, location parameter $\mu = \Sigma_{\mathbf{s}:\mathbf{x}} \Sigma_{\mathbf{x}}^{-1} \mathbf{z}$ and scale matrix

$$\tilde{\Sigma} = \frac{a_{\mathbf{x}}(\mathbf{z})}{k+1} \left(\Sigma_{\mathbf{s}} - \Sigma_{\mathbf{s}:\mathbf{x}} \Sigma_{\mathbf{x}}^{-1} \Sigma_{\mathbf{x}:\mathbf{s}} \right)$$

where

$$\Sigma_{(s,x)} = \begin{bmatrix} \Sigma_s & \Sigma_{s:x} \\ \Sigma_{x:s} & \Sigma_x \end{bmatrix}.$$

Example 3 continued: Extremal t process

The results are similar and generalize the case of Schlather processes. For $(s, x) \in \mathcal{X}^{m+k}$, $(u, z) \in \mathbb{R}^{m+k}$ we set $\mu = \Sigma_{s:x}\Sigma_x^{-1}z^{1/\nu}$ and

$$\tilde{\Sigma} = (k+\nu)^{-1}a_x(z,\nu)(\Sigma_s - \Sigma_{s:x}\Sigma_x^{-1}\Sigma_{x:s})$$

Then we have

$$\lambda_{s|x,z}(u) = \pi^{-m/2}(k+\nu)^{-m/2}\det(\tilde{\Sigma})^{-1/2}\cdots$$
$$\cdots\left\{1 + \frac{(u^{1/\nu} - \mu)^T\tilde{\Sigma}^{-1}(u^{1/\nu} - \mu)}{k+\nu}\right\}^{-(m+k+\nu)/2}\cdots$$
$$\cdots\frac{\Gamma\left(\frac{m+k+\nu}{2}\right)}{\Gamma\left(\frac{k+\nu}{2}\right)}\left\{\nu^{-m}\prod_{j=1}^{m}u_j^{-(\nu-1)/\nu}\right\}.$$

The last term in brackets in the previous equation corresponds to the Jacobian of the mapping $u \mapsto u^{1/\nu}$. Hence we recognize that the conditional intensity function is the density of the random vector T^ν where T is a Student random vector with $k + \nu$ degrees of freedom, mean μ and scale matrix $\tilde{\Sigma}$.

11.3 Conditional Sampling of Max-Stable Processes

We now consider the conditional sampling of max-stable processes following Dombry et al. (2013). We present a three-step sampling procedure that is quite straightforwardly derived from Theorem 11.2.2. We particularly focus on the first step and discuss a Gibbs sampling approach for the conditional hitting scenario.

11.3.1 A Three-Step Procedure for Conditional Sampling

Theorem 11.2.2 provides the conditional distribution of the point process Φ given the observation (11.1). In practice, we are rather interested in the conditional simulation of the max-stable process $Z(s)$ at some location $s = (s_1,\ldots,s_m) \in \mathcal{X}^m$. The connection is rather straightforward and is made explicit in the next proposition.

Proposition 11.3.1. *Assume that Λ is regular at $(x, s) \in \mathcal{X}^{k+m}$. Consider the three-step procedure:*

1. Draw a random partition $\theta \in \mathcal{P}_k$ with distribution (11.4);

2. *Given $\theta = (\tau_1, \ldots, \tau_\ell)$, draw ℓ independent random vectors $\varphi_j^+(\mathbf{s}), \ldots, \varphi_\ell^+(\mathbf{s})$ with distribution (11.6) and define the random vector*

$$Z^+(\mathbf{s}) = \max_{j=1,\ldots,\ell} \varphi_j^+(\mathbf{s}).$$

3. *Independently draw a Poisson point process $\{\zeta_i\}_{i \geqslant 1}$ on $(0, \infty)$ with intensity $\zeta^{-2}d\zeta$ and $\{Y_i\}_{i \geqslant 1}$ independent copies of Y, and define the random vector*

$$Z^-(\mathbf{s}) = \max_{i \geqslant 1} \zeta_i Y_i(\mathbf{s}) 1_{\{\zeta_i Y_i(\mathbf{x}) < \mathbf{z}\}}.$$

Then the random vector $\tilde{Z}(\mathbf{s}) = \max\{Z^+(\mathbf{s}), Z^-(\mathbf{s})\}$ follows the conditional distribution of $Z(s)$ given $Z(x) = z$.

This three-step procedure — which, in general, can also be applied in the non-regular case — is illustrated on Figure 11.2 on the simple case of 1-D Smith storm process with Gaussian shape.

In step 3, it is worth noting that $Z^-(\mathbf{s})$ follows the conditional distribution of $Z(\mathbf{s})$ given $Z(\mathbf{x}) < \mathbf{z}$. It is not difficult to show that the conditional cumulative distribution function of $Z(\mathbf{s})$ given $Z(\mathbf{x}) = \mathbf{z}$ is given by

$$\mathbb{P}[Z(\mathbf{s}) \leqslant \mathbf{a} \mid Z(\mathbf{x}) = \mathbf{z}] = \mathbb{P}[Z(\mathbf{s}) \leqslant \mathbf{a} \mid Z(\mathbf{x}) < \mathbf{z}] \sum_{\tau \in \mathcal{P}_k} \pi_{\mathbf{x},\mathbf{z}}(\tau) \prod_{j=1}^{\ell} F_{\tau,j}(\mathbf{a}),$$

where $\pi_{\mathbf{x},\mathbf{z}}(\tau) = \mathbb{P}[\theta = \tau \mid Z(\mathbf{x}) = \mathbf{z}]$ is given by Equation (11.4) and

$$
\begin{aligned}
F_{\tau,j}(\mathbf{a}) &= \mathbb{P}[\varphi_j^+(\mathbf{s}) \leqslant \mathbf{a} \mid Z(\mathbf{x}) = \mathbf{z}, \theta = \tau] \\
&= \frac{\int_{\{\mathbf{u}_j < \mathbf{z}_{\tau_j^c}, \mathbf{v} < \mathbf{a}\}} \lambda_{(\mathbf{s}, \mathbf{x}_{\tau_j^c}) \mid \mathbf{x}_{\tau_j}, \mathbf{z}_{\tau_j}}(\mathbf{v}, \mathbf{u}_j) d\mathbf{u}_j d\mathbf{v}}{\int_{\{\mathbf{u}_j < \mathbf{z}_{\tau_j^c}\}} \lambda_{\mathbf{x}_{\tau_j^c} \mid \mathbf{x}_{\tau_j}, \mathbf{z}_{\tau_j}}(\mathbf{u}_j) d\mathbf{u}_j}.
\end{aligned}
$$

11.3.2 Gibbs Sampling for the Conditional Hitting Scenario

In the above three-step procedure for conditional sampling, step 2 requires sampling the extremal functions. As we have seen in Examples 1, 2, and 3 where log-normal and Student distributions appear, these can be of a relatively simple structure although the imposed equality and inequality constraints may cause additional difficulties. Step 3 can be performed by an unconditional simulation of the max-stable process Z according to an appropriate spectral representation rejecting all those functions that do not respect the constraints given by the conditioning data. For details on unconditional simulation of max-stable processes we refer to the corresponding chapter. However, the conditional sampling of the hitting scenario (step 1) remains the most challenging step. According to Equation (11.4), it is given by

$$\pi_{\mathbf{x},\mathbf{z}}(\tau) = \mathbb{P}[\theta = \tau \mid Z(\mathbf{x}) = \mathbf{z}] = \frac{1}{C_{\mathbf{x},\mathbf{z}}} \prod_{j=1}^{\ell} \omega_{\tau_j}$$

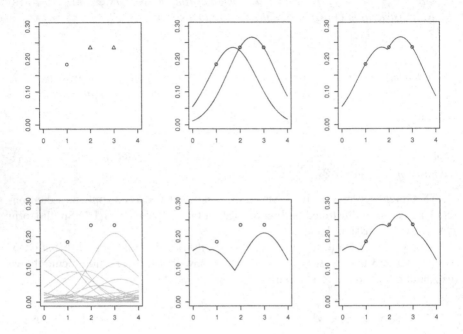

FIGURE 11.2
The three-step procedure for simulation sampling. First line: construction of the conditional hitting scenario (step 1, left), of the extremal functions (step 2, middle) and of the associated process Z^+ (right). Second line: construction of the subextremal functions (step 3, left), of the associated process Z^- (middle) and of the conditioned max-stable process $\tilde{Z} = \max(Z^+, Z^-)$ (right).

where

$$\omega_{\tau_j} = \lambda_{\mathbf{x}_{\tau_j}}(\mathbf{z}_{\tau_j}) \int_{\{\mathbf{u}_j < \mathbf{z}_{\tau_j^c}\}} \lambda_{\mathbf{x}_{\tau_j^c}|\mathbf{x}_{\tau_j},\mathbf{z}_{\tau_j}}(\mathbf{u}_j)\mathrm{d}\mathbf{u}_j.$$

The last integral is the multivariate cumulative distribution of the conditional intensity function $\lambda_{\mathbf{x}_{\tau_j^c}|\mathbf{x}_{\tau_j},\mathbf{z}_{\tau_j}}$. Hence this is a multivariate log-normal cdf in the Brown–Resnick model and a multivariate Student cdf in the Schlather or extremal-t models. In the following, we assume that the weights ω_{τ_j} can be accurately computed numerically and we focus on how to sample from $\pi_{\mathbf{x},\mathbf{z}}$, a discrete probability measure on the set $\mathcal{P}_k$ of partition of $\{x_1, \cdots, x_k\}$. The main difficulty is that the norming constant $C_{\mathbf{x},\mathbf{z}}$ is unknown and a naive computation of this constant requires the computation of the weight $\prod_{j=1}^{\ell} \omega_{\tau_j}$ for all $\tau \in \mathcal{P}_k$. The number of terms, or equivalently the cardinality of $\mathcal{P}_k$ is given by the so-called Bell number B_k. The first 10 Bell numbers are

k	1	2	3	4	5	6	7	8	9	10
B_k	1	2	5	15	52	203	877	4140	21147	115975

Hence determining the discrete probability $\pi_{\mathbf{x},\mathbf{z}}$ requires the computation of 52 weights for $k = 5$ and 115975 for $k = 10$. Even worse, $B_{15} \approx 1.4\,10^9$ and $B_{20} \approx 5.2\,10^{13}$ so that the storage of all partitions is beyond the memory capacities of a standard computer. This is the so-called phenomenon of combinatorial explosion and we need an alternative method to the naive discrete enumeration of all possibilities.

It is customary to use Monte Carlo Markov Chain sampling when the target distribution is known up to a multiplicative constant only. The aim is to construct a Markov chain with stationary distribution equal to the target distribution and with good mixing properties. Two main methods exist: the Metropolis–Hasting algorithm and the Gibbs sampler. We focus here on this second option.

For $\tau \in \mathcal{P}_k$ and $j \in \{1, \ldots, k\}$, let τ_{-j} be the restriction of τ to the set $\{x_1, \ldots, x_k\}$. As usual with Gibbs samplers, our goal is to simulate from

$$\mathbb{P}[\theta \in \cdot \mid \theta_{-j} = \tau_{-j}], \qquad (11.7)$$

where $\theta \in \mathcal{P}_k$ is a random partition which follows the target distribution $\pi_{\mathbf{x},\mathbf{z}}(\cdot)$ and τ is typically the current state of the Markov chain. It is easy to see that the number of possible updates according to (11.7) is always less than k, so that the combinatorial explosion is avoided. Indeed, the point x_j can be reallocated to any of the components of τ_{-j} or to a new component with a single point. We deduce that the number of possible updates $\tau^* \in \mathcal{P}_k$ such that $\tau^*_{-j} = \tau_{-j}$ is

$$b^+ = \begin{cases} \ell & \text{if } \{x_j\} \text{ is a partitioning set of } \tau, \\ \ell + 1 & \text{if } \{x_j\} \text{ is not a partitioning set of } \tau, \end{cases}$$

For illustration, consider the set $\{x_1, x_2, x_3\}$ and let $\tau = (\{x_1, x_2\}, \{x_3\})$. Then the possible partitions τ^* such that $\tau^*_{-2} = \tau_{-2}$ are

$$(\{x_1, x_2\}, \{x_3\}), \qquad (\{x_1\}, \{x_2\}, \{x_3\}), \qquad (\{x_1\}, \{x_2, x_3\}), \qquad (11.8)$$

while there exists only two partitions such that $\tau^*_{-3} = \tau_{-3}$, i.e.,

$$(\{x_1, x_2\}, \{x_3\}), \qquad (\{x_1, x_2, x_3\}).$$

For our work we use a random scan implementation of the Gibbs sampler (Liu et al., 1995), meaning that one iteration of the Gibbs sampler selects randomly an element of $j \in \{1, \ldots, k\}$ and then updates the current state τ according to the proposal distribution (11.7). For the sake of simplicity, we use the uniform random scan, i.e., j is selected according to the uniform distribution on $\{1, \ldots, k\}$. Figure 11.3 shows two successive iterations of this random scan Gibbs sampler.

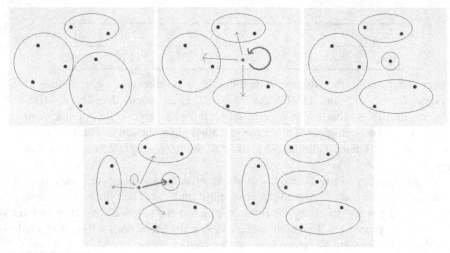

FIGURE 11.3
Two successive iterations of the Gibbs sampler for the conditional hitting scenario with $k = 8$ conditioning points: after choosing a random point, the arrows show the different possible reallocations, the bold arrow representing the chosen one.

The distribution (11.7) has nice properties. For all $\tau^* \in \mathscr{P}_k$ with $\tau^*_{-j} = \tau_{-j}$ we have

$$
\mathbb{P}[\theta = \tau^* \mid \theta_{-j} = \tau_{-j}] = \frac{\pi_{\mathbf{x},\mathbf{z}}(\tau^*)}{\displaystyle\sum_{\tilde{\tau} \in \mathscr{P}_k} \pi_{\mathbf{x},\mathbf{z}}(\tilde{\tau}) 1_{\{\tilde{\tau}_{-j} = \tau_{-j}\}}} \propto \frac{\prod_{k=1}^{|\tau^*|} w_{\tau^*_k}}{\prod_{k=1}^{|\tau|} w_{\tau_k}}. \tag{11.9}
$$

Since τ and τ^* share many components, it can be seen that many factors in the right-hand side of (11.9) cancel out except at most four of them. In the previous example (11.8), the corresponding weights are

$$
1, \quad \frac{w_{\{x_1,x_2\}}}{w_{\{x_1\}}w_{\{x_2\}}}, \quad \frac{w_{\{x_1,x_2\}}w_{\{x_3\}}}{w_{\{x_1\}}w_{\{x_2,x_3\}}} \quad \text{respectively.}
$$

This makes the Gibbs sampler especially convenient.

We finally provide some practical details on the computation of these conditional weights as well as a detailed algorithm; see Algorithm 11.1.

We first describe how each partition of $\{x_1, \ldots, x_k\}$ is stored. To illustrate consider the set $\{x_1, x_2, x_3\}$ and the partition $(\{x_1, x_2\}, \{x_3\})$. With our convention, this partition is defined as $(1, 1, 2)$ indicating that x_1 and x_2 belong to the same partitioning set labeled "1" and x_3 belongs to the partitioning set "2". There exist several equivalent notations for this partition: for example one can use $(2, 2, 1)$ or $(1, 1, 3)$. However there is a one-one mapping between $\mathscr{P}_k$ and the set

$$
\mathcal{P}^*_k = \left\{ (a_1, \ldots, a_k), \, \forall i \in \{2, \ldots, k\} : 1 = a_1 \leqslant a_i \leqslant \max_{1 \leqslant j < i} a_j + 1, \, a_i \in \mathbb{Z} \right\}.
$$

Algorithm 11.1 Computing the probabilities of (11.9).

Input: A partition $\tau \in \mathcal{P}_k^*$ of size ℓ and $j \in \{1, \ldots, k\}$

Output: The conditional weights $w^* = (w_1^*, \ldots, w_{b+}^*)$

// Identify which partitioning set x_j belongs to

$s \leftarrow \tau_j$;

// Compute the size of this partition set

$r_1 \leftarrow \sum_{m=1}^{k} \delta_{\tau_m = s}$;

// Create a new partition that will be updated

$\tau^* \leftarrow \tau$;

// Get the number of possible states for τ^*

$b^+ \leftarrow \ell + \delta_{r_1 \neq 1}$;

for $b \leftarrow 1$ **to** b^+ **do**

  // b refers to the partitioning set where x_j is moved to

  // Update the partition τ^*

  $\tau_j^* = b$;

  // Compute the conditional weights for this new partition

  $r_2 \leftarrow \sum_{m=1}^{k} \delta_{\tau_m = b}$;

  if $b = s$ **then**

    // This is case (11.10a)

    $w_b^* \leftarrow 1$;

    continue;

  end if

  Case $r_1 = 1$ and $r_2 \neq 0$ // This is case (11.10b)

    $w_b^* \leftarrow w_{\tau^*, b} / \{w_{\tau, b} w_{\tau, s}\}$;

  Case $r_1 \neq 1$ and $r_2 \neq 0$ // This is case (11.10c)

    $w_b^* \leftarrow w_{\tau^*, b} w_{\tau^*, s} / \{w_{\tau, b} w_{\tau, s}\}$;

  Case $r_1 \neq 1$ and $r_2 = 0$ // This is case (11.10d)

    $w_b^* \leftarrow w_{\tau^*, b} w_{\tau^*, s} / w_{\tau, s}$;

end for

Normalize the conditional weights to sum to 1;

return w^*

Consequently we shall restrict our attention to the partitions that live in $\mathcal{P}_k^*$ and going back to our example we see that $(1, 1, 2)$ is valid but $(2, 2, 1)$ and $(1, 1, 3)$ are not.

For $\tau \in \mathcal{P}_k^*$ of size ℓ, let $r_1 = \sum_{i=1}^{k} \delta_{\tau_i = a_j}$ and $r_2 = \sum_{i=1}^{k} \delta_{\tau_i = b}$, i.e., the number of conditioning locations that belong to the partitioning sets "a_j" and "b" where $b \in \{1, \ldots, b^+\}$ and

$$b^+ = \begin{cases} \ell & (r_1 = 1), \\ \ell + 1 & (r_1 \neq 1). \end{cases}$$

Then the conditional probability distribution $\mathbb{P}[\tau_j = b \mid \tau_i = a_i, \forall i \neq j]$ given by

Equation (11.9) is proportional to

$$
\begin{cases}
1 & (b = a_j), & (11.10\text{a}) \\
w_{\tau_*,b}/(w_{\tau,b}w_{\tau,a_j}) & (r_1 = 1, r_2 \neq 0, b \neq a_j), & (11.10\text{b}) \\
w_{\tau_*,b}w_{\tau_*,a_j}/(w_{\tau,b}w_{\tau,a_j}) & (r_1 \neq 1, r_2 \neq 0, b \neq a_j), & (11.10\text{c}) \\
w_{\tau_*,b}w_{\tau_*,a_j}/w_{\tau,a_j} & (r_1 \neq 1, r_2 = 0, b \neq a_j), & (11.10\text{d})
\end{cases}
$$

where $\tau_* = (a_1, \ldots, a_{j-1}, b, a_{j+1}, \ldots, a_k)$ and $w_{\tau,a_i} = w_{\{x_k: \tau_k = a_i\}}$. It is worth stressing that although τ_* does not necessarily belong to $\mathcal{P}_k^*$, it corresponds to a unique partition of $\mathcal{P}_k$ and we can use the bijection $\mathcal{P}_k \to \mathcal{P}_k^*$ to recode τ_* into an element of $\mathcal{P}_k^*$.

11.4   Simulation Study

11.4.1   Gibbs Sampler

In this section we check whether the Gibbs sampler is able to sample from $\pi_{\mathbf{x},\mathbf{z}}(\cdot)$. To illustrate a typical sample path, Figure 11.4 shows the trace plot of a simulated chain of length 2000 with $k = 5$ conditioning locations and a thinning lag of 5 and compares the theoretical probabilities $\{\pi_{\mathbf{x},\mathbf{z}}(\tau), \tau \in \mathcal{P}_k\}$ to the empirical ones estimated from the Markov chain. As mentioned in the previous section it can be seen that only a few states have a significant probability to occur and these states most often differs only slightly. As expected for this particular simulation the empirical probabilities match the theoretical ones.

To better assess if our uniform random scan Gibbs sampler is able to sample from $\pi_{\mathbf{x},\mathbf{z}}(\cdot)$ we simulate 250 Markov chains of length 1000—after removing a burn-in period and thinning the chain, and similarly to Figure 11.4, compare the theoretical probabilities $\{\pi_{\mathbf{x},\mathbf{z}}(\tau), \tau \in \mathcal{P}_k\}$ to the empirical ones using a χ^2 test for each of these chains. Since the computation of these theoretical probabilities is CPU intensive, the number of conditioning locations is at most 5 in our simulation study.

Figure 11.5 plots the sample p-values of these χ^2 tests against the quantiles of a $U(0, 1)$ distribution for a varying number of conditional locations. This figure corroborates what we saw in Figure 11.4 since the sample p-values seem to follow a $U(0, 1)$ distribution indicating that the sampler is able to sample from $\pi_{\mathbf{x},\mathbf{z}}(\cdot)$.

11.4.2   Conditional Simulations

In this section we check if our algorithm is able to produce realistic conditional simulations of Brown–Resnick processes with semi-variogram $\gamma(h) = (h/\lambda)^\nu$. In this case, the spectral process $Y(\cdot)$ in (11.2) is a fractional Brownian motion with Hurst index ν. To have a broad overview, we consider three different sample path properties as summarized below.

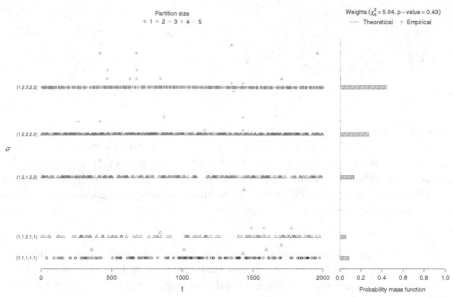

FIGURE 11.4
Left: Trace plot of one Markov chain generated using the uniform random scan Gibbs sampler and whose target distribution is $\pi_{\mathbf{x},\mathbf{z}}(\cdot)$ with $k = 5$ conditioning locations. The labels on the y-axis indicates the partitions having a probability greater than 0·01 to occur. Right: Comparison of the theoretical probabilities $\{\pi_{\mathbf{x},\mathbf{z}}(\tau), \tau \in \mathcal{P}_k\}$ and the empirical ones estimated from the simulated Markov chain with an insert showing the sample χ^2 statistic and the associated p-value.

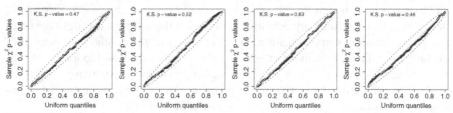

FIGURE 11.5
QQ-plots for the sample χ^2 test p-values against $U(0,1)$ quantiles with a varying number of conditioning locations k with an insert showing the p-values of a Kolmogorov–Smirnov test for uniformity — from left to right, $k = 2, 3, 4$ and 5. The dashed lines show the 95% confidence envelopes.

	Sample path properties		
	γ_1: Wiggly	γ_2: Smooth	γ_3: Very smooth
λ	25	54	69
ν	0·5	1·0	1·5

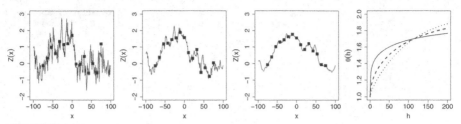

FIGURE 11.6

Three realizations of a Brown–Resnick process with standard Gumbel margins and semi-variograms γ_1, γ_2, and γ_3—from left to right. The squares correspond to the 15 conditioning values that will be used in the simulation study. The right panel shows the associated extremal coefficient functions where the solid, dashed, and dotted lines correspond respectively to γ_1, γ_2, and γ_3.

The variogram parameters are set to ensure that the extremal coefficient function satisfies $\theta(115) = 1.7$. Figure 11.6 shows one realization for each sample path configuration as well as the corresponding extremal coefficient function. These realizations will serve as the basis for our conditioning events.

In order to check if our sampling procedure is accurate, given a single conditional event $\{Z(\mathbf{x}) = \mathbf{z}\}$, we generated 1000 conditional realizations of a Brown–Resnick process with standard Gumbel margins and semi-variograms γ_j ($j = 1, 2, 3$).

Figure 11.7 shows the pointwise sample quantiles obtained from these 1000 simulated paths in comparison to unit Gumbel quantiles. As expected the conditional sample paths inherit the regularity driven by the Hurst index $\nu/2$ of the process $Y(\cdot)$ in (11.2) and there is less variability in regions close to some conditioning locations. On the opposite, although for this specific type of variogram $\theta(h) \to 2$ as $h \to \infty$, for regions far away from any conditioning location the sample quantiles converges to that of a standard Gumbel distribution indicating that the conditional event does not have any influence anymore. In addition the sample paths used to get the conditional events, see Figure 11.6, lay most of the time in the 95% pointwise confidence intervals corroborating that our sampling procedure seems to be accurate — the coverage ranges between $0 \cdot 93$ and $1 \cdot 00$ with a mean value of $0 \cdot 96$.

So far we have checked that the proposed sampling procedure yields the expected coverage as well as the right marginal properties as we move far away from any conditioning location. The last point to be fulfilled is to assess whether the simulation procedure honors the spatial dependence driven by the semi-variogram $\gamma(\cdot)$. To this aim we use the F-madogram (Cooley et al., 2006) to compare the pairwise extremal coefficient estimates to the theoretical extremal coefficient function. Since $Z(\cdot) \mid \{Z(\mathbf{x}) = \mathbf{z}\}$ is not max-stable the F-madogram cannot be used. However, by integrating out the conditional event we recover the original Brown–Resnick distribution and the max-stability property. So we generate independently 1000 conditional events $\{Z(\mathbf{x}) = \mathbf{z}\}$, $\mathbf{x}$ being fixed, and for each conditional event one conditional realization of a Brown–Resnick process. Figure 11.8 compares the pairwise extremal coefficient estimates based on these simulations to the theoretical extremal coeffi-

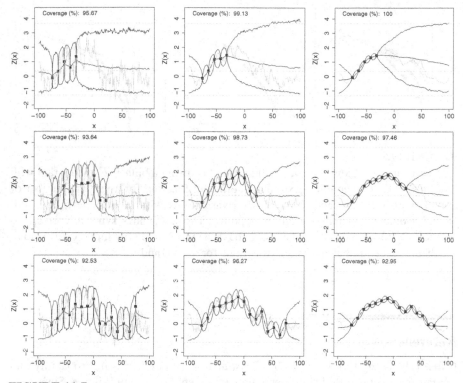

FIGURE 11.7

Pointwise sample quantiles estimated from 1000 conditional simulations of Brown–Resnick processes with standard Gumbel margins and semi-variograms γ_1, γ_2 and γ_3 (left to right) and with $k = 5, 10, 15$ conditioning locations—top to bottom. The solid black lines show the pointwise 0·025, 0·5, 0·975 sample quantiles and the dashed gray lines that of a standard Gumbel distribution. The squares show the conditional points $\{(x_i, z_i)\}_{i=1,\ldots,k}$. The solid gray lines correspond the simulated paths used to get the conditioning events. The inserts give the proportion of points lying in the 95% pointwise confidence intervals.

cient function. As expected, whatever the number of conditioning location and the semi-variograms are, the (binned) pairwise estimates match the theoretical curve indicating that the spatial dependence is honored.

Acknowledgements

M. Oesting and M. Ribatet acknowledge the ANR project McSim for financial support.

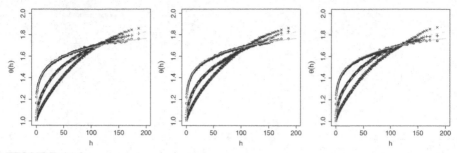

FIGURE 11.8

Comparison of the extremal coefficient estimates (using a binned F-madogram with 250 bins) and the theoretical extremal coefficient function for a varying number of conditioning locations and different semi-variograms with $k = 5$ (left), $k = 10$ (middle), or $k = 15$ (right) conditioning points. The o, +, and × symbols correspond respectively to γ_1, γ_2, and γ_3. The solid, dashed, and dotted gray lines correspond to the theoretical extremal coefficient functions for γ_1, γ_2, and γ_3.

References

Chils, J.-P. and P. Delfiner (1999). *Geostatistics*. Wiley Series in Probability and Statistics: Applied Probability and Statistics. New York: John Wiley & Sons.

Cooley, D., P. Naveau, and P. Poncet (2006). Variograms for spatial max-stable random fields. In P. Doukhan, G. Lang, D. Surgailis, and G. Teyssire (Eds.), *Dependence in Probability and Statistics*, Volume 187 of *Lecture Notes in Statististics*, pp. 373–390. New York: Springer.

Davis, R. A. and S. I. Resnick (1989). Basic properties and prediction of max-ARMA processes. *Advances in Applied Probability 21*(4), 781–803.

Davis, R. A. and S. I. Resnick (1993). Prediction of stationary max-stable processes. *The Annals of Applied Probability 3*(2), 497–525.

de Haan, L. (1984). A spectral representation for max-stable processes. *The Annals of Probability 12*(4), 1194–1204.

de Haan, L. and A. Ferreira (2006). *Extreme Value Theory*. Springer Series in Operations Research and Financial Engineering. New York: Springer. An introduction.

Dombry, C. and F. Eyi-Minko (2013). Regular conditional distributions of continuous max-infinitely divisible random fields. *Electronic Journal of Probability 18*(7), 1–21.

Dombry, C., F. Eyi-Minko, and M. Ribatet (2013). Conditional simulation of max-stable processes. *Biometrika 100*(1), 111–124.

Kabluchko, Z., M. Schlather, and L. de Haan (2009). Stationary max-stable fields associated to negative definite functions. *The Annals of Probability 37*(5), 2042–2065.

Liu, J. S., W. H. Wong, and A. Kong (1995). Covariance structure and convergence rate of the Gibbs sampler with various scans. *Journal of the Royal Statistical Society. Series B. Methodological 57*(1), 157–169.

Oesting, M. and M. Schlather (2014). Conditional sampling for max-stable processes with a mixed moving maxima representation. *Extremes. Statistical Theory and Applications in Science, Engineering and Economics 17*(1), 157–192.

Penrose, M. D. (1992). Semi-min-stable processes. *The Annals of Probability 20*(3), 1450–1463.

Ribatet, M. (2013). Spatial extremes: Max-stable processes at work. *Journal de la SFdS. Journal de la Socet Franuise de Statistique 154*(2), 156–177.

Schlather, M. (2002). Models for stationary max-stable random fields. *Extremes 5*(1), 33–44.

Stoyan, D., W. S. Kendall, and J. Mecke (1987). *Stochastic Geometry and Its Applications*. Berlin: Akademie-Verlag.

Wang, Y. and S. A. Stoev (2011). Conditional sampling for spectrally discrete max-stable random fields. *Advances in Applied Probability 43*(2), 461–483.

Weintraub, K. S. (1991). Sample and ergodic properties of some min-stable processes. *The Annals of Probability 19*(2), 706–723.

12

Composite Likelihood for Extreme Values

Huiyan Sang

Texas A&M University, United States

Abstract

Many problems involve dependent extremes, such as extreme precipitation, heavy snow, financial and insurance risk, to name a few. These problems motivate the need for the development of statistical modeling and inference tools for multivariate extreme values. Copula models and max-stable process models have become two popular modeling choices to characterize the dependence among multivariate extremes, especially for high-dimensional cases. Despite the sound mathematical properties of these models in terms of modeling tail dependence among multivariate extreme values, likelihood inference is challenging for such models because their corresponding joint likelihood functions are unavailable. Taking advantage of the availability of the low-dimensional marginal likelihoods, the composite likelihood approach has become a major inference tool for max-stable process models. In this chapter, we review the concepts of composite likelihoods and present some recent developments of composite likelihood approaches for the inference of max-stable process models with illustrative examples. We also discuss the use of composite likelihood for the inference of copulas for modeling multivariate extreme values.

12.1 Introduction

Models and inference methods for multivariate extreme values have gained increasing attention in recent years. In many applications, these methods are utilized as important statistical tools to investigate the dependence among multivariate extreme values. In particular, max-stable stochastic processes have emerged as a fundamental class of models to describe dependent extreme value phenomena in spatial statistics and time series. This class of models is viewed as a natural generalization of the univariate generalized extreme value (GEV) distribution in the infinite dimensional continuous spaces, providing an asymptotically justified approach to modeling process extremes incorporating spatial dependence. Despite many attractive properties of max-stable processes models, both classical and Bayesian inference encounter difficulties because closed-form expressions of the corresponding joint likelihoods are typically not available except for some trivial cases. This motivates the use of the composite likelihood (CL) approach constructed based on low-dimensional marginal likelihoods of the max-stable processes. It soon becomes a practical and reliable inference tool for fitting max-stable processes due to the well-established asymptotic theories of maximum composite likelihood estimators.

Padoan et al. (2010) are the first to suggest using a pairwise composite likelihood to estimate unknown parameters in the Gaussian max-stable processes models (Smith, 1990). Similar ideas are subsequently used in the context of extremal inference for other max-stable processes models (e.g., Blanchet and Davison, 2011; Davison et al., 2012). Recently, Genton et al. (2011) derived a closed-form expression for the higher-order likelihood function of a Gaussian max-stable process. Huser and Davison (2013) obtained the expression of the triplewise density function of a Brown–Resnick process. These results allow investigation and comparison of the maximum composite likelihood estimators based on both pairs and triples.

Weighted composite likelihood (WCL) has also been investigated in the context of max-stable processes, aiming to improve statistical efficiency or to reduce the computational burden associated with large datasets. Padoan et al. (2010) found that the composite likelihoods constructed only using neighboring sites can reduce the asymptotic variances of the model parameters in their simulation example of a Gaussian max-stable process. Sang and Genton (2014) proposed a tapered composite likelihood approach for the fitting of max-stable processes. They also showed through simulations studies that exclusion of likelihood contributions from sites very far apart leads to efficiency gain over the full composite likelihood.

Due to the availability of the asymptotic theories of maximum composite likelihood estimators, most of the work using composite likelihood for max-stable processes follows the frequentist paradigm. Very recently, Ribatet et al. (2012) proposed an approach to employ composite likelihood in the Markov chain Monte Carlo (MCMC) algorithms which allows Bayesian inference for max-stable processes.

The work on composite likelihood for extreme values has been mainly focused on max-stable process models. Nevertheless, the composite likelihood inference technique is applicable to other multivariate extreme value models in which lower-dimensional likelihood functions are available. In this chapter, we also consider several multivariate extreme value models under the copula framework. Both non-extreme value copulas and extreme value copulas have been developed to model multivariate extreme values (Ribatet and Sedki, 2013; Sang and Gelfand, 2010), but inference is not easy. For extreme value copulas there is a lack of closed form likelihood functions. And for some non-extreme value copulas such as Gaussian or multivariate t copulas, computational burden is also heavy in high-dimensional case. Very recently, Ribatet and Sedki (2013) applied a pairwise composite likelihood approach for the inference of several extreme value copula models.

We briefly review the CL approach in Section 12.2. We describe using CL for inferencing max-stable processes models in Section 12.3.1; in particular, we introduce the WCL and the Bayesian composite likelihood in Sections 12.3.2 and 12.3.3. We then introduce copula models and briefly discuss the application of CL inference methods for such models in Section 12.4. Finally, we illustrate the tapered pairwise and triplewise CL approach for several examples of max-stable processes with a simulation study in Section 12.5. In the supplementary files we give R examples that use the WCL for max-stable processes models.

12.2 Composite Likelihood

Consider a parametric statistical model with probability density function $\{f(z;\theta), z \in \mathcal{Z} \subseteq \mathbb{R}^K, \theta \in \Theta \subseteq \mathbb{R}^q\}$, and a set of marginal or conditional events $\{\mathcal{A}_i : \mathcal{A}_i \subseteq \mathcal{F}, i \in I \subseteq \mathbb{N}\}$, where $\mathcal{F}$ is some σ-algebra on $\mathcal{Z}$. The log composite likelihood (Lindsay, 1988) is defined as

$$\ell_c(\theta) = \sum_{i \in I} w_i \log f(z \in \mathcal{A}_i; \theta) \tag{12.1}$$

where $f(z \in \mathcal{A}_i; \theta)$ is the likelihood associated with the event $\mathcal{A}_i$, and $\{w_i, i \in I \subseteq \mathbb{N}\}$ is a set of weights. $\hat{\theta}_{wcl}$ is called the weighted maximum CL estimator if it is a global maximum of $\ell_c(\theta)$.

The definition in (12.1) allows for combinations of marginal and conditional densities (Cox and Reid, 2004). One example of the composite conditional likelihood inference is the pseudolikelihood proposed by Besag (1974) for approximate inference in spatial processes. This pseudolikelihood is the product of the conditional densities of a single observation given its neighbors

$$\ell_c(\theta) = \sum_{t=1}^{N} \sum_{i=1}^{K} \log f(z_i^{(t)} | z_j^{(t)} : z_j^{(t)} \text{ is neighbor of } z_i^{(t)}; \theta),$$

where $z_i^{(t)}$ is the sample of the t-th replicate at the i-th site, for $t = 1, \cdots, N$ and $i = 1, \cdots, K$. More recent variants of Besag's proposal involve using blocks of observations in both conditional and conditioned events (see, e.g., Stein et al., 2004; Vecchia, 1988).

Examples of the composite marginal likelihood inference include the pairwise and triplewise marginal likelihoods given below. The pairwise log CL is defined as

$$\ell_c(\theta) = \sum_{t=1}^{N} \sum_{i \neq j} w_{ij} \log f(z_i^{(t)}, z_j^{(t)}; \theta).$$

Analogously, we may define the triplewise log composite likelihood as

$$\ell_c(\boldsymbol{\theta}) = \sum_{t=1}^{N} \sum_{i \neq j \neq k} w_{ijk} \log f(z_i^{(t)}, z_j^{(t)}, z_k^{(t)}; \boldsymbol{\theta}).$$

The CL approach is known to yield estimators with sound asymptotic properties. This is mainly due to one attractive feature of the CL; the weighted composite score function $\nabla \ell_c(\boldsymbol{\theta}; z)$ is unbiased under standard regularity conditions (Lindsay, 1988) since each component in it is a likelihood object. Specifically, in the case of N independent and identically distributed observations $z^{(1)}, \ldots, z^{(N)}$ from the model $f(z; \boldsymbol{\theta})$ on $\mathbb{R}^K$ with K fixed, the maximized CL estimator $\widehat{\boldsymbol{\theta}}_c$ is a consistent and unbiased parameter estimator (Lindsay, 1988; Varin and Vidoni, 2005)

$$N^{1/2}(\widehat{\boldsymbol{\theta}}_c - \boldsymbol{\theta}) \to N_q\{0, G^{-1}(\boldsymbol{\theta})\} \tag{12.2}$$

in distribution as $N \to +\infty$, where $G(\boldsymbol{\theta}) = H(\boldsymbol{\theta})J(\boldsymbol{\theta})^{-1}H(\boldsymbol{\theta})$ is called the Godambe information matrix, $H(\boldsymbol{\theta}) = E\{-\nabla^2 \ell_c(\boldsymbol{\theta})\}$ and $J(\boldsymbol{\theta}) = \text{var}\{\nabla \ell_c(\boldsymbol{\theta})\}$.

Despite its sound asymptotic properties, the estimation method using CL typically results in loss of statistical efficiency compared with the maximum likelihood estimator counterpart. Indeed, CL can be viewed as a misspecified model, and hence may not attain the Cramér–Rao lower bound (Cox and Reid, 2004). Nevertheless, efficiency gain might be achieved by carefully designing ways to construct CL while keeping low computational cost (Bai et al., 2012; Lindsay et al., 2011). We will present one example of designing CL by tapering in Section 12.3.2.

Under the CL inference context, model selection is typically done by calculating the CL based information criterion (CLIC), defined as CLIC $= -2\{\ell_c(\widehat{\boldsymbol{\theta}}_c) - \text{tr}[\widehat{\mathbf{J}}(\widehat{\boldsymbol{\theta}}_c)\widehat{\mathbf{H}}(\widehat{\boldsymbol{\theta}}_c)^{-1}]\}$, where the second term is the composite log-likelihood penalty term for the effective number of parameters. The derivation of the CLIC is based on the expected Kullback–Leibler divergence between the true unknown model and the adopted misspecified model in the CL context (Davison, 2003, p.123). When the joint likelihood is used, the CLIC reduces to the usual Akaike criterion.

In practice, calculations of the Godambe information matrix require the estimations of $\mathbf{H}$ and $\mathbf{J}$ in (12.2). The Hessian matrix, $\mathbf{H}$, is usually estimated from the numerical maximization algorithm, whereas $\mathbf{J}$ is usually estimated from the variance of the observed score function. One may also use the Jackknife method to estimate the covariance matrix of the maximum CL estimates (Lipsitz et al., 1994). When N is not sufficiently large compared with the dimension of $\boldsymbol{\theta}$, this method may provide a more robust estimate of the covariance matrix than the asymptotic Godambe information matrix method provides. This method is also preferred in situations where the analytical derivatives of the score functions are difficult or tedious to derive. The Jackknife covariance matrix is given by

$$\text{var}\{\widehat{\boldsymbol{\theta}}_c\} = \frac{N-1}{N} \sum_{t=1}^{N} \{\widehat{\boldsymbol{\theta}}_c^{(-t)} - \widehat{\boldsymbol{\theta}}_c\}\{\widehat{\boldsymbol{\theta}}_c^{(-t)} - \widehat{\boldsymbol{\theta}}_c\}^{\mathsf{T}},$$

where $\widehat{\boldsymbol{\theta}}_c^{(-t)}$ is the CL estimator of $\boldsymbol{\theta}$ excluding the t-th sample.

12.3 Composite Likelihood for Extreme Values

Max-stable processes models have become a very popular modeling choice to characterize the dependence among high-dimensional multivariate extremes. The popularity of such models is gained partly due to the use of CL which makes the model inference feasible. Most of the recent developments on CL for extreme values have been focusing on max-stable process models. Very recently, the CL has also been introduced for inferencing multivariate extreme value copulas models. Below, we first describe CL methods for max-stable processes and then briefly discuss the use of CL for copulas models.

12.3.1 Composite Likelihood for Max-Stable Processes

Max-stable processes models have gained increasing popularity for the modeling of extreme values in space. They are natural extensions of the generalized extreme-value distribution in space and are asymptotically justified for maxima of independent replicates of spatial random fields. However, ordinary likelihood inference is infeasible because of the lack of analytical form of the joint distribution function. Only up to trivariate density functions have been derived. Therefore, marginal CL methods become natural choices for the inference in max-stable processes.

Let $\{\tilde{Z}(s)^{(i)}\}, s \in \mathbb{R}^d, i = 1, \ldots, n$, be n independent replicates of a continuous spatial stochastic process. A spatial stochastic process $Z(s)$ is *max-stable* if there exist continuous functions $a_n(s) > 0$ and $b_n(s)$ such that

$$Z(s) = \lim_{n \to +\infty} \frac{\max_{i=1}^n \tilde{Z}(s)^{(i)} - b_n(s)}{a_n(s)}.$$

By the above definition, a spatial max-stable process offers a natural choice for modeling spatial extremes. The marginal distributions of the random vector $\{Z(s_1), \ldots, Z(s_K)\}^T$ for any points $s_1, \ldots, s_K$ belong to the class of multivariate extreme value distributions (de Haan and Resnick, 1977). In particular, the univariate marginal distributions belong to the family of generalized extreme value (GEV) distributions. For univariate GEV distributions, there is no loss of generality in transforming the margins to the standard Fréchet distributions with distribution function $\exp(-1/z)$. Therefore, it is common to consider a *simple spatial max-stable process* $Z(s)$ with standard Fréchet marginals.

Schlather (2002) provided a spectral representation for simple spatial max-stable processes as follows. Let $\{r_i, i \in \mathbb{N}\}$ be the points of a Poisson process on $\mathbb{R}_+$ with intensity dr/r^2. Let $\{W(s), s \in \mathcal{S} \subset \mathbb{R}^d\}$ be a nonnegative stochastic process that satisfies $E[W(s)] = 1$ for all $s \in \mathcal{S}$ and let $\{W_i(s), i \in \mathbb{N}\}$, be a collection of independent copies of this process. Then

$$Z(s) = \max_i r_i W_i(s) \tag{12.3}$$

is a simple max-stable process.

Following the general representation in (12.3), there are several parametric formulations of $W(s)$ to construct a valid simple max-stable process. Below, we list three specific examples of max-stable process models and their corresponding bivariate or trivariate likelihood functions, which provides the basis for CL inference.

Example 1: A first max-stable process is referred to as the *Gaussian* max-stable process model, In a seminal unpublished University of Surrey 1990 technical report, R. L. Smith defined Gaussian max-stable processes and derived a closed form expression for the joint cumulative distribution function of the process Z for two spatial sites $s_1 \in \mathbb{R}^2$ and $s_2 \in \mathbb{R}^2$

$$-\log F(z_1, z_2) = \frac{1}{z_1} \Phi \left\{ \frac{\alpha}{2} + \frac{1}{\alpha} \log \left(\frac{z_2}{z_1} \right) \right\} + \frac{1}{z_2} \Phi \left\{ \frac{\alpha}{2} + \frac{1}{\alpha} \log \left(\frac{z_1}{z_2} \right) \right\}, \quad (12.4)$$

where Φ denotes the univariate standard normal cumulative distribution function, $\alpha^2(h) = h^\mathsf{T} \Sigma^{-1} h$, h is the separation vector between s_1 and s_2, and $\Sigma \in \mathbb{R}^{2 \times 2}$ is the covariance matrix with variances σ_{11}^2 and σ_{22}^2, and correlation ρ. The square roots of the eigenvalues of Σ control the range of the spatial dependence.

Example 2: A second max-stable process is referred to as the *Schlather* model, constructed by letting $W(s) = \max\{0, \epsilon(s)\}$, where $\epsilon(s)$ is taken to be independent copies of a stationary Gaussian process with unit variance and correlation function $\rho(h)$, scaled such that the condition $E[W(s)] = 1$ holds. Schlather (2002) showed that in this case the bivariate probability distribution takes the form

$$-\log F(z_1, z_2) = \frac{1}{2} \left(\frac{1}{z_1} + \frac{1}{z_2} \right) \left(1 + \left[1 - 2 \frac{\{\rho(h) + 1\} z_1 z_2}{(z_1 + z_2)^2} \right]^{1/2} \right).$$

Example 3: A third max-stable process is referred to as the *Brown–Resnick* model (see, e.g., Kabluchko et al., 2009), constructed by letting $W(s) = \exp\{\epsilon(s) - \gamma(s)\}$, where $\epsilon(s)$ denotes a zero mean Gaussian process with semi-variogram $\gamma(h)$ and with $\epsilon(0) = 0$ almost surely. In particular, when $\epsilon(s)$ is a Brownian process with semi-variogram $\gamma(h) = \|h\|^2 / 2$, the process corresponds to the Smith Gaussian max-stable process model. The closed form for the bivariate distribution function of the Brown–Resnick process has the same expression as that of the Smith model with $\alpha(h) = \{2\gamma(h)\}^{1/2}$.

We remark here that the bivariate distributions listed above are based on simple max-stable processes, that is, the marginal distributions are unit Fréchet. When the margins do not have a unit Fréchet distribution, the bijective transformation defining the GEV distribution can be used to straightforwardly modify the likelihood function. In the presence of GEV marginals with GEV parameters (μ_1, σ_1, ξ_1) and (μ_2, σ_2, ξ_2), the resulting bivariate density becomes:

$$f_{Y_1, Y_2}(y_1, y_2) = f_{Z_1, Z_2}(z_1, z_2) |J(y_1, y_2)|$$

where $f_{Y_1, Y_2}(y_1, y_2)$ is the bivariate density of a pair of variables (Y_1, Y_2) with GEV marginals, $f_{Z_1, Z_2}(z_1, z_2)$ is the bivariate density of (Z_1, Z_2) which are the transformation of (Y_1, Y_2) to ensure unit Fréchet marginals, and

$$|J(y_1, y_2)| = \frac{1}{\sigma_1 \sigma_2} \left(1 + \frac{\xi_1 (y_1 - \mu_1)}{\sigma_1}\right)_+^{1/\xi_1 - 1} \left(1 + \frac{\xi_2 (y_2 - \mu_2)}{\sigma_2}\right)_+^{1/\xi_2 - 1}.$$

The above expression allows simultaneous estimation of the dependence parameters and marginal GEV parameters for general max-stable processes.

Padoan et al. (2010) are the first to introduce the CL method for inference in Gaussian max-stable processes. They considered equal weight pairwise CL method. Utilizing the expression of the CL function after bijective transformation, they demonstrated the utility and flexibility of CL method for joint estimation of marginal and dependence parameters in the spatial context at a moderate computational cost. Pairwise CL soon became a popular tool for estimating unknown model parameters in other max-stable processes models in which pairwise density functions are available (e.g., Blanchet and Davison, 2011; Davison et al., 2012).

There are some recent works on deriving higher-order likelihood functions for max-stable processes, which allow a natural extension to triplewise CL inference and so on. Genton et al. (2011) suggested that there is substantial gain in efficiency of the maximum CL estimates from $p = 2$ to $p = 3$ sites in $\mathbb{R}^2$ for the Gaussian max-stable process models. Huser and Davison (2013) studied the triplewise CL for the Brown–Resnick process. Their findings through simulation studies also suggest that higher-order CL may lead to efficiency gains, especially for smoother max-stable processes.

12.3.2 Tapered Composite Likelihood for Max-Stable Processes

One issue with equally weighted CL is that the number of pairs can be enormous when the number of observations is large, making it computationally very expensive. Moreover, several investigations have suggested that statistical efficiency gain can be achieved by carefully selecting weights of the composite likelihoods in the context of time series and spatial statistics. Bevilacqua et al. (2012) suggested that down-weighting or excluding likelihood contributions from sites that are very far apart leads to efficiency gains over the full composite likelihood for spatial temporal Gaussian process models. In the context of time series, several works (e.g., Davis and Yau, 2011; Joe and Lee, 2009) have also shown that including unnecessary pairs can cause some loss of estimation efficiency. Padoan et al. (2010) found in a simulation study that the composite likelihoods constructed only using neighboring sites can reduce the asymptotic variances of the model parameters from a Gaussian extreme value process referred to as the *Smith* model (Smith, 1990).

Motivated by these findings, Sang and Genton (2014) proposed a tapered CL (TCL) approach to improve the efficiency of the estimates and to reduce the computational burden. For spatial max-stable processes, the weights of the TCL are set to depend on the distance between sites. Let γ_s denote the *taper range* in space. Let $D_2(\gamma_s)$ be a set containing all pairs of observations within γ_s units of space lag to exclude distant pairs. Given γ_s and D_2, the pairwise weight function is defined as a thresholding/tapering function, $w_{ij} = 1$ if $(i, j) \in D_2(\gamma_s)$, $w_{ij} = 0$

otherwise. For the Smith Gaussian extreme value process model and the Brown–Resnick process model, one can also construct a triplewise TCL function following the expressions of trivariate distribution functions derived in Genton et al. (2011) and Huser and Davison (2013). Similarly to the pairwise case, the triplewise weight function is defined as $w_{ijk} = 1$ if $(i, j, k) \in D_3(\gamma_s)$, $w_{ijk} = 0$ otherwise, where $D_3(\gamma_s) = \{(i, j, k); \max_{i_1, i_2 \in (i,j,k)} \|s_{i_1} - s_{i_2}\| \leqslant \gamma_s\}$.

Sang and Genton (2014) considered two criteria that measure statistical efficiency of TCL estimators to select a taper range. In particular, they proposed to use the determinant and the trace of the covariance matrix of TCL estimators. The former is often used as a measure of overall spread in a distribution (Mardia et al., 1979), while the latter equals the summation of the variances of the parameter estimates. Since the maximum TCL estimator is unbiased under certain regularity conditions, these two measures of the covariance matrix of estimates become natural scalar choices to compare the statistical efficiency. Both criteria defined above require estimation of the covariance matrix of the maximum TCL estimator, which can be calculated using the inverse of the Godambe information matrix based on the asymptotic results in (12.2). The optimal taper range is then chosen such that the trace or the determinant of the inverse of the Godambe information matrix is minimized.

12.3.3   Bayesian Composite Likelihood for Max-Stable Processes

Bayesian models have been extensively used for the statistical analysis of spatial extremes, due to their flexibility in describing dependence structures in complex processes. These models are primarily fitted using MCMC algorithms to enable inference for parameters and to provide spatial predictions. However, little work has been done in Bayesian inference for max-stable process again because full likelihood does not have analytical form. There were some attempts to employ the CL method in Bayesian analysis. One naive way is to replace the unknown joint likelihood function with the CL function in MCMC algorithms. However, the posterior inference based on this method often leads to overly precise inference, mainly because the CL treats the dependent events as mutually independent.

Recently, Ribatet et al. (2012) employed the CL method within Bayesian framework, which allows a practical inferential method for the fitting of max-stable processes to spatial data. This method is shown to be effective in terms of correcting biased inference when CL is used in a naive way in Bayesian inference.

They proposed to adjust the likelihood ratio of CL to obtain appropriate posterior inference based on the asymptotic results developed under the frequentist paradigm. Let $\theta = (\phi, \psi)$ denote an unknown parameter vector and $\theta_0 = (\phi_0, \psi_0)$ denote the true parameter value. Let $\widehat{\theta}$ be the maximum likelihood estimator that maximizes the full likelihood function, and let $\widehat{\theta}_c$ be the maximum CL estimator that maximizes the CL function. Fix ψ at the true parameter value ψ_0 of size q^*. Let $\tilde{\theta}$ be the restrictive maximum likelihood estimator that maximizes the full likelihood function of ϕ, and $\tilde{\theta}_c$ be the corresponding restrictive maximum CL estimator. As sample size $N \to \infty$,

$$\Lambda(\psi_0) = 2\{\ell(\widehat{\theta}) - \ell(\tilde{\theta})\} \xrightarrow{d} \chi_{q^*}^2$$

whereas for the CL

$$\Lambda_c(\psi_0) = 2\{\ell_c(\widehat{\boldsymbol{\theta}}_c) - \ell(\tilde{\boldsymbol{\theta}}_c)\} \xrightarrow{d} \sum_{i=1}^{q^*} \lambda_i X_i$$

where $(X_1, \cdots, X_{q^*})$ are $i.i.d$ chi-squared variables of degree 1, $\lambda_1, \cdots, \lambda_{q^*}$ are the eigen values of the matrix $\{\mathbf{H}(\boldsymbol{\theta}_0)^{-1}\mathbf{J}(\boldsymbol{\theta}_0)\mathbf{H}(\boldsymbol{\theta}_0)^{-1}\}_\psi[\{\mathbf{H}(\boldsymbol{\theta}_0)^{-1}\}_\psi]^{-1}$. Here $\mathbf{A}_\psi$ denotes the sub-matrix of a matrix $\mathbf{A}$ corresponding to the elements of ψ.

The authors propose to make two types of adjustments, namely the magnitude and curvature adjustments, to the log CL, aiming to recover convergence in distribution to the usual χ^2 distribution. Under such adjustments, the CL ratio behaves similarly in distribution to the full likelihood ratio as sample size goes to infinity.

The magnitude adjustment multiplies the CL by a multiplier k, that is, $\ell_{magn}(\boldsymbol{\theta}) = k\ell_c(\boldsymbol{\theta})$. Define

$$\Lambda_{magn}(\psi_0) = 2\{\ell_c(\widehat{\boldsymbol{\theta}}_{magn}) - \ell(\tilde{\boldsymbol{\theta}}_{magn})\} \xrightarrow{d} k \sum_{i=1}^{q^*} \lambda_i X_i$$

k is chosen such that the first moment of the distribution of $\Lambda_{magn}(\psi_0)$ matches with that of $\chi^2_{q^*}$ distribution.

The curvature adjustment modifies the curvature of the CL around the global maximum CL estimator. Define $\boldsymbol{\theta}_* = \widehat{\boldsymbol{\theta}}_c + \mathbf{B}(\boldsymbol{\theta} - \widehat{\boldsymbol{\theta}}_c)$, where $\mathbf{B}$ is a $q \times q$ matrix. Let $\ell_{curv}(\boldsymbol{\theta}) = \ell(\boldsymbol{\theta}^*)$. It clearly follows that $\widehat{\boldsymbol{\theta}}_c$ also maximizes ℓ_{curv}. $\mathbf{B}$ is chosen such that the curvature of ℓ_{curv} at $\widehat{\boldsymbol{\theta}}_c$ matches that of the asymptotic log-density of $\widehat{\boldsymbol{\theta}}_c$. In particular, the goal is to ensure that $-n^{-1}\nabla^2\ell_{curv}(\widehat{\boldsymbol{\theta}}_c)$ converges almost surely to the Godambe information matrix of the maximum CL estimator. Unlike the magnitude adjustment in which only the first moment of the CL is corrected, the curvature adjustment ensures that the adjusted CL ratio converges to a χ^2_q variable.

12.4 Composite Likelihood for Multivariate Extreme Value Distributions via Copulas

Copulas provide a convenient tool to construct a multivariate distribution with known marginals to describe the dependence structure across multivariate random variables. Consider a d-dimensional random vector $\mathbf{Y} = (Y_1, \cdots, Y_d)^T$ with joint cumulative distribution function (CDF) F and corresponding marginal CDFs F_i for each Y_i. According to Sklar's theorem (e.g., Sklar, 1996), $F(\mathbf{Y}) = C(F_1(Y_1), \cdots, F_d(Y_d)) = C(u_1, \cdots, u_d)$, where C is called a copula function defined on $[0, 1]^d$. When the marginals of $\mathbf{Y}$ are all continuous, the function C is uniquely defined.

By combining the univariate extreme value distribution theory for marginals and a valid copula model, one may construct a multivariate distribution for modeling multivariate extremes.

Multivariate Gaussian and multivariate t are two popular choices for copula functions since they can handle high-dimensional multivariate cases. The Gaussian copula function is defined as follows: $C(u_1, \cdots, u_d) = \Phi_\Sigma\left(\Phi^{-1}(u_1), \cdots, \Phi^{-1}(u_d)\right)$, where Φ denotes the standard normal CDF and Φ_Σ denotes the joint CDF of a multivariate Gaussian distribution with covariance matrix Σ. Similarly, the multivariate t copula is defined as $C(u_1, \cdots, u_d) = T_{\nu;\Sigma}\left(T_\nu^{-1}(u_1), \cdots, T_\nu^{-1}(u_d)\right)$, where $T_{\nu;\Sigma}$ denote for the CDF of a joint t distribution with ν degrees of freedom and covariance matrix Σ, and T_ν is the CDF of the marginal t distribution.

Despite the simplicity of multivariate Gaussian or multivariate t copulas and their utilities in handling high-dimensional data, these copulas may not appropriately describe dependence for extreme values. For example, the Gaussian copula is known to yield asymptotically independent random vectors. That motivates the need to define copula functions that respect dependence properties of multivariate extreme values. Extreme value copulas are one such class. A copula function C_* is called an extreme value copula if there exists a copula C such that

$$C(u_1^{1/n}, \cdots, u_d^{1/n})^n \to C_*(u_1, \cdots, u_d), \quad n \to \infty$$

for all $(u_1, \cdots, u_d)$. The above definition suggests that extreme value copulas are associated with componentwise maxima of n independent and identically distributed random vectors with copula C, and hence are asymptotically justified to model component-wise maxima.

Indeed, it can be shown that the class of extreme value copulas have max-stable property (de Haan and Resnick, 1977), that is,

$$C_*(u_1, \cdots, u_d)^n = C_*(u_1^n, \cdots, u_d^n), \quad n > 0$$

One well-known extreme value copula is the Gumbel–Hougaard copula which corresponds to the logistic family (Gumbel, 1960),

$$C_*(u_1, \cdots, u_d) = \exp[-\{\sum_{j=1}^{d}(-\log u_j)^{1/\alpha}\}^\alpha], \quad 0 < \alpha \leqslant 1$$

However, the logistic family is often considered to be too restrictive especially for high-dimensional variables since there is only one parameter α controlling the dependence.

Two other extreme value copulas are the Hüsler–Reiss and the extremal-t copulas based on the extreme value limits of Gaussian and t distributions, respectively (Hüsler and Reiss, 1989; Nikoloulopoulos et al., 2009). Consider a bivariate extreme value random vector. The Hüsler–Reiss copula is defined as

$$C_*(u_1, u_2) = \exp\left[\Phi\left\{\frac{\alpha}{2} + \frac{1}{\alpha}\log\left(\frac{\log u_2}{\log u_1}\right)\right\} \cdot \log u_1 \right.$$
$$\left. + \Phi\left\{\frac{\alpha}{2} + \frac{1}{\alpha}\log\left(\frac{\log u_1}{\log u_2}\right)\right\} \cdot \log u_2\right],$$

where α is a dependence parameter. The extremal-t copula is

$$C_*(u_1, u_2) = \exp\left[T_{\nu+1}\left\{ -\frac{\rho}{b} + \frac{1}{b}\log\left(\frac{\log u_2}{\log u_1}\right)^{1/\nu} \right\} \cdot \log u_1 \right.$$
$$\left. + T_{\nu+1}\left\{ -\frac{\rho}{b} + \frac{1}{b}\log\left(\frac{\log u_1}{\log u_2}\right)^{1/\nu} \right\} \cdot \log u_2 \right],$$

where ρ is the off-diagonal element in the correlation matrix of a standard bivariate t distribution and $b = \sqrt{(1 - \rho^2)/(\nu + 1)}$.

From the definition of copulas, the corresponding joint distribution function is easily obtained from $F(\mathbf{Y}) = C(u_1, \cdots, u_d)$. However, the full likelihood based inference requires differentiating the joint CDF, which typically involves an explosive number of terms as dimension d increases except for the Gaussian or multivariate t copulas. In fact, the inference for the Gaussian or Student t copulas also faces computational challenges when d is very large since the joint density function involves the inversion of a $d \times d$ covariance matrix.

To tackle this computational issue in inference, CL derived from lower-dimensional copulas can be used to replace the full likelihood. For example, assuming marginals are standard Fréchet distributions, a pairwise CL based on copulas can be written as:

$$\ell_c(\boldsymbol{\theta}) = \sum_{t=1}^{N} \sum_{i \neq j} \log f(z_i^{(t)}, z_j^{(t)}; \boldsymbol{\theta}) = \sum_{t=1}^{N} \sum_{i \neq j} \log c(u_i^{(t)}, u_j^{(t)}; \boldsymbol{\theta}).$$

where c is the bivariate marginal copula density derived from $\frac{\partial^2}{\partial u_i \partial u_j} C(u_i, u_j)$. Inference of copula model parameters is done by maximizing the above CL. The estimators obtained this way share the same asymptotic properties as in (12.2).

12.5 Simulation Study

Below, we present a simulation study to investigate the use of the TCL method for inference on the parameters of the three examples of max-stable process models, including the Smith model, the Schlather model, and the Brown–Resnick model.

For each of the 100 simulation replicates, we fix the sample size $N = 47$ and the sampling location at $K = 35$ rain gauging stations located in the north of the Alps and east of the Jura mountains in Switzerland, mimicking a real dataset of annual maximum precipitations from 1962 to 2008 provided by the national meteorological services of Switzerland, MeteoSuisse. For each max-stable process model, we obtain the equally weighted CL estimates and two TCL estimates, in which the optimal taper range is selected by minimizing the trace (denoted as T.Trc) and the determinant (denoted as T.Det) of the associated estimated covariance matrix, respectively. We

TABLE 12.1
Results of the TCL estimates for data simulated from the Smith model. The mean and standard deviations (in parentheses) of the estimates are for σ_{11}^2, σ_{12}, and σ_{22}^2 in Σ. Taper range gives the median and the 90% interval of the selected taper range. RE is the relative efficiency of the TCL estimators relative to the equally weighted pariwise CL (PCL) or tripplewise CL estimator. T.Trc, T.Det, and T.SSE are pairwise TCL estimators obtained by minimizing the trace of the estimated variance matrix, the determinant of the estimated variance matrix, and the sum of the squared errors, respectively. T.MahD and T.EucD are pairwise TCL estimators obtained by tapering the Mahalanobis distance and the Euclidean distance, respectively, to minimize the trace. C.Tri is the estimator minimizing the equally weighted tripplewise CL and T.Tri is the tapered triplewise CL estimator minimizing the trace. The practical distances h_- and h_+ are distances at which the extremal coefficient is 1.3 and 1.7, respectively, under the true models.

	Taper Range	Parameter Estimates			RE		
		σ_{11}^2	σ_{12}	σ_{22}^2	σ_{11}^2	σ_{12}	σ_{22}^2
True	$h_- = [11, 15]$	243	62	320			
	$h_+ = [30, 39]$						
PCL	81.78	245.8(36.6)	60.9(32.5)	337.6(64.7)	1.00	1.00	1.00
T.Trc	14.73(12.00,18.06)	243.0(26.4)	58.09(21.6)	329.5(33.1)	1.94	2.19	3.79
T.Det	13.91(11.28,16.95)	242.1(26.3)	59.0(21.3)	330.5(32.8)	1.94	2.28	3.78
T.SSE	14.00(9.23,19.66)	242.0(22.6)	58.5(17.4)	328.2(24.0)	2.63	3.34	6.96
C.Tri	81.78	243.1(0.9)	62.1(1.3)	319.6(1.3)	1.00	1.00	1.00
T.Tri	42.00	243.1(0.5)	62.0(0.6)	320.1(0.7)	3.48	5.17	3.44
True	$h_- = [8, 10]$	121.5	31.0	160.0			
	$h_+ = [21, 28]$						
PCL	81.78	123.9(13.0)	32.5(13.4)	165.9(21.9)	1.00	1.00	1.00
T.Trc	15.23(13.88,18.06)	123.5(10.3)	31.3(8.7)	161.0(14.2)	1.60	2.43	2.54
T.Det	14.73(13.17,18.06)	123.6(10.1)	31.3(8.4)	160.8(14.6)	1.64	2.58	2.41
T.SSE	15.58(11.25,22.66)	122.5(8.7)	31.2(7.5)	160.9(10.4)	2.29	3.26	4.74
True	$h_- = [16, 20]$	486	124	640			
	$h_+ = [42, 55]$						
PCL	81.78	496.0(87.5)	129.8(80.2)	670.3(140.2)	1.00	1.00	1.00
T.Trc	13.17(10.24,15.98)	489.4(66.3)	121.7(50.0)	642.9(76.1)	1.76	2.58	3.55
T.Det	13.17(10.64,17.39)	489.8(67.0)	123.8(50.1)	644.9(75.3)	1.72	2.58	3.61
T.SSE	12.71(7.64,16.00)	489.2(58.8)	122.06(41.58)	642.5(59.2)	2.24	3.73	5.87
True	$h_- = [8, 20]$	640	62	121			
	$h_+ = [22, 53]$						
PCL	81.78	649.9(90.5)	67.1(26.8)	127.2(20.6)	1.00	1.00	1.00
T.MahD	0.88(0.67,1.20)	644.8(71.3)	63.6(17.3)	124.57(11.63)	1.62	2.46	3.12
T.EucD	17.18(13.34,24.23)	640.1(74.5)	63.5(18.6)	124.74(12.04)	1.49	2.15	2.90

TABLE 12.2
Results of the TCL estimates for data simulated from the Schlather model. See Table 12.1 for the descriptions of the rows and the columns.

	Taper Range	Parameter Estimates		RE	
		ϕ	$\kappa \times 100$	ϕ	κ
True	$(h_-, h_+) = (6, 148)$	39.90	43		
PCL	81.78	43.99(18.92)	45(12.70)	1.00	1.00
T.Trc	21.34(10.36,54.15)	40.00(10.42)	53(9.80)	1.99	0.84
T.Det	27.12(15.81,40.10)	36.45(12.77)	48(9.79)	2.13	1.33
T.SSE	24.09(12.85,40.51)	36.92(6.66)	45(6.08)	7.03	3.83
True	$(h_-, h_+) = (15, 148)$	29.50	100		
PCL	81.78	31.10(11.01)	109(34.3)	1.00	1.00
T.Trc	21.93(10.36,34.52)	25.37(8.78)	226(768)	1.31	0.00
T.Det	24.62(15.81,32.58)	30.91(7.83)	205(852)	1.95	0.00
T.SSE	22.06(11.86,38.24)	29.31(3.54)	102(9.3)	9.85	13.55
True	$(h_-, h_+) = (20, 148)$	25.40	150		
PCL	81.78	27.15(14.22)	530(1620)	1.00	1.00
T.Trc	20.25(16.93,62.52)	21.16(9.12)	616(1660)	2.02	0.93
T.Det	22.06(16.93,62.09)	25.19(11.10)	593(1670)	1.67	0.93
T.SSE	18.07(12.85,34.02)	24.61(4.35)	349(1370)	10.48	1.45
True	$(h_-, h_+) = (3, 74)$	19.90	43		
PCL	81.78	24.55(17.42)	47(18.7)	1.00	1.00
T.Trc	20.80(16.93,35.96)	18.14(6.49)	51(15.7)	7.19	1.18
T.Det	23.00(16.93,31.08)	19.33(6.69)	48(12.2)	7.22	2.10
T.SSE	21.34(13.88,29.80)	19.56(4.82)	45(7.4)	13.96	6.13
True	$(h_-, h_+) = (12, 296)$	79.80	43		
PCL	81.78	85.72(30.53)	44(7.21)	1.00	1.00
T.Trc	22.69(16.93,57.56)	67.39(19.27)	48(6.22)	1.84	0.82
T.Det	27.34(17.61,42.82)	71.45(20.02)	47(5.79)	2.05	1.06
T.SSE	27.34(15.76,53.20)	74.33(11.31)	45(2.77)	6.12	4.71

also include the results of the TCL method by minimizing the sum of the squared errors (denoted as T.SSE) of the estimates as a benchmark for comparison purposes.

For the Smith model, we consider four scenarios. The true parameter values are given in Table 12.1. For each parameter set, we calculate their corresponding *practical range* (Davison et al., 2012), which is defined as a pair of distances (h_-, h_+) such that the extremal coefficients (see, e.g., Schlather and Tawn, 2003; Smith, 1990) are 1.3 and 1.7, respectively. In all the scenarios, the TCL estimators outperform the equally weighted CL estimators. The standard deviations of the TCL estimates are smaller than those using all pairs. The TCL also seems to reduce the biases in the estimations of the model parameters. We have observed that the longer the practical

TABLE 12.3

Results of the TCL estimates for data simulated from the Brown–Resnick model. See Table 12.1 for the descriptions of the rows and the columns.

	Taper Range	Parameter Estimates ϕ	$\kappa \times 100$	RE ϕ	κ
True	$(h_-, h_+) = (59, 83)$	30.00	75		
PCL	81.78	29.95(4.04)	74(8.95)	1.00	1.00
T.Trc	31.49(15.08,81.78)	27.81(3.38)	80(7.68)	1.01	0.93
T.Det	31.84(21.43,45.70)	28.93(3.57)	77(6.74)	1.18	1.64
T.SSE	31.20(17.61,81.78)	29.34(1.97)	75(4.85)	3.79	3.43
C.Tri	81.78	29.57(3.73)	75 (8.62)	1.00	1.00
T.Tri	31	29.43(3.56)	75(6.08)	1.09	2.01
True	$(h_-, h_+) = (11.39, 83)$	38.70	100		
PCL	81.78	39.87(5.44)	99(9.40)	1.00	1.00
T.Trc	28.64(12.83,69.16)	36.97(4.86)	105(8.43)	1.16	0.96
T.Det	28.64(16.93,41.53)	38.19(4.66)	102(7.35)	1.41	1.55
T.SSE	27.12(13.93,81.78)	38.65(3.00)	100(3.96)	3.43	5.65
True	$(h_-, h_+) = (22.25, 83)$	50.00	150		
PCL	81.78	51.15(5.85)	149(10.1)	1.00	1.00
T.Trc	30.89(17.58,41.21)	49.14(4.75)	153(7.38)	1.52	1.72
T.Det	25.44(18.07,34.15)	49.62(4.69)	152(7.13)	1.60	1.96
T.SSE	27.62(15.81,64.25)	49.97(2.82)	150(3.57)	4.45	8.20
True	$(h_-, h_+) = (22.25, 83)$	30.00	100		
PCL	81.78	30.89(3.44)	98(9.14)	1.00	1.00
T.Trc	26.79(13.93,69.02)	28.72(2.79)	105(6.37)	1.34	1.38
T.Det	27.48(18.01,43.21)	29.65(2.73)	102(5.89)	1.66	2.36
T.SSE	27.23(15.76,81.78)	29.81(1.59)	100(4.33)	4.90	4.66
True	$(h_-, h_+) = (22.25, 83)$	30.00	150		
PCL	81.78	30.90(2.61)	144(9.88)	1.00	1.00
T.Trc	25.95(14.84,41.19)	29.58(2.38)	151(7.75)	1.31	2.13
T.Det	25.44(17.58,33.14)	29.98(2.23)	149(6.79)	1.54	2.77
T.SSE	23.00(14.40,57.51)	29.97(1.18)	148(4.10)	5.46	6.76

range of the Smith max-stable process, the shorter the optimal taper range. In the last case, where the dependence is the most anisotropic, the use of the Mahalanobis distance for tapering achieves smaller MSE, leading to a higher efficiency gain over the equally weighted CL than the Euclidean distance counterpart. Moreover, it seems that the Mahalanobis distance results in fewer pairs selected for the TCL. We have also experimented with the triplewise TCL for the first parameter setting. Consistent with the findings in Genton et al. (2011), the triplewise equally weighted CL achieves much more accurate parameter estimations with drastically smaller standard errors than does the pairwise equally weighted CL. Since triplewise estimator is computationally expensive, we only searched the "optimal" range from a very small set with only 6 candidate taper ranges to minimize the trace of the estimated variance matrix. Therefore, the interval of the taper range is not reported. Efficiency is significantly further improved with the triplewise TCL.

For the Schlather model, we use a Matérn correlation function (Stein, 1999) with range parameter ϕ and smoothness parameter κ for the latent scaled Gaussian process. The results are presented in Table 12.2. We observe significant efficiency gains in most cases using the TCL method. Consistent with the findings of the first simulation with fixed smoothness, efficiency gains are stronger and the corresponding optimal taper ranges are shorter for the Schlather models with shorter ranges of dependence or smaller smoothness parameters. It is also noticeable that the efficiency gain for the range parameter is generally larger than that for the smoothness parameter. In fact, both equally weighted CL estimates and the TCL estimates for the smoothness tend to have serious upward biases for relatively large values of smoothness. Indeed, higher orders of smoothness are typically difficult to identify from real data.

For the Brown–Resnick model, we use the variogram function $2\gamma(h) = (|h|/\phi)^{\kappa}$, where ϕ is the range parameter and κ is the smoothness parameter. Five combinations of range and smoothness are considered. Table 12.3 shows that the TCL leads to more accurate parameter estimates with smaller biases and standard deviations compared with the equally weighted counterparts. And we notice that the benefit of using TCL is more prominent for smoother Brown–Resnick processes. But overall the efficiency gains from using the TCL for the Brown–Resnick model are not as significant as for the Smith model and the Schlather model. The selected taper range for the Brown–Resnick model is typically the longest among all three max-stable process models. The use of the determinant of the estimated inverse Godambe information matrix for taper range selection slightly improves efficiency over the method using the trace. This also seems to be true for the other two max-stable process models. The triplewise equally weighted CL leads to slightly more efficient parameter estimations than the equally weighted pairwise CL. But the efficiency gain from using the higher-order CL is not as prominent as from the Smith model, which has a much smoother max-stable field. With the taper range fixed at the same range chosen for the tapered pairwise CL, we observe further efficiency gains by using the tapered triplewise CL.

12.6 Discussion

In this chapter, we have reviewed CL methods in max-stable processes for spatial multivariate extremes, and also briefly discussed their application for the inference of copula models for extreme values. In particular, we focus on describing equal weight pairwise and triplewise CL, tapered CL, and Bayesian CL methods. Through a simulation study, we illustrate the capability of CL as a powerful and flexible tool in inferences for max-stable process models.

There are several ongoing and open research questions along this direction. Further statistical efficiency gains might be achieved by carefully designing CL. For instance, we considered only binary weights for the tapered CL inference. It is of interest to explore the use of smoothly decaying weights. One may also consider setting unequal weights to acknowledge the association between pairwise CL elements.

The CL approach for extreme values with large datasets faces critical computational challenges as the number of pairwise/triplewise CL elements grows explosively as sample size and the number of spatial location. Subsampling based CL approach is one potentially useful way to tackle this problem. In addition, taking advantage of the fact that CL essentially involves a number of separate calculations of likelihood elements, one natural strategy is to design computationally efficient algorithms that allow parallel computation and distributed inference.

Finally, one important gap that needs further study is the asymptotic theory in the case of increasing number of spatial locations with fixed or slowly increasing sample size, especially under fix domain infill asymptotic framework. The consistency property of the maximum CL estimators for max-stable processes and copula models is still largely unknown under various conditions on the number of locations and the sample size.

References

Bai, Y., P. X.-K. Song, and T. Raghunathan (2012). Joint composite estimating functions in spatiotemporal models. *Journal of the Royal Statistical Society: Series B (Statistical Methodology) 74*(5), 799–824.

Besag, J. (1974). Spatial interaction and the statistical analysis of lattice systems. *Journal of the Royal Statistical Society: Series B (Statistical Methodology) 36*(2), 192–236.

Bevilacqua, M., C. Gaetan, J. Mateu, and E. Porcu (2012). Estimating space and space-time covariance functions for large datasets: A weighted composite likelihood approach. *Journal of the American Statistical Association 107*(497), 268–280.

Blanchet, J. and A. Davison (2011). Spatial modeling of extreme snow depth. *The Annals of Applied Statistics 5*(3), 1699–1725.

Cox, D. R. and N. Reid (2004). A note on pseudolikelihood constructed from marginal densities. *Biometrika 91*, 729–737.

Davis, R. A. and C. Y. Yau (2011). Comments on pairwise likelihood in time series models. *Statistica Sinica 21*, 255–277.

Davison, A., S. Padoan, and M. Ribatet (2012). Statistical modeling of spatial extremes. *Statistical Science 27*(2), 161–186.

Davison, A. C. (2003). *Statistical Models*. Cambridge University Press.

de Haan, L. and S. I. Resnick (1977). Limit theory for multivariate sample extremes. *Probability Theory and Related Fields 40*(4), 317–337.

Genton, M. G., Y. Ma, and H. Sang (2011). On the likelihood function of Gaussian max-stable processes. *Biometrika 98*(2), 481–488.

Gumbel, E. J. (1960). Bivariate exponential distributions. *Journal of the American Statistical Association 55*(292), 698–707.

Huser, R. and A. C. Davison (2013). Composite likelihood estimation for the Brown–Resnick process. *Biometrika 100*, 511–518.

Hüsler, J. and R.-D. Reiss (1989). Maxima of normal random vectors: Between independence and complete dependence. *Statistics & Probability Letters 7*(4), 283–286.

Joe, H. and Y. Lee (2009). On weighting of bivariate margins in pairwise likelihood. *Journal of Multivariate Analysis 100*(4), 670–685.

Kabluchko, Z., M. Schlather, and L. de Haan (2009). Stationary max-stable fields associated to negative definite functions. *The Annals of Probability 37*(5), 2042–2065.

Lindsay, B. G. (1988). Composite likelihood methods. In N. U. Prabhu (Ed.), *Statistical Inference from Stochastic Processes*, pp. 221–239. Providence, RI: American Mathematical Society.

Lindsay, B. G., G. Y. Yi, and J. Sun (2011). Issues and strategies in the selection of composite likelihoods. *Statistica Sinica 21*(1), 71.

Lipsitz, S., K. Dear, and L. Zhao (1994). Jackknife estimators of variance for parameter estimates from estimating equations with applications to clustered survival data. *Biometrics 50*(3), 842–846.

Mardia, K. V., J. T. Kent, and J. M. Bibby (1979). *Multivariate Analysis*. Academic Press.

Nikoloulopoulos, A. K., H. Joe, and H. Li (2009). Extreme value properties of multivariate t copulas. *Extremes 12*(2), 129–148.

Padoan, S. A., M. Ribatet, and S. A. Sisson (2010). Likelihood-based inference for max-stable processes. *Journal of the American Statistical Association 105*(489), 263–277.

Ribatet, M., D. Cooley, and A. Davison (2012). Bayesian inference from composite likelihoods, with an application to spatial extremes. *Statistica Sinica 22*, 813–845.

Ribatet, M. and M. Sedki (2013). Extreme value copulas and max-stable processes. *Journal de la Société Française de Statistique 154*(1), 138–150.

Sang, H. and A. E. Gelfand (2010). Continuous spatial process models for spatial extreme values. *Journal of Agricultural, Biological, and Environmental Statistics 15*(1), 49–65.

Sang, H. and M. G. Genton (2014). Tapered composite likelihood for spatial max-stable models. *Spatial Statistics 8*, 86–103.

Schlather, M. (2002). Models for stationary max-stable random fields. *Extremes 5*(1), 33–44.

Schlather, M. and J. A. Tawn (2003). A dependence measure for multivariate and spatial extreme values: Properties and inference. *Biometrika 90*(1), 139–156.

Sklar, A. (1996). Random variables, distribution functions, and copulas: A personal look backward and forward. In L. Rüschendorf, B. Schweizer, and M. D. Taylor (Eds.), *Distributions with Fixed Marginals and Related Topics*, pp. 1–14. Hayward, CA: Institute of Mathematical Statistics.

Smith, R. L. (1990). Max-stable processes and spatial extremes. Unpublished manuscript.

Stein, M. (1999). *Interpolation of Spatial Data: Some Theory for Kriging*. Springer Verlag.

Stein, M. L., Z. Chi, and L. J. Welty (2004). Approximating likelihoods for large spatial data sets. *Journal of the Royal Statistical Society Series B* 66(2), 275–296.

Varin, C. and P. Vidoni (2005). A note on composite likelihood inference and model selection. *Biometrika* 92, 519–528.

Vecchia, A. V. (1988). Estimation and model identification for continuous spatial processes. *Journal of the Royal Statistical Society: Series B (Statistical Methodology)* 50(2), 297–312.

13

Bayesian Inference for Extreme Value Modelling

Alec Stephenson

CSIRO, Australia

Abstract

We present a practical introduction to the application of Bayesian inferential methods for commonly used parametric models for extreme event data. We discuss the general theory of Bayesian inference and the choice of a prior distribution for extreme value parameters. We also examine practical implementation issues using the statistical package R (R Core Team, 2013). We explain the main concepts through a series of examples. In Section 13.1 we provide the theoretical background and give two examples using simulated data. In Section 13.2 we provide two additional data examples. The first uses a non-stationary model for athletics data. The second, available on the book website, uses a bivariate model for monthly temperature maxima at two sites in the Australian state of Victoria. The algorithms and the data are also available on the book website. The R code provided on the website enables the reader to fully examine and reproduce all the examples given in this chapter.

13.1 Bayesian Theory

In this chapter we focus on practical applications of Bayesian inference in extreme value modelling. For demonstration purposes our examples use maxima or minima data, but the extension to threshold exceedance models is in many cases immediate. For Bayesian approaches to threshold selection, see Lee et al. (2015) and references therein. For the use of noninformative priors in threshold exceedance models, see, e.g., Castellanos and Cabras (2007) and Northrop and Attalides (2015).

Our aim is to discuss both theoretical approaches and practical implementation issues. In order to achieve this we present code that can be used to program algorithms within the statistical package R (R Core Team, 2013). In addition to the standard R distribution, the following R packages must be loaded to ensure that the code will work correctly: `KernSmooth` (Wand, 2013), `evd` (Stephenson, 2002), `mvtnorm` (Genz et al., 2013), and `coda` (Plummer et al., 2006).

Datasets and full R code that enables the reproduction of all the examples given here is available from the book website. Also available on the book website is a discussion of Bayesian model diagnostics and an example of modelling monthly temperature maxima. The code given here demonstrates only the basic ideas of implementation. We have included some advice on code optimization where appropriate, focusing on speed rather than memory; for time critical operations it may be necessary to use a compiled language either in preference to R or as a subroutine that is called from R. In our experience this typically results in a two-fold or three-fold speed increase over optimized R code for iterative simulation algorithms of the type used in Markov chain Monte Carlo simulation.

There are a large number of R packages that implement Bayesian methodology for particular classes of models or in particular application areas. An extensive list of these packages is available from the website of the Comprehensive R Archive Network (CRAN), which maintains a "Task View" for Bayesian inference. The list includes the package `evdbayes` for Bayesian analysis of extreme value models (Stephenson and Ribatet, 2012). There are also alternative high-level languages and environments designed specifically for Bayesian modelling. The BUGS (Bayesian Inference Using Gibbs Sampling) project distributes the widely used BUGS, Win-BUGS, and OpenBUGS software (Lunn et al., 2000). JAGS (Just Another Gibbs Sampler) is an alternative open source program (Plummer, 2003). Stan is a more recently developed environment that includes Hamiltonian Monte Carlo methods (Stan Development Team, 2014).

13.1.1   The Bayesian Paradigm

Bayesian inferences about a parameter or a vector of parameters θ are made in terms of probability statements that typically condition on the observed values of the data y. We write this as $f(\theta|y)$, which defines the conditional probability density of θ given y. In Bayesian inference we assume that, without reference to the data y, it is possible to formulate beliefs about θ that can be expressed as a probability distribution. For example, if $\theta \in (0, 1)$, and you believe that any value in $(0, 1)$ is equally likely, your belief can be expressed using the uniform probability distribution $\theta \sim U(0, 1)$. A distribution on θ, made without reference to the data, is called a prior distribution. The parameters of the prior distribution are called hyperparameters.

Let $f(\theta)$ denote the density of the prior distribution for θ. More generally, $f(\cdot)$ and $f(\cdot|\cdot)$ will denote arbitrary marginal and conditional densities with the arguments determined from the context. We use the terms "distribution" and "density" interchangeably, and we assume that the parameters θ are continuous. The joint probability distribution of θ and y is defined by $f(\theta, y) = f(\theta)f(y|\theta)$, where $f(y|\theta)$ is the data distribution. When viewed as a function of θ for fixed y, the data distribution

$f(y|\theta)$ is simply the likelihood function. Using the rules of conditional probability gives the posterior density

$$f(\theta|y) = \frac{f(\theta)f(y|\theta)}{f(y)} = \frac{f(\theta)f(y|\theta)}{\int f(\theta)f(y|\theta)\,d\theta}, \qquad (13.1)$$

where the integral is over all possible values of θ, and where $f(y)$ is the marginal distribution of y. Since $f(y)$ does not depend on θ it can, for fixed y, be considered a constant so that

$$f(\theta|y) \propto f(\theta)f(y|\theta) \qquad (13.2)$$

and therefore $f(\theta)f(y|\theta)$ is the unnormalized posterior density for θ given y.

The primary objective of an extreme value analysis is often prediction. Inferences about the parameter θ are derived through $f(\theta|y)$. We can make inferences about future observations $\tilde{y}$ or more generally about any unknown observable quantities. For example, suppose $y = (y_1, \ldots, y_n)$ is a vector of recorded annual maximum temperatures over n years, modelled as independent and identically distributed observations from a generalized extreme value (GEV) distribution with parameter vector $\theta = (\mu, \sigma, \xi)$. Then $\tilde{y}$ may be the yet to be recorded temperature maxima for the year $n + 1$. The distribution of $\tilde{y}$ is called the posterior predictive distribution. It is a predictive distribution because $\tilde{y}$ is observable in the future, and it is a posterior distribution because it is conditional on the observed y. The density function is

$$f(\tilde{y}|y) = \int f(\tilde{y},\theta|y)\,d\theta = \int f(\tilde{y}|\theta,y)f(\theta|y)\,d\theta = \int f(\tilde{y}|\theta)f(\theta|y)\,d\theta, \quad (13.3)$$

which follows from the conditional independence of y and $\tilde{y}$ given θ. The predictive distribution reflects the uncertainty in the model through $f(\theta|y)$ and the uncertainty due to the variability of future observations through $f(\tilde{y}|\theta)$. The predictive distribution can also be written as

$$f(\tilde{y}|y) = \mathbb{E}_{\theta|y}[f(\tilde{y}|\theta)],$$

so we can evaluate $f(\tilde{y}|y)$ by averaging over the different possible parameter values. If we do not observe any data y then our inferences about the future observable $\tilde{y}$ are based on the prior predictive density $\int f(\theta)f(\tilde{y}|\theta)\,d\theta$, which is equal to the denominator of Equation (13.1), expressed as a function of $\tilde{y}$.

Using Equation (13.3), the survivor function of the posterior predictive distribution is given by

$$\Pr(\tilde{Y} > \tilde{y}|y) = \int \Pr(\tilde{Y} > \tilde{y}|\theta)f(\theta|y)\,d\theta = \mathbb{E}_{\theta|y}[\Pr(\tilde{Y} > \tilde{y}|\theta)], \qquad (13.4)$$

which is often of interest in extreme value models where we are concerned with the probability of a future unknown observable exceeding some threshold. The survivor function of the prior predictive distribution is defined in a similar manner, replacing the posterior density $f(\theta|y)$ with the prior density $f(\theta)$.

The derivations given above are direct consequences of the axioms of probability and cannot be disputed. However, there are philosophical differences of opinion regarding the treatment of parameters θ as random variables, and to the necessity of adopting a prior distribution $f(\theta)$. We largely ignore these issues here, and focus on practical examples. In many cases the data overwhelm the prior and the choice of $f(\theta)$ is not particularly important, so long as it is non-zero over an appropriate range. Unfortunately, extreme value modelling is often based on small datasets, and conclusions may be sensitive to the prior distribution. In these cases it may be important to incorporate information external to the data within the prior. We discuss prior distributions and prior elicitation for GEV parameters at some length in Section 13.1.6.

Our view is that the ability to incorporate information external to the data y is a considerable benefit, particularly for extreme value modelling where the amount of information in the data is often low. Bayesian inference also provides computational advantages in complex models, such as the hierarchical models in Section 13.1.7, and easily allows for posterior predictions of a wide range of quantities of interest.

13.1.2   Markov Chain Monte Carlo Simulation

We assume in this chapter that the unnormalized density $f(\theta)f(y|\theta)$ in Equation (13.2) can be easily evaluated. In complex cases where the likelihood $f(y|\theta)$ is intractable, a technique known as approximate Bayesian computation (ABC) can be used; see Chapter 14.

Given the data y, the difficulty in obtaining inferences from $f(\theta|y)$ is in the calculation of the normalizing constant $\int f(\theta)f(y|\theta)\,d\theta$. For high-dimensional θ, this constant is often difficult to evaluate or approximate. In some cases $f(\theta)$ can be chosen so that $\int f(\theta)f(y|\theta)\,d\theta$ can be calculated analytically, but this is typically not possible for extreme value models of practical use. Distributions commonly used for extreme value modelling are not within the exponential family, and this prohibits the use of convenient analytical forms.

Fortunately, methods have been developed for sampling from arbitrary posterior distributions $f(\theta|y)$. If we could simulate N values $\theta^1, \theta^2, \theta^3, \ldots, \theta^N$ that are independent and identically distributed from $f(\theta|y)$, we can use them to estimate features of interest. For example, the sample means of the simulated values for each parameter can be used to estimate the marginal posterior means, and the histograms of the simulated values can be used to estimate the marginal posterior densities. Moreover, we can use these values to simulate a corresponding set of predictions $\tilde{y}^1, \tilde{y}^2, \tilde{y}^3, \ldots, \tilde{y}^N$ that are distributed according to the posterior predictive density $f(\tilde{y}|y)$.

Simulating an independent and identically distributed sample from $f(\theta|y)$ is usually not achievable, but we can use Markov chain Monte Carlo techniques to simulate a Markov chain $\theta^1, \theta^2, \theta^3, \ldots, \theta^N$ that converges to the target distribution $f(\theta|y)$. This means that the simulations $\theta^{B+1}, \ldots, \theta^N$ for some burn-in period B can be treated as a random sample from $f(\theta|y)$. The Markov property implies that the distribution of θ^t depends only on the distribution of the previous iteration θ^{t-1}. The samples are not independent, but they can still be used to estimate features of $f(\theta|y)$ and $f(\tilde{y}|y)$. The dependence between the samples influences the accuracy of posterior estimates. As the dependence becomes stronger, the run-length N must be larger in order to achieve the same precision. This dependence must therefore be taken into account when evaluating, e.g., the standard error of posterior means.

Alternative methods of inference for $f(\theta|y)$ are less frequently used in the context of extreme value modelling. If θ is not high dimensional, it may be feasible to calculate $\int f(\theta)f(y|\theta)\,d\theta$ numerically, leading to the use of more direct simulation methods such as discrete approximation or rejection sampling. If θ is high dimensional, alternative methods are available to approximate $f(\theta|y)$, such as mixtures of normal approximations, variational Bayes, and expectation propagation algorithms.

13.1.3 The Metropolis–Hastings Algorithm

The Metropolis–Hastings algorithm is a conceptually simple method to simulate a Markov chain $\theta^1, \theta^2, \theta^3, \ldots, \theta^N$ that converges (under simple regularity conditions) to the target distribution $f(\theta|y)$; see Algorithm 13.1.

Algorithm 13.1 The Metropolis–Hastings Algorithm.

1. Pick a starting point θ^0.
2. For $t = 1, \ldots, N$
   - (a) Sample a proposal θ^* from a proposal distribution $p_t(\theta^*|\theta^{t-1})$.
   - (b) Calculate the ratio

$$r = \frac{f(\theta^*|y)p_t(\theta^{t-1}|\theta^*)}{f(\theta^{t-1}|y)p_t(\theta^*|\theta^{t-1})} = \frac{f(\theta^*)f(y|\theta^*)p_t(\theta^{t-1}|\theta^*)}{f(\theta^{t-1})f(y|\theta^{t-1})p_t(\theta^*|\theta^{t-1})}$$

   - (c) Set

$$\theta^t = \begin{cases} \theta^* & \text{with probability } \min(r,1) \\ \theta^{t-1} & \text{otherwise.} \end{cases}$$

The denominator from Equation (13.1) cancels in the ratio r and therefore only the unnormalized density is required. The algorithm also requires a starting point θ^0 and a proposal distribution $p_t(\cdot|\cdot)$. The proposal distribution is often taken to be symmetric so that $p_t(\theta^*|\theta^{t-1}) = p_t(\theta^{t-1}|\theta^*)$, and r is then the ratio of the unnormalized densities. For example, $p_t(\theta^*|\theta^{t-1})$ could be a multivariate normal density with mean vector θ^{t-1}. In practice we often choose several starting points to ensure proper behaviour of the algorithm, as discussed in Section 13.1.5.

13.1.3.1 Simulated Data Example I

We simulate data from a Gumbel distribution with location parameter $\mu = 0$ and scale parameter $\sigma = 1$. Our parameter vector is $\theta = (\mu, \nu)$, where we take $\nu = \log(\sigma)$. For our prior distribution $f(\theta)$ we take μ and ν to be independent vague normal distributions with $\mu \sim N(0, 10^2)$ and $\nu \sim N(0, 10^2)$. We correctly model the data $y = (y_1, \ldots, y_n)$ as independent and identically distributed Gumbel so that $f(y|\theta)$ is the product over $i = 1, \ldots, n$ of the Gumbel probability density function evaluated at each y_i.

In practice we compute the unnormalized posterior density function $f(\theta)f(y|\theta)$

on the log scale to avoid problems with numerical underflow and overflow. In this example we employ the density functions dgumbel and dnorm to construct the unnormalized posterior density function as below.

```
log_post <- function(mu, logsig, data) {
    llhd <- sum(dgumbel(data, loc = mu, scale = exp(logsig), log = TRUE))
    lprior <- dnorm(mu, sd = 10, log = TRUE)
    lprior <- lprior + dnorm(logsig, sd = 10, log = TRUE)
    lprior + llhd
}
```

Numerical maximization of $f(\theta)f(y|\theta)$, and hence of $f(\theta|y)$, over θ using several starting values suggests that there is a single posterior mode at $\widehat{\theta} = (0.003, -0.023)$. We can construct a crude approximation of $f(\theta|y)$ by assuming a bivariate normal distribution. The matrix of second derivatives of the log posterior density function can be negated and then inverted to approximate the variance matrix. This yields

$$\theta|y \overset{\text{approx}}{\sim} N\left(\begin{bmatrix} 0.003 \\ -0.023 \end{bmatrix}, \begin{bmatrix} 0.011 & 0.003 \\ 0.003 & 0.006 \end{bmatrix}\right), \tag{13.5}$$

which provides a useful starting point and can be compared with our Markov chain outputs to identify any clear discrepancies that may derive from coding errors. It also provides a means of selecting our proposal distribution $p_t(\theta^*|\theta^{t-1})$. Theoretical results involving multivariate normal approximations suggest taking $p_t(\theta^*|\theta^{t-1}) = N(\theta^*|\theta^{t-1}, c^2\Sigma)$ where Σ is the variance matrix given in Equation (13.5) and $c \approx 2.4/\sqrt{d}$, with $d = 2$ being the number of parameters.

Using the code below we simulate a single Markov chain of run-length 2000, using the posterior mode $\widehat{\theta}$ as a starting point. The first lines of code specify the run-length, the starting value, and the variance matrix for the bivariate normal proposal distribution. Memory is allocated using a matrix with the starting values in the first row. The main simulation loop then follows.

```
varmat <- matrix(c(0.030,0.007,0.007,0.018), nrow = 2)
iter <- 2000; out <- matrix(NA, nrow = iter+1, ncol = 2)
dimnames(out) <- list(0:iter, c("mu", "logsig"))
out[1,] <- c(0.003, -0.023)
for(t in 1:iter) {
    prop <- rmvnorm(1, mean = out[t,], varmat)
    lpost_old <- log_post(out[t,1], out[t,2], data)
    lpost_prop <- log_post(prop[1], prop[2], data)
    r <- exp(lpost_prop - lpost_old)
    if(r > runif(1))
        out[t+1,] <- prop
    else out[t+1,] <- out[t,]
}
```

It is easy to adjust the above code in order to store the acceptance probabilities min(r, 1) at each iteration. In our case the mean acceptance rate was 0.34 which is a reasonable value for bivariate proposals. We discard the first 1000 iterations.

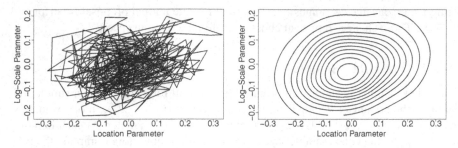

FIGURE 13.1
The left plot shows the sequence of the last 1000 iterations from a Markov chain Monte Carlo simulation, using Metropolis sampling. The simulated values can be treated as a correlated sample from the posterior distribution $f(\theta|y)$, where $\theta = (\mu, \nu)$ with $\nu = \log(\sigma)$. The right plot shows a bivariate kernel density estimate of $f(\theta|y)$, calculated using the simulated values.

Figure 13.1 displays the Markov chain sequence during the last 1000 iterations. This illustrates the movement of the Metropolis sampler throughout the parameter space. Figure 13.1 also displays a contour plot of a kernel density estimate for the posterior density $f(\theta|y)$, based on the sampled values. It is this density on which posterior inferences can be based.

Further examination of the code above reveals inefficiencies that will be important to correct in more complex cases. It is unnecessary to compute the `log_post` function twice on each iteration. Apart from the first iteration, the term `lpost_old` will always have been calculated previously. Also, the bivariate normal proposal involves the calculation of the matrix square root of the variance matrix `varmat`. This is unnecessarily calculated on every iteration. The following code solves the issues.

```
ev <- eigen(matrix(c(0.030,0.007,0.007,0.018), nrow = 2))
varmat <- ev$vector %*% diag(sqrt(ev$value)) %*% t(ev$vector)
iter <- 2000; out <- matrix(NA, nrow = iter+1, ncol = 2)
dimnames(out) <- list(0:iter, c("mu", "logsig"))
out[1,] <- c(0.003, -0.023)
lpost_old <- log_post(out[1,1], out[1,2], data)
for(t in 1:iter) {
  prop <- rnorm(2) %*% varmat + out[t,]
  lpost_prop <- log_post(prop[1], prop[2], data)
  r <- exp(lpost_prop - lpost_old)
  if(r > runif(1)) {
    out[t+1,] <- prop
    lpost_old <- lpost_prop
  }
  else out[t+1,] <- out[t,]
}
```

The term `lpost_old` is now calculated at the starting values and is updated to equal `lpost_prop` whenever the proposal is accepted. The matrix `varmat` is now the matrix square root of the variance matrix, calculated outside of the main loop using the eigenvalue decomposition, and the bivariate normal proposal is performed directly. These changes represent about a ten-fold increase in speed.

13.1.4 The Gibbs Sampler

The Gibbs sampler is another algorithm that is useful for Markov chain simulation. Suppose the parameter θ is divided into d subvectors $(\theta_1, \ldots, \theta_d)$. In our examples, each subvector θ_j for $j = 1, \ldots, d$ is typically a single parameter. At each iteration $t = 1, \ldots, N$, the Gibbs sampler cycles through the d subvectors, sampling the sub-vector θ_j^t conditional on both the data y and the remaining subvectors θ_{-j}^{t-1} at their current values. We therefore have $\theta_{-j}^{t-1} = (\theta_1^t, \ldots, \theta_{j-1}^t, \theta_{j+1}^{t-1}, \ldots, \theta_d^{t-1})$ with each θ_j^t sampled from $f(\theta_j | \theta_{-j}^{t-1}, y)$.

The Gibbs sampler depends on being able to simulate from $f(\theta_j | \theta_{-j}^{t-1}, y)$. In many cases, direct simulation is not possible. However, we can apply the Metropolis–Hastings algorithm to $f(\theta_j | \theta_{-j}^{t-1}, y)$, giving Algorithm 13.2.

Algorithm 13.2 The Gibbs Sampler.

---

1. Pick a starting point θ^0.
2. For $t = 1, \ldots, N$
   For $j = 1, \ldots, d$
   (a) Sample a proposal θ_j^* from a proposal distribution $p_{t,j}(\theta_j^* | \theta_j^{t-1})$.
   (b) Calculate the ratio

$$r = \frac{f(\theta_j^* | \theta_{-j}^{t-1}, y) p_{t,j}(\theta_j^{t-1} | \theta_j^*)}{f(\theta_j^{t-1} | \theta_{-j}^{t-1}, y) p_{t,j}(\theta_j^* | \theta^{t-1})}$$

$$= \frac{f(\theta_1^t, \ldots, \theta_{j-1}^t, \theta_j^*, \theta_{j+1}^{t-1}, \ldots, \theta_d^{t-1} | y) p_{t,j}(\theta_j^{t-1} | \theta_j^*)}{f(\theta_1^t, \ldots, \theta_{j-1}^t, \theta_j^{t-1}, \theta_{j+1}^{t-1}, \ldots, \theta_d^{t-1} | y) p_{t,j}(\theta_j^* | \theta^{t-1})}$$

   (c) Set

$$\theta_j^t = \begin{cases} \theta_j^* & \text{with probability } \min(r, 1) \\ \theta_j^{t-1} & \text{otherwise.} \end{cases}$$

If we are able to perform the simulation directly then we can take $p_{t,j}(\theta_j^* | \theta_j^{t-1}) = f(\theta_j^* | \theta_{-j}^{t-1}, y)$ so that $r = 1$.

The proposal distribution is often symmetric, so that $p_{t,j}(\theta_j^* | \theta_j^{t-1}) = p_{t,j}(\theta_j^{t-1} | \theta_j^*)$. In our examples θ_j is often univariate and $p_{t,j}(\theta_j^* | \theta_j^{t-1})$ is typically taken to be a univariate normal distribution with mean θ_j^{t-1} and standard deviation σ_j. If σ_j is too large, then r will be too small and θ_j^* will be rarely accepted. If σ_j is too small, then the proposal will often be accepted but θ_j^* will be very close to θ_j^{t-1}.

When θ_j is univariate, setting σ_j so that the average probability of acceptance is between 0.4 and 0.5 typically yields Markov chains with desirable properties. It can be difficult to set each σ_j to achieve average acceptance probabilities between 0.4 and 0.5 for all parameters. In our examples we use a trial-and-error approach, where we simply re-sample with different values of σ_j in those cases where the average acceptance probabilities are not as anticipated. An alternative approach is to use an adaptive algorithm, where the σ_j values are altered as appropriate in the earlier iterations of the main for loop. This leads to more elaborate code, but may be advantageous for more complex models.

13.1.4.1 Simulated Data Example II

We continue the example of Section 13.1.3.1, but we now apply the Gibbs sampling methodology given above. Recall that our parameter vector is $\theta = (\mu, \nu)$, where we take $\nu = \log(\sigma)$, with independent vague normal distributions $\mu \sim N(0, 10^2)$ and $\nu \sim N(0, 10^2)$. In the notation given for the Gibbs sampler, we split θ into two subvectors of size one, namely the two individual parameters μ and ν, so that $d = 2$ with $\theta_1 = \mu$ and $\theta_2 = \nu$.

The code to perform the sampling is given below. The code is similar to that given in Section 13.1.3.1, again using the function log_post to calculate the logarithm of the unnormalized density posterior density function $f(\theta)f(y|\theta)$. We again simulate a single Markov chain of run-length 2000, using the posterior mode $\widehat{\theta}$ as a starting point. For our proposal distributions $p_{t,j}(\theta_j^* | \theta_j^{t-1})$ we take univariate normal distributions with mean θ_j^{t-1}. The proposal variances are taken to equal the diagonal entries in the variance matrix of Equation (13.5) multiplied by the factor c^2, where $c \approx 2.4/\sqrt{d}$ with $d = 1$.

```
propsd <- sqrt(c(0.061, 0.036))
iter <- 2000; out < matrix(NA, nrow - iter+1, ncol = 2)
dimnames(out) <- list(0:iter, c("mu", "logsig"))
out[1,] <- c(0.003, -0.023)
lpost_old <- log_post(out[1,1], out[1,2], data)
for(t in 1:iter) {
  prop1 <- rnorm(1, mean = out[t,1], propsd[1])
  lpost_prop <- log_post(prop1, out[t,2], data)
  r <- exp(lpost_prop - lpost_old)
  if(r > runif(1)) {
    out[t+1,1] <- prop1
    lpost_old <- lpost_prop
  }
  else out[t+1,1] <- out[t,1]

  prop2 <- rnorm(1, mean = out[t,2], propsd[2])
  lpost_prop <- log_post(out[t+1,1], prop2, data)
  r <- exp(lpost_prop - lpost_old)
  if(r > runif(1)) {
    out[t+1,2] <- prop2
    lpost_old <- lpost_prop
  }
  else out[t+1,2] <- out[t,2]
}
```

The first lines of code are similar to those given in Section 13.1.3.1, but we now have a vector propsd giving the standard deviations of the univariate normal proposal distributions. The main loop now updates each parameter individually, but the implementation is similar to the earlier Metropolis algorithm. It is easy to adjust the above code in order to store the acceptance probabilities min(r, 1) at each iteration

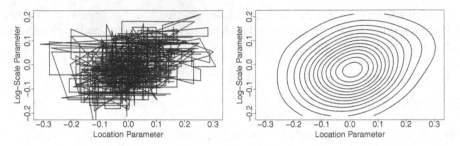

FIGURE 13.2
The left plot shows the sequence of the last 1000 iterations from a Markov chain
Monte Carlo simulation, using Gibbs sampling with Metropolis steps for each pa-
rameter. The simulated values can be treated as a correlated sample from the poste-
rior distribution $f(\theta|y)$, where $\theta = (\mu, \nu)$ with $\nu = \log(\sigma)$. The right plot shows a
bivariate kernel density estimate of $f(\theta|y)$, calculated using the simulated values.

and for each parameter. In our case the mean acceptance rates were 0.43 for μ and
0.42 for $\nu = \log(\sigma)$ which are reasonable values for univariate proposals.

We discard the first 1000 iterations. Figure 13.2 displays the Markov chain sequence during the last 1000 iterations. If a particular iteration accepts a proposal for
exactly one of the two parameters, the movement of the chain is in a parallel direction to the axes. This is unlike the bivariate proposal of Figure 13.1, and therefore the
two figures display a different behaviour in the movement throughout the parameter
space. However, the target distribution $f(\theta|y)$ is the same, and therefore the kernel
density estimates of $f(\theta|y)$ in Figures 13.1 and 13.2 are similar.

If we consider the details of our implementation it becomes apparent that we are
unnecessarily calculating terms that eventually cancel in the acceptance ratio r. We
can improve the speed of implementation by replacing the pre-built density functions
in `log_post` by code that explicitly calculates only the required terms. It is easy to
show that we can replace the original function by the following, giving a two-fold
increase in the speed of each iteration.

```
log_post <- function(mu, logsig, data) {
  data <- (data - mu)/exp(logsig)
  llhd <- -length(data)*logsig  - sum(data) - sum(exp(-data))
  lprior <- -mu*mu/200 - logsig*logsig/200
  lprior + llhd
}
```

Another approach we could use for implementation is to construct different
`log_post` functions for each parameter, or more generally for each subvector θ_j.
The advantage of this approach is that we could remove from each function any unnecessary terms that depend only on the other subvectors θ_{-j}. The disadvantage is
that we could not store values of the function for use in subsequent updates of other
subvectors, as we have done in our code. For examples where large cancellation factors are involved, this alternative approach should be preferred.

The generalized extreme value (GEV) distribution, with parameter vector $\theta = (\mu, \sigma, \xi)$, extends the Gumbel distribution that is used in the above example. The code above can easily be extended to incorporate the additional shape parameter ξ. The only additional consideration is that the GEV distribution has finite support. When $\xi > 0$ there is a finite lower end-point at $\mu - \sigma/\xi$. When $\xi < 0$ there is a finite upper end-point, also at $\mu - \sigma/\xi$. One way of dealing with this is to explicitly return the R language negative infinity construct `-Inf` from the function `log_post` whenever any of $1 + \xi(y_i - \mu)/\sigma$ are non-positive. If the posterior density is not zero at the chosen starting values, this will ensure that any proposals outside the parameter space result in rejection with $r = 0$, and the algorithm will continue to work correctly. We recommend explicitly returning an error condition to ensure that the posterior density is not zero at the starting values. It may also be useful to store the values returned in each call to `log_post`, as this can help to identify cases where there is a large posterior mass near this complex parameter boundary.

13.1.5 Simulation Inferences

Simulations $\theta^{B+1}, \ldots, \theta^N$ for some burn-in period B and run-length N can be treated as a correlated sample from $f(\theta|y)$. The burn-in period B should be chosen to ensure that convergence to $f(\theta|y)$ has been reached, and N should be chosen to ensure sufficient accuracy in the quantities being estimated. There is much literature on the choices of B and N, and there are many output diagnostics that evaluate one or more simulated chains to ensure convergence and an appropriate run-length (see Cowles and Carlin (1996) and Brooks and Roberts (1998) for reviews). We briefly sketch our approach here, based on Gelman and Rubin (1992), which we then apply to the data examples in Section 13.2.

Simulating several chains from multiple starting points is the only way to ensure that the chains have converged and fully explored all regions of high probability, particularly when the target distribution is complex. We simulate four chains using starting values sampled from a starting distribution $p_0(\theta)$, which is over-dispersed relative to the target distribution $f(\theta|y)$. Since $f(\theta|y)$ is unknown, a crude multivariate normal approximation is used to specify $p_0(\theta)$. We discard the first half of each chain, taking $B = N/2$. This is a fairly conservative approach and will often discard too many simulations, but in our examples the simulations are relatively fast and therefore this is not a major concern.

To ensure that B is large enough we use a diagnostic that compares within and between chain variances to simultaneously identify poor mixing behaviour and non-stationarity. The last $N/2$ of each of the four chains are split in half, giving eight chains of length $N/4$. The eight chains are used to produce an estimated "potential scale reduction factor." Values substantially above 1 indicate a lack of convergence. This diagnostic assumes that the initial values are sampled from a distribution $p_0(\theta)$ that is over-dispersed relative to the target distribution. It also relies on a normal approximation to the samples of each parameter within each chain; we typically transform parameters so that a normal distribution is more appropriate. If no lack of convergence is indicated, we use a run-length diagnostic based on a criterion of accuracy of estimation of a quantile. Depending on the run-length indications, we either pool the eight chains to produce a sample of size $2N$, or we repeat the process with a larger value of N.

For the choice of run-length N, it is important to recognize that most output diagnostics are based on means and variances, and examine the estimation of quantities in the center of the target distribution. If we wish to, e.g., produce density estimates for extreme upper quantiles of heavy-tailed distributions, we may need N to be far larger than the values indicated by typical output diagnostics. Standard run-length diagnostics may be useful indicators, but for extreme value modelling, much larger sample sizes may be needed. If N is very large and computer storage is a problem, it is possible to thin the output by storing only every kth iteration following convergence. This will reduce the precision of posterior estimates, but may represent a necessary computational saving.

Once we have pooled our chains to form $\theta^1, \ldots, \theta^{2N}$ we can estimate features of the posterior distribution $f(\theta|y)$ in the usual manner. In particular, we can estimate the probability of a future unknown observable exceeding some fixed threshold $\tilde{y}$, as given in Equation (13.4), using

$$\widehat{\Pr}(\tilde{Y} > \tilde{y}|y) = \frac{1}{2N} \sum_{t=1}^{2N} \Pr(\tilde{Y} > \tilde{y}|\theta_t). \tag{13.6}$$

13.1.6   Prior Elicitation for the GEV Model

The prior distribution $f(\theta)$ may not be of great importance if the dataset is large, because the information contained in the dataset is then dominant. For large datasets posterior inferences will not be particularly sensitive to the choice of $f(\theta)$ so long as $f(\theta)$ is non-zero over an appropriate range. However, extreme value modelling is often based on small datasets, and therefore it may be important to incorporate information external to the data y within the prior distribution. We therefore discuss prior specification in some detail here.

The methods used for prior construction will depend on the model, the parameters θ, and the information that is available for any particular application. We will discuss prior elicitation for the generalized extreme value (GEV) distribution, with parameter vector (μ, σ, ξ). Several extreme value models are either based on the GEV distribution or use parameters with similar interpretations, so this discussion may serve as a reasonable starting point for a wide range of practical applications.

In Bayesian statistics it is necessary to specify the prior distribution $f(\theta)$, but we often have little information other than what is contained in the data y. This leads to a desire to construct prior distributions that represent a lack of knowledge so that they play only a minimal role in posterior inferences. There is a large body of literature on the construction of these noninformative distributions. Noninformative distributions may be improper, meaning that the integral of $f(\theta)$ over the parameter space is not finite. For example, one common approach is to employ the improper prior $f(\mu, \sigma, \xi) \propto 1/\sigma$, which is equivalent to $f(\mu, \nu, \xi) \propto 1$ where $\nu = \log(\sigma)$.

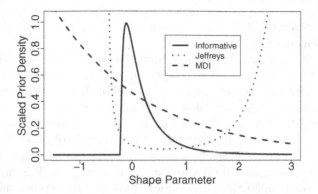

FIGURE 13.3
Scaled prior density functions for the Jeffreys prior (dotted line), the MDI prior (dashed line), and an informative prior (solid line), plotted against the shape parameter ξ at fixed values of (μ, σ). The solid line denotes a proper informative prior distribution constructed on the probability space, for fixed quantiles. It uses the construction of Section 13.1.6.2 with quantiles $q = (37, 39, 43)$ millimetres and hyperparameters $\alpha = (4, 2.5, 2, 0.5)$.

General principles of construction exist for deriving noninformative priors based on the likelihood $f(y|\theta)$. We give two common constructions here, but for the GEV model neither should be used without adjustment, as discussed subsequently. The Jeffreys prior is specified to be proportional to the square root of the determinant of the Fisher information matrix (see Prescott and Walden, 1980). It produces a prior that is invariant to reparameterization, but it typically has a complex form for extreme value models. For the GEV model, the Jeffreys prior for $\theta = (\mu, \sigma, \xi)$ exists only when $\xi > -0.5$ and is given by $f(\theta) \propto \phi(\xi)/\sigma^2$, where $\phi(\xi)$ is a function only of ξ. The calculation of $\phi(\xi)$ involves gamma and digamma functions but is otherwise easy to compute. We omit the mathematical details; a plot of $\phi(\xi)$ is given in Figure 13.3. An alternative criterion that leads to a much simpler form is to take the prior to be proportional to $\exp(\mathbb{E}_\theta[\log\{f(y|\theta)\}])$, which is known as the maximal data information (MDI) prior. For the GEV model this gives $f(\theta) \propto \exp(-\gamma\xi)/\sigma$ for $\theta = (\mu, \sigma, \xi)$, where $\gamma \approx 0.577$ is the Euler–Mascheroni constant. A plot of the MDI prior for fixed σ is also given in Figure 13.3.

It is valid to use an improper prior as long as the posterior distribution is proper, which will occur when $\int f(\theta|y)\,d\theta$ is finite, for a given dataset y. We assume here that the sample size is greater than the number of parameters. Showing that the posterior distribution is proper is not necessarily an easy task, and this cannot typically be identified from Markov chain Monte Carlo output. For the GEV model it has been shown that (Northrop and Attalides, 2015), for any sample size, both the Jeffreys and MDI priors give improper posterior distributions and they should not be used without truncation of the shape parameter. For the Jeffreys prior $f(\theta) \to \infty$ as $\xi \to \infty$, and truncation of ξ from above leads to a proper posterior distribution. For the MDI prior

$f(\theta) \to \infty$ as $\xi \to -\infty$, and we require truncation from below. Specifying, e.g., $\xi < 2.5$ for the Jeffreys prior and $\xi > -1$ for the MDI prior are reasonable values in practice (see Figure 13.3).

An alternative approach to the construction of noninformative priors is to employ vague proper priors, such as specifying a prior normal distribution with a large variance. The variance could be very large for cases where there is no information, or it could reflect a small amount of real-world information derived from the context of the problem. For the GEV model, we could take the prior distributions for the parameters (μ, ν, ξ) to be independent, each with vague normal distributions. The improper prior $f(\mu, \nu, \xi) \propto 1$, or equivalently $f(\mu, \sigma, \xi) \propto 1/\sigma$, can be seen as the limit of this vague independent normal prior as the variances increase. The improper prior $f(\mu, \nu, \xi) \propto 1$ does lead to a proper posterior distribution for the GEV model, and in practice there may be little difference between the improper and vague independent normal approaches, particularly if the sample size is not small.

An obvious extension to taking independent normal priors on (μ, ν, ξ) is to take (μ, ν, ξ) to be trivariate normal with a specified mean vector and variance matrix (Coles and Powell, 1996; Nolde and Joe, 2013). This is a flexible approach, involving nine hyperparameters, but these hyperparameters can be difficult to specify appropriately, even if there is external information with which to formulate a dependence structure. If we wish to use an informative prior that incorporates expert opinion external to the data, it can be easier to transform the GEV parameters to quantities that are more directly interpretable. Sections 13.1.6.1 and 13.1.6.2 give two different constructions. Section 13.1.6.1 constructs priors on the quantile space, for fixed probabilities, using gamma distributions. Section 13.1.6.2 constructs priors on the probability space, for fixed quantiles, using beta distributions. These constructions induce a natural dependence structure using only a small number of hyperparameters. Moreover, they enable the elicitation of information using familiar quantities.

13.1.6.1 Gamma Distributions for Quantile Differences

Let $\Pr(Y > q_p) = p$, where Y has a GEV distribution, with parameter vector $\theta = (\mu, \sigma, \xi)$. It follows that

$$q_p = \mu + \sigma(x_p^{-\xi} - 1)/\xi,$$

where $x_p = -\log(1 - p)$. A prior distribution can be constructed in terms of the quantiles $(q_{p_1}, q_{p_2}, q_{p_3})$ for specified probabilities $p_1 > p_2 > p_3$. For example, we could take $p_i = 10^{-i}$ for $i = 1, 2, 3$. Since $q_{p_1} < q_{p_2} < q_{p_3}$ it is easier to work with the differences $(\tilde{q}_{p_1}, \tilde{q}_{p_2}, \tilde{q}_{p_3})$, so that $\tilde{q}_{p_i} = q_{p_i} - q_{p_{i-1}}$ for $i = 1, 2, 3$, where q_{p_0} is the physical lower end-point of the process variable. The measurement scale can always be transformed to make the lower end-point zero, so we take $q_{p_0} = 0$ without loss of generality. We can take priors on the quantile differences to be independent, with

$$\tilde{q}_{p_i} \sim \text{gamma}(\alpha_i, \beta_i), \qquad \alpha_i, \beta_i > 0,$$

for $i = 1, 2, 3$. The differences $(\tilde{q}_{p_2}, \tilde{q}_{p_3})$ only depend on the scale and shape parameters (σ, ξ), and therefore the prior information on the location parameter μ arises only through $\tilde{q}_{p_1}$. The hyperparameters $(\alpha_1, \alpha_2, \alpha_3)$ and $(\beta_1, \beta_2, \beta_3)$, and the probabilities $p_1 > p_2 > p_3$, must all be specified. This construction (Coles and Tawn, 1996) leads to the prior density

$$\pi(\theta) \propto \mathrm{J}(\theta) \prod_{i=1}^{3} \tilde{q}_{p_i}^{\alpha_i - 1} \exp\{-\tilde{q}_{p_i}/\beta_i\}, \tag{13.7}$$

provided that $q_{p_1} < q_{p_2} < q_{p_3}$, where $\mathrm{J}(\theta)$ is the Jacobian of the transformation from $(q_{p_1}, q_{p_2}, q_{p_3})$ to $\theta = (\mu, \sigma, \xi)$. It can be shown that

$$\mathrm{J}(\theta) = \sigma/\xi^2 \left| \sum_{\substack{i,j \in \{1,2,3\} \\ i<j}} (-1)^{i+j} (x_i x_j)^{-\xi} \log(x_j/x_i) \right|,$$

where $x_i = -\log(1 - p_i)$ for $i = 1, 2, 3$.

13.1.6.2 Beta Distributions for Probability Ratios

Let $\Pr(Y > q) = p_q$, where Y has a GEV distribution, with parameter vector $\theta = (\mu, \sigma, \xi)$. It follows that

$$p_q = 1 - \exp\left\{ -\left[1 + \xi (q - \mu)/\sigma\right]_+^{-1/\xi} \right\},$$

where $h_+ = \max(h, 0)$. A prior distribution can be constructed in terms of the probabilities $(p_{q_1}, p_{q_2}, p_{q_3})$ for specified quantiles $q_1 < q_2 < q_3$. Define $p_{q_0} = 1$ and $p_{q_4} = 0$. Since $p_{q_1} > p_{q_2} > p_{q_3}$ it is easier to work with the ratios $(\tilde{p}_{q_1}, \tilde{p}_{q_2}, \tilde{p}_{q_3})$, where $\tilde{p}_{q_i} = p_{q_i}/p_{q_{i-1}}$ for $i = 1, 2, 3$.

We can take priors on the probability ratios to be independent, with

$$\tilde{p}_{q_i} \sim \mathrm{beta}\left(\sum_{j=i+1}^{4} \alpha_j, \alpha_i \right), \qquad i = 1, 2, 3,$$

where $(\alpha_1, \alpha_2, \alpha_3, \alpha_4)$ are positive hyperparameters. This construction (Crowder, 1992) leads to the prior density

$$\pi(\theta) \propto \mathrm{J}(\theta) \prod_{i=1}^{4} (p_{q_{i-1}} - p_{q_i})^{\alpha_i - 1}, \tag{13.8}$$

provided that $p_{q_1} > p_{q_2} > p_{q_3}$ and that $1 + \xi(q_i - \mu)/\sigma$ is positive for each $i = 1, 2, 3$. $\mathrm{J}(\theta)$ is the Jacobian of the transformation from $(p_{q_1}, p_{q_2}, p_{q_3})$ to $\theta = (\mu, \sigma, \xi)$. It can be shown that

$$\mathrm{J}(\theta) = \sigma/\xi^2 \left\{ \prod_{i=1}^{3} g(q_i) \right\} \left| \sum_{\substack{i,j \in \{1,2,3\} \\ i<j}} (-1)^{i+j} (x_i x_j)^{-\xi} \log(x_j/x_i) \right|,$$

where $x_i = -\log(1 - p_{q_i})$ for $i = 1, 2, 3$, and $g(q_i) = x_i^{1+\xi} e^{-x_i}/\sigma$ is the density function of the GEV distribution.

For a specific example of the above construction, suppose we have a dataset y of annual maximum daily rainfall observations in millimetres at a particular location, and suppose we take $q_1 = 37$, $q_2 = 39$ and $q_3 = 43$. The corresponding probabilities are denoted by $p_{37} > p_{39} > p_{43}$, and the probability ratios are given by $\tilde{p}_{37} = p_{37}$, $\tilde{p}_{39} = p_{39}/p_{37}$ and $\tilde{p}_{43} = p_{43}/p_{39}$. Suppose that we first elicit a beta(5,4) prior distribution for $\tilde{p}_{37} = p_{37}$, which is the probability that the annual maximum rainfall will exceed 37 millimetres. This gives $\alpha_1 = 4$ and $\alpha_2 + \alpha_3 + \alpha_4 = 5$. This distribution has mean $5/9$, so the prior opinion is that the annual maximum rainfall will exceed 37 millimetres just over half of the time. Suppose it is also thought that half of the annual maxima that exceed 37 millimetres will also exceed 39 millimetres, but that only one-fifth of the annual maxima that exceed 39 millimetres will also exceed 43 millimetres. We can equate the means of $\tilde{p}_{39} \sim \text{beta}(\alpha_3 + \alpha_4, \alpha_2)$, and $\tilde{p}_{43} \sim \text{beta}(\alpha_4, \alpha_3)$ to these ratios, giving $\alpha_3 + \alpha_4 = 0.5 \times 5 = 2.5$ and $\alpha_4 = 0.2 \times 2.5 = 0.5$. This yields $\alpha = (\alpha_1, \alpha_2, \alpha_3, \alpha_4) = (4, 2.5, 2, 0.5)$. The parameter vector α can now be used to construct the prior distribution. The mode of this prior distribution is calculated using numerical optimization to be $\widehat{\theta} = (\widehat{\mu}, \widehat{\sigma}, \widehat{\xi}) = (37.01, 1.34, -0.11)$. Figure 13.3 shows this prior plotted against ξ for fixed $\mu = \widehat{\mu}$ and $\sigma = \widehat{\sigma}$. The lower bound of the plotted prior density derives from the condition that $\xi > -\widehat{\sigma}/(q_3 - \widehat{\mu}) = -0.22$.

13.1.7 Hierarchical Models

Hierarchical models for extreme events are most commonly applied in a spatial setting. We assume again that our data y are modelled according to $f(y|\theta)$ where θ is a parameter vector, but we now assume that θ is distributed according to $f(\theta|\phi)$ for some parameter vector ϕ, with prior distribution $f(\phi)$. We therefore have

$$f(\theta, \phi, |y) \propto f(\theta, \phi) f(y|\theta, \phi) = f(\theta, \phi) f(y|\theta) = f(\phi) f(\theta|\phi) f(y|\theta),$$

and therefore $f(\phi) f(\theta|\phi) f(y|\theta)$ is the unnormalized posterior density for (θ, ϕ) given y. Markov chain Monte Carlo algorithms can proceed as before with no additional difficulties, using the target density $f(\theta, \phi|y)$. There typically exist large cancellation factors in the acceptance ratio r that can be taken advantage of for implementation purposes. For example, when performing a Metropolis step on a subvector of ϕ, it is unnecessary to compute $f(y|\theta)$. The Markov chain algorithms produce iterations (θ^t, ϕ^t) for $t = 1, \ldots, N$. For examples of hierarchical modelling for spatial extremes see e.g., Cooley et al. (2007), Gaetan and Grigoletto (2007), Sang and Gelfand (2008), Cooley and Sain (2010), Apputhurai and Stephenson (2012), and Stephenson et al. (2014).

As an example of the above, suppose we have temperature maxima data at d sites in a region, with the maxima at site j modelled as a random sample from a GEV distribution with parameter vector (μ_j, σ, ξ). We also take $\mu = (\mu_1, \ldots, \mu_d)$ to be distributed according to a multivariate normal distribution with a known variance matrix Σ and an unknown mean vector λ. In the notation given above, we have $\theta = \mu$ and $\phi = \lambda$. We additionally have the parameters (σ, ξ) which are not part of the hierarchy and are taken to be *a priori* independent of λ. The posterior distribution of the parameters is then given by $f(\lambda)f(\sigma,\xi)f(\mu|\lambda)f(y|\mu,\sigma,\xi)$, where $f(\lambda)$ and $f(\sigma,\xi)$ are prior densities, $f(\mu|\lambda)$ is the multivariate normal density and $f(y|\mu,\sigma,\xi)$ is the likelihood function.

13.2 Data Examples

This section provides two data examples; due to space constraints the second example is only available on the book website. These examples are more complex than the simulated examples of Sections 13.1.3.1 and 13.1.4.1. The provided R code is therefore designed for clarity rather than efficiency. We use built-in distribution and simulation functions where available, and we often calculate unnecessary terms that cancel in the acceptance ratio. It is left as an exercise for the reader to optimize the code using the methods applied in Sections 13.1.3.1 and 13.1.4.1.

The first example gives a model for athletics data, focusing on techniques for Markov chain Monte Carlo simulation. The second example uses a bivariate model for meteorological data, focusing on issues of model choice and model diagnostics.

13.2.1 A Model for Athletics Data

Figure 13.4 shows the annual best times taken to run the men's 1500 metres event, for the period 1912–2010. The data includes right-censored observations, which derive from the knowledge that an unknown annual best time must be greater than the known existing world record. Using annual best times gives us more information than would be the case if we only used world records, and hence allows for more accurate inferences that can incorporate information from fast performances that were not necessarily record breaking. The data is part of a larger dataset across different race distances analyzed by Stephenson and Tawn (2013).

For this section only, we let GEV denote the generalized extreme value distribution for minima. If X has a generalized extreme value distribution for minima with parameters (μ,σ,ξ), then $-X$ had a generalized extreme distribution for maxima with parameters $(-\mu,\sigma,\xi)$. Let X_t denote the annual best time taken, in seconds, to run the 1500 metre event for years $t = 1912,\ldots,2010$. We then model X_t as GEV(μ_t,σ_t,ξ_t), where the specifications of (μ_t,σ_t,ξ_t) for each $t = 1912,\ldots,2010$ are to be determined. We assume that the X_t are independent over different years.

It is common to take the shape parameter ξ_t to be constant across time, since the data provides little information about the shape of the distribution relative to the location μ_t and the scale σ_t. The annual best times seem to display an initial linear decay over time, with the rate of reduction then levelling off. Thus an appropriate model for μ_t is an exponential function which we parametrize so that a linear model is a special case. We also see evidence of the variation around the trend over time decreasing in a similar fashion to the trend location, and therefore we take σ_t/μ_t to

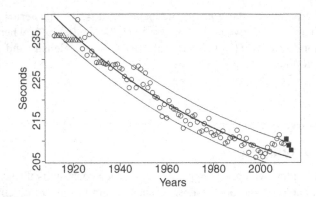

FIGURE 13.4

The annual best running times for the men's 1500 metres event, for the period 1912–2013. Triangular symbols denote right-censored observations. The curves give the median, the 5% quantile, and the 95% quantile from the predictive posterior distribution of the model fitted in this section. Filled squares denote recent annual best running times that were not used in the fitted model.

be constant over time. This yields

$$\mu_t = \alpha - \frac{\beta}{\gamma} \{1 - \exp(-\gamma t^*)\}, \tag{13.9}$$

with $\sigma_t = e^{-\delta} \mu_t > 0$ and $\xi_t = \xi$, and where $t^* = (t - 2000)/100$ is the rescaled year.

If $\gamma = 0$ then Equation (13.9) is defined in the limit so that $\mu_t = \alpha - \beta t^*$ becomes a linear function in t. If $\gamma > 0$ then μ_t tends to a limiting value $\alpha - \beta/\gamma$. The parameter γ represents the curvature of μ_t as a function of t. The parameter α represents the annual best time in the year $t = 2000$. It would be possible to remove the γ term from the ratio β/γ in Equation (13.9), but the parametrization used here has a more intuitive interpretation and reduces the correlation in the Markov chain output.

The vector of parameters is given by $\theta = (\alpha, \beta, \gamma, \delta, \xi)$. We do not have any external information with which to construct a prior distribution, so we take independent vague normal distributions on each parameter. We specify a positive prior mean for α to take into account the physical impossibility of running the event in under, e.g., 10 seconds per 100 metres.

We follow Sections 13.1.3.1 and 13.1.4.1 by defining a function `log_post` that computes the unnormalized posterior density function $f(\theta)f(y|\theta)$. However, we need to take into account the existence of right-censored observations. A data point that is right censored at x contributes to the likelihood through $1 - G(x)$, where $G(\cdot)$ is the generalized extreme value distribution for minima. It would also be possible to take into account the rounding of the recorded times by incorporating interval censoring, but for simplicity we ignore this aspect of the data.

```
log_post <- function(alpha, beta, gamma, delta, xi, data, cs) {
  theta <- c(alpha, beta, gamma, delta, xi)
  tt <- (1912:2010 - 2000)/100
  mu <- alpha - beta * (1 - exp(-gamma*tt))/gamma
  sig <- exp(-delta) * mu
  if(any(sig <= 0)) return(-Inf)
  llhd1 <- sum(dgev(-data[!cs], -mu[!cs], sig[!cs], xi, TRUE))
  llhd2 <- sum(log(pgev(-data[cs], -mu[cs], sig[cs], xi)))
  mnpr <- c(200,0,0,0,0); sdpr <- c(40,40,10,10,10)
  lprior <- sum(dnorm(theta, mean = mnpr, sd = sdpr, TRUE))
  lprior + llhd1 + llhd2
}
```

The functions pgev and dgev correspond to the generalized extreme value distribution for maxima, and therefore we need to manipulate the arguments to these functions so that they give the correct result for minima. The dataset consists of the race times data and a censoring indicator cs which is TRUE when the race time is right censored.

Numerical optimization suggests that the function log_post obtains a global maximum at $\hat{\theta} = (209.1, 21.1, 1.15, 4.63, -0.39)$. The function log_post can be used to simulate a Markov chain by expanding the code given in Section 13.1.4.1. The only difference here is that we need to extend the code to five parameters. We also use a starting value that is randomly selected from a distribution that is over-dispersed relative to the target distribution. The code below gives the calculations prior to the main for loop. The proposal standard deviations were initially specified using the same methodology as in Section 13.1.4.1, but this was seen to produce mean acceptance rates that were too low. We therefore used a trial-and-error approach to decrease the proposal standard deviations to the values given in the code below. We simulated the starting value out[1,] using a crude multivariate normal approximation (see the book website for details).

```
propsd <- c(0.6,0.7,0.07,0.15,0.1)
iter <- 10000; out <- matrix(NA, nrow = iter+1, ncol = 5)
pars <- c("alpha", "beta", "gamma", "delta", "xi")
dimnames(out) <- list(0:iter, pars)
out[1,] <- st <- c(208.02,22.80,1.03,5.06,-0.24)
lpost_old <- log_post(st[1],st[2],st[3],st[4],st[5], data, cs)
```

As discussed in Section 13.1.5, our approach was to simulate four different chains of length 10 000, and then to combine the last half of each chain and use the combined 20 000 iterations for estimation purposes. We used the four chains to ensure proper mixing behaviour; the largest potential scale reduction factor equals 1.04 (for the β parameter), which is well below the commonly used threshold of 1.10. The relationship between the parameters (β, γ) leads to fairly strong autocorrelations and cross-correlations in the simulated β and γ values, and therefore much larger sample sizes are required than for the examples in Sections 13.1.3.1 and 13.1.4.1.

Figure 13.5 compares two of the four individual chains for the parameter δ, which does not have high autocorrelations and appears to burn-in relatively quickly. Similar behaviour is observed for ξ. Figure 13.6 compares two of the four individual chains

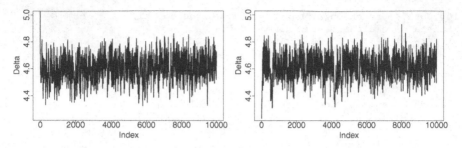

FIGURE 13.5

Two Markov chain trace plots for the parameter δ, using different randomly selected starting values. The autocorrelations for δ were the smallest of the five parameters.

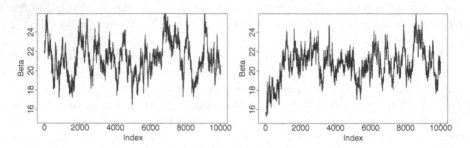

FIGURE 13.6

Two Markov chain trace plots for the parameter β, using different randomly selected starting values. The autocorrelations for β were the largest of the five parameters.

for the parameter β, which has high autocorrelations and therefore larger sample sizes are needed to ensure that it mixes properly.

As discussed in Section 13.1.2, the 20 000 combined iterations can be used to estimate features of the posterior distribution of $\theta = (\alpha, \beta, \gamma, \delta, \xi)$. Figure 13.7 displays kernel density estimates of the bivariate marginal posterior densities for (α, β) and for (δ, ξ). The marginal posterior density for (δ, ξ) appears to be fairly elliptical, whereas the relationship between the parameters (α, β) is more complex. Table 13.1 gives quantile estimates for marginal posterior distributions for each of the five parameters.

Posterior density estimates can be easily calculated for any function of the parameters. In particular, posterior density estimates can be calculated for any generalized extreme value (for minima) quantile for any given year t and any lower tail probability p. This includes the finite lower end-point which is obtained at $p = 0$.

The code below uses the Markov chain values to simulate from the posterior predictive distribution, using the method given in Section 13.1.2. These simulations can be used to estimate features of the posterior predictive distribution. We extend the years to include the more recent period 2011–2013. The last line of code prints the values that are used to produce the curves in Figure 13.4.

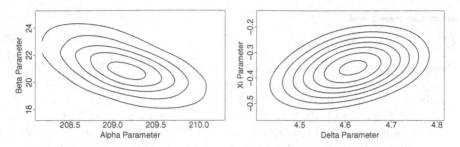

FIGURE 13.7
Bivariate kernel density estimates for the marginal posterior distributions of (from left to right) the parameters (α, β) and (δ, ξ).

TABLE 13.1
Estimated quantiles for marginal posterior distributions for each of the five parameters $(\alpha, \beta, \gamma, \delta, \xi)$.

	0.025	0.25	Median	0.75	0.975
Alpha	208.31	208.83	209.11	209.38	209.90
Beta	17.96	19.99	21.04	22.21	24.69
Gamma	0.84	1.05	1.16	1.28	1.52
Delta	4.42	4.55	4.61	4.67	4.76
Xi	-0.51	-0.41	-0.36	-0.31	-0.21

```
tt <- (1912:2013 - 2000)/100; ny <- length(tt)
repl <- matrix(NA, nrow(out), ny)
for(t in 1:nrow(out)) {
  mu <- out[t,1] - out[t,2] * (1 - exp(-out[t,3]*tt))/out[t,3]
  repl[t,] <- -rgev(ny, -mu, exp(-out[t,4]) * mu, out[t,5])
}
apply(repl, 2, function(x) quantile(x, c(0.05,0.5,0.95)))
```

The three annual best times within the period 2011–2013 were not used in the model, and are plotted as filled squares in Figure 13.4. All three points lie above the median of the predictive distribution, with one point lying above the 95% quantile. They therefore tend to be slower than predicted by the model, and are more similar to the recorded annual bests of 2006–2010. This may indicate a weakness in the model, and therefore suggest potential improvements. For example, the data may reflect the increased effectiveness of drug testing in recent years. Additionally, the assumption of independence may be inappropriate since the same athlete is often responsible for multiple annual bests. It would be possible to extend the model to incorporate these features (see Stephenson and Tawn, 2013).

13.3   Further Reading

There are a number of books on the subject of general Bayesian inference. The approach we have taken here typically follows that given in Gelman et al. (2013). More introductory texts include Bolstad (2007) and Lee (2012). The books by Lunn et al. (2012) and Kruschke (2010) have a more practical emphasis, using BUGS and R software. Gamerman and Lopes (2006) focus on Markov chain Monte Carlo simulation. Gilks et al. (1995) and Brooks et al. (2011) contain theoretical and practical examples of Markov chain Monte Carlo methodology from several authors.

Section 9.1 of Coles (2001) provides an introduction to Bayesian inference in extremes, and gives an example using the GEV model for annual maximum sea-levels. He uses the algorithm of Section 13.1.4 with vague independent normal priors on $(\mu, \log \sigma, \xi)$. Chapter 11 of Beirlant et al. (2004) gives another introduction, providing examples for both GEV and generalized Pareto (GP) models. They apply the GEV model to monthly maximal wind gust measurements, using weakly informative independent prior normal distributions on $(\mu, \log \sigma, \log \xi)$, giving zero prior probability for $\xi < 0$. For the GP model they use the MDI prior of Section 13.1.6.

References

Apputhurai, A. and A. G. Stephenson (2012). Spatiotemporal hierarchical modelling of extreme precipitation in Western Australia using anisotropic Gaussian random fields. *Environmental and Ecological Statistics 20*(4), 667–677.

Beirlant, J., Y. Geogebeur, J. Segers, and J. Teugels (2004). *Statistics of Extremes*. Chichester: John Wiley & Sons.

Bolstad, W. M. (2007). *Introduction to Bayesian Statistics* (Second ed.). New Jersey: John Wiley & Sons.

Brooks, S., A. Gelman, G. Jones, and M. Xiao-Li (2011). *Handbook of Markov Chain Monte Carlo*. Boca Raton: Chapman & Hall/CRC.

Brooks, S. P. and G. O. Roberts (1998). Convergence assessment techniques for Markov chain Monte Carlo. *Statistics and Computing 8*, 319–335.

Castellanos, M. and S. Cabras (2007). A default Bayesian procedure for the generalised Pareto distribution. *Journal of Statistical Planning and Inference 137*, 473–483.

Coles, S. G. (2001). *An Introduction to Statistical Modeling of Extreme Values*. London: Springer–Verlag.

Coles, S. G. and E. A. Powell (1996). Bayesian methods in extreme value modelling: a review and new developments. *International Statistical Review 64*, 119–136.

Coles, S. G. and J. A. Tawn (1996). A Bayesian analysis of extreme rainfall data. *Applied Statistics 45*, 463–478.

Cooley, D., D. Nychka, and P. Naveau (2007). Bayesian spatial modeling of extreme precipitation return levels. *Journal of the American Statistical Association 102*, 824–840.

Cooley, D. and R. S. Sain (2010). Spatial hierarchical modeling of precipitation extremes from a regional climate model. *Journal of Agricultural, Biological, and Environmental Statistics 15*, 381–402.

Cowles, M. K. and B. P. Carlin (1996). Markov chain Monte Carlo convergence diagnostics: A comparative review. *Journal of the American Statistical Association 91*, 883–904.

Crowder, M. (1992). Bayesian priors based on a parameter transformation using the distribution function. *Annals of Institute of Statistical Mathematics 44*, 405–416.

Gaetan, C. and M. Grigoletto (2007). A hierarchical model for the analysis of spatial rainfall extremes. *Journal of Agricultural, Biological, and Environmental Statistics 12*, 434–449.

Gamerman, D. and H. F. Lopes (2006). *Markov Chain Monte Carlo: Stochastic Simulation for Bayesian Inference* (Second ed.). Boca Raton: CRC Press.

Gelman, A., J. B. Carlin, H. S. Stern, D. B. Dunson, A. Vehtari, and D. B. Rubin (2013). *Bayesian Data Analysis* (Third ed.). Boca Raton: CRC Press.

Gelman, A. and D. B. Rubin (1992). Inference from iterative simulation using multiple sequences. *Statistical Science 7*, 457–511.

Genz, A., F. Bretz, T. Miwa, X. Mi, F. Leisch, F. Scheipl, and T. Hothorn (2013). *mvtnorm: Multivariate Normal and t Distributions*. R package version 0.9-9996.

Gilks, W. R., S. Richardson, and D. Spiegelhalter (1995). *Markov Chain Monte Carlo in Practice*. Boca Raton: Chapman & Hall/CRC.

Kruschke, J. K. (2010). *Doing Bayesian Data Analysis: A Tutorial with R and BUGS*. Oxford: Academic Press Press.

Lee, J., Y. Fan, and S. A. Sisson (2015). Bayesian threshold selection for extremal models using measures of surprise. *Computational Statistics and Data Analysis 85*, 84–99.

Lee, P. M. (2012). *Bayesian Statistics: An Introduction* (4th ed.). West Sussex: John Wiley & Sons.

Lunn, D., C. Jackson, N. Best, A. Thomas, and D. Spiegelhalter (2012). *The BUGS Book: A Practical Introduction to Bayesian Analysis*. Boca Raton: CRC Press.

Lunn, D., A. Thomas, N. Best, and D. Spiegelhalter (2000). WinBUGS – a Bayesian modelling framework: Concepts, structure, and extensibility. *Statistics and Computing 10*, 325–337.

Nolde, N. and H. Joe (2013). A Bayesian extreme value analysis of debris flows. *Water Resources Research 49*, 7009–7022.

Northrop, P. J. and N. Attalides (2015). Posterior propriety in Bayesian extreme value analyses using reference priors. *Statistica Sinica*. In press.

Plummer, M. (2003, March). JAGS: A program for analysis of Bayesian graphical models using Gibbs sampling. In K. Hornik, F. Leisch, and A. Zeileis (Eds.), *Proceedings of the 3rd International Workshop on Distributed Statistical Computing*, Vienna, Austria.

Plummer, M., N. Best, K. Cowles, and K. Vines (2006). CODA: Convergence diagnosis and output analysis for MCMC. *R News* 6(1), 7–11.

Prescott, P. and A. T. Walden (1980). Maximum likelihood estimation of the parameters of the generalized extreme value distribution. *Biometrika 67*, 723–724.

R Core Team (2013). *R: A Language and Environment for Statistical Computing.* Vienna, Austria: R Foundation for Statistical Computing.

Sang, H. and A. E. Gelfand (2008). Continuous spatial process models for spatial extreme values. *Journal of Agricultural, Biological and Environmental Statistics 15*, 49–65.

Stan Development Team (2014). Stan: A C++ library for probability and sampling, Version 2.2. http://mc-stan.org/.

Stephenson, A. G. (2002). evd: Extreme value distributions. *R News* 2(2), 31–32.

Stephenson, A. G. and M. Ribatet (2012). *evdbayes: Bayesian Analysis in Extreme Value Theory.* R package version 1.1-0.

Stephenson, A. G. and J. A. Tawn (2013). Determining the best track performances of all time using a conceptual population model for athletics records. *Journal of Quantitative Analysis in Sports 9*, 67–76.

Stephenson, A. G. Shaby, B. J., B. J. Reich, and A. L. Sullivan (2014). Estimating spatially varying severity thresholds of the forest fire danger rating system using max-stable extreme event modelling. *Journal of Applied Meteorology and Climatology 54*, 395–407.

Wand, M. (2013). *KernSmooth: Functions for Kernel Smoothing for Wand & Jones (1995).* R package version 2.23-10.

14

Modelling Extremes Using Approximate Bayesian Computation

Robert Erhardt

Wake Forest University, United States

Scott A. Sisson

University of New South Wales, Australia

Abstract

By the nature of their construction, many statistical models for extremes result in likelihood functions that are computationally prohibitive to evaluate. This is consequently problematic for the purposes of likelihood-based inference. With a focus on the Bayesian framework, this chapter examines the use of approximate Bayesian computation (ABC) techniques for the fitting and analysis of statistical models for extremes. After introducing the ideas behind ABC algorithms and methods, we demonstrate their application to extremal models in stereology and spatial extremes.

14.1   Introduction

Suppose interest is in modelling the extremes of a multivariate random process. A useful example to hold in mind might be measurements of temperature y sampled at locations $x_1, ..., x_D$, where there is dependence among the D locations due to their proximity to one another. The extremal dependence may, in general, differ from the dependence of non-extremes, and so the model should target the extremes only and not allow the bulk of non-extreme data to overwhelm the model fit. Models for extremes are useful when trying to estimate the risks associated with rare but influential events.

Call y_{obs} the observed data from a random process $f(y|\theta)$, where θ is the parameter, and call $\mathcal{A}$ an event or set of events of particular concern. The probability of such an event (with notation to remind of the influence from the parameter θ) is

$$p_\theta = Pr(y_{obs} \in A) = \int_A f(y|\theta)dy.$$

The reciprical p_θ^{-1} describes the expected number of independent realizations of y needed until event A occurs. If this process is temporal and the realizations come over time (as in $f(y(t) \mid \theta, t)$), then p_θ^{-1} is the *return period* of the event A. A return period of twenty years means that the event occurs roughly once every twenty years. Estimating return periods of extremes is crucial to designing systems capable of handling such extremes. Examples include flood walls that can withstand the 100-year flood, infrastructure that can withstand extreme heat or cold, insurance companies that can remain solvent after losses of a particular magnitude, and so forth. Given the sparsity of data on extreme events, information on θ can be minimal, and this can translate into large parameter uncertainty in θ. Coles and Powell (1996) advocate the Bayesian approach to inference, arguing that expert opinion incorporated through a prior distribution $p(\theta)$ could be of tremendous value given the natural scarcity of observed data extremes; see Chapter 13 for a review. Alternatively, if one chooses a vague prior $p(\theta)$ and obtains a posterior $\pi(\theta|y_{obs})$, then the parameter uncertainty in θ would naturally be incorporated into calculations, such as predictive return levels.

Suppose that $Y_1, ..., Y_n$ are univariate, independent and identically distributed replicates from some distribution function F, and define $M_n = \max(Y_1, ..., Y_n)$ as the maximum of the n random variables. The distribution of M_n can be obtained exactly assuming F is known. In practice, then one could estimate F from all of the data $Y_1, ..., Y_n$ and estimate the distribution of M_n as $P(M_n \leqslant z) = F^n(z)$, but this approach has two drawbacks. The first is that even minor discrepancies in estimating F result in large discrepancies in F^n, particularly in the tails of F. Put another way, why should a model which fits the bulk of data also be a good fit in the tails? A second drawback concerns the limit as $n \to \infty$: if the support of Y has a finite upper bound E, then F^n converges to a point mass at E; alternatively if the support of Y is not bounded, then F^n does not converge to a non-degenerate distribution (and can be thought of to converge to the degenerate distribution with a point mass at ∞). Instead, one may model renormalized maxima $(M_n - b_n)/a_n$ for sequences $a_n > 0$ and b_n. If there exist sequences $a_n > 0$ and b_n such that

$$\lim_{n \to \infty} P\left(\frac{M_n - b_n}{a_n} \leqslant z\right) \to G(z)$$

for some non-degenerate distribution function G, then G is a member of the *generalized extreme value* (GEV) family, with distribution function

$$G(z) = \exp\left[-\left(1 + \xi \frac{z - \mu}{\sigma}\right)_+^{-1/\xi}\right], \tag{14.1}$$

where $a_+ = \max(a, 0)$, and μ, σ, and ξ are the location, scale, and shape parameters, respectively (Coles, 2001; Gnedenko, 1944). The sign of the shape parameter ξ determines that G corresponds to one of the three classical extreme values distributions:

$\xi > 0$ is Fréchet with support $z \in [\mu - \sigma/\xi, +\infty)$, $\xi < 0$ is Weibull with support $z \in (-\infty, \mu - \sigma/\xi]$, and $\xi \to 0$ is Gumbel with support $z \in (-\infty, +\infty)$. The GEV distribution G has the property of max-stability, understood as follows: if $Y_1, ..., Y_n$ are i.i.d. draws from G, then $\max(Y_1, ..., Y_n)$ also has distribution G, meaning

$$G^n(z) = G(A_n z + B_n)$$

for appropriate sequences $A_n > 0$ and B_n. In fact, a distribution is max-stable if and only if it is a member of the GEV family (Leadbetter et al., 1983). If block maxima are taken over a block size large enough to allow the GEV to be a valid approximation, then if one further increased the block size (from monthly to annual maxima, for example) the GEV model would still hold, with only a change in the three parameters. While these results for the GEV family assume i.i.d. data, this assumption can be relaxed and the limiting distribution still holds so long as certain mixing conditions are satisfied (Leadbetter et al., 1983).

A useful member of the GEV family is the unit-Fréchet distribution, with distribution function $P(Z \leqslant z) = \exp(-1/z)$. The simplicity of the unit-Fréchet distribution function is helpful when one considers multivariate and ultimately spatial extremes. Any member of the GEV family may be transformed to have unit-Fréchet margins as follows: if Z has a GEV distribution, and a new variable U is defined as

$$U = \left(1 + \xi \frac{Z - \mu}{\sigma}\right)^{1/\xi}, \tag{14.2}$$

then U is unit-Fréchet. This transformation assumes that the parameters are known. If the parameters are unknown, they may first be estimated and then the transformation to U is taken. For extreme values data in a spatial setting, the first step is often to transform data at each location to unit-Fréchet by fitting all marginal distributions. Then the second step is to analyze the spatial dependence among sites once every location has been transformed. Thus, there is no loss of generality when one assumes unit-Fréchet margins.

This approach can be extended to handle multivariate extremes. Let $(X_{i1}, ..., X_{iD})$, $i = 1, ..., n$ be a D-dimensional random vector and let $M_n = (M_{n1}, ..., M_{nD})$ be the vector of componentwise maxima, where $M_{nd} = \max(X_{1d}, ..., X_{nd})$ for $d = 1, ..., D$. It is worth noting that M_n will not appear in the data record unless the occurrence times of each element's block maximum happen to coincide. In a spatial context, this vector M_n might refer to the annual maxima of some variable at D locations. A non-degenerate limit for M_n exists if there exist sequences $a_{nd} > 0$ and $b_{nd}, d = 1, ..., D$ such that

$$\lim_{n \to \infty} P\left(\frac{M_{n1} - b_{n1}}{a_{n1}} \leqslant z_1, \ldots, \frac{M_{nD} - b_{nD}}{a_{nD}} \leqslant z_D\right) = G(z_1, \ldots, z_D).$$

Then G is a multivariate extreme value distribution (MEVD), and is max-stable if there exist sequences $A_{nd} > 0$, $B_{nd}, d = 1, ..., D$ such that, for any $n > 1$

$$G^n(z_1, \ldots, z_D) = G(A_{n1}z_1 + B_{n1}, \ldots, A_{nD}z_D + B_{nD}).$$

Finally, a max-stable process is defined as the infinite dimensional generalization of multivariate extreme value distribution. Let $Z(x), x \in X \subseteq \mathbb{R}^p$ be a spatial process. If for all $n \geq 1$, there exist sequences $a_n(x), b_n(x)$, $x \in X$ such that for any $x_1, ..., x_D \in X$,

$$P^n \left(\frac{Z(x_d) - b_n(x_d)}{a_n(x_d)} \leqslant z(x_d), d = 1, ..., D \right) \to G_{x_1, ..., x_D}(z(x_1), ..., z(x_D))$$

then $G_{x_1, ..., x_D}$ is a multivariate extreme value distribution. If the above holds for all possible subsets $x_1, ..., x_D \in X$ for any $D \geq 1$, then the process is termed a max-stable process.

The univariate marginal distributions of a multivariate extreme value distribution are all necessarily GEV distributions. Thus, for each margin one can define a transformation with parameter (μ_d, σ_d, ξ_d) and transform to unit-Fréchet using Equation (14.2). Since all GEV distributions can be transformed into unit-Fréchet, all MEVD can be transformed into multivariate unit-Fréchet, and thus one may assume, without loss of generality, that all MEVD have unit Fréchet margins. For D fixed locations $x_1, ..., x_D$, the joint distribution function can be written as

$$P(Z(x_1) \leqslant z_1, ..., Z(x_D) \leqslant z_D) = \exp(-V(z_1, ..., z_D)) \qquad (14.3)$$

where $V(z_1, ..., z_D)$ is the exponent measure first described by Pickands (1981). This function takes the form

$$V(z) = D \cdot \int_{\Delta_D} \max_{d=1,...,D} \frac{w_d}{z_d} H(dw) \qquad (14.4)$$

where $\Delta_D = \{w \in \mathbb{R}_+^D \mid w_1 + ... + w_D = 1\}$ is the $D - 1$ dimensional simplex, and the angular (or spectral) measure H is a probability measure on Δ_D which determines the dependence structure of the random vector. Due to the common marginal distributions, H has moment conditions $\int_{\Delta_D} w_d H(w) = 1/D$ for $d = 1, ..., D$.

Max-stability implies that for all N,

$$P(Z_1 \leqslant z_1, ..., Z_D \leqslant z_D)^N = \exp(-N \cdot V(z_1, ..., z_D))$$
$$= \exp(-V(z_1/N, ..., z_D/N))$$

with the final equality following from the homogeneity property of the exponent measure. The measure also satisfies two bounds: if all locations are independent, $V(z_1, ..., z_D) = 1/z_1 + ... + 1/z_D$; if all locations are totally dependent, $V(z_1, ..., z_D) = \max(1/z_1, ..., 1/z_D)$. Thus, we always have

$$\max(1/z_1, ..., 1/z_D) \leqslant V(z_1, ..., z_D) \leqslant 1/z_1 + ... + 1/z_D.$$

There are two challenges to working with the spectral representation of the joint distribution function shown in Equation (14.3). First, even if a closed form for the exponent measure can be found by solving Equation (14.4), the joint density function undergoes a combinatorial explosion as the dimension D increases. Differentiating $\exp(-V)$ with respect to the values $z_1, ..., z_D$ leads to a rapid growth in terms:

$$\frac{\partial}{\partial z_1} \exp(-V) = -V_1 \exp(-V)$$

$$\frac{\partial^2}{\partial z_1 \partial z_2} \exp(-V) = (V_1 V_2 - V_{12}) \exp(-V)$$

$$\frac{\partial^3}{\partial z_1 \partial z_2 \partial z_3} \exp(-V) = (-V_1 V_2 V_3 + V_{12} V_3 + V_{13} V_2 + V_{23} V_1 - V_{123}) \exp(-V)$$

$$\ldots$$

where V_i is the partial derivative of V with respect to z_i. Thus even if a reasonable choice for V can be found, as the dimension D increases one is left with an unwieldy likelihood function, which may be difficult to maximize. More common, though, is the situation where closed-form expressions for the exponent measure cannot be obtained by solving Equation (14.4).

As a result, the lack of a closed-form likelihood presents a stumbling block for modelling high-dimensional extremes data. While there are a number of procedures that may permit some form of statistical inference in this setting (e.g., the composite likelihood method, Chapter 12), the remainder of this chapter will demonstrate how approximate Bayesian computing can be implemented as one solution to this problem in the Bayesian framework.

14.2 A Primer on Approximate Bayesian Computation

Suppose that interest is in performing a standard Bayesian analysis for a parameter $\theta \in \Theta$ which has a prior distribution $p(\theta)$. The model is defined through the likelihood function $L(y|\theta)$, which is assumed to be a candidate for the true data generating process that produced the observed dataset $y_{obs} \in \mathcal{Y}$. The posterior distribution of the model parameter θ, having now observed the data y_{obs} through the likelihood function, is expressed as $\pi(\theta|y_{obs}) \propto L(y_{obs}|\theta)p(\theta)$. In the present setting, we could have, e.g., $\theta = (\mu, \sigma, \xi)$ as the parameters of a GEV distribution. The posterior distribution contains all the information that is needed to perform inference on the model (e.g., Gelman et al., 2003; O'Hagan and Forster, 2004).

As the posterior distribution is rarely available in closed form, it is common to base subsequent analysis on a Monte Carlo approximation to the posterior (Brooks et al., 2011). In this manner, if $\theta^{(1)}, \ldots, \theta^{(N)} \sim \pi(\theta|y_{obs})$ are N samples drawn from the posterior distribution, then the posterior expectation of some function $a(\theta)$ can be approximated by

$$\mathbb{E}_\pi[a(\theta)] = \int_\Theta a(\theta)\pi(\theta|y_{obs})d\theta \approx \frac{1}{N}\sum_{i=1}^N a(\theta^{(i)}),$$

Algorithm 14.1 A simple importance sampling algorithm, based on a single large sample of size N.

Input:
An observed dataset, y_{obs}.
A desired number of samples $N > 0$.
A sampling distribution $g(\theta)$, with $g(\theta) > 0$ if $p(\theta) > 0$.

Iterate: For $i = 1, \ldots, N$:
1. Sample a parameter vector from sampling distribution $\theta^{(i)} \sim g(\theta)$.
2. Weight each sample $\theta^{(i)}$ by $w^{(i)} \propto \pi(\theta^{(i)}|y_{obs})/g(\theta^{(i)})$.

Output:
A set of N weighted samples $(w^{(1)}, \theta^{(1)}), \ldots, (w^{(N)}, \theta^{(N)})$ drawn from $\pi(\theta|y_{obs})$.

---

where $\mathbb{E}_\pi$ denotes expectation under π. There is a wide variety of algorithms available to draw samples from the posterior distribution (Brooks et al., 2011; Chen et al., 2000; Doucet et al., 2001). One of the simplest of these is importance sampling, as illustrated in Algorithm 14.1, which produces weighted samples $(w^{(1)}, \theta^{(1)}), \ldots, (w^{(N)}, \theta^{(N)})$ from $\pi(\theta|y_{obs})$ based on samples $\theta^{(1)}, \ldots, \theta^{(N)}$ from a sampling distribution $g(\theta)$. In this setting, and writing $w(\theta) = \pi(\theta|y_{obs})/g(\theta)$, then

$$
\mathbb{E}_g[w(\theta)a(\theta)] = \int_\Theta w(\theta)a(\theta)g(\theta)d\theta = \int_\Theta a(\theta)\pi(\theta|y_{obs})d\theta = \mathbb{E}_\pi[a(\theta)]
$$
$$
\approx \sum_{i=1}^N w(\theta^{(i)})a(\theta^{(i)}).
$$

That is, appropriately weighted samples from $g(\theta)$ can be used as samples from $\pi(\theta|y_{obs})$.

Almost all posterior simulation algorithms need to be able to evaluate the likelihood function $L(y|\theta)$ in order to be correctly implemented. In the importance sampling algorithm (Algorithm 14.1) this occurs in the weight evaluation $w^{(i)} \propto L(y_{obs}|\theta^{(i)})p(\theta^{(i)})/g(\theta^{(i)})$. In the present setting, the natural construction of many useful statistical models for extremes results in the likelihood function being computationally prohibitive to evaluate (see Section 14.1). This computational intractability means that for these classes of models, an alternative procedure is needed to sample from the posterior distribution *without* directly evaluating the likelihood function. One class of procedures that has been developed to achieve this is known as approximate Bayesian computation (Beaumont, 2010; Bertorelle et al., 2010; Csilléry et al., 2010; Marin et al., 2012; Sisson and Fan, 2011).

14.2.1   Approximate Bayesian Computation Basics

All approximate Bayesian computation (ABC) procedures operate on the following heuristic argument, that was first developed in the population genetics literature

(Pritchard et al., 1999; Tavaré et al., 1991). Suppose that we have a parameter vector $\theta^{(i)}$ that is a candidate draw from the posterior distribution $\pi(\theta|y_{obs})$. Further suppose that we can quickly generate an auxiliary dataset from the model, conditional on this candidate parameter vector, so that $y^{(i)} \sim L(y|\theta^{(i)})$. Now, the argument states that if $y^{(i)}$ and y_{obs} are "close" to each other (in a general sense that will be made more precise below), then it is credible that the parameter vector $\theta^{(i)}$ could also have generated the observed dataset y_{obs}. In which case, the parameter vector $\theta^{(i)}$ should be retained as an approximate sample from $\pi(\theta|y_{obs})$. Conversely, if $y^{(i)}$ and y_{obs} are not "close" to each other, then $\theta^{(i)}$ is unlikely to be able to generate the observed dataset, and so it should be discarded as not being a draw from the posterior. Repeating this procedure will produce samples that are either approximate draws from the posterior $\pi(\theta|y_{obs})$, or exact draws from some as yet unspecified approximation to the posterior distribution. Either way, direct numerical evaluation of the computationally intractable likelihood function has been avoided.

The principles behind the above heuristic method can be made more precise. Suppose that we generate our candidate parameter vectors from the prior $\theta^{(i)} \sim p(\theta)$. (As part of an importance sampling algorithm, the candidate parameter vectors $\theta^{(i)}$ may be generated from $g(\theta)$ and then reweighted.) Then, given this, a dataset is generated from the likelihood. In this way, the pair

$$(\theta^{(i)}, y^{(i)}) \sim L(y|\theta)p(\theta)$$

has been generated from the prior predictive distribution $p(y, \theta) = L(y|\theta)p(\theta)$. In general, the auxiliary and observed dataset can be compared via some distance metric $\|y^{(i)} - y_{obs}\|$, such as Euclidean or Mahalanobis distance. The retention and discarding of "close" and not "close" auxiliary datasets can be mimicked by computing an importance weight $K_h(\|y^{(i)} - y_{obs}\|)$ where $K_h(u) = K(u/h)/h$ is a standard smoothing kernel with scale parameter $h > 0$. For example, if K_h corresponds to the uniform kernel over $(-h, h)$, then $\theta^{(i)}$ will receive the weight 1 if $\|y^{(i)} - y_{obs}\| \leqslant h$ (i.e., it is retained if $y^{(i)}$ and y_{obs} are sufficiently "close") and the weight 0 if $\|y^{(i)} - y_{obs}\| > h$ (i.e., it is rejected). Other choices of kernel K_h will produce continuous weights.

The resulting weighted samples $(w^{(i)}, \theta^{(i)}, y^{(i)})$ are then draws from

$$\pi_h^{ABC}(\theta, y|y_{obs}) \propto K_h(\|y - y_{obs}\|)L(y|\theta)p(\theta) \qquad (14.5)$$

(Reeves and Pettitt, 2005; Wilkinson, 2013). Distribution (14.5) is the joint posterior distribution of model parameter θ and auxiliary dataset y such that y and y_{obs} are "close" in a specific sense. If we are just interested in the resulting distribution of the parameter vector θ as an approximation of the posterior distribution $\pi(\theta|y_{obs})$, then

$$\pi_h^{ABC}(\theta|y_{obs}) \propto \int_y K_h(\|y - y_{obs}\|)L(y|\theta)p(\theta)dy. \qquad (14.6)$$

Equation (14.6) is the ABC approximation to the true posterior distribution $\pi(\theta|y_{obs})$.

The quality of this approximation is determined by the kernel scale parameter $h > 0$. Consider what happens to $\pi_h^{ABC}(\theta|y_{obs})$ as h gets small. We have

$$
\begin{aligned}
\lim_{h\to 0} \pi_h^{ABC}(\theta|y_{obs}) &\propto \lim_{h\to 0} \int_{\mathcal{Y}} K_h(\|y - y_{obs}\|)L(y|\theta)p(\theta)dy \\
&= \int_{\mathcal{Y}} 1(y = y_{obs})L(y|\theta)p(\theta)dy \\
&= L(y_{obs}|\theta)p(\theta) \\
&\propto \pi(\theta|y_{obs}),
\end{aligned}
$$

where $1(A) = 1$ if A is true, and 0 otherwise. That is, if we only accept those candidate parameter values $\theta^{(i)}$ that exactly reproduce the observed dataset, then we will exactly recover the true posterior distribution $\pi(\theta|y_{obs})$. In practice however, unless y_{obs} is discrete, this can never occur (and in fact, will be unlikely to occur practically for discrete y_{obs} in all but trivial analyses). Hence h will typically be greater than zero in practice.

In the other extreme

$$
\begin{aligned}
\lim_{h\to\infty} \pi_h^{ABC}(\theta|y_{obs}) &\propto \lim_{h\to\infty} \int_{\mathcal{Y}} K_h(\|y - y_{obs}\|)L(y|\theta)p(\theta)dy \\
&\propto \int_{\mathcal{Y}} L(y|\theta)p(\theta)dy \\
&= p(\theta),
\end{aligned}
$$

assuming that $K_h(u) \propto 1$ as $h \to \infty$. That is, sampling from the prior $p(\theta)$ and then not showing any discrimination in favour of auxiliary data closely matching the observed data will simply result in all samples being equally weighted, and the ABC approximation to the posterior distribution being given by the prior $p(\theta)$. It should be clear that increasing fidelity toward reproducing the observed data, as measured by decreasing h, defines a smooth transition from prior $p(\theta)$ to posterior $\pi(\theta|y_{obs})$, and that in practice the actually attainable ABC posterior approximation $\pi_h^{ABC}(\theta|y_{obs})$ lies somewhere between the two. Given that it results in greater closeness to $\pi(\theta|y_{obs})$, for inferential purposes lower h is desirable.

Aside from the approximation to the posterior linked to h, in practice a second level of approximation is commonly introduced in an ABC analysis, and one which can have a greater impact on the quality of the ABC approximation of $\pi(\theta|y_{obs})$ than h. To understand the motivation for this, consider the form of the likelihood component of $\pi_h^{ABC}(\theta|y_{obs})$ (14.6) where, for simplicity, both y_{obs} and θ are univariate. Using a Taylor expansion and the substitution $u = (y - y_{obs})/h$, then

$$
\int_{\mathcal{Y}} K_h(\|y - y_{obs}\|)L(y|\theta)dy \approx L(y_{obs}|\theta) + \frac{1}{2}h^2 L''(y_{obs}|\theta)\int u^2 K(u)du,
$$

assuming the usual kernel function properties of $\int K(u)du = 1$, $\int uK(u)du = 0$ and $K(u) = K(-u)$. That is, the ABC simulation procedure is simply performing a form of conditional kernel density estimation, by estimating a smoothed likelihood function, and then using this as part of a regular Bayesian analysis (Blum, 2010).

Under this interpretation, a limitation of the proposed ABC method becomes apparent. Kernel density estimation is well known to suffer from the curse of dimensionality, and is arguably impractical in more than two dimensions. Here, the appropriate dimensionality is in $y - y_{obs}$ (although it is masked by the univariate measure $\|y - y_{obs}\|$) within the kernel function K_h. That is, as the dimension of y_{obs} increases, the performance of the ABC posterior approximation $\pi_h^{ABC}(\theta|y_{obs})$ will rapidly deteriorate (Blum, 2010). In effect, it becomes increasingly unlikely to be able to reproduce the observed dataset y_{obs} by randomly sampling y, as the dimension of y_{obs} increases, thereby forcing the practitioner to increase h for a fixed number of samples N.

A simple solution to this is to reduce the dimensionality of y_{obs} by reducing it to a vector of summary statistics $s_{obs} = S(y_{obs})$ and then performing the ABC procedure as before, but using s_{obs} rather than y_{obs} (Beaumont et al., 2002; Pritchard et al., 1999; Tavaré et al., 1991). In this manner, (14.5) and (14.6) become

$$\pi_h^{ABC}(\theta, s|s_{obs}) \quad \propto \quad K_h(\|s - s_{obs}\|)L(s|\theta)p(\theta) \tag{14.7}$$

$$\pi_h^{ABC}(\theta|s_{obs}) \quad \propto \quad \int_S K_h(\|s - s_{obs}\|)L(s|\theta)p(\theta)ds, \tag{14.8}$$

where $S = \{S(y) : y \in \mathcal{Y}\}$ is the image of $\mathcal{Y}$ under S, and where $L(s|\theta)$ corresponds to the likelihood function of the summary statistic, which is also assumed to be computationally intractable. In the case where $S(y) = y$ then (14.7) and (14.8) reduce to (14.5) and (14.6).

Note that once a decision has been made on the summary statistics, then the most accurate possible ABC posterior approximation is given by $\pi(\theta|s_{obs})$ as

$$\begin{aligned}
\lim_{h \to 0} \pi_h^{ABC}(\theta|s_{obs}) \quad &\propto \quad \lim_{h \to 0} \int_S K_h(\|s - s_{obs}\|)L(s|\theta)p(\theta)ds \\
&= \quad \int_S \mathbf{1}(s = s_{obs})L(s|\theta)p(\theta)ds \\
&= \quad L(s_{obs}|\theta)p(\theta) \\
&\propto \quad \pi(\theta|s_{obs}),
\end{aligned}$$

in the same manner as before. Consequently it is important that due consideration is given to the summary statistics aspect of the model.

In the case of sufficient statistics, there is no loss of information and so $\pi(\theta|s_{obs}) = \pi(\theta|y_{obs})$. When $S(y)$ is not sufficient, then there is some information loss, and $\pi(\theta|s_{obs})$ will be less precise than $\pi(\theta|y_{obs})$. In turn, the ABC posterior approximation $\pi_h^{ABC}(\theta|s_{obs})$ will be less precise than $\pi(\theta|s_{obs})$. While the reduction in dimension from y_{obs} to s_{obs} will allow for increased precision by permitting a reduced kernel scale parameter h, this must be offset by any loss of information in the construction of the summary statistics (Blum et al., 2013).

A precise importance sampling algorithm to generate from $\pi_h^{ABC}(\theta|s_{obs})$ (14.8)

Algorithm 14.2 A simple ABC importance sampling algorithm, based on a single large sample of size N.

Input:

An observed dataset, y_{obs}.
A desired number of samples $N > 0$.
A sampling distribution $g(\theta)$, with $g(\theta) > 0$ if $p(\theta) > 0$.
A smoothing kernel K_h and scale parameter $h > 0$.
A low-dimensional vector of summary statistics $s = S(y)$.
Compute $s_{obs} = S(y_{obs})$.

Iterate: For $i = 1, \ldots, N$:

1. Sample a parameter vector from sampling distribution $\theta^{(i)} \sim g(\theta)$.
   Simulate a dataset from the likelihood given parameter vector $\theta^{(i)}$ as $y^{(i)} \sim L(y|\theta^{(i)})$.
   Compute the summary statistics $s^{(i)} = S(y^{(i)})$.
2. Weight each sample $\theta^{(i)}$ by $w^{(i)} \propto K_h(\|s^{(i)} - s_{obs}\|)p(\theta^{(i)})/g(\theta^{(i)})$.

Output:

A set of $i = 1, \ldots, N$ weighted samples $(w^{(i)}, \theta^{(i)}, s^{(i)})$, drawn from $\pi_h^{ABC}(\theta, s|s_{obs})$ (14.7). Or a set of $i = 1, \ldots, N$ weighted samples $(w^{(i)}, \theta^{(i)})$, drawn from $\pi_h^{ABC}(\theta|s_{obs})$ (14.8).

---

is given by Algorithm 14.2. To see how this works while avoiding evaluation of the intractable likelihood function $L(s|\theta)$, note that this algorithm targets the joint posterior approximation $\pi_h^{ABC}(\theta, s|s_{obs})$ – marginalising over s so that the draws come from $\pi_h^{ABC}(\theta|s_{obs})$ is achieved by simply discarding the $s^{(i)}$ in the Monte Carlo output. Hence our sampling distribution must span (θ, s). In the present setting, it is natural to use

$$(\theta^{(i)}, s^{(i)}) \sim L(s|\theta)g(\theta).$$

From which the importance weight becomes

$$
\begin{aligned}
w^{(i)} &\propto \frac{\pi_h^{ABC}(\theta^{(i)}|s_{obs})}{L(s^{(i)}|\theta^{(i)})g(\theta^{(i)})} \propto \frac{K_h(\|s^{(i)} - s_{obs}\|)L(s^{(i)}|\theta^{(i)})p(\theta^{(i)})}{L(s^{(i)}|\theta^{(i)})g(\theta^{(i)})} \\
&= \frac{K_h(\|s^{(i)} - s_{obs}\|)p(\theta^{(i)})}{g(\theta^{(i)})},
\end{aligned}
$$

which is conveniently free of computationally intractable likelihood terms.

14.2.2   A Simple Example

As an illustration of ABC methods in a simple setting, we analyse annual maximum daily rainfall (measured in millimetres) in the years 1951–1999, recorded at Maiquetia International Airport, on the central coast of Venezuela. This dataset is particularly notable as the annual maximum rainfall was around 50–150mm from 1950–1998, but in 1999 the maximum more than doubled to 410mm. The unusually extreme daily rainfall events in December 1999 caused the worst environmentally related tragedy in Venezuelan history, and one of the largest historical rainfall-induced debris flows documented in the world (Wieczorek et al., 2001). The resulting human, infrastructure, and economic impacts have become known as the *Vargas Tragedy*. These data have been analysed by Coles and Pericchi (2003) and Coles et al. (2003).

As these data are univariate annual maxima, they may be approximately modelled by the GEV distribution (14.1). As the GEV is computationally tractable, a gold standard result is available to compare with an ABC posterior approximation. Four specifications of the vector of summary statistics are considered for the dataset $y = (y_1, \ldots, y_n)'$:

$$
\begin{aligned}
s_1(y) &= (y_1, \ldots, y_n)' = y \\
s_2(y) &= (y_{(1)}, \ldots, y_{(n)})' \\
s_3(y) &= (\widehat{\mu}_L, \widehat{\sigma}_L, \widehat{\xi}_L)' \\
s_4(y) &= (\widehat{\mu}, \widehat{\sigma}, \widehat{\xi})'.
\end{aligned}
$$

Here, $s_1(y) = y$ is the original full dataset, whereas $s_2(y)$ is the vector of n order statistics of y such that $y_{(1)} \leqslant \ldots \leqslant y_{(n)}$. The vector $s_3(y)$ consists of the three L-moments estimates (following Hosking (1990)) of each GEV parameter, and $s_4(y)$ is formed from the standard maximum likelihood estimates based on (14.1). Note that $s_1(y)$, $s_2(y)$, and $s_4(y)$ are sufficient statistics for this analysis, and so $\pi(\mu, \sigma, \xi | s_{obs}) = \pi(\mu, \sigma, \xi | y_{obs})$ for these choices. However, both $s_1(y)$ and $s_2(y)$ are high-dimensional ($n = 49$), so in practice, the resulting ABC posterior approximation may suffer from the curse of dimensionality. Both $s_3(y)$ and $s_4(y)$ are low dimensional, but $s_3(y)$ is not sufficient, leading to some loss of information. Overall it might be expected that $s_4(y)$ will perform the best, as it is a minimal sufficient statistic. However, $s_4(y)$ will not typically be available in an ABC analysis.

For convenience, the prior is set as $p(\mu, \sigma, \xi) \propto 1$ over the support of the parameter space. To implement Algorithm 14.2 a sampling distribution, $g(\mu, \sigma, \xi)$, is required. In the current setting, we exploit the fact that we know the actual location of the posterior distribution and specify $g(\mu, \sigma, \xi) = U(30, 70) \times U(5, 45) \times U(-0.3, 1.5)$. See, e.g., Fearnhead and Prangle (2012) for a general method to usefully restrict the sampling distribution. We define

$$
\|s - s_{obs}\| = \left[(s - s_{obs})' \widehat{\Sigma}^{-1} (s - s_{obs}) \right]^{1/2}, \tag{14.9}
$$

where $\widehat{\Sigma}$ is an estimate of the covariance of s_{obs}. In practice the estimate of the covariance matrix can be approximate, and so can be determined by identifying some point in parameter space (μ_0, σ_0, ξ_0) that is likely in an area of high posterior density, generating a number of summary statistic vectors $s^{(i)} \sim L(s | \mu_0, \sigma_0, \xi_0)$ conditional on this point, and then computing the sample covariance matrix of these vectors (Luciani et al., 2009; Sisson and Fan, 2011). Here, (μ_0, σ_0, ξ_0) are set as the maximum likelihood estimates for convenience.

The following results are based on $N = 1$ million samples, where $K_h(u)$ is uniform over $(-h, h)$, and h is determined as the 0.15, 0.05, and 0.005 quantile of

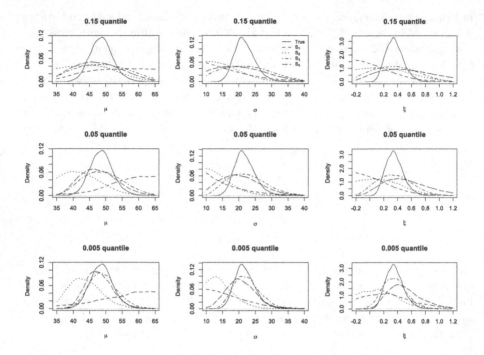

FIGURE 14.1
Estimates of the marginal posterior distributions of the GEV parameters μ (left panels), σ (centre) and ξ (right) based on the Vargas dataset. Each panel shows the true posterior marginal distribution (solid line), and the four ABC posterior approximations based on $s_1(y), \ldots, s_4(y)$. Rows indicate the quantile of $\| s_j^{(i)} - s_j(y_{obs}) \|$ used to determine the kernel scale parameter h.

$\| s_j^{(i)} - s_j(y_{obs}) \|$ for each summary vector $j = 1, \ldots, 4$. This results in approximate posterior $150,000$, $50,000$, and $5,000$ samples, respectively. Note that this retrospective definition of h is not quite in line with Algorithm 14.2, but is fairly common in practice (Beaumont et al., 2002; Blum et al., 2013).

Figure 14.1 illustrates the estimated posterior marginal distributions of μ, σ, and ξ for both the true marginal posterior (solid line) and the four ABC marginal posterior approximations based on $s_1(y), \ldots, s_4(y)$. Figure rows correspond to the different values of kernel scale parameter h. Clearly all ABC approximations based on $s_1(y) = y$ (dashed lines) are very poor, regardless of the size of h. This should not be a surprise, as $\dim(s_1(y)) = 49$, and the chance of generating such a vector of independent observations, $s_1^{(i)}$, such that $\| s_1^{(i)} - s_{1,obs} \| \leqslant h$ becomes vanishingly small as h decreases. Even taking h as the 0.005 quantile was not enough to produce any reasonable accuracy in this case. Hence, it is not viable to use $s_1(y)$ in practice.

The posterior approximations based on $s_2(y)$ (dotted lines) are as poor as those based on $s_1(y)$ for high values of h. However, there is some evidence that the density estimates for μ in particular improve when h is reduced to the 0.005 quantile of $\|s_2^{(i)} - s_{2,obs}\|$. The improved performance of $s_2(y)$ over $s_1(y)$ is due to the induced dependence of the summary statistics. However, even with the performance gain, $s_2(y)$ is clearly not useful in practice.

However, the 3-dimensional summary statistics $s_3(y)$ (dot-dash line) and $s_4(y)$ (long dashed line) both produce more practically useful ABC posterior approximations. The broad adherence of the approximate and true posterior densities is apparent in Figure 14.1 even at the 0.15 quantile, with increasing accuracy as h decreases. At the 0.005 quantile, either posterior is almost good enough to use in practice, although the L-moments based ABC approximation does appear to slightly better approximate the true posterior in this case, despite the information loss. It seems likely that decreasing h further would result in a viable ABC posterior approximation based on $s_3(y)$. While in theory using $s_4(y)$ should be slightly more efficient than using $s_3(y)$, this choice of summary statistics will not be available in practice.

Overall, the main concepts of ABC have been illustrated by this simple analysis. ABC methods are themselves a simple and intuitive procedure that can produce viable approximations to posterior distribution without the need to evaluate the likelihood function. They work best when the vector of summary statistics is both highly informative for the model parameters, and is low dimensional. However, even then, ABC methods can have large computational overheads if high accuracy is desired. This is the price to pay for not being able to evaluate the likelihood function.

14.2.3 Other Useful ABC Methods

As previously discussed, standard ABC methods suffer from the curse of dimensionality (Blum, 2010), so that the kernel scale parameter h must increase for practical purposes as the number of summary statistics increases. As a result, there is still often a large discrepancy between $s^{(i)}$ and s_{obs} in the final sample from $\pi_h^{ABC}(\theta, s|s_{obs})$, whereas greatest accuracy of the ABC posterior approximation is obtained if $s^{(i)} \approx s_{obs}$ (i.e., if $h \to 0$). Regression-adjustment techniques aim to reduce this discrepancy by explicitly modelling the relationship between the sampled $\theta^{(i)}$ and $s^{(i)}$, and then adjusting $(\theta^{(i)}, s^{(i)}) \to (\theta^{(i)*}, s^{(i)*})$ so that $s^{(i)*} = s_{obs}$. The resulting (weighted) samples $\theta^{(i)*}$ will then form an improved approximation to $\pi(\theta|s_{obs})$ over those (weighted) $\theta^{(i)}$ from $\pi_h^{ABC}(\theta|s_{obs})$, if the regression model is correct.

The simplest form of a model for this is a homoscedastic regression in the region of s_{obs}, so that

$$\theta^{(i)} = m(s^{(i)}) + e^{(i)},$$

where $m(s^{(i)}) = \mathbb{E}[\theta|s = s^{(i)}]$ is the mean function, and the $e^{(i)}$ are zero mean, common variance random variates. In the case of a local-linear regression we have the model $m(s^{(i)}) = \alpha + \beta' s^{(i)}$. Beaumont et al. (2002) estimated this model by minimising the least squares criterion $\sum_{i=1}^{N} \omega^{(i)} \|m(s^{(i)}) - \theta^{(i)}\|^2$ where $\omega^{(i)} = K_h(\|s^{(i)} - s_{obs}\|)$. The regression-adjusted weighted sample $(\theta^{(i)*}, w^{(i)})$

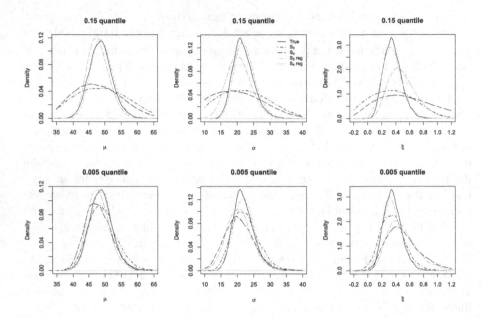

FIGURE 14.2
Regression-adjusted estimates of the marginal posterior distributions of the GEV parameters μ, σ, and ξ based on the Vargas dataset. Panels shows the true posterior marginal distribution (solid line) and the two ABC approximations to the posterior based on the vectors of summary statistics $s_3(y)$ and $s_4(y)$. ABC approximations without regression adjustment are illustrated with black lines, and those with a subsequent regression-adjustment are in gray.

drawn approximately from $\pi(\theta|s_{obs})$ is then obtained as

$$\theta^{(i)*} = \widehat{m}(s_{obs}) + (\theta^{(i)} - \widehat{m}(s^{(i)})), \qquad (14.10)$$

where $\widehat{m}(s) = \widehat{\alpha} + \widehat{\beta}'s$ is the fitted mean function. Variations on this approach include a non-linear, heteroscedastic regression-adjustment (Blum and François, 2010) and a ridge regression-adjustment (Blum et al., 2013). A qualitatively different, but also useful form of marginal adjustment to improve ABC posterior approximation accuracy, that can be used in conjunction with the regression-adjustment is described by Nott et al. (2014).

Figure 14.2 illustrates the ABC approximations to the posterior marginal distributions of the GEV analysis of Section 14.2.2, focusing on $s_3(y)$ and $s_4(y)$. The black lines indicate the original approximations, whereas the gray lines show the result of the regression adjustment (14.10). In all cases, the regression-adjustment has made a noticeable improvement, indicating that regression-adjustment methods should be routinely used in ABC analyses. Note that in this case, the regression adjustment produces effectively the same approximation whether the initial samples $(\theta^{(i)}, s^{(i)})$ are obtained using relatively high (top row) or low (bottom row) values of kernel scale parameter h. As a result, computational savings can be made by using the larger value of h, in this case.

Summary statistic identification itself is essentially a problem of dimension-reduction with respect to sufficiency for the target model. This is a huge research area in its own right. See Blum et al. (2013) for a recent comparative review. Improved techniques in this area are still being developed, and there is no single "best" method. The optimum approach remains careful consideration of the specific model and analysis at hand.

Finally, the ABC method listed in Algorithm 14.2 is based on importance sampling. However there are many other standard algorithms for generating draws from posterior distributions, such as those based on MCMC and sequential Monte Carlo. For further information on MCMC algorithms for ABC see, e.g., Bortot et al. (2007); Marjoram et al. (2003); Sisson and Fan (2011), and for sequential Monte Carlo algorithms refer to, e.g., Beaumont et al. (2009); Del Moral et al. (2012); Drovandi and Pettitt (2011); Sisson et al. (2007).

14.3 ABC for Stereological Extremes

In the production of clean steels, microscopically small particles termed *inclusions* are introduced during the production process. Metallurgic considerations indicate that the strength of the steel is directly related to the size of the largest inclusion in the block, and so inference on the largest inclusion size is important. Commonly, the sampling of inclusions involves measuring the maximum cross-sectional slice of each observed inclusion, $y_{obs} = (y_{1,obs}, \ldots, y_{n,obs})$, obtained from a two-dimensional planar slice through the steel block. Each cross-sectional inclusion size is greater than some measurement threshold, $y_{i,obs} > u$. The inferential problem is to analyse the unobserved distribution of the largest inclusion in the block, based on the information in the cross-sectional slice, y_{obs}. The focus on the size of the largest inclusion means that this is an extreme value variation on the standard stereological problem (Baddeley and Jensen, 2004).

Anderson and Coles (2002) proposed a mathematical model for those observed cross-sectional measurements. The proposed model assumed that the inclusions were spherical with diameters V, that their centres followed a homogeneous Poisson process with rate $\lambda > 0$ in the volume of steel, and that the inclusion diameters were mutually independent and independent of inclusion location. The distribution of the largest inclusion diameters, $V|V > v_0$, (i.e., those conditional on exceeding some threshold, v_0) was assumed to be well approximated by a generalised Pareto distribution, with distribution function

$$\Pr(V \leqslant v | V > v_0) = 1 - \left[1 + \frac{\xi(v - v_0)}{\sigma} \right]_+^{-1/\xi}, \qquad (14.11)$$

for $v > v_0$, where $[a]_+ = \max\{0, a\}$, and $\sigma > 0$ and $-\infty < \xi < \infty$ are scale and shape parameters, following standard extreme value theory arguments (Coles, 2001). Accordingly the parameters of the full spherical inclusion model are $\theta = (\lambda, \sigma, \xi)$.

Each observed cross-sectional inclusion diameter, $y_{i,obs}$, is associated with an unobserved true inclusion diameter V_i. However, note that the probability of observing the cross-sectional diameter size $y_{i,obs}$ (where $y_{i,obs} \leqslant V_i$) is dependent on the value of V_i, as larger inclusion diameters give a greater chance that the inclusion will be observed in the two-dimensional planar cross-section. The number of observed inclusions, n, is also a random variable. In these terms, interest is in the distribution of the largest inclusion diameters, $V_1, \dots, V_n$, given the observed cross-sectional measurements, $y_{1,obs}, \dots, y_{n,obs}$.

Anderson and Coles (2002) were able to construct a likelihood function for this model by adapting the solution to Wicksell's corpuscle problem (Wicksell, 1925). They also overcome numerical difficulties with the resulting likelihood function of the model by treating the unobserved V_i as latent variables in a Bayesian hierarchical formulation. However, while their model assumptions of a Poisson process and inclusion independence are not unreasonable, the assumption that the inclusions are spherical is not plausible in practice.

Because of this, a generalisation of the spherical inclusion model was proposed by Bortot et al. (2007), who considered a family of ellipsoidal inclusions. In addition to the previous Poisson process and independence assumptions, this new model considered inclusions to be ellipsoidal and randomly oriented in space, with principal diameters (V^1, V^2, V^3). As before, the distribution function of $V^3 | V^3 > v_0$ is specified as the generalised Pareto distribution (14.11), where V^3 is defined as the largest ellipsoidal diameter. The other two principal diameters are defined as $V^j = U_j V^3$ for $j = 1, 2$, where U_1 and U_2 are independent uniform $U(0, 1)$ variables. Finally, in order to avoid ambiguity, the observed cross-sectional measurement $s_{i,obs}$ is assumed to be the largest principal diameter of the ellipse generated by the planar section of an ellipsoidal inclusion.

While the ellipsoidal inclusion is more realistic than the spherical inclusion model, there are analytic and computational difficulties in extending likelihood-based inference to more general families of inclusion (Baddeley and Jensen, 2004; Bortot et al., 2007). As a result ABC methods are a good candidate procedure to approximate the posterior distribution $\pi(\lambda, \sigma, \xi | y_{obs})$ under the ellipsoidal model.

In the following analysis, the prior $p(\lambda, \sigma, \xi) \propto 1$ is specified as uniform over the support of the model parameters. For the purposes of implementing Algorithm 14.2 total of $N = 2$ million samples were drawn from the sampling distribution $g(\lambda, \sigma, \xi) = U(10, 80) \times U(0, 10) \times U(-3, 3)$ for the spherical model and $g(\lambda, \sigma, \xi) = U(60, 130) \times U(0, 10) \times U(-3, 3)$ for the ellipsoidal model based on a

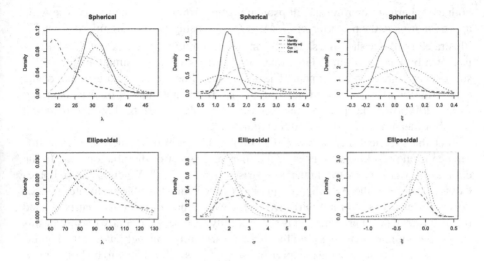

FIGURE 14.3

Marginal posterior estimates of λ, σ and ξ for the stereological extremes analysis. Top and bottom rows respectively correspond to the spherical and ellipsoidal inclusion models. Panels show the true posterior marginal distribution (solid line – spherical model only), and the ABC approximations to the posterior based on the summary statistics (14.12). Dashed lines indicate ABC posterior approximation using the identity matrix $\widehat{\Sigma} = I$ in the Mahalanobis distance, dotted lines use an estimate of $\widehat{\Sigma} = \mathrm{Cov}(s|\lambda_0, \sigma_0, \xi_0)$. Gray versions of each line indicate a subsequent regression-adjusted estimate. Points show the marginal posterior means estimated by Bortot et al. (2007).

pilot analysis. A 7-dimensional vector of summary statistics was specified as

$$S(y) = (q_{0.5}(y), q_{0.7}(y), q_{0.9}(y), q_{0.95}(y), q_{0.99}(y), q_1(y), n') \qquad (14.12)$$

where $q_a(y)$ denotes the a-th quantile of the dataset y (with interpolation if necessary), and n' is the number of observations in y. The Mahalanobis distance (14.9) is used to compare auxiliary and observed summary statistics $\|s - s_{obs}\|$, where the covariance matrix $\widehat{\Sigma}$ is either the identity matrix I, or an estimate of $\mathrm{Cov}(s|\lambda_0, \sigma_0, \xi_0)$ based on 1,000 samples $s \sim L(s|\lambda_0, \sigma_0, \xi_0)$, where $(\lambda_0, \sigma_0, \xi_0) = (30, 1.5, -0.05)$ for the spherical model and $(\lambda_0, \sigma_0, \xi_0) = (95, 1.9, -0.1)$ for the ellipsoidal model. The values for $(\lambda_0, \sigma_0, \xi_0)$ were based on preliminary analyses for each model. The smoothing kernel $K_h(u)$ is the uniform kernel over $(-h, h)$ and the kernel scale parameter h is determined as the 0.001 quantile of the N samples $\|s^{(i)} - s_{obs}\|$, resulting in 2,000 samples from the approximate posterior $\pi_h^{ABC}(\lambda, \sigma, \xi|s_{obs})$. The observed dataset s_{obs} is derived from a set of 112 inclusion diameters previously analysed by Anderson and Coles (2002) and Bortot et al. (2007).

Figure 14.3 illustrates various marginal posterior estimates of $\pi(\lambda, \sigma, \xi|s_{obs})$ for both the spherical (top row) and ellipsoidal (bottom row) inclusion models. The ABC approximation to the posterior $\pi_h^{ABC}(\lambda, \sigma, \xi | s_{obs})$ using the identity matrix $\widehat{\Sigma} = I$ is illustrated by the dashed lines, and the approximation using $\widehat{\Sigma} = \text{Cov}(s | \lambda_0, \sigma_0, \xi_0)$ is shown by the dotted lines. Gray lines denote the same approximations but with a subsequent regression adjustment. The solid black line indicates an estimate of the true posterior $\pi(\lambda, \sigma, \xi | y_{obs})$ following the latent variable model of Anderson and Coles (2002), for the spherical inclusion model only. The dots show the marginal posterior means estimated by Bortot et al. (2007).

For the spherical inclusions model (Figure 14.3, top row), there is a clear difference in the resulting estimates of the marginal posterior distributions due to the choice of covariance matrix in the comparison $\| s^{(i)} - s_{obs} \|$. When $\widehat{\Sigma}$ approximates $\text{Cov}(s | \lambda_0, \sigma_0, \xi_0)$, the ABC approximation (dotted lines) is able to locate the true posterior density (solid line) fairly well. However, when the identity matrix is used (dashed lines), the approximation is substantially worse. The primary reason for this is that the summary statistics with the highest variability – namely n' and the highest quantiles of y – dominate the comparison, so $\| s^{(i)} - s_{obs} \|$ is likely to be large unless these highly variable statistics happen to be close to those in s_{obs}. When summary statistic scaling (and correlation) is taken into consideration, the relative closeness of $s^{(i)}$ to s_{obs} can be better measured.

In addition, the implementation of a regression-adjustment also makes a substantial improvement in the quality of the ABC approximation, even for the very poor estimates using $\widehat{\Sigma} = I$. The best results are clearly obtained when $\widehat{\Sigma} \approx \text{Cov}(s | \lambda_0, \sigma_0, \xi_0)$ is used, followed by a regression adjustment. Indeed, the inclusion rate parameter λ is almost perfectly estimated marginally. More detailed analysis (not shown) suggests that the results may improve further if the kernel scale parameter h could be lowered further. However, this would require a larger number of initial simulations, N.

For the ellipsoidal inclusions model (Figure 14.3, bottom row), while estimates of the true marginal posterior densities of $\pi(\lambda, \theta, \xi | y_{obs})$ are not available, qualitatively similar conclusions to the spherical inclusions model can be made. The best performing ABC approximation is likely to be when accounting for the scale and correlation of the summary statistics, followed by a regression adjustment. This approximation agrees well with the previous estimates of marginal posterior means for these data given by Bortot et al. (2007)

A primary difference between the parameter estimates from the two models is that the number (rate) of inclusions is much higher for the ellipsoidal model. This reflects that the overall dimensions of an ellipsoid are smaller than a sphere with diameter the same as the largest principal diameter of the ellipsoid. Hence, given the observed planar intersections, a smaller rate of inclusion is predicted under the spherical inclusion model than under the ellipsoidal inclusion model. This analysis suggests that measures of inclusion impact that depend strongly on the rate of extreme inclusions are likely to be strongly sensitive to assumptions on inclusion shape.

While it is possible to perform Bayesian model selection through the computation of Bayes factors in the ABC framework (Marin et al., 2014; Robert, 2011), in the present setting it is doubtful if the current data – the maximum cross-sectional slice of each observed inclusion – is informative for this quantity. To proceed further along the road of model choice, further measurements on each cross-sectional inclusion, such as the minimum and maximum diameters, would be required.

14.4 ABC and Max-Stable Processes

Suppose that there is interest in using ABC methods to fit max-stable processes to extreme temperature data in the context of actuarial risk estimation for a class of financial products known as weather derivatives. Weather derivatives are contracts that specify a weather reporting station, a time period, and payments corresponding to certain pre-determined weather events. The intention is for the buyer of this contract to be compensated by the seller if certain undesirable weather events occur (low snowfall for ski resorts, low rainfall for farms, high temperatures for electricity producers, etc.), but in practice, weather derivatives can also be bought and sold for purely speculative reasons.

Denoting the weather random variable by Y, the resulting payment is also a random variable $P(Y)$. Here we are interested in derivatives with payments triggered by events $\{\max Y \geq u\}$ or $\{\min Y \leq u\}$ where Y is the variable of interest and u is some threshold meant to target extremes. For example, a payment P can be triggered if the maximum temperature in July exceeds 100 degrees Fahrenheit, with payment of $1000 for each degree in excess. This is shown mathematically as

$$P(Y) = \max\left(1000 \cdot \left(\max_{Y \in July}(Y) - 100\right), 0\right).$$

Because weather random variables Y can be positively correlated due to close proximity, financial outcomes of weather derivatives can also be positively correlated. When one considers the total payment from a collection of D weather derivatives $P = P_1 + \ldots + P_D$, recognition of the dependence is essential for fully characterising the distribution of P. The use of max-stable processes is one way to incorporate both spatial dependence and target extremes.

de Haan (de Haan, 1984; de Haan and Ferreira, 2006) introduced a point-process approach for constructing max-stable processes, and Smith (1990) introduced the Gaussian max-stable process and gave a convenient interpretation of this process. Here we consider the process constructed by Schlather (2002), who allowed $Y(x)$ to be any stationary Gaussian process on $\mathbb{R}^p$ with correlation function $\rho(\cdot)$ and finite mean $\mu = \mathbb{E}[\max(0, Y(x))] \in (0, \infty)$. With s_i as a Poisson process on $(0, \infty)$ with intensity measure $\mu^{-1}s^{-2}ds$, the quantity

$$Z(x) = \max_i s_i \max(0, Y_i(x))$$

is a stationary max-stable process with unit-Fréchet margins. The bivariate distribution function is

$$P(Z_1 \leqslant z_1, Z_2 \leqslant z_2) = \exp\left[-\frac{1}{2}\left(\frac{1}{z_1} + \frac{1}{z_2}\right)\left(1 + \sqrt{1 - 2(\rho(h) + 1)\frac{z_1 z_2}{(z_1 + z_2)^2}}\right)\right] \tag{14.13}$$

where $\rho(h)$ is the correlation of the underlying Gaussian process Y and $h = ||x_1 - x_2||$. The correlation function may be chosen from one of the valid families of correlations for Gaussian processes, with one common choice being the Whittle–Matérn correlation,

$$\rho(h) = c_1 \frac{2^{1-\nu}}{\Gamma(\nu)}\left(\frac{h}{c_2}\right)^{\nu} K_{\nu}\left(\frac{h}{c_2}\right), \quad 0 \leqslant c_1 \leqslant 1,\ c_2 > 0,\ \nu > 0, \tag{14.14}$$

where c_1 is the nugget, c_2 is the range and ν is the smooth parameter.

To quickly summarize, we have a method for simulating realizations from max-stable processes, along with closed-form expressions for the bivariate distribution. Max-stable processes sampled at D locations form a $D-$dimensional MEVD, whose joint likelihood function can be expressed in terms of an exponent measure as in (14.3). If we were to evaluate the probability that the process is below z at all locations $x_1, ..., x_D$, we get

$$P(Z(x_1) \leqslant z, ..., Z(x_D) \leqslant z) = \exp\left\{-\frac{\phi(x_1, ..., x_D)}{z}\right\},$$

where $\phi(x_1, ..., x_D) = V(1, ..., 1)$ is the *extremal coefficient* for the D locations. For a pair of locations, using the bivariate distribution function in (14.13) one can relate this to the spatial dependence as

$$\phi(h) = 1 + \left\{\frac{1 - \rho(h; \theta)}{2}\right\}^{1/2}, \tag{14.15}$$

where $h = ||x_1 - x_2||$. Following Smith (1990) and Coles and Dixon (1999), if the field $Z(\cdot)$ has been transformed to unit-Fréchet, then $1/Z(\cdot)$ is unit exponential. This means $1/\max(Z(x_1), Z(x_2))$ is exponential with mean $1/\phi(x_1, x_2)$, and so estimating ϕ is equivalent to estimating the parameter of an exponential distribution. Taking this idea further, Erhardt and Smith (2012) defined the tripletwise extremal coefficient with estimator

$$\widehat{\phi}(x_j, x_k, x_l) = \frac{n}{\sum_{i=1}^{n} 1/\max(z_i(x_j), z_i(x_k), z_i,(x_l))}. \tag{14.16}$$

where i is the index for the block. For D locations the number of tripletwise extremal coefficients is $\binom{D}{3}$, which grows quite rapidly as D increases. Hence, a clustering step can be added to group these coefficients into a fixed number $K << \binom{D}{3}$ (see Erhardt and Smith (2012) for full details on this step). The $\binom{D}{3}$ triplet extremal coefficients may be estimated for the observed data using (14.16), and then these values are averaged within the K clusters. The result is the summary of the observed data, $s = (\bar{\phi}_1, ..., \bar{\phi}_K)$.

Following Algorithm 14.2, independent draws from the prior $\theta^{(i)} \sim p(\theta)$ are taken. For each draw from the prior, a max-stable process $Z^{(i)}$ with unit-Fréchet margins is simulated on the same locations and for the same number of years as the observed data. For this data $Z^{(i)}$, all tripletwise extremal coefficients are estimated, and averages within each cluster group are taken to produce $s^{(i)} = (\bar{\phi}_1, ..., \bar{\phi}_K)$. To compare summaries, the sum of absolute deviations is used as a distance measure, so that

$$\|s^{(i)} - s_{obs}\| = \sum_{k=1}^{K} |s_k^{(i)} - s_{k,obs}|$$

where $s_k^{(i)}$ and $s_{k,obs}$ denote the k-th element of $s^{(i)}$ and s_{obs}, respectively. As before, the smoothing kernel $K_h(u)$ is uniform over $(-h, h)$, and the kernel scale parameter h is determined as the 0.01 quantile of the N samples $\|s^{(i)} - s_{obs}\|$.

An application of this model concerns annual maxima daily temperatures taken from 39 locations in the midwestern United States with complete summer (June 1 – August 31) temperature records from 1895 to 2009 (Erhardt and Smith, 2014). All sites are located between longitudes 93 and 103 degrees west, and latitudes 37 to 45 degrees north.

Each of the 39 marginal distributions was transformed to unit-Fréchet through ordinary maximum likelihood estimation. Based on the dependence model (14.14) with no nugget effect ($c_1 = 1$), the prior for the spatial dependence parameter vector $\theta = (c_2, \nu)$ is specified as $[0, 7] \times [0, 7]$. This choice aims to represent vague non-informativeness and also place positive support on a large range of possible parameter values. Based on candidate draws $\theta^{(i)} \sim p(\theta)$, $i = 1, \dots, N$, max-stable processes $Z^{(i)} \sim f(Z|\theta^{(i)})$ were simulated for 115 years at the same 39 locations.

The ABC posterior approximation $\pi_h^{ABC}(\theta|y_{obs})$ based on $N = 100,000$ samples is shown in Figure 14.4. The approximate posterior has correctly identified the credible regions of the parameter space from these data, retaining only draws from a crescent-shaped subspace. Inversely changing values of c_2 and ν can produce similarly shaped correlation functions $\rho(h)$ — hence the crescent shape for the posterior approximation. The posterior predictive distribution of spatial correlation functions $\rho(h)$ is illustrated in the lower panel of Figure 14.4. This approximate posterior may now be used to fully incorporate the parameter uncertainty of θ into estimates of the distribution of total loss $P = P_1 + \dots + P_D$ and associated actuarial risk measures.

14.5 Discussion

This chapter has outlined how approximate Bayesian statistical modeling of extremes can be viable in situations where the likelihood function either cannot be evaluated numerically, or written in closed form. Sections 14.3 and 14.4 illustrate ABC methods as applied to two different models in extremes, with various model and data complexities, in which an intractable likelihood is a common issue. All that ABC

ABC Approximate Posterior

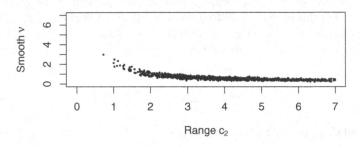

Posterior Predictive Correlation Function

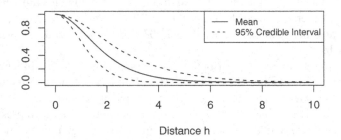

FIGURE 14.4
Top: Samples drawn from the ABC approximate posterior distribution based on fitting a max-stable process. The uniform prior for (c_2, ν) is the full range displayed on the scatterplot. Bottom: Posterior predictive distribution of the corresponding spatial correlation function $\rho(h)$ following Equation (14.14). The solid line shows the pointwise posterior mean, and the dashed lines show the pointwise 2.5% and 97.5% quantiles.

methods require in order to be implemented is a fast algorithm to simulate realisations from the intractable likelihood, and a suitable vector of summary statistics. The former requirement is not always a trivial procedure for extremal models, and commonly needs some subtle theory and computational tricks. However, procedures for accurately simulating observations from extremal models continues to be an area of active research.

Perhaps the most important aspect of any ABC analysis is the identification of a low-dimensional vector of summary statistics that are highly informative for the model parameters. Once the summary statistic is chosen, the best possible ABC approximation to the posterior is $\pi(\theta|s_{obs}) \approx \pi(\theta|y_{obs})$, and this occurs when the kernel scale parameter $h \to 0$. If the summary statistic is poorly chosen, the rest of the ABC procedure (e.g., Algorithm 14.2) cannot recover the loss of information. In fact, as $h > 0$ is a necessity in practice, the best ABC posterior approximation achievable is $\pi_h^{ABC}(\theta|s_{obs})$ which is in general a worse approximation than $\pi(\theta|s_{obs})$. Hence, the choice of summary statistics is critical. To date, probably the most useful and principled general approach to identify summary statistics is developed by Fearnhead and Prangle (2012), whereas a method to derive summary statistics from the score function of a related composite model (such as a pairwise likelihood) could be of particular interest for modelling spatial extremes (Ruli et al., 2015).

Coupled with the choice of summary statistics is the issue of computational overheads. ABC methods are a classic example of a computation versus accuracy trade-off. This implies that any developments in improving the efficiency of an ABC analysis can be converted into producing a more accurate inference for the same computational cost. Some more sophisticated algorithms than importance sampling are detailed in Section 14.2.3, but other ideas, such as terminating the generation of a dataset early if it is likely to be rejected (Prangle, 2014) are also starting to be developed. Improving the "approximation" (in approximate Bayesian computation) while reducing the "computation" is an active research area in ABC.

It is worth stating that the development of a tractable, closed-form joint likelihood for a stochastic process would entirely obviate the need for ABC methods by allowing for a more conventional Bayesian analysis based on direct usage of the likelihood. A flavour of this evolution can be found in both Genton et al. (2011) and Aandahl et al. (2014). After a period of roughly two decades in which researchers could write down only the bivariate joint distribution function for the Gaussian max-stable process, Genton et al. (2011) extended the closed-form solution to include the trivariate joint likelihood. In the setting of population genetics, Aandahl et al. (2014) show that models that were previously only available for analysis using ABC, have subsequently become tractable with further analytical development. These are useful reminders that what is considered intractable today may not remain intractable in the future. In the meantime, however, the statistician's toolbox remains empowered through the availability of ABC methods.

References

Aandahl, R. Z., T. Stadler, S. A. Sisson, and M. M. Tanaka (2014). Exact vs. approximation computation: Reconciling different estimates of Mycobacterium tuberculosis epidemiological parameters. *Genetics 196*, 1227–1230.

Anderson, C. W. and S. G. Coles (2002). The largest inclusions in a piece of steel. *Extremes 5*, 237–252.

Baddeley, A. and E. B. V. Jensen (2004). *Stereology for Statisticians*. Chapman & Hall/CRC.

Beaumont, M. A. (2010). Approximate Bayesian computation in evolution and ecology. *Annual Review of Ecololgy, Evolution, and Systematics 41*(1), 379–406.

Beaumont, M. A., J.-M. Marin, J. M. Corunet, and C. P. Robert (2009). Adaptivity for ABC algorithms: The ABC-PMC scheme. *Biometrika 96*, 983–990.

Beaumont, M. A., W. Zhang, and D. J. Balding (2002). Approximate Bayesian computation in population genetics. *Genetics 162*, 2025–2035.

Bertorelle, G., A. Benazzo, and S. Mona (2010). ABC as a flexible framework to estimate demography over space and time: Some cons, many pros. *Molecular Ecology 19*(13), 2609–2625.

Blum, M. G. B. (2010). Approximate Bayesian computation: A nonparametric perspective. *Journal of the Amercan Statistical Association 105*, 1178–1187.

Blum, M. G. B. and O. François (2010). Non-linear regression models for approximate Bayesian computation. *Statistics and Computing 20*, 63–73.

Blum, M. G. B., M. A. Nunes, D. Prangle, and S. A. Sisson (2013). A comparative review of dimension reduction methods in approximate Bayesian computation. *Statistical Science 28*(189-208), 189–208.

Bortot, P., S. G. Coles, and S. A. Sisson (2007). Inference for stereological extremes. *Journal of the Amercan Statistical Association 102*, 84–92.

Brooks, S. P., A. Gelman, G. L. Jones, and X.-L. Meng (Eds.) (2011). *Handbook of Markov Chain Monte Carlo*. CRC Press.

Chen, M.-H., Q.-M. Shao, and J. G. Ibrahim (2000). *Monte Carlo Methods in Bayesian Computation*. Springer.

Coles, S. (2001). *An Introduction to Statistical Modeling of Extreme Values*, Volume 208. Springer.

Coles, S. G. and M. J. Dixon (1999). Likelihood-based inference for extreme value models. *Extremes 2*(1), 5–23.

Coles, S. G. and L. R. Pericchi (2003). Anticipating catastrophes through extreme value modelling. *Journal of the Royal Statistical Society, Series C (Applied Statistics) 52*, 405–416.

Coles, S. G., L. R. Pericchi, and S. A. Sisson (2003). A fully probabilistic approach to extreme value modelling. *Journal of Hydrology 273*, 35–50.

Coles, S. G. and E. A. Powell (1996). Bayesian methods in extreme value modelling: A review and new developments. *International Statistical Review 64*, 119–136.

Csilléry, K., M. G. B. Blum, O. E. Gaggiotti, and O. François (2010). Approximate Bayesian computation (ABC) in practice. *Trends in Ecology and Evolution 25*(7), 410–418.

de Haan, L. (1984). A spectral representation for max-stable processes. *The Annals of Probability 12*, 1194–1204.

de Haan, L. and A. Ferreira (2006). *Extreme Value Theory: An Introduction*. Springer.

Del Moral, P., A. Doucet, and A. Jasra (2012). An adaptive sequential Monte Carlo method for approximate Bayesian computation. *Statistics and Computing 22*, 1009–1020.

Doucet, A., N. de Freitas, and N. Gordon (Eds.) (2001). *Sequential Monte Carlo Methods in Practice*. Springer.

Drovandi, C. C. and A. N. Pettitt (2011). Likelihood-free Bayesian estimation of multivariate quantile distributions. *Computational Statistics & Data Analysis 55*, 2541–2556.

Erhardt, R. J. and R. L. Smith (2012). Approximate bayesian computing for spatial extremes. *Computational Statistics & Data Analysis 56*(6), 1468–1481.

Erhardt, R. J. and R. L. Smith (2014). Weather derivative risk measures for extreme events. *North American Actuarial Journal 18*(3), 379–393.

Fearnhead, P. and D. Prangle (2012). Constructing summary statistics for approximate Bayesian computation: Semi-automatic approximate bayesian computation. *Journal of the Royal Statistical Society, Series B (Statistical Methodology) 74*, 419–474.

Gelman, A., J. B. Carlin, H. S. Stern, and D. B. Rubin (2003). *Bayesian Data Analysis* (2 ed.). Chapman & Hall/CRC.

Genton, M. G., Y. Ma, and H. Sang (2011). On the likelihood function of Gaussian max-stable processes. *Biometrika 98*(2), 481–488.

Gnedenko, B. V. (1944). Limit theorems for sums of independent random variables. *Uspekhi Matematicheskikh Nauk* (10), 115–165.

Hosking, J. R. M. (1990). L-moments: Analysis and estimation of distributions using linear combinations of order statistics. *Journal of the Royal Statistical Society, Series B (Statistical Methodology) 52*, 105–124.

Leadbetter, M. R., G. Lindgren, and H. Rootzén (1983). *Extremes and Related Properties of Random Sequences and Processes*. Berlin: Springer.

Luciani, F., S. A. Sisson, H. Jiang, A. R. Francis, and M. M. Tanaka (2009). The epidemiological fitness cost of drug resistance in *mycobacterium tuberculosis*. *Proceedings of the National Academy of Sciences of the USA 106*, 14711–14715.

Marin, J.-M., N. Pillai, C. P. Robert, and J. Rousseau (2014). Relevant statistics for Bayesian model choice. *Journal of the Royal Statistical Society, Series B (Statistical Methodology) 76*, 833–859.

Marin, J.-M., P. Pudlo, C. P. Robert, and R. J. Ryder (2012). Approximate Bayesian computational methods. *Statistics and Computing 22*, 1167–1180.

Marjoram, P., J. Molitor, V. Plagnol, and S. Tavaré (2003). Markov chain Monte Carlo without likelihoods. *Proceedings of the National Academy of Sciences of the USA 100*, 15324–15328.

Nott, D. J., Y. Fan, L. Marshall, and S. A. Sisson (2014). Approximate Bayesian computation and Bayes linear analysis: Towards high-dimensional ABC. *Journal of Computational and Graphical Statistics 23*, 65–86.

O'Hagan, A. and J. Forster (2004). *Bayesian Inference* (2 ed.), Volume 2B of *Kendall's Advanced Theory of Statistics*. Arnold.

Pickands, J. (1981). Multivariate extreme value distributions. In *Proceedings 43rd Session International Statistical Institute*, Volume 2, pp. 859–878.

Prangle, D. (2014). Lazy ABC. *Statistics and Computing*. In press.

Pritchard, J. K., M. T. Seielstad, A. Perez-Lezaun, and M. W. Feldman (1999). Population growth of human Y chromosomes: A study of Y chromosome microsatellites. *Molecular Biology Evolution 16*, 1791–1798.

Reeves, R. and A. N. Pettitt (2005). A theoretical framework for approximate Bayesian computation. In A. R. Francis, K. M. Matawie, A. Oshlack, and G. K. Smythe (Eds.), *Statistical Solutions to Modern Problems: Proceedings of the 20th International Workshop on Statistical Modelling, Sydney, 10-15 July, 2005*. University of Western Sydney Press, Sydney.

Robert, C. P. (2011). Lack of confidence in approximate Bayesian computation model choice. *Proceedings of the National Academy of Sciences of the USA 108*, 15112–15117.

Ruli, E., N. Sartori, and L. Ventura (2015). Approximate Bayesian computation with composite score functions. *Statistics and Computing*. In press.

Schlather, M. (2002). Models for stationary max-stable random fields. *Extremes 5*(1), 33–44.

Sisson, S. A. and Y. Fan (2011). Likelihood-free Markov chain Monte Carlo. In S. P. Brooks, A. Gelman, G. Jones, and X.-L. Meng (Eds.), *Handbook of Markov Chain Monte Carlo*, pp. 319–341. Chapman and Hall/CRC Press.

Sisson, S. A., Y. Fan, and M. M. Tanaka (2007). Sequential Monte Carlo without likelihoods. *Proceedings of the National Academy of Sciences of the USA 104*, 1760–1765. Errata (2009), 106, 16889.

Smith, R. L. (1990). Max-stable processes and spatial extremes. Unpublished manuscript.

Tavaré, S., D. J. Balding, R. C. Griffiths, and P. Donnelley (1991). Infering coalescent times from DNA sequence data. *Genetics 145*, 505–518.

Wicksell, S. D. (1925). The corpsucle problem: A mathematical study of a biometric problem. *Biometrika 17*, 84–99.

Wieczorek, G. F., M. C. Larsen, L. S. Eaton, B. A. Morgan, and J. L. Blair (2001). Debris-flow and flooding hazards associated with the December 1999 storm in coastal Venezuela and strategies for mitigation. Open File Report 01-0144, U. S. Geological Survey. Available at http://pubs.usgs.gov/of/2001/ofr-01-0144/.

Wilkinson, R. D. (2013). Approximate Bayesian computation (ABC) gives exact results under the assumption of model error. *Statistical Applications in Genetics and Molecular Biology 12*(2), 129–141.

15

Estimation of Extreme Conditional Quantiles

Huixia Judy Wang

George Washington University, United States

Deyuan Li

Fudan University, China

Abstract

The estimation of extreme quantiles of the response distribution is of great interest in many areas. Extreme value theory provides a useful tool for estimating extreme quantiles. However, current extreme value literature focuses primarily on the extreme quantiles of a univariate variable. In this chapter, we provide a survey of available methods, including parametric, nonparametric, semiparametric, and quantile-regression-based approaches, for estimating the extreme conditional quantiles of the quantity of interest when some covariates are recorded simultaneously. A simulation study is carried out to assess the performance of various methods.

15.1 Introduction

Estimation of tail quantiles is of great interest in many studies of rare events that happen infrequently but have heavy consequences. Extreme value theory provides a useful tool for modeling rare events and estimating extreme quantiles. The current extreme value literature focuses primarily on the tail quantiles of a univariate variable. However, in many applications, the conditional extreme quantiles of the response variable Y given some covariates X are of interest, for instance, high quantiles of tropical cyclone intensity given time or certain climate variables (Jagger and Elsner, 2008), localized high precipitation conditional on global climate model projections (Friederichs, 2010), low conditional quantiles of a portfolio's future return given the past or assumptions on future interest rate changes (Engle and Manganelli, 2004), low quantiles of birth weight given maternal behavior (Abrevaya, 2001), and so on.

In this chapter, we provide a survey of methods for estimating extreme conditional quantiles. Without loss of generality, we focus on the estimation of conditional high quantiles, because a low quantile of Y can be viewed as a high quantile of $-Y$.

Throughout, let Y denote the univariate response of interest, and $\mathbf{X}$ be the p-dimensional covariate vector. In addition, let $\xi(\mathbf{x})$ denote the conditional extreme value index of Y given $\mathbf{X} = \mathbf{x}$, which determines the rate of tail decay of the conditional distribution of Y. Suppose that we observe a random sample $\{(y_i, \mathbf{x}_i), i = 1, \ldots, n\}$ of $(Y, \mathbf{X})$. Our main interest is in estimating the τ_n-th conditional quantile of Y given $\mathbf{X} = \mathbf{x}$, $Q_Y(\tau_n|\mathbf{x})$, which satisfies $P\{Y > Q_Y(\tau_n|\mathbf{x})|\mathbf{x}\} = \tau_n$, where $\tau_n \to 1$ as $n \to \infty$. The conditional quantile $Q_Y(\tau_n|\mathbf{x})$ can also be interpreted as the $1/(1 - \tau_n)$ return level of Y given that the covariate $\mathbf{X} = \mathbf{x}$.

The rest of this chapter is organized as follows. In Section 15.2, we review the commonly used methods for estimating unconditional extreme quantiles. In Section 15.3, we discuss four classes of approaches for estimating conditional extreme quantiles: (1) parametric methods, (2) semiparametric methods, (3) nonparametric methods, and (4) quantile-regression-based methods. We present some numerical comparison of different estimation methods in Section 15.4 and some final remarks in Section 15.5.

15.2 Estimation of Extreme Unconditional Quantiles

We first review some classic methods for estimating extreme quantiles of a univariate response distribution without considering the covariate information. Let $\{y_1, \ldots, y_n\}$ be a random sample of Y with cumulative distribution function F, and $y_{1,n} \leqslant y_{2,n} \leqslant \cdots \leqslant y_{n,n}$ be the order statistics. Denote $Q(\tau) = F^{\leftarrow}(\tau) = \inf\{y : F(y) \geqslant \tau\}$ as the τ-th quantile of Y. We are interested in estimating the high quantile $Q(\tau_n)$ when $\tau_n \to 1$ as $n \to \infty$.

For a general distribution F, we assume that F belongs to the maximum domain of attraction of an extreme value distribution G_ξ with the extreme value index (EVI) $\xi \in \mathbb{R}$ that measures the heaviness of the tail of F, denoted by $F \in D(G_\xi)$. This means there exist $a_n > 0$ and $b_n \in \mathbb{R}$ such that

$$\lim_{n \to \infty} \{F(a_n y + b_n)\}^n = G_\xi(y) = \exp\{-(1 + \xi y)^{-1/\xi}\}, \quad 1 + \xi y > 0.$$

This condition is equivalent to

$$\lim_{t \to \infty} \frac{U(tx) - U(t)}{a(t)} = \frac{x^\xi - 1}{\xi}, \quad x > 0,$$

where $U(t) = F^{\leftarrow}(1 - 1/t) = Q(1 - 1/t)$ and $a(\cdot)$ is some positive function. There are some other equivalent conditions for $F \in D(G_\xi)$, for example see Theorems 1.1.6 and 1.1.8 in de Haan and Ferreira (2006). Based on the above relation, $Q(\tau_n)$

can be estimated by

$$\widehat{Q}(\tau_n) = y_{n-k,n} + \widehat{a}(n/k)\left\{\left(\frac{k}{np_n}\right)^{\widehat{\xi}} - 1\right\}\widehat{\xi}^{-1},$$

where $p_n = 1 - \tau_n$, $k = k_n$ is a positive integer such that $k \to \infty$ and $k/n \to 0$, $\widehat{\xi}$ and $\widehat{a}(\cdot)$ are some estimators of ξ and $a(\cdot)$, respectively. The asymptotic normality of $\widehat{Q}(\tau_n)$ can be obtained under some second-order conditions. For example, Dekkers et al. (1989) and de Haan and Rootzén (1993) established the asymptotical properties of $\widehat{Q}(\tau_n)$ based on the moment estimator of (ξ, a). In general, if the estimator $(\widehat{\xi}, \widehat{a})$ is asymptotically normal (for example, the maximum likelihood estimator, see Drees et al. (2004)), then the asymptotic properties of $\widehat{Q}(\tau_n)$ can be obtained by applying Theorem 4.3.1 in de Haan and Ferreira (2006).

For a heavy-tailed F, one common assumption is that for some $\xi > 0$,

$$1 - F(y) = y^{-1/\xi}l(y), \quad \text{as } y \to \infty, \tag{15.1}$$

where $l(\cdot)$ is a slowly varying function that satisfies the condition $l(ty)/l(y) \to 1$ as $y \to \infty$ for all $t > 0$. The condition (15.1) is equivalent to the following condition on the quantile function:

$$Q(1 - 1/y) = y^{\xi}L(y), \tag{15.2}$$

where $L(\cdot)$ is also a slowly varying function and is related to $l(\cdot)$. Consequently, as $\tau \to 1$ and $\tau_n \to 1$, $Q(\tau_n)/Q(\tau) \sim \{(1 - \tau)/(1 - \tau_n)\}^{\xi}$. This is the basis of the popular Weissman estimator (Weissman, 1978),

$$\widehat{Q}(\tau_n) = y_{n-k,n}\left[k/\{n(1 - \tau_n)\}\right]^{\widehat{\xi}},$$

where $\widehat{\xi}$ is some estimator of ξ, for instance, the Hill estimator (Hill, 1975) $\widehat{\xi} = k^{-1}\sum_{i=1}^{k}\log(y_{n-i+1,n}/y_{n-k,n})$. Under some second-order conditions, the asymptotic normality of the Weissman estimator $\widehat{Q}(\tau_n)$ based on some asymptotically normal estimator $\widehat{\xi}$ is presented in Theorem 4.3.8 of de Haan and Ferreira (2006).

The Weissman estimator of extreme quantiles was also adapted to Weibull-tail distributions in Diebolt et al. (2008) and Gardes and Girard (2005). Recently, some bias-reduced extreme quantile estimation methods for heavy-tailed distributions have been developed; see for instance Gomes and Figueiredo (2006), Gomes and Pestana (2007), Li et al. (2010), and references therein. In addition, Drees (2003) discussed the estimation of extreme quantiles for dependent random variables.

15.3 Estimation of Extreme Conditional Quantiles

In this section, we focus on the estimation of extreme high conditional quantile of Y given covariate $\mathbf{x}$, $Q_Y(\tau_n|\mathbf{x})$, where $\tau_n \to 1$ as $n \to \infty$. We discuss four different classes of approaches: parametric, semiparametric, nonparametric, and quantile-regression-based methods. The focus of this chapter differs from that in Smith (1994), Portnoy and Jurečková (1999), which studied extreme quantile regression with quantile level $\tau = 0$ or 1.

15.3.1   Parametric Methods

To incorporate the covariate information in modeling extremes, the first class of work fit parametric models such as the generalized extreme value (GEV) distribution based on block maximum data or the generalized Pareto distribution (GPD) based on exceedances over high thresholds, where the location, shape, and scale parameters are assumed to depend on covariates parametrically.

One parametric model is based on block maximum data, for example, the annual maximum of daily precipitation. Suppose that Y is the block maximum variable. The basic model assumes that the conditional distribution $F_Y(\cdot|\mathbf{x})$ can be approximated by the GEV distribution, that is,

$$F_Y(y|\mathbf{x}) \approx H\{y; \mu(\mathbf{x}), \sigma(\mathbf{x}), \xi(\mathbf{x})\} = \exp\left[-\left\{1 + \xi(\mathbf{x})\frac{y - \mu(\mathbf{x})}{\sigma(\mathbf{x})}\right\}_+^{-1/\xi(\mathbf{x})}\right], \quad (15.3)$$

where $\mu(\mathbf{x})$, $\sigma(\mathbf{x})$ and $\xi(\mathbf{x})$ are the location, scale, and shape parameters, respectively, and $1 + \xi(\mathbf{x})\{y - \mu(\mathbf{x})\}/\sigma(\mathbf{x}) > 0$. The GEV approximation is based on the result of Fisher and Tippett (1928). Under this model assumption, the τ-th conditional quantile of Y given $\mathbf{x}$ is

$$Q_Y(\tau|\mathbf{x}) = \begin{cases} \mu(\mathbf{x}) + \frac{\sigma(\mathbf{x})}{\xi(\mathbf{x})}\{(-\log\tau)^{-\xi(\mathbf{x})} - 1\}, & \xi(\mathbf{x}) \neq 0, \\ \mu(\mathbf{x}) - \sigma(\mathbf{x})\log(-\log\tau), & \xi(\mathbf{x}) = 0. \end{cases} \quad (15.4)$$

To capture the dependence of the distribution of Y on $\mathbf{x}$, one common practice is to model $\mu(\mathbf{x})$, $\sigma(\mathbf{x})$, and $\xi(\mathbf{x})$ as some linear functions of $\mathbf{x}$ after known link transformations, that is, assume

$$\mu(\mathbf{x}) = \Lambda_\mu(\mathbf{x}^T\boldsymbol{\gamma}), \quad \sigma(\mathbf{x}) = \Lambda_\sigma(\mathbf{x}^T\boldsymbol{\beta}), \quad \xi(\mathbf{x}) = \Lambda_\xi(\mathbf{x}^T\boldsymbol{\theta}),$$

where Λ_μ, Λ_σ, and Λ_ξ are some known link functions. We can then estimate $\boldsymbol{\gamma}$, $\boldsymbol{\beta}$, $\boldsymbol{\theta}$, and the extreme conditional quantile $Q_Y(\tau_n|\mathbf{x})$ by using existing estimation methods such as maximum likelihood estimation. This GEV modeling approach has been considered in Sang and Gelfand (2009), Coles (2001, Chapter 6), Friederichs and Thorarinsdottir (2012), to name a few.

One limitation of the GEV modeling approach based on maximum data is its inefficient use of the available data. This problem can be remedied by using the observations exceeding a high threshold (Davison and Smith, 1990; Smith, 1989). Let u be some high threshold, and $Z = Y - u|Y > u$ be the positive exceedance. Motivated by the GPD approximation result from Pickands (1975), the method assumes that for $z > 0$,

$$F_Z(z|\mathbf{x}) = \frac{F_Y(u + z|\mathbf{x}) - F_Y(u|\mathbf{x})}{1 - F_Y(u|\mathbf{x})} \approx G\{z; \sigma(\mathbf{x}), \xi(\mathbf{x})\}, \quad (15.5)$$

where $G(z; \sigma, \xi) = 1 - (1 + \xi z/\sigma)^{-1/\xi}$ is the cumulative distribution function of GPD, $\sigma(\mathbf{x}) > 0$ and $\xi(\mathbf{x})$ are the scale and shape parameters satisfying $1 + z\xi(\mathbf{x})/\sigma(\mathbf{x}) > 0$. Similar to the method based on maximum data, $\sigma(\mathbf{x})$ and $\xi(\mathbf{x})$ can be modeled parametrically by

$$\sigma(\mathbf{x}) = \Lambda_\sigma(\mathbf{x}^T \beta), \quad \xi(\mathbf{x}) = \Lambda_\xi(\mathbf{x}^T \theta),$$

where Λ_ξ and Λ_σ are some known functions. Let $(\widehat{\theta}, \widehat{\beta})$ be the estimator of (θ, β) based on the sample $\{(z_i, \mathbf{x}_i), i = 1, \ldots, n\}$ with $z_i = y_i - u > 0$, for instance, the maximum likelihood estimator (Smith, 1985) or the method of moments estimator (Hosking and Wallis, 1987). Consequently, $Q_Y(\tau_n | \mathbf{x})$ can be estimated by

$$\widehat{Q}_Y(\tau_n | \mathbf{x}) = u + \frac{\Lambda_\sigma(\mathbf{x}^T \widehat{\beta})}{\Lambda_\xi(\mathbf{x}^T \widehat{\theta})} \left[\left\{ \frac{1 - F_Y(u | \mathbf{x})}{1 - \tau_n} \right\}^{\Lambda_\xi(\mathbf{x}^T \widehat{\theta})} - 1 \right].$$

15.3.2 Semiparametric Methods

Instead of assuming an exact distribution form for $Y | \mathbf{x}$ as in the parametric methods discussed in Section 15.3.1, some researchers (Beirlant and Goegebeur, 2003; Wang and Tsai, 2009) considered semiparametric approaches that model the tail of $Y | \mathbf{x}$ as a Pareto-type distribution with parameters depending on $\mathbf{x}$ in a parametric way.

The basic assumption is that the conditional distribution of Y given $\mathbf{x}$ is heavy-tailed or Pareto-type, that is, there exists a $\xi(\mathbf{x}) > 0$ such that

$$1 - F_Y(y | \mathbf{x}) = y^{-1/\xi(\mathbf{x})} l(y; \mathbf{x}), \quad y > 0, \tag{15.6}$$

where $l(\cdot; \mathbf{x})$ is an unknown slowly varying function at infinity, which means that for any $y > 0$, $l(ty; \mathbf{x})/l(t; \mathbf{x}) \to 1$ as $t \to \infty$. The extreme value index $\xi(\mathbf{x})$ is modeled parametrically. For instance, Beirlant and Goegebeur (2003) and Wang and Tsai (2009) assumed that $\xi(\mathbf{x}) = \exp(\mathbf{x}^T \beta)$ for some unknown parameter β.

Suppose that the first element of $\mathbf{x}$ is 1. Write $\mathbf{x} = (1, \tilde{\mathbf{x}}^T)^T$, and $\beta = (\beta_0, \beta_1^T)^T$ with β_0 denoting the coefficient corresponding to the intercept. Beirlant and Goegebeur (2003) assumed that the transformation $R = R(\beta_1) = Y^{\exp(-\tilde{\mathbf{x}}^T \beta_1)}$ removes the dependence of ξ and l on $\mathbf{x}$ completely, so that $1 - F_R(r | \mathbf{x}) = r^{-1/\xi_0} l(r)$, where $\xi_0 = \exp(\beta_0)$.

Define $Z_j = j(\log R_{n-j+1, n} - \log R_{n-j, n})$, $j = 1, \ldots, n$, where $R_{1, n} \leqslant \cdots \leqslant R_{n, n}$ are the order statistics of the so-called generalized residuals $\{R_1, \ldots, R_n\}$. Under a so-called slow variation with remainder condition on the slowly varying function $l(\cdot)$, Beirlant and Goegebeur (2003) proposed the following exponential regression model:

$$Z_j = \left\{ \xi_0 + b_{n,k} \left(\frac{j}{k+1} \right)^{-\rho} \right\} F_j, \quad j = 1, \ldots, k,$$

where $F_1, \ldots, F_k$ denote independent standard exponential random variables, $\rho < 0$, and $b_{n,k} = b\{(n+1)/(k+1)\}$ with $b(\cdot)$ a rate function satisfying $b(t) \to 0$ as $t \to \infty$. The authors then proposed a maximum likelihood estimation procedure to estimate $\xi_0, \rho, b_{n,k}$ and consequently β.

Wang and Tsai (2009) proposed an alternative approximate maximum likelihood estimator for β. They assumed that as $y \to \infty$, the slowly varying function $l(y; \mathbf{x})$ converges to a constant $c(\mathbf{x})$ with a reasonably fast speed. Under this assumption, the distribution of Y given $\mathbf{x}$ can be approximated by an exponential distribution, that is, for sufficiently large y, $f_Y(y|\mathbf{x}) \approx c(\mathbf{x})/\xi(\mathbf{x})y^{-1/\xi(\mathbf{x})-1}$. Therefore, the approximate maximum likelihood estimator of β is defined as

$$\widehat{\beta} = \underset{\beta}{\operatorname{argmin}} \sum_{i=1}^{n} \left\{ \exp(-\mathbf{x}_i^T \beta) \log(y_i/\omega_n) + \mathbf{x}_i^T \beta \right\} I(y_i > \omega_n),$$

where ω_n is the threshold. Under some second-order conditions, Wang and Tsai (2009) established the asymptotic normality of $\widehat{\beta}$.

Once β is estimated, by adapting the Weissman estimator, the extreme conditional quantile $Q_Y(\tau_n|\mathbf{x})$ can be estimated by

$$\widehat{Q}_Y(\tau_n|\mathbf{x}) = \widehat{Q}_Y(1 - k/n|\mathbf{x}) \left\{ \frac{k}{n(1 - \tau_n)} \right\}^{\exp(\mathbf{x}^T \widehat{\beta})},$$

where $\widehat{Q}_Y(1 - k/n|\mathbf{x})$ is some estimation of the $(1 - k/n)$-th conditional quantile of Y given $\mathbf{x}$, for instance, $\left\{ \widehat{R}_{n-k,n} \right\}^{\exp(\tilde{\mathbf{x}}^T \widehat{\beta}_1)}$ with $\widehat{R}_{n-k,n}$ representing the $(k+1)$-th largest order statistic of the generalized residuals based on $\widehat{\beta}_1$.

15.3.3   Nonparametric Methods

The parametric and semiparametric methods all model the dependence of the distributional parameters (location, scale, and shape) on covariates parametrically, which are often restrictive and may not describe the data well. As an alternative, researchers have considered nonparametric modeling of the distributional parameters, which are more flexible and can be used for exploratory data analysis or for checking the adequacy of a parametric model.

In the current literature, there exist three main classes of nonparametric methods. By focusing on either maximum data or exceedances, the first class of work is based on a likelihood assumption of either GEV distribution or GPD, which allow the parameters to depend on covariates in a nonparametric way. The second class of work is based on a local two-step estimation, where in the first step a subset of data within a neighborhood of $\mathbf{x}$ of interest is selected and then univariate extreme value theory is applied to y_i in the neighborhood to estimate $\xi(\mathbf{x})$ and $Q_Y(\tau_n|\mathbf{x})$ in the second step. In the third class of work, the intermediate conditional quantiles are first obtained by inverting the kernel estimation of the conditional distribution function and then extrapolated to the high tails to estimate extreme conditional quantiles.

15.3.3.1 Likelihood-Based Methods

In Section 15.3.1, we discussed parametric methods that assume either the GEV distribution for block maximum data or the GPD for exceedances over high thresholds, where the form of the dependence of the distributional parameters on $\mathbf{x}$ is fully specified. In many applications, however, the dependence on $\mathbf{x}$ is more complex than what a simple parametric model could accommodate; see Hall and Tajvidi (2000) for examples. To allow more flexibility, we can model the parameters in the GEV distribution or GPD to be nonparametric functions of $\mathbf{x}$. For instance, Davison and Ramesh (2000) assumed the GEV distribution (15.3) for block maximum data, and proposed a local polynomial estimator of $\mu(t)$, $\sigma(t)$ and $\xi(t)$, where t is the univariate time variable. Beirlant and Goegebeur (2004) proposed a local polynomial estimator by fitting the GPD to exceedances over high thresholds. To estimate $Q_Y(\tau_n|\mathbf{x})$ at a given $\mathbf{x}$, the method uses covariate-dependent thresholds $u_\mathbf{x}$ and assumes that the positive exceedances $z_i = y_i - u_\mathbf{x}$ are independent following the GPD as in (15.5). Focusing on the case with a univariate covariate x, the authors established the consistency and asymptotic normality of the proposed local polynomial estimator, and also suggested a leave-one-out cross validation procedure for choosing the bandwidth h and threshold u_x. Using a similar GPD approximation to exceedances over high thresholds, Chavez-Demoulin and Davison (2005) proposed an alternative smoothing spline estimator obtained by maximizing the penalized GPD likelihood, and studied the finite sample properties of the estimator.

15.3.3.2 Two-Step Local Estimation

Gardes and Girard (2010) developed a nearest-neighbor method and Gardes et al. (2010) developed a moving window approach for estimating the extreme conditional quantiles of heavy-tailed distributions. The main idea of the two methods is to first select observations in a neighborhood of $\mathbf{x}$ of interest, and then apply the univariate extreme value methods to the neighborhood data to estimate the conditional quantiles of Y given $\mathbf{x}$.

Suppose that the design points $\mathbf{x}_1, \ldots, \mathbf{x}_n$ are nonrandom. Let E be a metric space associated to a metric d. Assume that for all $\mathbf{x} \in E$, $F_Y(\cdot|\mathbf{x})$ is a heavy-tailed distribution with EVI $\xi(\mathbf{x}) > 0$. In the first step of the estimation, Gardes and Girard (2010) proposed to first select $m_{n,\mathbf{x}} = m_\mathbf{x}$ nearest covariates of $\mathbf{x}$ (with respect to the distance d), where $m_\mathbf{x}$ is a sequence of integers such that $1 < m_\mathbf{x} < n$. On the other hand, to accommodate functional covariates, Gardes et al. (2010) proposed to form the neighborhood covariates by including the $m_{n,\mathbf{x}}$ covariates that belong to the ball $B(\mathbf{x}, h_\mathbf{x}) = \{\mathbf{t} \in E, d(\mathbf{t}, \mathbf{x}) \leqslant h_\mathbf{x}\}$, where $h_\mathbf{x}$ is a positive sequence tending to zero as $n \to \infty$, and $m_\mathbf{x} = \sum_{i=1}^{n} I\{\mathbf{x}_i \in B(\mathbf{x}, h_\mathbf{x})\}$. Denote the covariates in the selected neighborhood by $\{\mathbf{x}_1^*, \ldots, \mathbf{x}_{m_\mathbf{x}}^*\}$, and the associated observations taken from $\{y_1, \ldots, y_n\}$ by $\{z_1, \ldots, z_{m_\mathbf{x}}\}$. In the second step, univariate extreme value methods are applied to the order statistics $z_{1,m_\mathbf{x}} \leqslant \cdots \leqslant z_{m_\mathbf{x},m_\mathbf{x}}$ to estimate the EVI $\xi(\mathbf{x})$ and $Q_Y(\tau_n|\mathbf{x})$. For instance, Gardes and Girard (2010) considered the following estimator of $\xi(\mathbf{x})$ based on weighted rescaled log-spacings:

$$\widehat{\xi}(\mathbf{x}; a, \lambda) = \sum_{i=1}^{k_\mathbf{x}} \{w(i/k_\mathbf{x}, a, \lambda) i (\log z_{m_\mathbf{x}-i+1,m_\mathbf{x}} - \log z_{m_\mathbf{x}-i,m_\mathbf{x}})\} / \sum_{i=1}^{k_\mathbf{x}} w(i/k_\mathbf{x}, a, \lambda),$$

where $k_{\mathbf{x}} = k_{n,\mathbf{x}}$ is a sequence of integers such that $1 \leqslant k_{\mathbf{x}} \leqslant m_{\mathbf{x}}$, and

$$w(s, a, \lambda) = \frac{\lambda^{-a}}{\Gamma(a)} s^{1/\lambda - 1} (-\log s)^{a-1}, \text{ for } s \in (0, 1), a \leqslant 1, 0 \leqslant \lambda \leqslant 1$$

is the density of log-gamma distribution defined in Consul and Jain (1971). The parameters (a, λ) in the weighting function determine the weights assigned to different extreme order statistics. A special case $a = \lambda = 1$ leads to the Hill estimator (Hill, 1975), and $(a, \lambda) = (2, 1)$ leads to the Zipf estimator (Kratz and Resnick, 1996; Schultze and Steinebach, 1996). Adopting the unconditional quantile estimator proposed by Weissman (1978), based on the local EVI estimator of $\xi(\mathbf{x})$, $Q_Y(\tau_n|\mathbf{x})$ can be estimated by

$$\widehat{Q}_Y(\tau_n|\mathbf{x}) = z_{m_{\mathbf{x}} - k_{\mathbf{x}} + 1, m_{\mathbf{x}}} \left\{ \frac{k_{\mathbf{x}}}{m_{\mathbf{x}}(1 - \tau_n)} \right\}^{\widehat{\xi}(\mathbf{x})},$$

which can be viewed as an extrapolation from the $(1 - k_{\mathbf{x}}/m_{\mathbf{x}})$-th conditional quantile of Y. Suppose that $k_{\mathbf{x}}$ is an intermediate sequence such that $k_{\mathbf{x}} \to \infty$ and $k_{\mathbf{x}}/m_{\mathbf{x}} \to 0$ as $n \to \infty$. Under the second-order condition and some regularity conditions, Gardes and Girard (2010) and Gardes et al. (2010) have established the asymptotic distribution of $\widehat{Q}_Y(\tau_n|\mathbf{x})$.

15.3.3.3 Kernel Estimation

Based on the kernel estimation of $F_Y(\cdot|\mathbf{x})$, Daouia et al. (2011) and Gardes and Girard (2011) proposed kernel-type estimators of the extreme conditional quantile $Q_Y(\tau_n|\mathbf{x})$ for heavy-tailed distributions. Assume that $F_Y(\cdot|\mathbf{x})$ belongs to the Fréchet maximum domain of attraction with EVI $\xi(\mathbf{x})$. For any $(\mathbf{x}, y) \in \mathbb{R}^p \times \mathbb{R}$, Daouia et al. (2011) defined the kernel estimator of $F_Y(y|\mathbf{x})$ as

$$\widehat{F}_Y(y|\mathbf{x}) = 1 - \left\{ \sum_{i=1}^n K_h(\mathbf{x} - \mathbf{x}_i) I(y_i > y) \right\} / \sum_{i=1}^n K_h(\mathbf{x} - \mathbf{x}_i),$$

where h is the bandwidth such that $h \to 0$ as $n \to \infty$, and $K_h(t) = K(t/h)/h^p$ with K being a p-dimensional kernel function. For any $\tau \in (0, 1)$, the kernel estimator of $Q_Y(\tau|\mathbf{x})$ is defined via the generalized inverse of $\widehat{F}_Y(\cdot|\mathbf{x})$:

$$\widehat{Q}_Y(\tau|\mathbf{x}) = \inf\{t : \widehat{F}_Y(t|\mathbf{x}) \geqslant \tau\}.$$

Daouia et al. (2011) showed that the kernel estimator $\widehat{Q}_Y(\tau_n|\mathbf{x})$ still has the asymptotic normality for intermediate quantiles such that $n(1 - \tau_n) > \{\log(n)\}^p$. However, for extreme order of quantiles, for instance $\tau_n \to 1$ at a rate faster than $1/n$, the kernel estimation is not feasible as it cannot extrapolate beyond the maximum observation in the ball centered at $\mathbf{x}$ with radius h. To overcome this difficulty, Daouia et al. (2011) proposed a Weissman-type estimator of the extreme conditional quantile $Q_Y(\tau_n|\mathbf{x})$:

$$\widehat{Q}_Y(\tau_n|\mathbf{x}) = \widehat{Q}_Y(\alpha_n|\mathbf{x}) \left(\frac{1 - \alpha_n}{1 - \tau_n} \right)^{\widehat{\xi}(\mathbf{x})},$$

where α_n is an intermediate quantile level, $(1 - \tau_n)/(1 - \alpha_n) \to 0$ as $n \to \infty$ and $\widehat{\xi}(\mathbf{x})$ is an estimator of the conditional EVI $\xi(\mathbf{x})$, for instance, a kernel version of the Hill estimator (Hill, 1975)

$$\widehat{\xi}(\mathbf{x}) = \sum_{j=1}^{J} \left(\log \left[\widehat{Q}_Y \{1 - w_j(1 - \alpha_n) | \mathbf{x} \} \right] - \log \left\{ \widehat{Q}_Y(\alpha_n | \mathbf{x}) \right\} \right) / \sum_{j=1}^{J} \log(1/w_j),$$

where $w_1 > w_2 > \ldots > w_J > 0$ is a decreasing sequence of weights and J is a positive integer. This extrapolation allows the estimation of extreme conditional quantile with $\tau_n \to 1$ arbitrarily fast.

The estimation procedure in Gardes and Girard (2011) is similar but the authors considered a different double-kernel estimator of $F_Y(y | \mathbf{x})$:

$$\widehat{F}_Y(y | \mathbf{x}) = 1 - \left[\sum_{i=1}^{n} K_h(\mathbf{x} - \mathbf{x}_i) G\{(y_i - y)/\lambda\} \right] / \sum_{i=1}^{n} K_h(\mathbf{x} - \mathbf{x}_i),$$

where $G(t) = \int_{-\infty}^{t} g(s) ds$ with $g(\cdot)$ being a univariate kernel function, and λ is the bandwidth parameter associated with $G(\cdot)$.

15.3.4 Quantile Regression Methods

Quantile regression, first introduced by Koenker and Bassett (1978), focuses on studying the impact of covariates on the quantiles of the response variable and thus provides a natural alternative to estimating conditional tail quantiles. Researchers have applied quantile regression for estimating tail quantiles in different areas of studies. For instance, Bremnes (2004a) and Bremnes (2004b) used a local quantile regression method to predict the conditional quantiles of precipitation and wind power given outputs from numerical weather prediction models. To account for zero precipitation, Friederichs and Hense (2007) applied a censored linear quantile regression method to estimate the high quantiles of precipitation conditional on the NCEP (National Centers for Environmental Prediction) reanalysis variables. Jagger and Elsner (2008) applied linear quantile regression to study the conditional quantiles of tropical cyclone wind speeds given climate variables. Taylor (2008) proposed an exponentially weighted quantile regression method to estimate the value at risk, which corresponds to the tail quantile of financial returns conditional on the current information. In the above work, conventional parametric or nonparametric quantile regression was directly applied even when the interests are at the extreme tails. However, due to data sparsity, direct estimation from quantile regression is often unstable or infeasible at the extreme tails.

To estimate extreme conditional quantiles in the very far tails with few or no observations available, additional conditions or models for the tails are needed. Chernozhukov and Du (2008), Wang et al. (2012), and Wang and Li (2013) proposed new estimating methods for extreme conditional quantiles that combine linear quantile regression and extreme value theory.

Let $0 < \tau_L < 1$ be a fixed constant that is close to one. Consider the following linear quantile regression model:

$$Q_Y(\tau|\mathbf{x}) = \alpha(\tau) + \mathbf{x}^T\beta(\tau), \ \tau \in [\tau_L, 1], \tag{15.7}$$

where $\alpha(\tau) \in \mathbb{R}$ and $\beta(\tau) \in \mathbb{R}^p$ are the unknown quantile coefficients. Given the random sample $\{(y_i, \mathbf{x}_i), i = 1, \ldots, n\}$, the quantile coefficients can be estimated by

$$(\widehat{\alpha}(\tau), \widehat{\beta}(\tau)) = \underset{\alpha, \beta}{\operatorname{argmin}} \sum_{i=1}^n \rho_\tau(y_i - \alpha - \mathbf{x}_i^T\beta), \tag{15.8}$$

where $\rho_\tau(u) = \{\tau - I(u < 0)\}u$ is the quantile loss function.

At the extreme quantiles such that $\tau_n \to 1$ as $n \to \infty$, the conventional quantile regression estimators $\widehat{\alpha}(\tau)$ and $\widehat{\beta}(\tau)$ are often not precise due to data sparsity. The basic idea of the estimation methods in Chernozhukov and Du (2008), Wang et al. (2012), and Wang and Li (2013) is to first estimate less extreme quantiles through conventional quantile regression, and then extrapolate these quantile estimates to the high end based on different assumptions on the tail behavior of the conditional response distribution. We will focus on the estimation for heavy-tailed distributions.

15.3.4.1   Estimation Based on the Common-Slope Assumption

We first consider a common-slope assumption, which assumes that the quantile slope coefficient $\beta(\tau)$ in model (15.7) is constant in the upper quantiles, that is, $\beta(\tau) = \beta$ for $\tau \in [\tau_L, 1]$. In addition, assume that $F_Y(\cdot|\mathbf{x} = 0)$ belongs to the maximum domain of attraction with extreme value index $\xi > 0$. Let $\widehat{e}_i = y_i - \mathbf{x}_i^T\widehat{\beta}$, $i = 1, \ldots, n$, where $\widehat{\beta}$ is a consistent estimator of β. For instance, we can take $\widehat{\beta} = \widehat{\beta}(\tau_L)$ or the composite estimator proposed in Koenker (1984) and Zou and Yuan (2008), which is obtained by pooling information across a sequence of quantiles $\tau_L = \tau_1 < \ldots < \tau_l = \tau_U$ with $\tau_L < \tau_U < 1$ and $l \geqslant 1$. Let $\widehat{e}_{1,n} \leqslant \cdots \leqslant \widehat{e}_{n,n}$ be the order statistics of $\{\widehat{e}_1, \ldots, \widehat{e}_n\}$. Wang et al. (2012) showed that the upper order statistics of $\widehat{e}_i$ are asymptotically equivalent to those of $Q_Y(u_i|\mathbf{x} = 0)$, where $\{u_1, \ldots, u_n\}$ is a random sample from $U(0, 1)$. Therefore, the order statistics of $\{\widehat{e}_1, \ldots, \widehat{e}_n\}$ can be used to estimate the EVI ξ by existing estimating methods, for instance, the Hill estimator (Hill, 1975),

$$\widehat{\xi} = \frac{1}{k} \sum_{j=1}^k \log \frac{\widehat{e}_{n-j+1,n}}{\widehat{e}_{n-k,n}},$$

where k is an integer such that $k = k_n \to \infty$ and $k/n \to 0$ as $n \to \infty$. A Weissman-type extrapolation estimator for $Q_Y(\tau_n|\mathbf{x})$ can be constructed by

$$\widehat{Q}_Y(\tau_n|\mathbf{x}) = \mathbf{x}^T\widehat{\beta} + \left(\frac{k/n}{1 - \tau_n}\right)^{\widehat{\xi}} \widehat{e}_{n-k,n},$$

where $1 - \tau_n = o(k/n)$.

15.3.4.2 Estimation without the Common-Slope Assumption

We next discuss an estimation method proposed by Chernozhukov and Du (2008) based on a more relaxed assumption that allows the quantile slope coefficient $\beta(\tau)$ in model (15.7) to vary across τ. In addition to model (15.7), assume that after being transformed by some auxiliary regression line, the response variable Y has regularly varying tails with EVI $\xi > 0$. More specifically, suppose that there exists an auxiliary slope β_e such that the following tail-equivalence relationship holds as $\tau \to 1$,

$$Q_Y(\tau|\mathbf{x}) - \mathbf{x}^T \beta_e \sim F_0^{\leftarrow}(\tau), \text{ uniformly in } \mathbf{x}, \tag{15.9}$$

where $F_0(\cdot)$ is a distribution that belongs to the maximum domain of attraction with EVI $\xi > 0$. The tail-equivalence condition (15.9) implies that the covariate $\mathbf{x}$ affects the extreme quantiles of Y through β_e approximately.

Under model (15.7) and the tail-equivalence condition (15.9), Chernozhukov (2005) showed that for intermediate order sequences $\tau_n \to 1$ and $n(1 - \tau_n) \to \infty$, $a_n\{\widehat{\boldsymbol{\theta}}(\tau) - \boldsymbol{\theta}(\tau)\}$ converges to a normal distribution with mean zero, where $\boldsymbol{\theta}(\tau) = (\alpha(\tau), \beta(\tau)^T)^T$, $\widehat{\boldsymbol{\theta}}(\tau) = (\widehat{\alpha}(\tau), \widehat{\beta}(\tau)^T)^T$ and $a_n = \{(1 - \tau_n)n\} / [(1, E(\mathbf{X})^T)^T \{\boldsymbol{\theta}(\tau_n) - \boldsymbol{\theta}(1 - m(1 - \tau_n))\}]$ with $m > 1$. This suggests that we can estimate the intermediate conditional quantiles by conventional quantile regression, and then extrapolate these estimates to the high tail to estimate extreme conditional quantiles. With this idea, Chernozhukov and Du (2008) proposed to estimate the EVI ξ by the Hill estimator

$$\widehat{\xi} = \{n(1 - \tau_{0n})\}^{-1} \sum_{i=1}^{n} \log \left(\frac{y_i}{\widehat{\alpha}(\tau_{0n}) + \mathbf{x}_i^T \widehat{\beta}(\tau_{0n})} \right)_+,$$

where $\log(u)_+ = \log(u)I(u > 0)$, $\tau_{0n} \to 1$ and $n(1 - \tau_{0n}) \to \infty$. For $1 - \tau_n = o(1 - \tau_{0n})$, the Weissman-type extrapolation estimator of $Q_Y(\tau_n|\mathbf{x})$ thus can be constructed as

$$\widehat{Q}_Y(\tau_n|\mathbf{x}) = \{\widehat{\alpha}(\tau_{0n}) + \mathbf{x}^T \widehat{\beta}(\tau_{0n})\} \left(\frac{1 - \tau_{0n}}{1 - \tau_n} \right)^{\widehat{\xi}}. \tag{15.10}$$

15.3.4.3 Three-Stage Estimation

The methods in Chernozhukov and Du (2008) and Wang et al. (2012) are based on two main assumptions: (1) the conditional quantiles of Y are linear in $\mathbf{x}$ at the upper quantiles; (2) the conditional distribution $F_Y(\cdot|\mathbf{x})$ is tail equivalent across $\mathbf{x}$ with a common EVI ξ. In many applications, the covariate may affect the heaviness of the tail distribution of Y and thus the EVI $\xi(\mathbf{x})$ is dependent on $\mathbf{x}$. It would be interesting to construct a covariate-dependent EVI estimator while still being able to adopt linear quantile regression to borrow information across multidimensional covariates. However, Proposition 2.1 in Wang and Li (2013) suggests that in situations where the EVI $\xi(\mathbf{x})$ varies with $\mathbf{x}$, it is rarely the case that the conditional high quantiles of Y are still linear in $\mathbf{x}$. This result suggests that to accommodate covariate-dependent EVI, we have to consider nonparametric quantile regression, which, however, is known to be unstable at tails in finite samples especially when the dimension of $\mathbf{x}$ is high. Wang and Li (2013) showed that in some cases with covariate-dependent EVI, the quantiles of Y may still be linear in $\mathbf{x}$ after some appropriate transformation such as log transformation. Motivated by this, Wang and Li (2013) considered a power-transformed quantile regression model:

$$Q_{\Lambda_\lambda(Y)}(\tau|\mathbf{x}) = \mathbf{z}^T\boldsymbol{\theta}(\tau), \tau \in [\tau_L, 1], \tag{15.11}$$

where $\mathbf{z} = (1, \mathbf{x}^T)^T$ and $\Lambda_\lambda(y) = \{(y^\lambda - 1)/\lambda\}I(\lambda \neq 0) + \log(y)I(\lambda = 0)$ denotes the family of power transformations (Box and Cox, 1964).

For estimating the extreme conditional quantiles of Y, Wang and Li (2013) proposed a three-stage estimating procedure. In the first stage, the power transformation parameter λ is estimated by

$$\widehat{\lambda} = \operatorname*{argmin}_{\lambda \in \mathbb{R}} \sum_{i=1}^{n} \{R_n(\mathbf{x}_i, \lambda; \tau_L)\}^2, \tag{15.12}$$

where $R_n(\mathbf{t}, \lambda; \tau) = n^{-1}\sum_{j=1}^{n} I(\mathbf{x}_j \leqslant \mathbf{t})\left[\tau - I\{\Lambda_\lambda(y_j) - \mathbf{z}_j^T\widehat{\boldsymbol{\theta}}(\tau; \lambda) \leqslant 0\}\right]$ is a residual cusum process that is often used in lack-of-fit tests, $\mathbf{x} \leqslant \mathbf{t}$ means that each component of $\mathbf{x}$ is less than or equal to the corresponding component of $\mathbf{t} \in \mathbb{R}^p$, and $\widehat{\boldsymbol{\theta}}(\tau; \lambda) = \operatorname{argmin}_{\boldsymbol{\theta}} \sum_{i=1}^{n} \rho_\tau \{\Lambda_\lambda(y_i) - \mathbf{z}_i^T\boldsymbol{\theta}\}$.

In the second stage, the conditional quantiles of Y at a sequence of intermediate quantile levels are estimated by first fitting model (15.11) on the transformed scale and then transforming the estimates back to the original scale. Specifically, define a sequence of quantile levels $\tau_L < \tau_{n-k} < \ldots < \tau_m \in (0, 1)$, where $k = k_n \to \infty$ and $k/n \to 0$, $m = n - [n^\eta]$ with $\eta \in (0, 1)$ as some small constant satisfying $n^\eta < k$, and $\tau_j = j/(n+1)$. The trimming parameter η is introduced for technical purposes, more specifically, to obtain the Bahadur representation of $\widehat{\boldsymbol{\theta}}(\tau_m; \lambda)$. In practice we can choose $\eta = 0.1$. For each $j = n - k, \ldots, m$, $Q_{\Lambda_\lambda(Y)}(\tau_j|\mathbf{x})$ can be estimated by $\mathbf{z}^T\widehat{\boldsymbol{\theta}}(\tau_j; \widehat{\lambda})$. By the equivariance property of quantiles to monotone transformations, we can estimate $Q_Y(\tau_j|\mathbf{x})$ by $\widehat{Q}_Y(\tau_j|\mathbf{x}) = \Lambda_{\widehat{\lambda}}^{-1}\left\{\mathbf{z}^T\widehat{\boldsymbol{\theta}}(\tau_j; \widehat{\lambda})\right\}$. For a given $\mathbf{x}$, $\{\widehat{Q}_Y(\tau_j|\mathbf{x}), j = n - k, \ldots, m\}$ can be roughly regarded as the extreme order statistics of a sample from $F_Y(\cdot|\mathbf{x})$.

In the third stage, extrapolation from the intermediate quantile estimates is performed to estimate $Q_Y(\tau_n|\mathbf{x})$ with $1 - \tau_n = o(k/n)$ as

$$\widehat{Q}_Y(\tau_n|\mathbf{x}) = \widehat{Q}_Y(\tau_{n-k}|\mathbf{x})\left(\frac{1 - \tau_{n-k}}{1 - \tau_n}\right)^{\widehat{\xi}(\mathbf{x})}, \tag{15.13}$$

where $\widehat{\xi}(\mathbf{x}) = (k - [n^\eta])^{-1}\sum_{j=[n^\eta]}^{k} \log\left\{\widehat{Q}_Y(\tau_{n-j}|\mathbf{x})/\widehat{Q}_Y(\tau_{n-k}|\mathbf{x})\right\}$ is the Hill estimator based on the pseudo order statistics of a sample from $F_Y(\cdot|\mathbf{x})$.

The method in Wang and Li (2013) allows the EVI $\xi(\mathbf{x})$ to depend on $\mathbf{x}$ and thus provides more flexibility. However, due to lack of information, the covariate-dependent EVI estimator could be unstable in regions where $\mathbf{x}$ is sparse. In situations where $\xi(\mathbf{x})$ is constant across $\mathbf{x}$ (or in some region of $\mathbf{x}$), we can estimate the common ξ by the pooled estimator $\widehat{\xi}_p = n^{-1} \sum_{i=1}^{n} \widehat{\xi}(\mathbf{x}_i)$. Numerical studies in Wang and Li (2013) showed that the pooled EVI estimator often leads to more stable and efficient estimation of the extreme conditional quantiles when $\xi(\mathbf{x})$ is indeed constant or varies little across $\mathbf{x}$. To identify the commonality of the EVI, Wang and Li (2013) proposed a test statistic $T_n = n^{-1} \sum_{i=1}^{n} \{\widehat{\xi}(\mathbf{x}_i) - \widehat{\xi}_p\}^2$ and established the asymptotic distribution of T_n under $H_0 : \xi(\mathbf{x}) = \xi$ for all $\mathbf{x}$ in its support. Suppose that $E(\mathbf{X}) = \mathbf{0}_p$. For two special cases: (1) homogenous case such as the location-shift model (Koenker, 2005); (2) the EVI of $\Lambda_\lambda(Y)$ is $\xi^* = 0$, it was shown that $kT_n \overset{d}{\to} \xi^2 \chi^2(p-1)$ under H_0, so the test can be easily carried out.

The quantile-regression-based methods discussed in this section can be regarded as semiparametric methods since they make no parametric distributional assumptions but assume that the conditional upper quantiles of the response (or some transformation thereof) are linear in covariates. This quantile linearity assumption allows us to model the effect of covariates $\mathbf{x}$ across the entire range of $\mathbf{x}$ and thus borrow information across $\mathbf{x}$ to estimate the extreme conditional quantiles of the response, and to avoid the curse of dimensionality issue faced by the nonparametric methods.

15.4 Numerical Comparison

We carry out a simulation study to compare different methods for estimating extremely high conditional quantiles. The data are generated from the following four different models.

- Model 1: $y_i = 2 + 2x_{i1} + 2x_{i2} + (2 + 1.6x_{i1})\epsilon_i$, $\epsilon_i \sim$ Pareto(0.5), $i = 1, \ldots, n$.
- Model 2: $y_i | x_i \sim$ Pareto with $\xi(\mathbf{x}_i) = \exp(-1 + x_i)$, $i = 1, \ldots, n$.
- Model 3: $\log(y_i) = 2 + x_{i1} + x_{i2} + (0.5 + 0.25x_{i1})\epsilon_i$, and ϵ_i are i.i.d. random variables with quantile function $Q(\tau) = \tau - 1 - \log(1 - \tau)$ for $\tau \in (0, 1)$, $i = 1, \ldots, n$. In this case, the conditional distribution of Y is in the domain of attraction with EVI $\xi(x_{i1}, x_{i2}) = 0.5 + 0.25x_{i1}$.
- Model 4: $y_i | x_i \sim$ Fréchet distribution with distribution function $F_Y(y | x_i) = \exp\{-y^{-1/\xi(x_i)}\}$, where $\xi(x) = 1/2[1/10 + \sin\{\pi(x + 1)/2\}]\{11/10 - 1/2\exp(-16x^2)\}$, $i = 1, \ldots, n$.

In the four models, $x_{i1}, x_{i2}, x_i, i = 1, \ldots, n$, are independent random variables from Uniform$(-1, 1)$. The sample size is set as $n = 2000$. For each model, the simulation is repeated 500 times.

We compare five estimators: (1) the parametric method assuming GPD for the exceedances with scale $\sigma(\mathbf{x}) = \exp(\mathbf{x}^T \boldsymbol{\beta})$ and shape $\xi(\mathbf{x}) = \exp(\mathbf{x}^T \boldsymbol{\theta})$; (2) the semiparametric tail index regression (TIR) method of Wang and Tsai (2009); (3) the nonparametric kernel method (KER) of Daouia et al. (2011); (4) the quantile-

TABLE 15.1

The integrated bias (IBias) and root integrated mean squared error (RIMSE) of different estimators of $Q_Y(\tau_n|\mathbf{x}_i)$ at $\tau_n = 0.99$ and 0.995. Values in the parentheses are the standard errors. The results of KER and 3Stage are taken from Tables 1–2 of Wang and Li (2013).

Method	IBias		RIMSE	
	$\tau_n = 0.99$	$\tau_n = 0.995$	$\tau_n = 0.99$	$\tau_n = 0.995$
	Model 1 ($p = 2$, constant EVI)			
GPD	0.27 (0.16)	0.75 (0.42)	3.53 (4.47)	9.33 (9.33)
TIR	-0.72 (0.07)	-1.87 (0.14)	1.78 (0.50)	3.63 (1.04)
KER	0.77 (0.55)	2.08 (2.74)	12.23 (15.06)	61.29 (161.39)
CD	0.04 (0.06)	-0.26 (0.10)	1.28 (0.37)	2.30 (0.66)
3Stage	-0.52 (0.07)	-1.62 (0.13)	1.75 (0.53)	3.26 (0.87)
	Model 2 ($p = 1$, Pareto distribution)			
GPD	-2.63 (0.53)	-5.65 (1.02)	12.09 (5.19)	23.45 (10.38)
TIR	-2.60 (0.58)	-5.90 (1.09)	13.25 (5.17)	25.02 (10.13)
KER	1.66 (0.81)	3.97 (2.32)	18.19 (13.35)	52.04 (58.19)
CD	-3.66 (0.74)	-9.05 (1.52)	16.97 (6.37)	35.15 (13.48)
3Stage	-1.67 (0.42)	-3.45 (1.11)	9.58 (9.74)	24.96 (29.69)
	Model 3 ($p = 2$, EVI linear in $\mathbf{x}$)			
GPD	1.52 (4.89)	16.29 (11.84)	109.34 (83.00)	265.27 (171.47)
TIR	6.13 (6.53)	17.07 (9.85)	146.16 (60.98)	221.01 (94.47)
KER	18.31 (15.20)	74.73 (85.72)	340.28 (483.88)	1918.17 (4441.34)
CD	-15.55 (8.09)	-43.81 (13.89)	181.47 (75.09)	313.72 (131.05)
3Stage	-14.01 (3.82)	-41.09 (7.90)	86.47 (90.71)	181.27 (203.67)
	Model 4 ($p = 1$, Fréchet distribution)			
GPD	0.77 (0.16)	1.65 (0.31)	3.57 (1.61)	7.04 (3.63)
TIR	0.48 (0.12)	0.83 (0.18)	2.75 (0.61)	4.10 (0.90)
KER	0.26 (0.08)	0.67 (0.19)	1.90 (0.85)	4.22 (2.48)
CD	0.46 (0.12)	0.78 (0.18)	2.74 (0.60)	4.12 (0.90)
3Stage	0.13 (0.12)	0.07 (0.19)	2.79 (0.63)	4.27 (1.01)

regression-based method of Chernozhukov and Du (2008), denoted by CD; (5) the three-stage estimator (3Stage) of Wang and Li (2013).

The extreme value index $\xi(\mathbf{x})$ is a constant in Model 1, while it depends on the covariates in different ways in Models 2–4. The CD method is based on the assumption of linear conditional quantiles of Y, which is satisfied in Model 1 but violated in Models 2–4. Since the conditional quantiles of Y are linear in $\mathbf{x}$ after log transformation in Model 3, the model assumption required by the 3Stage method is satisfied in Models 1 and 3 but violated in Models 2 and 4. Both the GPD and TIR methods assume that $\log\{\xi(\mathbf{x})\}$ is linear in $\mathbf{x}$, and this assumption is satisfied only in Models 1 and 2. The KER method is most flexible and it works in all four cases.

Table 15.1 summarizes the performance of different estimators for estimating $Q_Y(\tau_n|\mathbf{x})$ at $\tau_n = 0.99$ and 0.995. The IBias is the integrated bias defined as the average of $n^{-1}\sum_{i=1}^{n}\{\widehat{Q}_Y(\tau_n|\mathbf{x}_i) - Q_Y(\tau_n|\mathbf{x}_i)\}$, and RIMSE is the root integrated mean squared error defined as the square root of the average of $n^{-1}\sum_{i=1}^{n}\{\widehat{Q}_Y(\tau_n|\mathbf{x}_i) - Q_Y(\tau_n|\mathbf{x}_i)\}^2$ across 500 simulated data. The GPD and TIR methods rely on the correct specification of the EVI function; they perform competitively well in Model 2 when $\xi(\mathbf{x})$ is correctly specified but they are slightly less efficient than the 3Stage method in Model 3 when the function form is misspecified. In Model 1 with a constant EVI, the GPD and TIR methods underperform the CD and 3Stage methods due to the noise involved in estimating the zero parameters in $\xi(\mathbf{x}) = \mathbf{x}^T\boldsymbol{\theta}$. The CD estimator is slightly more efficient than 3Stage in Model 1 when the conditional quantiles of untransformed Y are linear in covariates but the latter is more flexible and performs competitively well or slightly better in the other three models even in Models 2 and 4 where the power-transformation model (15.11) is violated. As observed in Wang and Li (2013), the nonparametric method KER is the most flexible and can capture complicated dependence of EVI on the covariates for instance as in Model 4. However, the KER method gives unstable estimation especially in Models 1 and 3 with two predictors due to the data sparsity and curse of dimensionality.

15.5 Final Remarks

Estimation of conditional extreme quantiles has drawn much attention in recent years. We surveyed and compared various types of estimation methods based on different model assumptions. As for most statistical problems, there is a tradeoff between the model flexibility and stability. The nonparametric methods are more flexible but are subject to the curse of dimensionality and reduced effective sample size with local estimation. The parametric methods are sensitive to the misspecification of models. Semiparametric methods aim to achieve a better balance between model flexibility and parsimony. Regarded as also semiparametric, the quantile-regression-based methods discussed in Section 15.3.4 are relatively newer to the extreme value literature but they serve as useful alternative tools in cases where the conditional tail quantiles of the response after some transformation appear to be linear in covariates. In practice, we would suggest first use nonparametric methods as exploratory tools to examine the dependence of extreme value index and the conditional tail quantiles on the covariates, which may help identify a reasonable parametric or semiparametric model to carry out analysis with less variability.

Acknowledgments

The research of Wang is partially supported by the National Science Foundation CAREER award DMS-1149355. The research of Li is partially supported by the National Natural Science Foundation of China grant 11171074.

References

Abrevaya, J. (2001). The effect of demographics and maternal behavior on the distribution of birth outcomes. *Empirical Economics 26*, 247–259.

Beirlant, J. and Y. Goegebeur (2003). Regression with response distributions of Pareto-type. *Computational Statistics and Data Analysis 42*, 595–619.

Beirlant, J. and Y. Goegebeur (2004). Local polynomial maximum likelihood estimation for Pareto-type distributions. *Journal of Multivariate Analysis 89*, 97–118.

Box, G. E. P. and D. R. Cox (1964). An analysis of transformations. *Journal of the Royal Statistical Society, Series B 26*, 211–252.

Bremnes, J. B. (2004a). Probabilistic forecasts of precipitation in terms of quantiles using NWP model output. *Monthly Weather Review 132*, 338–347.

Bremnes, J. B. (2004b). Probabilistic wind power forecasts using local quantile regression. *Wind Energy 7*, 47–54.

Chavez-Demoulin, V. and A. C. Davison (2005). Generalized additive modelling of sample extremes. *Journal of the Royal Statistical Society, Series C 54*, 207–222.

Chernozhukov, V. (2005). Extremal quantile regression. *Annals of Statistics 33*, 806–839.

Chernozhukov, V. and S. Du (2008). Extremal quantiles and value-at-risk. In S. N. Durlauf and L. E. Blume (Eds.), *The New Palgrave Dictionary of Economics*. Basingstoke: Palgrave Macmillan.

Coles, S. (2001). *An Introduction to Statistical Modeling of Extreme Values*. Springer.

Consul, P. C. and G. C. Jain (1971). On the log-gamma distribution and its properties. *Statistische Hefte 12*, 100–106.

Daouia, A., L. Gardes, S. Girard, and A. Lekina (2011). Kernel estimators of extreme level curves. *Test 20*, 311–333.

Davison, A. C. and N. I. Ramesh (2000). Local likelihood smoothing of sample extremes. *Journal of the Royal Statistical Society, Series B, 62*, 191–208.

Davison, A. C. and R. L. Smith (1990). Models for exceedances over high thresholds. *Journal of the Royal Statistical Society. Series B 52*, 393–442.

de Haan, L. and A. Ferreira (2006). *Extreme Value Theory: An Introduction*. Springer.

de Haan, L. and H. Rootzén (1993). On the estimation of high quantiles. *Journal of Statistical Planning and Inference 35*, 1–13.

Dekkers, A., J. Einmahl, and L. de Haan (1989). A moment estimator for the index of an extreme-value distribution. *Annals of Statistics 17*, 1833–1855.

Diebolt, J., L. Gardes, S. Girard, and A. Guillou (2008). Bias-reduced extreme quantiles estimators of weibull distributions. *Journal of Statistical Planning and Inference 138*, 1389–1401.

Drees, H. (2003). Extreme quantile estimation for dependent data with applications to finance. *Bernoulli 9*, 617–657.

Drees, H., A. Ferreira, and L. de Haan (2004). On maximum likelihood estimation of the extreme value index. *Annals of Apllied Probability 14*, 1179–1201.

Engle, R. F. and S. Manganelli (2004). CAViaR: Conditional autoregressive value at risk by regression quantiles. *Journal of Business and Economic Statistics 22*, 367–381.

Fisher, R. A. and L. H. C. Tippett (1928). Limiting forms of the frequency distribution in the largest particle size and smallest number of a sample. *Proceedings of the Cambridge Philosophical Society 24*, 180–190.

Friederichs, P. (2010). Statistical downscaling of extreme precipitation events using extreme value theory. *Extremes 13*, 109–132.

Friederichs, P. and A. Hense (2007). Statistical downscaling of extreme precipitation events using censored quantile regression. *Monthly Weather Review 135*, 2365–2378.

Friederichs, P. and T. L. Thorarinsdottir (2012). Forecast verification for extreme value distributions with an application to probabilistic peak wind prediction. *Environmetrics 23*, 579–594.

Gardes, L. and S. Girard (2005). Estimating extreme quantiles of weibull tail-distributions. *Communications in Statistics–Theory and Methods 34*, 1065–1080.

Gardes, L. and S. Girard (2010). Conditional extremes from heavy-tailed distributions: An application to the estimation of extreme rainfall return levels. *Extremes 13*, 177–204.

Gardes, L. and S. Girard (2011). Functional kernel estimators of conditional extreme quantiles. In *Recent Advances in Functional Data Analysis and Related Topics Contributions to Statistics*, pp. 135–140. Springer.

Gardes, L., S. Girard, and A. Lekina (2010). Functional nonparametric estimation of conditional extreme quantiles. *Journal of Multivariate Analysis 101*, 419–433.

Gomes, M. and F. Figueiredo (2006). Bias reduction in risk modelling: Semiparametric quantile estimation. *Test 15*, 375–396.

Gomes, M. I. and D. Pestana (2007). A sturdy reduced-bias extreme quantile (VaR) estimator. *Journal of the American Statistical Association 102*, 280–292.

Hall, P. and N. Tajvidi (2000). Nonparametric analysis of temporal trend when fitting parametric models to extreme-value data. *Statistical Science 15*, 153–167.

Hill, B. M. (1975). A simple general approach to inference about the tail of a distribution. *Annals of Statistics 3*, 1163–1174.

Hosking, J. R. M. and J. R. Wallis (1987). Parameter and quantile estimation for the generalized Pareto distribution. *Technometrics 339*, 339–349.

Jagger, T. H. and J. B. Elsner (2008). Modeling tropical cyclone intensity with quantile regression. *International Journal of Climatology 29*, 1351–1361.

Koenker, R. (1984). A note on l-estimates for linear models. *Statistics and Probability Letters 2*, 323–325.

Koenker, R. (2005). *Quantile Regression*. Cambridge University Press, Cambridge.

Koenker, R. and G. Bassett (1978). Regression quantiles. *Econometrica 46*, 33–50.

Kratz, M. and S. Resnick (1996). The QQ-estimator and heavy tails. *Stochastic Models 12*, 699–724.

Li, D., L. Peng, and J. Yang (2010). Bias reduction for high quantiles. *Journal of Statistical Planning and Inference 140*, 2433–2441.

Pickands, J. (1975). Statistical inference using extreme order statistics. *Annals of Statistics 3*, 119–131.

Portnoy, S. and Jurečková (1999). On extreme regression quantiles. *Extremes 2*, 227–243.

Sang, H. and A. E. Gelfand (2009). Hierarchical modeling for extreme values observed over space and time. *Environmental and Ecological Statistics 16*, 407–426.

Schultze, J. and J. Steinebach (1996). On least squares estimates of an exponential tail coefficient. *Statistics and Decisions 14*, 353–372.

Smith, R. L. (1985). Maximum likelihood estimation in a class of nonregular cases. *Biometrika 72*, 67–92.

Smith, R. L. (1989). Extreme value analysis of environmental time series: an application to trend detection in ground-level ozone. *Statistical Science 4*, 367–377.

Smith, R. L. (1994). Nonregular regression. *Biometrika 81*, 173–183.

Taylor, J. W. (2008). Using exponentially weighted quantile regression to estimate value at risk and expected shortfall. *Journal of Financial Econometrics 6*, 382–406.

Wang, H. and D. Li (2013). Estimation of extreme conditional quantiles through power trnsformation. *Journal of the American Statistical Association 108*, 1062–1074.

Wang, H., D. Li, and X. He (2012). Estimation of high conditional quantiles for heavy-tailed distributions. *Journal of the American Statistical Association 107*, 1453–1464.

Wang, H. and C. L. Tsai (2009). Tail index regression. *Journal of the American Statistical Association 104*, 1233–1240.

Weissman, I. (1978). Estimation of parameters and large quantiles based on the k largest observations. *Journal of the American Statistical Association 73*, 812–815.

Zou, H. and M. Yuan (2008). Composite quantile regression and the oracle model selection theory. *The Annals of Statistics 36*, 1108–1126.

16

Extreme Dependence Models

Boris Beranger

Pierre and Marie Curie University, France
University of New South Wales, Australia

Simone Padoan

Bocconi University, Italy

Abstract

Extreme values of real phenomena are events that occur with low frequency, but can have a large impact on real life. These are, in many practical problems, high dimensional by nature (e.g., Coles and Tawn, 1991; Tawn, 1990). To study these events is of fundamental importance. For this purpose, probabilistic models and statistical methods are in high demand. There are several approaches to modelling multivariate extremes as described in Falk et al. (2011), linked to some extent. We describe an approach for deriving multivariate extreme value models and we illustrate the main features of some flexible extremal dependence models. We compare them by showing their utility with a real data application, in particular analyzing the extremal dependence among several pollutants recorded in the city of Leeds, UK.

16.1 Introduction

Statistical analyses of extreme events are of crucial importance for risk assessment in many areas such as the financial market, telecommunications, industry, environment, and health. For example governments and insurance companies need to statistically quantify the frequency of natural disasters in order to plan risk management and take preventive actions.

Several examples of univariate analysis are available, for instance in Coles (2001). Two main approaches are used in applications, the block-maximum and the peak over a threshold. These are based on the generalized extreme value (GEV) distribution and the generalized Pareto distribution (GPD), which are milestones of the extreme value theory, see e.g., Coles (2001, Ch. 3–4) and the references therein.

Many practical problems in finance, the environment, etc., are high dimensional by nature, for example when analyzing the air quality in an area, the amount of pollution depends on the levels of different pollutants and the interaction between them. Today the extreme value theory provides a sufficiently mature framework for applications in the multivariate case. Indeed a large number of theoretical results and statistical methods and models are available; see for instance the monographs Resnick (2007), de Haan and Ferreira (2006), Falk et al. (2011), Beirlant et al. (2004), Coles (2001), and Kotz and Nadarajah (2000). In this article we review some basic theoretical results on the extreme values of multivariate variables (multivariate extremes for brevity). With the block-maximum approach we explain what type of dependence structures can be described. We discuss the main features of some families of parametric extremal dependence models. By means of real data analysis we show the utility of these extremal dependence models when assessing the dependence of multivariate extremes. Their utility is also illustrated when estimating the probabilities that multivariate extreme events occur.

The analysis of real phenomena such as heavy rainfall, heat waves, and so on is a challenging task. The first difficulty is the complexity of the data, i.e., observations are collected over space and time. In this case, theory deals with extremes of temporal- or spatial-processes (e.g., de Haan and Ferreira, 2006, Ch. 9). Examples of such statistical analysis are Davison et al. (2012), Davison and Gholamrezaee (2012), and for a simple review see Padoan (2013c). This theory is closely linked to that of multivariate extremes presented here. The second difficulty is that the dependence of multivariate extremes is not always well captured by the models illustrated here. Ledford and Tawn (1996, 1997) have shown that in some applications a more suitable dependence structure is described by the so-called *asymptotic independence*. This framework has been recently extended to continuous processes (e.g., De Haan and Zhou, 2011; Padoan, 2013a; Wadsworth and Tawn, 2012). These motivations make the multivariate extreme value theory a very active research field at present.

The chapter is organized as follows. In Section 16.2 a definition of multivariate extremes is provided and the main characteristics are presented. In Section 16.3 some of the most popular extremal dependence models are described. In Section 16.4 some estimation methods are discussed and in Section 16.5 the analysis of the extremes of multiple pollutants is performed.

16.2   Multivariate Extremes

Applying the block-maximum approach to every component of a multivariate random vector gives rise to a definition of multivariate extremes. Specifically, for $d \in \mathbb{N}$, let $I = \{1, \ldots, d\}$ be an index set and $X = (X_1, \ldots, X_d)$ be an $\mathbb{R}^d$-valued random vector with joint (probability) distribution function F and marginal distribution functions $F_j = F(\infty, \ldots, x_j, \ldots, \infty)$, $j \in I$. Suppose that $X_1, \ldots, X_n$ are n independent and identically distributed (i.i.d.) copies of X. The sample vector of componentwise maxima (sample maxima for brevity) is $M_n = (M_{n,1}, \ldots, M_{n,d})$, where $M_{n,j} = \max(X_{1,j}, \ldots, X_{n,j})$.

Typically, in applications the distribution F is unknown and so the distribution of the sample maxima is also unknown. A possible solution is to study the asymptotic distribution of M_n as $n \to \infty$ and to use it as an approximation for a large but finite sample size, resulting in an approximate distribution for multivariate extremes. At first glance, this notion of multivariate extremes may seem too simple to provide a useful approach for applications. However, a number of theoretical results justify its practical use. For example, with this definition of multivariate extremes, the dependence that arises is linked to the dependence that all the components of X are simultaneously large. Thus, by estimating these dependence structures we are also able to estimate the probabilities that multiple exceedances occur.

16.2.1 Multivariate Extreme Value Distributions

The asymptotic distribution of M_n is derived with a similar approach to the univariate case. Assume there are sequences of normalizing constants $a_n = (a_{n1}, \ldots, a_{nd}) > 0$, with $0 = (0, \ldots, 0)$, and $b_n = (b_{n1}, \ldots, b_{nd}) \in \mathbb{R}^d$ such that

$$\mathrm{pr}\left(\frac{M_n - b_n}{a_n} \leqslant x \right) = F^n(a_n x + b_n) \to G(x), \qquad n \to \infty, \qquad (16.1)$$

for all the continuity points x of a nondegenerate distribution G. The class of the limiting distributions in (16.1) is called *multivariate extreme value distributions* (MEVDs) (Resnick, 2007, pp. 263). A distribution function F that satisfies the convergence result (16.1) is said to be in the *(maximum) domain of attraction* of G (de Haan and Ferreira, 2006, pp. 226–229). An attractive property of MEVDs is the *max-stability*. A distribution G on $\mathbb{R}^d$ is max-stable if for every $n \in \mathbb{N}$, there exists sequences $a_n > 0$ and $b_n \in \mathbb{R}^d$ such that

$$G(a_n x + b_n) = G^{1/n}(x), \qquad (16.2)$$

(Resnick, 2007, Proposition 5.9). As a consequence, G is such that G^a is a distribution for every $a > 0$. A class of distributions that satisfies such a property is named *max-infinitely divisible* (max-id). More precisely, a distribution G on $\mathbb{R}^d$ is max-id, if for any $n \in \mathbb{N}$ there exists a distribution F_n such that $G = F_n^n$ (Resnick, 2007, pp. 252). This means that G can always be defined through the distribution of the sample maxima of n i.i.d. random vectors.

In order to characterize the class of MEVDs we need to specify: a) the form of the marginal distributions, b) the form of the dependence structure.

a) To illustrate the first feature is fairly straightforward. If F converges, then so too does the marginal distributions F_j for all $j \in I$. Choosing a_{jn} and b_{jn} for all $j \in I$ as in de Haan and Ferreira (2006, Corollary 1.2.4), implies that each marginal distribution of G is a generalized extreme value (GEV), i.e.

$$G(\infty, \ldots, x_j, \ldots, \infty) = \exp\left[-\left\{1 + \xi_j\left(\frac{x_j - \mu_j}{\sigma_j}\right)\right\}_+^{-1/\xi_j}\right], \quad j \in I,$$

where $(x)_+ = \max(0, x)$, $-\infty < \mu_j, \xi_j < \infty$, $\sigma_j > 0$ (de Haan and Ferreira, 2006, pp. 208–211). Because the marginal distributions are continuous then G is also continuous.

b) The explanation of the dependence form is more elaborate, although it is not complicated. The explanation is based on three steps: 1) G is transformed so that its marginal distributions are equal, 2) a Poisson point process (PPP) is used to represent the standardised distribution, 3) the dependence form is made explicit by means of a change of coordinates. Here are the steps.

1) Let $U_j(a) = F_j^{\leftarrow}(1 - 1/a)$, with $a > 1$, be the left-continuous inverse of F_j, for all $j \in I$. The sequences a_{nj} and b_{nj} in (16.1) are such that for all $y_j > 0$,

$$\lim_{n \to \infty} \frac{U_j(ny_j) - b_n}{a_n} = \frac{\sigma_j(y_j^{\xi_j} - 1)}{\xi_j} + \mu_j, \quad j \in I,$$

and therefore

$$\lim_{n \to \infty} F^n\{U_1(ny_1), \ldots, U_d(ny_d)\}$$

$$= G\left(\frac{\sigma_1(y_1^{\xi_1} - 1)}{\xi_1} + \mu_1, \ldots, \frac{\sigma_d(y_d^{\xi_d} - 1)}{\xi_d} + \mu_d\right) \equiv G_0(y), \quad (16.3)$$

for all continuity points $y > 0$ of G_0 (see de Haan and Ferreira, 2006, Theorems 1.1.6, 6.1.1). G_0 is a MEVD with identical unit Fréchet marginal distributions.

Now, for all $y > 0$ such that $0 < G_0(y) < 1$, by taking the logarithm on the right and left side of (16.3) and using a first-order Taylor expansion of $\log F\{U_1(ny_1), \ldots, U_d(ny_d)\}$, as $n \to \infty$, it follows that

$$\lim_{n \to \infty} n[1 - F\{U_1(ny_1), \ldots, U_d(ny_d)\}] = -\log G_0(y) \equiv V(y). \quad (16.4)$$

The function V, named *exponent (dependence) function*, represents the dependence structure of multiple extremes (extremal dependence for brevity). According to (16.4) the derivation of V depends on the functional form of F. In most of the practical problems the latter is unknown. A possible solution is obtained exploiting the max-id property of G_0, which says that every max-id distribution permits a PPP representation; see Resnick (2007, pp. 257–262) and Falk et al. (2011, pp. 141–142).

2) Let $N_n(\cdot)$ be a PPP defined by

$$N_n(\mathcal{A}) := \sum_{i=1}^{\infty} \mathbb{I}_{\{P_i\}}(\mathcal{A}), \quad \mathbb{I}_{\{P_i\}}(\mathcal{A}) = \begin{cases} 1, & P_i \in \mathcal{A}, \\ 0, & P_i \notin \mathcal{A}, \end{cases}$$

where $\mathcal{A} \subset \mathbb{A}$ with $\mathbb{A} := (0, \infty) \times \mathbb{R}_+^d$,

$$P_i = \left[\frac{i}{n}, \left\{ 1 + \xi_1 \left(\frac{X_{i1} - b_{n1}}{a_{n1}} \right) \right\}^{\frac{1}{\xi_1}}, \ldots, \left\{ 1 + \xi_1 \left(\frac{X_{id} - b_{nd}}{a_{nd}} \right) \right\}^{\frac{1}{\xi_d}} \right],$$

for every $n \in \mathbb{N}$ and X_i, $i = 1, 2, \ldots$ are i.i.d random vectors with distribution F. The intensity measure is $\zeta \times \eta_n$ where ζ is the Lebesgue measure and for every $n \in \mathbb{N}$ and all critical regions defined by $\mathcal{B}_y := \mathbb{R}_+^d \setminus [\mathbf{0}, y]$ with $y > 0$,

$$\eta_n(\mathcal{B}_y) = n[1 - F\{U_1(ny_1), \ldots, U_d(ny_d)\}],$$

is a finite measure. If the limit in (16.3) holds, then N_n converges weakly to N as $n \to \infty$, i.e., a PPP with intensity measure $\zeta \times \eta$ where

$$\eta(\mathcal{B}_y) = \eta\{(v \in \mathbb{R}_+^d : v_1 > y \text{ or} \ldots \text{or } v_d > y_d)\} \equiv V(y), \quad y > 0,$$

is a fine measure, named *exponent measure* (see de Haan and Ferreira, 2006, Theorems 6.1.5, 6.1.11). Observe that η must concentrate on $\overline{\mathbb{R}} = \mathbb{R}_+^d \setminus \{\mathbf{0}\}$ in order to be uniquely determined. Also, η must satisfy $\eta(\infty) = 0$; see Falk et al. (2011, pp. 143) for details.

This essentially means that numbering the rescaled observations that fall in a critical region, e.g., see the shaded sets in the left panels of Figure 16.1, where at least one coordinate is large, makes it possible for (16.3) to be computed using the void probability of N, that is

$$\begin{aligned} G_0(y) &= \text{pr}[N\{(0, 1] \times \mathcal{B}_y\} = 0] \\ &= \exp(-[\zeta\{(0, 1]\} \times \eta(\mathcal{B}_y)]) \\ &= \exp\{-V(y)\} \quad y > 0. \end{aligned} \quad (16.5)$$

From Figure 16.1 we see that in the case of strong dependence (top-left panel) all the coordinates of the extremes are large, while in the case of weak dependence (bottom-left panels) only one coordinate of the extremes is large.

At this time it remains to specify the structure of the exponent measure. This task is simpler to fulfill when working with pseudopolar coordinates.

3) With unit Fréchet margins, the stability property (16.2) can be rephrased by $G_0^a(ay) = G_0(y)$ for any $a > 0$, implying that η satisfies the homogeneity property

$$\eta(a\mathcal{B}_y) = \eta(\mathcal{B}_y)/a, \quad (16.6)$$

for all $\mathcal{B}_y \subset \overline{\mathbb{R}}$, where $\mathcal{B}_y := \overline{\mathbb{R}} \setminus (\mathbf{0}, y]$ with $y > 0$. Note that for a Borel set $\mathcal{B} \subset \overline{\mathbb{R}}$ we have $a\mathcal{B} = \{av : v \in \mathcal{B}\}$ and $\mathcal{B}_{ay} = a\mathcal{B}_y$. Now, let

$$\mathbb{W} := (v \in \overline{\mathbb{R}} : v_1 + \ldots + v_d = 1),$$

be the unit simplex on $\overline{\mathbb{R}}$ (simplex for brevity), where $d - 1$ variables are free to vary and one is fixed, e.g., $v_d = 1 - (v_1 + \cdots + v_{d-1})$. For any $v \in \mathbb{R}_+^d$, with the sum-norm, $\|v\| = |v_1| + \cdots + |v_d|$, we measure the distance of v from $\mathbf{0}$. Other norms can also be considered (e.g., Resnick, 2007, pp. 270–274). We consider the one-to-one transformation $Q : \overline{\mathbb{R}} \to (0, \infty) \times \mathbb{W}$, given by

$$(r, \boldsymbol{w}) := Q(\boldsymbol{v}) = (\|\boldsymbol{v}\|, \|\boldsymbol{v}\|^{-1}\boldsymbol{v}), \quad \boldsymbol{v} \in \overline{\mathbb{R}}.$$

By means of this, the induced measure is $\psi := \eta * Q$, i.e., $\psi(\mathcal{W}_r) = \eta\{Q^{\leftarrow}(\mathcal{W}_r)\}$ for all sets $\mathcal{W}_r = r \times \mathcal{W}$ with $r > 0$ and $\mathcal{W} \subset \mathbb{W}$, is generated. Then, from the property (16.6) it follows that

$$\begin{aligned}
\psi(\mathcal{W}_r) &= \eta\{(\boldsymbol{v} \in \overline{\mathbb{R}} : \|\boldsymbol{v}\| > r, \boldsymbol{v}/\|\boldsymbol{v}\| \in \mathcal{W})\} \\
&= \eta\{(r\boldsymbol{u} \in \overline{\mathbb{R}} : \|\boldsymbol{u}\| > 1, \boldsymbol{u}/\|\boldsymbol{u}\| \in \mathcal{W})\} \\
&= r^{-1}H'(\mathcal{W}),
\end{aligned}$$

where $H'(\mathcal{W}) := \eta\{(\boldsymbol{u} \in \overline{\mathbb{R}} : \|\boldsymbol{u}\| > 1, \boldsymbol{u}/\|\boldsymbol{u}\| \in \mathcal{W})\}$. The benefit of transforming the coordinates into pseudopolar is that the measure η becomes a product of two independent measures: the *radial measure* $(1/r)$ and *spectral measure* or *angular measure* (H') (e.g., Falk et al., 2011, pp. 145). The first measures the intensity (or distance) of the points from the origin and the second measures the angular spread (or direction) of the points. This result is known as the *spectral decomposition* (de Haan and Resnick, 1977). Hereafter we will use the term angular measure.

The density of ψ is $\mathrm{d}\psi(r, \boldsymbol{w}) = r^{-2}\mathrm{d}r \times \mathrm{d}H'(\boldsymbol{w})$ for all $r > 0$ and $\boldsymbol{w} \in \mathbb{W}$, by means of which we obtain the explicit form

$$\begin{aligned}
\eta(\mathcal{B}_{\boldsymbol{y}}) &= \psi\{Q(\boldsymbol{v} \in \overline{\mathbb{R}} : v_1 > y_1 \text{ or} \ldots \text{or } v_d > y_d)\} \\
&= \psi[\{(r, \boldsymbol{w}) \in (0, \infty) \times \mathbb{W} : r > \min(y_j/w_j, j \in I)\}] \\
&= \int_{\mathbb{W}} \int_{\min(y_j/w_j, j\in I)}^{\infty} r^{-2}\mathrm{d}r\mathrm{d}H'(\boldsymbol{w}) \qquad\qquad (16.7) \\
&= \int_{\mathbb{W}} \max_{j \in I}(w_j/y_j)\,\mathrm{d}H'(\boldsymbol{w}).
\end{aligned}$$

In pseudopolar coordinates, extremes are the values whose radial component is higher than a high threshold, see the black points in the middle panels of Figure 16.1. The angular components are concentrated around the center of the simplex, in the case of strong dependence (middle-top panel), while they are concentrated around the vertices of the simplex (middle-bottom panel), in the case of weak dependence.

The measure H' can be any finite measure on $\mathbb{W}$ satisfying the first moment conditions

$$\int_{\mathbb{W}} w_j\,\mathrm{d}H'(\boldsymbol{w}) = 1, \quad \forall j \in I.$$

This guarantees that the marginal distributions of G_0 are unit Fréchet. If H' satisfies the first moment conditions, then the total mass is equal to

$$H'(\mathbb{W}) = \int_{\mathbb{W}} (w_1 + \cdots + w_d)\mathrm{d}H'(\boldsymbol{w}) = \sum_{j\in I} \int_{\mathbb{W}} w_j\mathrm{d}H'(\boldsymbol{w}) = d.$$

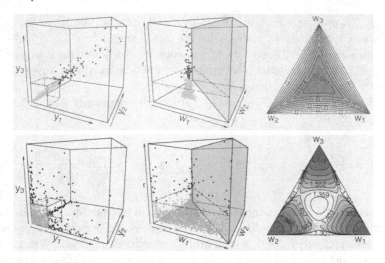

FIGURE 16.1
Examples of critical regions in $\mathbb{R}_+^3$ (left-panels) and its representation in pseudopolar coordinates (middle-panels). Black points are the extremes with strong (top-panels) and weak (bottom-panels) dependence. Right panels display the angular densities on the simplex.

So setting $H := H'/H'(\mathbb{W})$, then H is a probability measure satisfying

$$\int_{\mathbb{W}} w_j \mathrm{d}H(\boldsymbol{w}) = 1/d, \quad \forall\, j \in I. \tag{16.8}$$

Concluding, combining (16.3), (16.4), (16.5), and (16.7) all together, we have that a MEVD with unit Fréchet margins is equal to

$$G_0(\boldsymbol{y}) = \exp\left\{ -d \int_{\mathbb{W}} \max_{j \in I} (w_j/y_j)\, \mathrm{d}H(\boldsymbol{w}) \right\}. \tag{16.9}$$

16.2.2 Angular Densities

The measure H can place mass on the interior as well as on other subspaces of the simplex, such as the edges and the vertices. Thus H can have several densities that lie on these sets, which are named *angular densities*. Coles and Tawn (1991) described a way to derive the angular densities when G is absolutely continuous (see also Resnick, 2007, Example. 5.13).

Specifically, let $\mathbb{S} := \mathbb{P}(I) \backslash \varnothing$, where $\mathbb{P}(I)$ is the power set of I and $\mathcal{S}$ be the index set that takes values in $\mathbb{S}$. Given fixed d, the sets

$$\mathbb{W}_{d,\mathcal{S}} = (\boldsymbol{w} \in \mathbb{W} : w_j = 0,\; \text{if } j \notin \mathcal{S};\; w_j > 0 \text{ if } j \in \mathcal{S}),$$

for all $S \in \mathbb{S}$ provide a partition of $\mathbb{W}$ in $2^d - 1$ subsets. Similar to the simplex, there are $k - 1$ variables w_j in $\mathbb{W}_{d,S}$ that are free to vary, where $j \in S$ and $k = |S|$ denotes the size of S. We denote by $h_{d,S}$ the density that lies on the subspace $\mathbb{W}_{d,S}$, where $S \in \mathbb{S}$. When the latter is a vertex e_j of the simplex $\mathbb{W}$, for any $j \in I$, then the density is a point mass, that is $h_{d,S} = H(\{e_j\})$.

Let $S = \{i_1, \ldots, i_k\} \subset I$, when G_0 is absolutely continuous the angular density for any $y \in \mathbb{R}_+^d$ is

$$h_{d,S}\left(\frac{y_{i_1}}{\sum_{i \in S} y_i}, \cdots, \frac{y_{i_{k-1}}}{\sum_{i \in S} y_i}\right) = -\left(\sum_{i \in S} y_i\right)^{(k+1)} \lim_{\substack{y_j \to 0, \\ j \notin S}} \frac{\partial^k V}{\partial y_{i_1} \cdots \partial y_{i_k}}(y).$$

(16.10)

Two examples of a tridimensional angular density in the interior of the simplex are reported in the right panels of Figure 16.1. These are the densities of a symmetric logistic model (Gumbel, 1960) with a strong and weak dependence. When $S = \{i\}$ for any $i \in I$ the angular density $h_{d,S}$ represents the mass of H at the vertex e_j with $j = i$, thus (16.10) reduces into

$$h_{d,S} = H(\{e_i\}) = -y_i^{(2)} \lim_{y_j \to 0, j \notin S} \frac{\partial V}{\partial y_i}(y).$$

(16.11)

In the bivariate case these results are equal to the ones obtained by Pickands (1981). Kotz and Nadarajah (2000) discussed the bivariate case in the following terms. With $d = 2$ the unit simplex $\mathbb{W} = [0, 1]$ can be partitioned into

$$\mathbb{W}_{2,\{1\}} = \{(1,0)\}, \quad \mathbb{W}_{2,\{2\}} = \{(0,1)\}, \quad \mathbb{W}_{2,\{1,2\}} = \{(w, 1-w), w \in (0,1)\}.$$

The densities that lie on them are

$$h_{2,\{1\}} = H(\{0\}) = -y_1^2 \lim_{y_2 \to 0} \frac{\partial V}{\partial y_1}(y_1, y_2),$$

$$h_{2,\{2\}} = H(\{1\}) = -y_2^2 \lim_{y_1 \to 0} \frac{\partial V}{\partial y_2}(y_1, y_2),$$

and

$$h_{2,\{1,2\}}(w) = -\frac{\partial^2 V}{\partial y_1 \, \partial y_2}(w, 1-w).$$

respectively, for any $y_1, y_2 > 0$. The first two densities describe the case when extremes are only observed in one variable, while the third density describes the case when extremes are observed in both variables.

16.2.3   Extremal Dependence

From (16.5) it emerges that the extremal dependence is expressed through the exponent function. This is a map from $\mathbb{R}_+^d$ to $(0, \infty)$ satisfying the properties:
1. is a continuous function and homogeneous of order -1, the latter meaning that
   $V(ay) = a^{-1}V(y)$ for all $a > 0$;

2. is a convex function, that is $V(a\mathbf{y} + (1-a)\mathbf{y}') \leqslant aV(\mathbf{y}) + (1-a)V(\mathbf{y}')$, for $a \in [0,1]$ and $\mathbf{y}, \mathbf{y}' \in \mathbb{R}^d_+$;

3. $\max(1/y_1, \ldots, 1/y_d) \leqslant V(\mathbf{y}) \leqslant (1/y_1 + \ldots + 1/y_d)$, with the lower and upper limits representing the complete dependence and complete independence cases, respectively.

See de Haan and Ferreira (2006, pp. 223–226) for details. In summary, let $\mathbf{Y}$ be a random vector with distribution (16.9). When H places the total mass 1 on the center of the simplex $(1/d, \ldots, 1/d)$, then $Y_1 = Y_2 = \cdots = Y_d$ almost surely and hence $G_0(\mathbf{y}) = \exp\{\max(1/y_1, \ldots, 1/y_d)\}$. When H places mass $1/d$ on $\mathbf{e}_j$ for all $j \in I$, i.e., the vertices of the simplex, then $Y_1, \ldots, Y_d$ are independent and hence $G_0(\mathbf{y}) = \exp(1/y_1 + \ldots + 1/y_d)$. This rephrased for a random vector $\mathbf{X}$ with distribution (16.1) becomes

$$\min\{G_1(x_1), \ldots, G_d(x_d)\} \leqslant G(\mathbf{x}) \leqslant G_1(x_1) \cdot \ldots \cdot G_d(x_d), \quad \mathbf{x} \in \mathbb{R}^d.$$

In order to visualise the exponent function more easily, its restriction in the simplex is usually considered. This is a function $A : \mathbb{W} \to [1/d, 1]$, named the *Pickands dependence* function (Pickands, 1981), defined by

$$A(\mathbf{t}) := d \int_{\mathbb{W}} \max_{j \in I} (w_j t_j) \, dH(\mathbf{w}),$$

where $z_j = 1/y_j$, $j \in I$, $t_j = z_j/(z_1 + \cdots + z_d)$ with $j = 1, \ldots, d-1$ and $t_d = 1 - (t_1 + \cdots + t_{d-1})$. A inherits the above properties from V with the obvious modifications. In particular, $1/d \leqslant \max(t_1, \ldots, t_d) \leqslant A(\mathbf{t}) \leqslant 1$, where lower and upper bounds represent the complete dependence and independence cases, and for the homogeneity property of A the exponent function can be rewritten as

$$V(\mathbf{z}) = (z_1 + \cdots + z_d)A(t_1, \ldots, t_d), \quad \mathbf{z} \in \mathbb{R}^d_+.$$

The exponential function can be profitably used in several ways. First, an important summary of the extremal dependence is given by

$$\vartheta = V(1, \ldots, 1) = d \int_{\mathbb{W}} \max_{j \in I}(w_j) dH(\mathbf{w}). \tag{16.12}$$

This is named the *extremal coefficient* (Smith, 1990) and it represents the (fractional) number of independent components of the random vector $\mathbf{Y}$. The coefficient takes values in $[1, d]$, depending on whether the measure H concentrates near the center or the vertices of the simplex. The bounds regard the cases of complete dependence and independence.

Second, for any $\mathbf{y} > 0$ and failure region

$$\mathcal{F}_{\mathbf{y}} = (\mathbf{v} \in \overline{\mathbb{R}} : v_1 > y_1 \text{ and} \ldots \text{and } v_d > y_d), \tag{16.13}$$

the *tail dependence* function (de Haan and Ferreira, 2006; Nikoloulopoulos et al., 2009, pp. 225) is defined by

$$R(\mathbf{y}) := \eta\{(\mathbf{v} \in \overline{\mathbb{R}} : v_1 > y_1 \text{ and} \ldots \text{and } v_d > y_d)\} \equiv \eta(\mathcal{F}_{\mathbf{y}}), \quad \mathbf{y} > 0.$$

This counts the number of observations that fall in the failure region, i.e., all their coordinates are simultaneously large. The tail dependence function is related to the exponent function by the inclusion-exclusion principle. Using similar arguments to those in (16.7) and (16.8) it follows that

$$R(\boldsymbol{y}) = d \int_{\mathbb{W}} \min_{j \in I}(w_j/y_j) \mathrm{d}H(\boldsymbol{w}) \quad \boldsymbol{y} > \boldsymbol{0}. \tag{16.14}$$

By means of the tail dependence function, another important summary of the dependence between the components of $\boldsymbol{Y}$ is obtained. The *coefficient of upper tail dependence* is given by

$$\chi = R(1, \dots, 1) = d \int_{\mathbb{W}} \min_{j \in I}(w_j) \mathrm{d}H(\boldsymbol{w}). \tag{16.15}$$

It measures the strength of dependence in the tail of the distribution of $\boldsymbol{Y}$ or in other terms the probability that all the components of $\boldsymbol{Y}$ are simultaneously large. This coefficient was introduced in the bivariate case by Joe (1997, Ch. 2) and extended to the multivariate case by Li (2009). When H concentrates near the center or on the vertices of the simplex, then $\chi > 0$ or $\chi = 0$, respectively. In these cases we say that $\boldsymbol{Y}$ is upper tail dependent or independent.

In addition, the exponent and the tail dependence functions can be used for approximating the probability that certain types of extreme events will occur. Specifically, let $\boldsymbol{Y}$ be a random vector with unit Pareto margins. F is in the domain of attraction of a MEVD with Fréchet margins. From (16.4) and for the homogeneity property of V we have that $\{1 - F(n\boldsymbol{y})\} \approx V(n\boldsymbol{y})$ for large n. Then, for the relations (16.7) and (16.8), the approximating result follows

$$\mathrm{pr}(Y_1 > y_1 \text{ or } \dots \text{ or } Y_d > y_d) \approx d \int_{\mathbb{W}} \max_{j \in I}(w_j/y_j) \, \mathrm{d}H(\boldsymbol{w}), \tag{16.16}$$

when $y_1, \dots, y_d$ are high enough thresholds. Furthermore, with similar arguments to those in Section 16.2.1 we have that

$$\lim_{n \to \infty} n\bar{F}(ny_1, \dots, ny_d) = R(\boldsymbol{y}),$$

where $\bar{F}$ is the survivor function of $\boldsymbol{Y}$. R has the same homogeneity property of V. Hence, $\bar{F}(n\boldsymbol{y}) \approx R(n\boldsymbol{y})$ for large n. Then, for the relation (16.14), the approximating result also follows

$$\mathrm{pr}(Y_1 > y_1 \text{ and} \dots \text{and } Y_d > y_d) \approx d \int_{\mathbb{W}} \min_{j \in I}(w_j/y_j) \, \mathrm{d}H(\boldsymbol{w}), \tag{16.17}$$

when $y_1, \dots, y_d$ are high enough thresholds.

Lastly, when $\chi = 0$ the elements of $\boldsymbol{Y}$ are independent in the limit. However, they may still be dependent for large but finite samples. Ledford and Tawn (1996) proposed another dependence measure in order to capture this feature. For brevity, we focus on the bivariate case. Suppose that $\bar{F}$ for $y \to \infty$ satisfies the condition

$$\bar{F}(y, y) \approx y^{-1/\tau} \mathcal{L}(y), \quad 0 < \tau \leqslant 1,$$

where $\mathcal{L}$ is a slowly function, i.e., $\mathcal{L}(ay)/\mathcal{L}(y) \to 1$ as $y \to \infty$ for any $a > 0$. Then for large y, assuming $\mathcal{L}$ constant, different tail behaviours are covered. The case $\chi > 0$ is reached when $\tau = 1$ and so the variables are asymptotically dependent. When $1/2 < \tau < 1$ this means that $\chi = 0$ and so the variables are asymptotically independent, but they are still positively associated and the value of τ expresses the degree (see Ledford and Tawn, 1996, for details).

16.3   Parametric Models for the Extremal Dependence

From the previous sections, it emerges that both the exponent and tail dependence functions depend on the angular measure. There is no unique angular measure that generates the extremal dependence; any finite measure that satisfies the first moment conditions is suitable. In order to represent the extremal dependence, in principle it is insufficient to use a parametric family of models for the distribution function of the angular measure. However, flexible classes of parametric models can still be useful for applications; e.g., see Tawn (1990), Coles and Tawn (1991), and Boldi and Davison (2007), to name a few. To this end, in previous years different parametric extremal dependence models have been introduced in the literature. A fairly comprehensive overview can be found in Kotz and Nadarajah (2000, Section 3.4), Coles (2001, Section 8.2.1), Beirlant et al. (2004, Section 9.2.2), and Padoan (2013b). In the next sections we describe some of the most popular models.

16.3.1   Asymmetric Logistic Model

The multivariate asymmetric logistic model is an extension of the symmetric, introduced by Tawn (1990) (see also Coles and Tawn, 1991) for modelling extremes in complex environmental applications.

Let $\mathbb{S}$ and $\mathcal{S}$ as in Section 16.2.2 and $N_{\mathcal{S}}$ be a Poisson random variable with rate $1/\tau_{\mathcal{S}}$. This describes the number of storm events, $n_{\mathcal{S}}$, that take place on the sites $\mathcal{S}$ in a time interval. Given $n_{\mathcal{S}}$, for any site $j \in \mathcal{S}$, let $\{X_{j,\mathcal{S};i}, i = 1, \ldots, n_{\mathcal{S}}\}$ be a sequence of i.i.d. random variables that describe an environmental episode such as rain. For a fixed i, $\{X_{j,\mathcal{S};i}\}_{j \in \mathcal{S}}$ is assumed to be a dependent sequence. The maximum amount of rain observed at j is $X_{j,\mathcal{S}} = \max_{i=1...,n_{\mathcal{S}}}\{X_{j,\mathcal{S};i}\}$. Let $A_{\mathcal{S}}$ be a random effect with a positive stable distribution and stability parameter $\alpha_{\mathcal{S}} \geqslant 1$ (Nolan, 2015), representing an unrecorded additional piece of information on storm events. Assume $\{X_{j,\mathcal{S}}\}_{j \in \mathcal{S}}|\alpha_{\mathcal{S}}$ as an independent sequence. Define $Y_j = \max_{\mathcal{S} \in \mathbb{S}_j}\{X_{j,\mathcal{S}}\}$, where $\mathbb{S}_j \subset \mathbb{S}$ contains all nonempty sets including j and so the maximum is over all the storm events involving j. Then, the exponent function of the joint survival function of $(Y_1, \ldots, Y_d)$, after transforming the margins into unit exponential variables, is

$$V(\boldsymbol{y}; \boldsymbol{\theta}) = \sum_{\mathcal{S} \in \mathbb{S}} \left\{ \sum_{j \in \mathcal{S}} (\beta_{j,\mathcal{S}} y_j^{-1})^{\alpha_{\mathcal{S}}} \right\}^{1/\alpha_{\mathcal{S}}}, \quad \boldsymbol{y} \in \mathbb{R}_+^d,$$

where $\boldsymbol{\theta} = \{\alpha_S, \beta_{j,S}\}_{S \in \mathbb{S}}$, $\alpha_S \geqslant 1$, $\beta_S = \tau_S / \sum_{S \in \mathbb{S}_j} \tau_S$ and $\beta_{j,S} = 0$ if $j \notin S$, and for $j \in I$, $0 \leqslant \beta_{j,S} \leqslant 1$ and $\sum_{S \in \mathbb{S}} \beta_{j,S} = 1$. The parameter $\beta_{j,S}$ represents the probability that the maximum value observed at j is attributed to a storm event involving the sites of S. The number of the model parameters is $2^{d-1}(d+2) - (2d+1)$.

In this case the angular measure places mass on all the subspaces of the simplex. From (16.10) it follows that the angular density is, for every $S \in \mathbb{S}$ and all $w \in \mathbb{W}_{d,S}$ equal to

$$h_{d,S}(\boldsymbol{w}; \boldsymbol{\theta}) = \prod_{i=1}^{k-1}(i\alpha_S - 1) \prod_{j \in S} \beta_{j,S}^{\alpha_S} w_j^{-(\alpha_S+1)} \left\{ \sum_{j \in S} (\beta_{j,S}/w_j)^{\alpha_S} \right\}^{1/\alpha_S - k}.$$

When $S = I$, $\alpha_S = \alpha$, $\beta_{j,S} = \beta_j$ and so the angular density on the interior of the simplex simplifies to

$$h(\boldsymbol{w}; \boldsymbol{\theta}) = \prod_{i=1}^{d-1}(i\alpha - 1) \prod_{j \in I} \beta_j^{\alpha} w_j^{-(\alpha+1)} \left\{ \sum_{j \in I} (\beta_j/w_j)^{\alpha} \right\}^{1/\alpha - d}, \quad \boldsymbol{w} \in \mathbb{W}.$$

When $S = \{j\}$, for all $j \in I$, then from (16.11) it follows that the point mass at each extreme point of the simplex is $h_{d,S} = \beta_{j,S}$.

For example in the bivariate case, the conditions on the parameters are $\beta_{1,\{1\}} + \beta_{1,\{1,2\}} = 1$ and $\beta_{2,\{2\}} + \beta_{2,\{1,2\}} = 1$, so the masses at the corners of $\mathbb{S}_2 = [0, 1]$ are given by $h_{2,\{1\}} = 1 - \beta_1$ and $h_{2,\{2\}} = 1 - \beta_2$, where for simplicity $\beta_{1,\{1,2\}} = \beta_1$ and $\beta_{2,\{1,2\}} = \beta_2$, while the density in the interior of the simplex, for $0 < w < 1$, is

$$h_{2,\{1,2\}}(w) = (\alpha - 1)(\beta_1 \beta_2)^{\alpha} \{w(1 - w)\}^{\alpha-2} [(\beta_1(1 - w))^{\alpha} + (\beta_2 w)^{\alpha}]^{1/\alpha-2}.$$

The top row of Figure 16.2 illustrates some examples of trivariate angular densities for different values of the parameters $\boldsymbol{\theta} = (\alpha, \beta_1, \beta_2, \beta_3)$, where the subscript of the index set $S = \{1, 2, 3\}$ has been omitted for simplicity. The values of the parameters are, from left to right $\{(5.75, 0.5, 0.5, 0.5); (1.01, 0.9, 0.9, 0.9); (1.25, 0.5, 0.5, 0.5); (1.4, 0.7, 0.15, 0.15)\}$. The first panel shows that with large values of α and equal values of the other parameters, the case of strong dependence among the variables is obtained. The mass is mainly concentrated towards the center of the simplex. The second panel shows that when α is close to 1 and the other parameters are equal, the case of weak dependence is obtained. The mass is concentrated on the vertices of the simplex. The third panel shows the case of a symmetric dependence structure with the mass near the corners of the simplex but not along the edges. Finally, the fourth panel shows a case of an asymmetric dependence structure where the mass tends to be closer to the components whose corresponding values of β are high.

16.3.2   Tilted Dirichlet Model

Extremal dependence models with an angular measure that places mass on the interior, vertices, and edges of the simplex are more flexible than those with a measure

FIGURE 16.2
Examples of trivariate angular densities for the asymmetric logistic, tilted Dirichlet, pairwise beta, Hüsler–Reiss, and extremal-t models from top to bottom.

that concentrates only on the interior. An example is the asymmetric logistic model versus the symmetric. However, the former has too many parameters to estimate, so parsimonious models may be preferred. In order to derive a parametric model for the angular density whose mass concentrates on the interior of the simplex, Coles and Tawn (1991) proposed the following method. Consider a continuous function $h' : \mathbb{W} \to [0, \infty)$ such that $m_j = \int_{\mathbb{S}_d} v_j h'(v) dv < \infty$ for all $j \in I$. Then, the function

$$h(w) = d^{-1}(m_1 w_1 + \cdots + m_d w_d)^{-(d+1)} h'\{mw/(m_1 w_1 + \cdots + m_d w_d)\}, w \in \mathbb{W}$$

is a valid angular density. It satisfies the first moment conditions (16.8) and its mass is centered at $(1/d, \ldots, 1/d)$ and integrates to one. For example, if h' is the density of the Dirichlet distribution, then we obtain the angular density

$$h(w; \theta) = \frac{\Gamma(\sum_{j \in I} \alpha_j + 1)}{d(\sum_{j \in I} \alpha_j w_j)^{d+1}} \prod_{j=1}^{d} \frac{\alpha_j}{\Gamma(\alpha_j)} \left(\frac{\alpha_j w_j}{\sum_{j \in I} \alpha_j w_j} \right)^{\alpha_j - 1}, w \in \mathbb{W}, \quad (16.18)$$

where $\theta = \{\alpha_j > 0\}_{j \in I}$. This density is asymmetric and it becomes symmetric when $\alpha_1 = \cdots = \alpha_d$. Extremes are independent or completely dependent when for all $j \in I$ the limiting cases $\alpha_j \to 0$ and $\alpha_j \to \infty$ arise. The dependence parameters $\alpha_j, j \in I$, are not easy to interpret. However, Coles and Tawn (1994) draw attention to the quantities $r_1 = (\alpha_i - \alpha_j)/2$ and $r_2 = (\alpha_i + \alpha_j)/2$ which can be interpreted as the asymmetry and intensity of the dependence between pairs of variables.

In this case, the exponent function cannot be analytically computed; nonetheless it can still be evaluated numerically.

The second row of Figure 16.2 illustrates some examples of trivariate angular densities obtained with different sets of the parameters $\theta = (\alpha_1, \alpha_2, \alpha_3)$. The plots from left to right have been obtained using the parameter sets $\{(2, 2, 2); (0.5, 0.5, 0.5); (2, 2.5, 30); (0.1, 0.25, 0.95)\}$. The first panel shows that, when values of the parameters are equal and greater than 1, the mass concentrates in the center of the simplex leading to strong dependence. The second panel shows the opposite, when values of α are equal and less than 1, it yields to the case of weak dependence as the mass concentrates on the vertices of the simplex. The third panel shows the case of an asymmetric dependence structure and this is obtained when the values of the parameters are all greater than one. In this specific case the mass tends to spread towards the bottom and top left edges. The fourth panel illustrates another case of an asymmetric dependence structure, in this case obtained with all the values of the parameters that are less than 1, leading to a mass that concentrates along the top right edge and vertices.

16.3.3 Pairwise Beta Model

The tilted Dirichlet model has been successfully used for applications (e.g., Coles and Tawn, 1991), although it suffers from a lack of interpretability of the parameters. Cooley et al. (2010) proposed a similar model but with easily interpretable parameters. The definition of their model is based on a geometric approach. Specifically, they considered the symmetric pairwise beta function

$$h^*(w_i, w_j) = \frac{\Gamma(2\beta_{i,j})}{\Gamma^2(\beta_{i,j})} \left(\frac{w_i}{w_i + w_j} \right)^{\beta_{i,j} - 1} \left(\frac{w_j}{w_i + w_j} \right)^{\beta_{i,j} - 1}, i, j \in I,$$

where w_i and w_j are two elements of w and $\beta_{i,j} > 0$. This function has its center at the point $(1/d, \ldots, 1/d)$ and it verifies the first moment conditions (16.8). Then, the angular pairwise beta density is defined by summing together all the $d(d-1)/2$ possible pairs of variables, namely

$$h(w; \theta) = \frac{2(d-3)!\Gamma(\alpha d + 1)}{d(d-1)\Gamma(2\alpha+1)\Gamma\{\alpha(d-2)\}} \sum_{i,j \in I, i<j} h(w_i, w_j), \quad w \in \mathbb{W},$$

where

$$h(w_i, w_j) = (w_i + w_j)^{2\alpha - 1}\{1 - (w_i + w_j)\}^{\alpha(d-2)-d+2} h^*(w_i, w_j)$$

and $\theta = (\alpha, \{\beta_{i,j}\}_{i,j \in I})$ with $\alpha > 0$. Each parameter $\beta_{i,j}$ controls the level of dependence between the ith and the jth components and the dependence increases for increasing values of $\beta_{i,j}$. The function h^* is introduced to guarantee that the dependence ranges between weak and strong dependence. The parameter α controls the dependence of all the variables, when it increases the overall dependence increases.

Also in this case the exponent function cannot be computed in closed form and hence it can only be evaluated numerically.

The third row of Figure 16.2 provides some examples of trivariate angular densities obtained with different values of the parameters $\theta = (\alpha, \beta_{1,2}, \beta_{1,3}, \beta_{2,3})$. The plots from left to right have been obtained using the parameter sets $\{(4,2,2,2); (0.5,1,1,1); (1,2,4,15); (1,10,10,10)\}$. The first panel shows a case of symmetric density obtained with all equal parameters $\beta_{i,j}\ i,j \in I$. A large value of the overall dependence parameter α pulls the mass towards the center of the simplex, indicating a strong dependence between the variables. On the contrary, the second panel shows that when the overall dependence parameter is close to zero then the mass concentrates on the vertices of the simplex, indicating weak dependence among the variables. The third panel illustrates a case of asymmetric angular density with strong dependence between the second and third variables that is due to a large value of $\beta_{2,3}$. Although the value of the global dependence parameter α is not large, it is enough to slightly push the mass towards the center of the simplex. The fourth panel shows a case of symmetric angular density, which is obtained with large values of the pairwise dependence parameters and an average value of the global dependence parameter. The mass is mainly concentrated on the center of the simplex and some mass tends to lie near the centers of the edges.

16.3.4 Hüsler–Reiss Model

One of the most popular models is the Hüsler–Reiss (Hüsler and Reiss, 1989). Let $X_1, \ldots, X_n$ be n i.i.d. copies of a zero-mean unit variance Gaussian random vector. Assume that for all $i, j \in I$ the pairwise correlation $\rho_{i,j;n}$ satisfies the condition

$$\lim_{n \to \infty} \log n(1 - \rho_{i,j;n}) = \lambda_{i,j}^2 \in [0, \infty).$$

Then, the exponent function of the limit distribution of $b_n(M_n - b_n)$ for $n \to \infty$, where $b_n = (b_n, \ldots, b_n)$ is a vector of real sequences (see Resnick, 2007, pp. 71–

72), is

$$V(\boldsymbol{y};\boldsymbol{\theta}) = \sum_{j=1}^{d} \frac{1}{y_j} \Phi_{d-1}\left\{\left(\lambda_{i,j} + \frac{\log y_i/y_j}{2\lambda_{i,j}}\right)_{i\in I_j} ; \bar{\Lambda}_j\right\}, \quad \boldsymbol{y} \in \mathbb{R}_+^d, \qquad (16.19)$$

where $\boldsymbol{\theta} = \{\lambda_{i,j}\}_{i,j\in I}, I_j := I\backslash\{j\}$, Φ_{d-1} is $d-1$ dimensional Gaussian distribution with partial correlation $\bar{\Lambda}_j$. For all $j \in I$, the elements of $\bar{\Lambda}_j$ are $\lambda_{k,i;j} = (\lambda_{k,j}^2 + \lambda_{i,j}^2 - \lambda_{k,i}^2)/(2\lambda_{k,j}\lambda_{i,j})$, for $k,i \in I_j$. The parameter $\lambda_{i,j}$, $i,j \in I$, controls the dependence between the ith and j^{th} elements of a vector of d extremes. These are completely dependent when $\lambda_{ij} = 0$ and become independent as $\lambda_{ij} \to \infty$.

In this case the angular measure concentrates on the interior of the simplex. Applying (16.10) it can be checked (Engelke et al., 2015) that the angular density is

$$h(\boldsymbol{w};\boldsymbol{\theta}) = \phi_{d-1}\left\{\left(\lambda_{i,1} + \frac{\log w_i/w_1}{2\lambda_{i,1}}\right)_{i\in I_1} ; \bar{\Lambda}_1\right\}\left\{w_1^2 \prod_{i=2}^{d}(w_i 2\lambda_{i,1})\right\}^{-1}, \boldsymbol{w} \in \mathbb{W},$$

where ϕ_{d-1} is $d-1$ dimensional Gaussian density with partial correlation matrix $\bar{\Lambda}_1$.

The second last row of Figure 16.2 provides some examples of trivariate angular densities obtained with different values of the parameters $\boldsymbol{\theta} = (\lambda_{1,2}, \lambda_{1,3}, \lambda_{2,3})$. The plots from left to right have been obtained using the parameter sets $\{(0.3, 0.3, 0.3), (1.4, 1.4, 1.4), (1.7, 0.7, 1.1), (0.52, 0.71, 0.52)\}$. The first panel shows that with small and equal values of parameters the case of strong dependence among all the variables is obtained. In this case the mass concentrates around the center of the simplex. On the contrary, the second panel shows that with large and equal values of the parameters the case of weak dependence is obtained. In this case the mass is placed close to the vertices of the simplex. The third panel shows that an asymmetric dependence structure is obtained when the parameter values are different. In this case the mass tends to concentrate around the vertices and edges that are concerned with the smaller values of the parameters. The fourth panel shows that a symmetric dependence structure, with respect to the second component is obtained setting the values of two parameters to be equal. In this case the mass is equally divided up towards the two vertices and edges that are concerned with the smaller values of the parameters.

16.3.5 Extremal-t Model

The extremal-t model (Nikoloulopoulos et al., 2009) is more flexible than the Hüsler–Reiss but it is still simple enough. It is easily interpretable and useful in practical applications (see Davison et al., 2012). Let $\boldsymbol{X}_1, \ldots, \boldsymbol{X}_n$ be n i.i.d. copies of a zero-center unit scale Student-t random vector with dispersion matrix Σ and $\nu > 0$ degrees of freedom (d.f.). Then, the exponent function of the limiting distribution of $\boldsymbol{M}_n/\boldsymbol{a}_n$ for $n \to \infty$, where $\boldsymbol{a}_n = (a_n \ldots, a_n)$ is a vector of positive sequences (see

Demarta and McNeil, 2005), is

$$V(\boldsymbol{y};\boldsymbol{\theta}) = \sum_{j=1}^{d} \frac{1}{y_j} T_{d-1,\nu+1} \left\{ \left[\sqrt{\frac{\nu+1}{1-\rho_{i,j}^2}} \left\{ \left(\frac{y_i}{y_j}\right)^{\frac{1}{\nu}} - \rho_{i,j} \right\} \right]_{i \in I_j} ; \bar{\Sigma}_j \right\}, \quad (16.20)$$

for all $\boldsymbol{y} \in \mathbb{R}_+^d$, where $\boldsymbol{\theta} = (\{\rho_{i,j}\}_{i,j \in I}, \nu)$ and $T_{d-1,\nu+1}$ is a $d-1$ dimensional Student-t distribution with $\nu + 1$ d.f. and partial correlation matrix $\bar{\Sigma}_j$. The correlation parameter $\rho_{i,j}$, $i, j \in I$, drives the dependence between pairs of variables with the dependence that increases with the increasing of $\rho_{i,j}$. The parameter ν controls the overall dependence among all the variables. For decreasing values of ν the dependence increases and vice versa.

The Hüsler–Reiss model is a special case of the extremal-t. Indeed, for all $i, j \in I$ if the correlation parameters of the extremal-t distribution are equal to $\rho_{i,j;\nu} = 1 - \lambda_{i,j}^2/\nu$, then this distribution converges weakly, as $\nu \to \infty$, to the Hüsler–Reiss (see Nikoloulopoulos et al., 2009).

In this case the angular measure places mass on all the subspaces of the simplex. When $\mathcal{S} = I$, then applying (16.10) we obtain that the angular density is

$$h(\boldsymbol{w};\boldsymbol{\theta}) = \frac{t_{d-1,\nu+1}\left(\left[\sqrt{\frac{\nu+1}{1-\rho_{i,1}^2}} \left\{ (w_i/w_1)^{1/\nu} - \rho_{i,1} \right\} \right]_{i \in I_1} ; \bar{\Sigma}_1 \right)}{\nu^{d-1} w_1^{d+1} \left\{ \prod_{i=2}^{d} \sqrt{\frac{\nu+1}{1-\rho_{i,1}^2}} (w_i/w_1)^{(\nu-1)/\nu} \right\}^{-1}}, \quad \boldsymbol{w} \in \mathbb{W},$$

where $t_{d-1,\nu+1}$ is $d-1$ dimensional Student-t density with partial correlation matrix $\bar{\Sigma}_1$ (e.g., Ribatet, 2013). When $\mathcal{S} = \{j\}$, then applying (16.11) we obtain that the mass on the extreme points of the simplex is

$$h_{d,\mathcal{S}} = T_{d-1,\nu+1} \left[\left\{ -\rho_{i,j}(\nu+1)^{1/2}/(1-\rho_{i,j}^2)^{1/2} \right\}_{i \in I_j} ; \bar{\Sigma}_j \right], \quad j \in I.$$

The last row of Figure 16.2 provides some examples of the trivariate angular densities obtained with different values of the parameters $\boldsymbol{\theta} = (\rho_{1,2}, \rho_{1,3}, \rho_{2,3}, \nu)$. From left to right the plots are obtained using the parameter values $\{(0.95, 0.95, 0.95, 2); (-0.3, -0.3, -0.3, 5); (0.52, 0.71, 0.52, 3); (0.52, 0.71, 0.52, 2)\}$. The first panel shows that when the scale parameters ρ_{ij} are all equal and close to one and the d.f. ν are small, then the mass concentrates around the center of the simplex and therefore the dependence is strong. The second panel shows the opposite, when the correlations are close to zero and the d.f. are high, the mass concentrates around the vertices of the simplex and hence the dependence is weak. The third panel shows that when two scale parameters are equal then the dependence structure is symmetric with respect to the second component and the mass tends to concentrate on the top vertex and the bottom edge and vertices. The fourth panel shows that with the same setting but with smaller d.f. the mass is pushed towards the center of the simplex and hence the dependence is stronger.

16.4　Estimating the Extremal Dependence

Several inferential methods have been explored for inferring the extremal dependence. Nonparametric and parametric approaches are available. In the first case recent advances are Gudendorf and Segers (2011), Gudendorf and Segers (2012), and Marcon et al. (2014); see also the references therein. Both likelihood based and Bayesian inferential methods have been widely investigated. Examples of likelihood based methods are the approximate likelihood (e.g., Coles and Tawn, 1994; Cooley et al., 2010; Engelke et al., 2015) and the composite likelihood (e.g., Davison and Gholamrezaee, 2012; Padoan et al., 2010). Examples of Bayesian techniques are Apputhurai and Stephenson (2011), Sabourin et al. (2013), Sabourin and Naveau (2014).

For comparison purposes in the next section the real data analysis is performed using the maximum approximate likelihood estimation method and the approximate Bayesian method based on the approximate likelihood. Here is a brief description.

From the theory in Sections 16.2.1, if $Y_1, \ldots, Y_n$ are i.i.d. copies of Y on $\mathbb{R}_+^d$ with a distribution in the domain of attraction of an MEVD, then the distribution of the sequence $\{R_i/n, W_i, i = 1, \ldots, n\}$, where $R_i = Y_{i,1} + \cdots + Y_{i,d}$ and $W_i = Y_i/R_i$, converges as $n \to \infty$ to the distribution of a PPP with density $d\psi(r, w) = r^{-2} dr \times dH(w)$.

Assume that $x_1, \ldots, x_n$ are i.i.d. observations from a random vector with an unknown distribution. Since the aim is estimating the extremal dependence, we transform the data into the sample $y_1, \ldots, y_n$ with unit Fréchet marginal distributions. This is done by applying the probability integral transform, after fitting the marginal distributions. Next, the coordinates of the datapoints are changed from Euclidean into pseudopolar by the transformation

$$r_i = y_{i,1} + \cdots + y_{i,d} \quad w_i = y_i/r_i, \quad i = 1, \ldots, n.$$

Then, the sequence $\{(r_i, w_i), i = 1, \ldots, n : r_i > r_0\}$, where $r_0 > 0$ is a large threshold, comes approximately from a Poisson point process with intensity measure ψ. Let $\mathcal{W}_{r_0} = \{(r, w) : r > r_0\}$ be the set of points with a radial component larger than r_0, then the number of points falling in $\mathcal{W}_{r_0}$ is given by $N(\mathcal{W}_{r_0}) \sim \text{Pois}\{1/\psi(\mathcal{W}_{r_0})\}$. Conditionally to $N(\mathcal{W}_{r_0}) = m$, the points $\{(r_{(i)}, w_{(i)}), i = 1, \ldots, m\}$ are i.i.d. with common density $d\psi(r, w)/\psi(\mathcal{W}_{r_0})$. If we assume that H is known apart from a vector of unknown parameters $\theta \in \Theta \subset \mathbb{R}^p$, then the approximate likelihood of the excess is

$$L(\theta; (r_{(i)}, w_{(i)}), i = 1, \ldots, m) = \frac{e^{-\psi(\mathcal{W}_{r_0})} \psi(\mathcal{W}_{r_0})^m}{m!} \prod_{i=1}^m \frac{d\psi(r_{(i)}, w_{(i)})}{\psi(\mathcal{W}_{r_0})}$$

$$\propto \prod_{i=1}^m h(w_{(i)}, \theta), \tag{16.21}$$

where h is a parametric angular density function; see, Beirlant et al. (e.g., 2004, pp.

170–171) and Engelke et al. (2015). In the next section the angular density models described in Section 16.3 are fitted to the data by the maximization of the likelihood (16.21). For brevity the asymmetric logistic model is not considered since it has too many parameters. The likelihood (16.21) is proportional to the product of angular densities, therefore the maximizer of (16.21) is obtained equivalently by maximizing the log-likelihood

$$\ell(\boldsymbol{\theta}) = \sum_{i=1}^{m} \log h(\boldsymbol{w}_{(i)}, \boldsymbol{\theta}). \qquad (16.22)$$

Denote by $\widehat{\boldsymbol{\theta}}$ the maximizer of ℓ and by $\ell'(\boldsymbol{\theta}) = \nabla_{\boldsymbol{\theta}} \ell(\boldsymbol{\theta})$ the score function. Since (16.21) provides an approximation of the true likelihood, then from the theory on model misspecification (e.g., Davison, 2003, pp. 147–148) it follows that

$$\sqrt{n}(\widehat{\boldsymbol{\theta}} - \boldsymbol{\theta}) \overset{d}{\to} \mathcal{N}_p(0, J(\boldsymbol{\theta})^{-1} K(\boldsymbol{\theta}) J(\boldsymbol{\theta})^{-1}), \quad n \to \infty,$$

where $\mathcal{N}_p(\boldsymbol{\mu}, \Sigma)$ is the p-dimensional normal distribution with mean $\boldsymbol{\mu}$ and covariance Σ, $\boldsymbol{\theta}$ is the true parameter and

$$J(\boldsymbol{\theta}) = -\mathrm{E}\{\nabla_{\boldsymbol{\theta}} \ell'(\boldsymbol{\theta})\}, \qquad K(\boldsymbol{\theta}) = \mathrm{Var}_{\boldsymbol{\theta}}\{\ell'(\boldsymbol{\theta})\},$$

are the sensitivity and variability matrices, respectively (Varin et al., 2011). In the case of misspecified models, model selection can be performed by computing the Takeuchi information criterion (TIC) (e.g., Sakamoto et al., 1986), that is

$$\mathrm{TIC} = -2 \left[\ell(\widehat{\boldsymbol{\theta}}) - \mathrm{tr}\{K(\widehat{\boldsymbol{\theta}}) J^{-1}(\widehat{\boldsymbol{\theta}})\} \right],$$

where the log-likelihood and the variability and sensitive matrices are evaluated at $\widehat{\boldsymbol{\theta}}$. The model with the smallest value of the TIC is preferred.

In order to derive an approximate posterior distribution for the parameters of an angular density, the approximate likelihood (16.21) can be used within the Bayesian paradigm (see Sabourin et al., 2013). Briefly, let $q(\boldsymbol{\theta})$ be a prior distribution on $\boldsymbol{\theta}$, then the posterior distribution of the angular density's parameters is

$$q(\boldsymbol{\theta}|\boldsymbol{w}) = \frac{\prod_{i=1}^{m} h(\boldsymbol{w}_{(i)}, \boldsymbol{\theta}) q(\boldsymbol{\theta})}{\int_{\Theta} \prod_{i=1}^{m} h(\boldsymbol{w}_{(i)}, \boldsymbol{\theta}) q(\boldsymbol{\theta}) \, d\boldsymbol{\theta}}. \qquad (16.23)$$

With the angular density models in Section 16.3 the analytical expression of $q(\boldsymbol{\theta}|\boldsymbol{w})$ cannot be derived. Therefore, we use a Markov chain Monte Carlo method for sampling from an approximation of $q(\boldsymbol{\theta}|\boldsymbol{w})$. Specifically, we use a Metropolis–Hastings simulating algorithm (e.g., Hastings, 1970). With the pairwise beta models we use the prior distributions described by Sabourin et al. (2013). With the tilted Dirichlet and Hüsler–Reiss model we use independent zero-mean normal prior distributions with standard deviations equal to 3 for $\log \alpha_j$ and $\log \lambda_{i,j}$ with $i, j \in I$. For the extremal-t model we use independent zero-mean normal prior distributions with standard deviations equal to 3 for $\mathrm{sign}(\rho_{ij})\mathrm{logit}(\rho_{ij}^2)$ with $i, j \in I$, where $\mathrm{sign}(x)$ is the sign of x for $x \in \mathbb{R}$ and $\mathrm{logit}(x) = \log(x/(1-x))$ for $0 \leqslant x \leqslant 1$, and a zero-mean normal prior distribution with standard deviations equal to 3 for $\log \nu$. Similar to Sabourin et al. (2013), for each models' parameter we select a sample of 50×10^3 observations from the approximate posterior, after a burn-in period of length 30×10^3. These sizes have been determined using the Geweke convergence diagnostics (Geweke, 1992) and the Heidelberger and Welch test (Heidelberger and Welch, 1981) respectively.

Model selection is performed using the Bayesian information criterion (BIC) (e.g., Sakamoto et al., 1986), that is

$$\mathrm{BIC} = -2\,\ell(\widehat{\boldsymbol{\theta}}) + p\{\log m + \log(2\pi)\},$$

where p is the number of parameters and m is the sample size. The model with the smallest value of the BIC is preferred.

16.5 Real Data Analysis: Air Quality Data

We analyze the extremal dependence of the air quality data, recorded in the city centre of Leeds, UK. The aim is to estimate the probability that multiple pollutants will be simultaneously high in the near future. This dataset has been previously studied by Heffernan and Tawn (2004), Boldi and Davison (2007), and Cooley et al. (2010). The data are the daily maximum of five air pollutants: particulate matter (PM10), nitrogen oxide (NO), nitrogen dioxide (NO2), ozone (03), and sulfur dioxide (SO2). Levels of the gases are measured in parts per billion, and those of PM10 in micrograms per cubic meter. We focus our analysis on the winter season (from November to February) from 1994 to 1998.

A preliminary analysis focuses on the data of triplets of variables. For brevity we only report the results of the most dependent triplets: PM10, NO, SO2 (PNS), NO2, SO2, NO (NSN) and PM10, NO, NO2 (PNN). For each variable, the empirical distribution function is estimated with the data below the 0.7 quantile and a GPD is fitted to the data above the quantile (Cooley et al., 2010). Then, each marginal distribution is transformed into a unit Fréchet. The coordinates of the datapoints are transformed to radial distances and angular components. For each triplet, the 100 observations with the largest radial distances are retained. The angular density models in Section 16.3 are fitted to the data using the methods in Section 16.4.

The results are presented in Table 16.1. Maximum likelihood estimates are similar to the estimated posterior means and the estimated posterior standard deviations are typically larger than the standard errors. For PNS we obtain the same maximum likelihood estimates as Cooley et al. (2010) with the pairwise beta model; however we use (16.4) to compute the variances of the estimates and so we attain larger standard errors than they do. Both the TIC and BIC lead to the same model selection. The Hüsler–Reiss model provides the best fit for all the groups of pollutants.

From top to bottom, Figure 16.3 displays the angular densities, computed with the posterior means. From left to right the Hüsler–Reiss, the tilted Dirichlet, and the

TABLE 16.1

Summary of the extremal dependence models fitted to the UK air pollution data. For each angular density model the estimation results of the triplets of pollutants are reported. Four models are TD (tilted Dirichlet), PB (pairwise beta), HR (Hüsler–Reiss), and ET (extremal-t). Two methods are L (approximate likelihood) and B (Bayesian). The Bayesian estimates are posterior means. Standard errors of the estimates are in parentheses.

Model	Method	Estimates				$\ell(\widehat{\theta})$	TIC/BIC
TD		$\widehat{\alpha}_1$	$\widehat{\alpha}_2$	$\widehat{\alpha}_3$			
PNS	L	1.20(0.24)	0.67(0.07)	0.41(0.08)		199.63	−399.21
	B	1.22(0.25)	0.68(0.11)	0.42(0.09)			−379.90
NSN	L	0.85(0.12)	0.39(0.08)	0.90(0.11)		200.84	−401.63
	B	0.86(0.15)	0.39(0.09)	0.81(0.15)			−382.32
PNN	L	1.43(0.28)	1.55(0.31)	1.28(0.20)		186.35	−372.64
	B	1.45(0.30)	1.57(0.28)	1.29(0.23)			−353.36
PB		$\widehat{\beta}_{1,2}$	$\widehat{\beta}_{1,3}$	$\widehat{\beta}_{2,3}$	$\widehat{\alpha}$		
PNS	L	3.21(0.70)	0.47(0.05)	0.45(0.04)	0.68(0.06)	95.95	−191.87
	B	3.31(1.13)	0.48(0.11)	0.46(0.10)	0.68(0.09)		−166.10
NSN	L	0.40(0.03)	3.74(1.77)	0.50(0.05)	0.64(0.05)	102.59	−205.13
	B	0.40(0.09)	4.00(1.72)	0.51(0.12)	0.64(0.08)		−179.36
PNN	L	3.75(1.38)	0.71(0.09)	3.18(1.21)	1.35(0.18)	84.31	−168.55
	B	3.83(1.75)	0.72(0.16)	3.70(1.80)	1.37(0.20)		−142.66
HR		$\widehat{\lambda}_{1,2}$	$\widehat{\lambda}_{1,3}$	$\widehat{\lambda}_{2,3}$			
PNS	L	0.65(0.06)	0.90(0.04)	0.98(0.03)		234.51	−468.93
	B	0.65(0.04)	0.90(0.04)	0.98(0.04)			−449.67
NSN	L	1.00(0.04)	0.56(0.04)	0.96(0.04)		251.80	−503.54
	B	1.00(0.04)	0.57(0.03)	0.97(0.04)			−484.25
PNN	L	0.60(0.05)	0.70(0.04)	0.51(0.03)		198.23	−396.38
	B	0.60(0.03)	0.70(0.04)	0.51(0.03)			−377.11
ET		$\widehat{\rho}_{1,2}$	$\widehat{\rho}_{1,3}$	$\widehat{\rho}_{2,3}$	$\widehat{\nu}$		
PNS	L	0.87(0.02)	0.74(0.03)	0.66(0.03)	3.89(0.51)	152.13	−304.18
	B	0.87(0.02)	0.77(0.02)	0.72(0.01)	4.02(0.35)		−275.13
NSN	L	0.58(0.04)	0.87(0.02)	0.64(0.03)	3.50(0.01)	141.92	−283.80
	B	0.72(0.01)	0.89(0.02)	0.73(0.02)	4.00(0.33)		−242.50
PNN	L	0.88(0.02)	0.82(0.02)	0.89(0.01)	3.70(0.78)	180.74	−361.38
	B	0.86(0.02)	0.78(0.03)	0.87(0.02)	3.21(0.43)		−330.33

pairwise beta densities are reported. With PNS, we see that there are many observations along the edge that link PM10 and NO, revealing strong dependence between these two pollutants. There are also several observations on the SO2 vertex, reflecting that this pollutant is mildly dependent with the other two. There are also some data in the middle of the simplex, indicating that there is mild dependence among the pollutants. Similarly, with NSN we see that there is strong dependence between

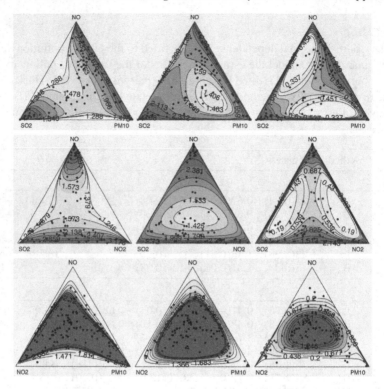

FIGURE 16.3
Estimated angular densities in logarithm scale. Dots represent the largest 100 observations.

NO and NO2, because there are many observations along the edge that link them. There is a mild dependence between SO2 and the other pollutants, because there is a considerable amount of data on the O3 vertex. Overall, there is mild dependence among the pollutants, because there is a small amount of data in the middle of the simplex. With PNN we see that most of the observations are placed on the middle of the simplex revealing an overall strong dependence among the pollutants. There is a small amount of data along the edge that link NO2 and NO and on the PM10 vertex. This reflects more dependence between NO2 and NO than between NO2 and PM10 and PM10 and NO. All these features are well captured by the angular densities estimated using the Hüsler–Reiss model.

With this analysis we found that O3 is only weakly dependent with the other pollutants. This result was also found by Heffernan and Tawn (2004). Then, the second part of the analysis focuses only on PM10, NO, NO2, and SO2. Now, because a larger number of parameters needs to be estimated, then the 200 observations with the largest radial distances are selected (see Cooley et al., 2010). Table 16.2 presents the estimation results. For brevity we only report the maximum value of the log-

TABLE 16.2
Summary of the extremal dependence models fitted to the UK air pollution data.

	Tilted Dirichlet	Pairwise Beta	Hüsler–Reiss	extremal-t
$\ell(\widehat{\boldsymbol{\theta}})$	654.3	402.5	762.7	532.3
TIC	−1308.6	−805.0	−1525.3	−1064.5
BIC	−1280.0	−753.4	−1475.5	−974.7

likelihood, the TIC and the BIC. The Hüsler–Reiss model provides the smallest values of the TIC, and BIC, revealing again that it better fits the pollution data. Accordingly hereafter calculations will be made using this model and the estimates obtained with the Bayesian approach.

We summarize the extremal dependence of the four variables using the extremal coefficient (16.12) and the coefficient of tail dependence (16.15). Specifically, $\widehat{\vartheta} = 2.267$ with a 95% credible interval is equal to $(1.942, 2.602)$ and $\widehat{\chi} = 0.242$ with a 95% credible interval is $(0.150, 0.361)$. These results suggest a strong extremal dependence among the pollutants. The estimated extremal dependence can be used in turn to estimate the probability that multiple pollutants exceed a high threshold. Consider a value y whose radial component is a high threshold r_0. Then, the probability of falling in the failure region (16.13) is approximately equal to the right-hand side of (16.17). Because the exponent function is related to the tail function by the inclusion-exclusion principle, then using (16.19) we have

$$\mathrm{pr}\{Y_1 > y_1, \ldots, Y_d > y_d\} \approx \sum_{j=1}^{d} \frac{1}{y_j} \bar{\Phi}_{d-1}\left\{\left(\lambda_{k,j} + \frac{\log y_k/y_j}{2\lambda_{k,j}}\right)_{k \in I_j}; \bar{\Lambda}_j\right\}, \quad (16.24)$$

where $\bar{\Phi}_{d-1}$ is the survival function of the multivariate normal distribution (Nikoloulopoulos et al., 2009). Similar to Cooley et al. (2010) we define three extreme events: $\{PM10 > 95, NO > 270, SO2 > 95\}$, $\{NO2 > 110, SO2 > 95, NO > 270\}$, and $\{PM10 > 95, NO > 270, NO2 > 110, SO2 > 95\}$. Then, we compute probability (16.24) using in place of the parameters their estimates. Table 16.3 reports the results. For the three events the estimates fall inside the 95% confidence intervals highlighting the ability of the model to estimate such extreme events.

The right-hand side of (16.24) can also be used for estimating joint return levels. In the univariate case see Coles (2001, pp.49–50). In the multivariate case different definitions of return levels may be available (Johansen, 2004). Let $J \subset I$, $\{x_i, i \in I \setminus J\}$ be a sequence of fixed high thresholds and $p \in (0,1)$ be a fixed probability. Given a return period $1/p$, we define *joint return levels* the quantiles $\{y_{j;p}, j \in J\}$ that satisfy the equation

$$p = \mathrm{pr}(Y_j > y_{j;p}, Y_i > x_i, j \in J, i \in I \setminus J).$$

TABLE 16.3
Probability estimates of excesses. The first row reports the number of excess and the sample size. The second row reports the empirical estimates and between brackets the 95% confidence intervals obtained with the normal approximation. The third row reports the model estimates.

	Event 1	Event 2	Event 3
Excess / n	18/528	14/562	12/528
Emp. Est.	0.034 $(0.019, 0.050)$	0.025 $(0.012, 0.038)$	0.023 $(0.010, 0.035)$
Mod. Est.	0.038	0.030	0.030

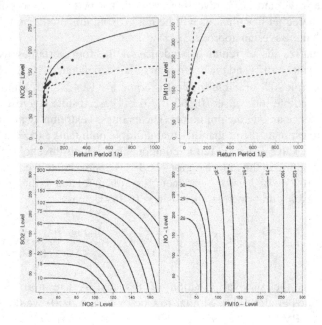

FIGURE 16.4
Joint return level plots of single components NO2 and PM10 and of the two components (SO2, NO2) and (NO, PM10).

Figure 16.4 displays univariate and bivariate *joint return level plots*. When $J = \{j\}$, with $j \in I$, then the joint return level plot displays $y_{j;p}$ against $1/p$ for different values of p. When $J = \{i, j\}$, with $i, j \in I$, then for different values of $1/p$ the contour levels of $(y_{i;p}, y_{j;p})$ are displayed. With solid lines, the top-left and right panels of Figure 16.4 report the estimated return levels of NO2 and PM10 jointly to the extreme events $\{SO2 > 95, NO > 270\}$ and $\{NO > 270, NO2 > 110, SO2 > 95\}$, respectively. The dots are the empirical estimates and the dashed lines are the pointwise 95% confidence intervals. These are computed using the normal approximation when $p > 0.02$ and using exact binomial confidence intervals when $p < 0.02$. The bottom-left and right panels report the contour levels of the return levels for (NO2, SO2) and (PM10, NO) jointly to events $\{NO > 270\}$ and $\{NO2 > 110, SO2 > 95\}$ respectively.

The joint return level can be interpreted as follows. For example, from the top-right panel we have that the 50 years joint return level of PM10 is 166. Concluding, we expect that PM10 will exceed the level 166 together with the event that NO, SO2, and NO2 simultaneously exceed the levels 270, 95, and 110, respectively, on average every 50 years.

16.6 Computational Details

The figures and the estimation results have been obtained using the free software R (R Core Team, 2015) and in particular the package ExtremalDep, available at https://r-forge.r-project.org/projects/extremaldep/. Bayesian estimation was obtained using and extending some routines of the package BMAmev. The left and middle panels of Figure 16.1 were obtained using the routines scatter3d and polygon3d of the package plot3D.

References

Apputhurai, P. and A. Stephenson (2011). Accounting for uncertainty in extremal dependence modeling using Bayesian model averaging techniques. *Journal of Statistical Planning and Inference 141*(5), 1800–1807.

Beirlant, J., Y. Goegebeur, J. Segers, and J. Teugels (2004). *Statistics of Extremes: Theory and Applications.* John Wiley & Sons.

Boldi, M.-O. and A. C. Davison (2007). A mixture model for multivariate extremes. *Journal of the Royal Statistical Society: Series B (Statistical Methodology) 69*(2), 217–229.

Coles, S. (2001). *An Introduction to Statistical Modeling of Extreme Values.* Springer Series in Statistics. London: Springer-Verlag.

Coles, S. G. and J. A. Tawn (1991). Modelling extreme multivariate events. *Journal of the Royal Statistical Society. Series B (Methodological) 53*(2), pp. 377–392.

Coles, S. G. and J. A. Tawn (1994). Statistical methods for multivariate extremes: An application to structural design. *Journal of the Royal Statistical Society: Series C (Applied Statistics) 43*(1), pp. 1–48.

Cooley, D., R. A. Davis, and P. Naveau (2010). The pairwise beta distribution: A flexible parametric multivariate model for extremes. *Journal of Multivariate Analysis 101*(9), 2103–2117.

Davison, A. C. (2003). *Statistical Models*. Cambridge.

Davison, A. C. and M. M. Gholamrezaee (2012). Geostatistics of extremes. *Proceedings of the Royal Society of London Series A: Mathematical and Physical Sciences 468*, 581–608.

Davison, A. C., S. A. Padoan, and M. Ribatet (2012). Statistical Modeling of Spatial Extremes. *Statistical Science 27*, 161–186.

de Haan, L. and A. Ferreira (2006). *Extreme Value Theory: An Introduction*. Springer Series in Operations Research and Financial Engineering. New York: Springer.

de Haan, L. and S. I. Resnick (1977). Limit theory for multivariate sample extremes. *Z. Wahrscheinlichkeitstheorie und Verw. Gebiete 40*(4), 317–337.

De Haan, L. and C. Zhou (2011). Extreme residual dependence for random vectors and processes. *Advances in Applied Probability 43*(1), 217–242.

Demarta, S. and A. J. McNeil (2005). The t copula and related copulas. *International Statistical Review 73*(1), 111–129.

Engelke, S., A. Malinowski, Z. Kabluchko, and M. Schlather (2015, July). Estimation of Hüsler–Reiss distributions and Brown–Resnick processes. *Journal of the Royal Statistical Society: Series B (Statistical Methodology) 77*(1), 239–265.

Falk, M., J. Hüsler, and R.-D. Reiss (2011). *Laws of Small Numbers: Extremes and Rare Events* (extended ed.). Birkhäuser/Springer Basel AG, Basel.

Geweke, J. (1992). Evaluating the accuracy of sampling-based approaches to the calculation of posterior moments. In J. M. Bernado, J. O. Berger, A. P. Dawid, and A. F. M. Smith (Eds.), *Bayesian Statistics*, Volume 4. Oxford, UK: Claredon Press.

Gudendorf, G. and J. Segers (2011). Nonparametric estimation of an extreme-value copula in arbitrary dimensions. *Journal of Multivariate Analysis 102*, 37–47.

Gudendorf, G. and J. Segers (2012). Nonparametric estimation of multivariate extreme-value copula. *Journal of Statistical Planning and Inference 142*, 3073–3085.

Gumbel, E. J. (1960). Distributions des valeurs extremes en plusieurs dimensions. *Publications de l'Institut de statistique de l'Universit de Paris 9*, 171–173.

Hastings, W. K. (1970). Monte Carlo sampling methods using Markov chains and their applications. *Biometrika 57*(1), 97–109.

Heffernan, J. E. and J. A. Tawn (2004). A conditional approach for multivariate extreme values (with discussion). *Journal of the Royal Statistical Society: Series B (Statistical Methodology) 66*(3), 497–546.

Heidelberger, P. and P. D. Welch (1981). A spectral method for confidence interval generation and run length control in simulations. *Communications of the ACM 24*(4), 233–245.

Hüsler, J. and R.-D. Reiss (1989). Maxima of normal random vectors: Between independence and complete dependence. *Statistics & Probability Letters 7*(4), 283–286.

Joe, H. (1997). *Multivariate models and dependence concepts*, Volume 73 of *Monographs on Statistics and Applied Probability*. London: Chapman & Hall.

Johansen, S. S. (2004). *Bivariate frequency analysis of flood characteristics in Glomma and Gudbrandsdalslagen*. Ph. D. thesis, University of Oslo, Department of Geosciences, Section of Geohazards and Hydrology.

Kotz, S. and S. Nadarajah (2000). *Extreme Value Distributions: Theory and Applications*. London: Imperial College Press.

Ledford, A. W. and J. A. Tawn (1996). Statistics for near independence in multivariate extreme values. *Biometrika 83*(1), 169–187.

Ledford, A. W. and J. A. Tawn (1997). Modelling dependence within joint tail regions. *Journal of the Royal Statistical Society: Series B (Statistical Methodology) 59*(2), 475–499.

Li, H. (2009). Orthant tail dependence of multivariate extreme value distributions. *Journal of Multivariate Analysis 100*(1), 243–256.

Marcon, G., S. A. Padoan, P. Naveau, and P. Muliere (2014). Multivariate nonparametric estimation of the Pickands dependence function using Bernstein polynomials. arXiv preprint arXiv:1405.5228.

Nikoloulopoulos, A. K., H. Joe, and H. Li (2009). Extreme value properties of multivariate t copulas. *Extremes 12*(2), 129–148.

Nolan, J. (2015). *Stable Distributions: Models for Heavy-Tailed Data*. Boston: Birkhauser. In progress.

Padoan, S. A. (2013a). Extreme dependence models based on event magnitude. *Journal of Multivariate Analysis 122*, 1–19.

Padoan, S. A. (2013b). Extreme value analysis. In A.-H. El-Shaarawi and W. Piegorsch (Eds.), *Encyclopedia of Environmetrics*, Volume 2. Chichester: John Wiley & Sons.

Padoan, S. A. (2013c). Max-stable processes. In A.-H. El-Shaarawi and W. Piegorsch (Eds.), *Encyclopedia of Environmetrics*, Volume 4. Chichester: John Wiley & Sons.

Padoan, S. A., M. Ribatet, and S. A. Sisson (2010). Likelihood-based inference for max-stable processes. *Journal of the American Statistical Association 105*(489), 263–277.

Pickands, III, J. (1981). Multivariate extreme value distributions. In *Proceedings 43rd Session International Statistical Institute*, Volume 2, pp. 859–878. Amsterdam: International Statistical Institute.

R Core Team (2015). *R: A Language and Environment for Statistical Computing*. Vienna, Austria: R Foundation for Statistical Computing.

Resnick, S. I. (2007). *Extreme Values, Regular Variation, and Point Processes*. Springer.

Ribatet, M. (2013). Spatial extremes: Max-stable processes at work. *Journal de la Société Francaise de Statistique 154*, 156–177.

Sabourin, A. and P. Naveau (2014). Bayesian Dirichlet mixture model for multivariate extremes: A re-parametrization. *Computational Statistics and Data Analysis 71*, 542–567.

Sabourin, A., P. Naveau, and A.-L. Fougères (2013). Bayesian model averaging for multivariate extremes. *Extremes 16*(3), 325–350.

Sakamoto, Y., M. Ishiguro, and G. Kitagawa (1986). *Akaike Information Criterion Statistics*. D. Reidel Publishing Company.

Smith, R. L. (1990). Max-stable processes and spatial extremes. University of Surrey.

Tawn, J. A. (1990). Modelling multivariate extreme value distributions. *Biometrika 77*(2), pp. 245–253.

Varin, C., N. Reid, and D. Firth (2011). An overview of composite likelihood methods. *Statist. Sinica 21*(1), 5–42.

Wadsworth, J. L. and J. A. Tawn (2012). Dependence modelling for spatial extremes. *Biometrika 99*, 253–272.

17

Nonparametric Estimation of Extremal Dependence

Anna Kiriliouk
Université Catholique de Louvain, Belgium

Johan Segers
Université Catholique de Louvain, Belgium

Michał Warchoł
Université Catholique de Louvain, Belgium

Abstract

There is an increasing interest to understand the dependence structure of a random vector not only in the center of its distribution but also in the tails. Extreme-value theory tackles the problem of modelling the joint tail of a multivariate distribution by modelling the marginal distributions and the dependence structure separately. For estimating dependence at high levels, the stable tail dependence function and the spectral measure are particularly convenient. These objects also lie at the basis of nonparametric techniques for modelling the dependence among extremes in the max-domain of attraction setting. In case of asymptotic independence, this setting is inadequate, and more refined tail dependence coefficients exist, serving, among others, to discriminate between asymptotic dependence and independence. Throughout, the methods are illustrated on financial data.

17.1 Introduction

Consider a financial portfolio containing three stocks: JP Morgan, Citibank, and IBM. We download stock prices from http://finance.yahoo.com between January 1, 2000, and December 31, 2013, and convert them to weekly negative log-

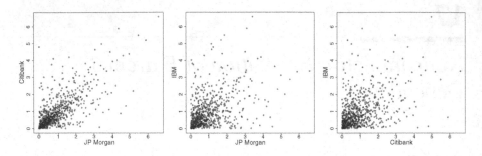

FIGURE 17.1
Scatterplots of negative weekly log-returns of stock prices of JP Morgan, Citibank, and IBM, plotted on the exponential scale.

returns: if $P_1, \ldots, P_n$ is a series of stock prices, the negative log-returns are

$$X_i = -\log(P_i/P_{i-1}), \qquad i = 1, \ldots, n.$$

These three series of negative log-returns will be denoted by the vectors $X_i = (X_{i1}, X_{i2}, X_{i3})$ for $i = 1, \ldots, 729$. By taking negative log-returns we force (extreme) losses to be in the upper tails of the distribution functions. It is such extreme values, and in particular their simultaneous occurrence, that are the focus of this chapter.

Figure 17.1 shows the scatterplots of the three possible pairs of negative log-returns on the exponential scale. Specifically, we transformed the negative log-returns to unit-Pareto margins through

$$\widehat{X}_{ij}^* := \frac{1}{1 - \widehat{F}_j(X_{ij})}, \qquad i \in \{1, \ldots, n\}, \; j \in \{1, \ldots, d\}, \qquad (17.1)$$

(with $d = 3$) and plotted them on the logarithmic scale; the empirical distribution functions $\widehat{F}_j$ evaluated at the data are defined as

$$\widehat{F}_j(X_{ij}) := \frac{R_{ij,n} - 1}{n}, \qquad i \in \{1, \ldots, n\}, \; j \in \{1, \ldots, d\}, \qquad (17.2)$$

where $R_{ij,n} = \sum_{l=1}^{n} \mathbf{1}\{X_{lj} \leqslant X_{ij}\}$ is the rank of X_{ij} among $X_{1j}, \ldots, X_{nj}$.

From Figure 17.1, we observe that joint occurrences of large losses are more frequent for JP Morgan and Citibank (left) than for JP Morgan and IBM (middle) or for Citibank and IBM (right). Given the fact that JP Morgan and Citibank are financial institutions while IBM is an IT company, this is no surprise. In statistical parlance, the pair JP Morgan versus Citibank exhibits the strongest degree of upper *tail dependence*.

The purpose of this chapter is to describe a statistical framework for modelling such tail dependence. Specifically, we wish to model the tail dependence structure of a d-variate random variable $X = (X_1, \ldots, X_d)$, with continuous marginal distribution functions $F_j(x_j) = \mathbb{P}(X_j \leq x_j)$ and joint distribution function $F(x) = \mathbb{P}(X_1 \leq x_1, \ldots, X_d \leq x_d)$.

Preliminary marginal transformations of the variables or changes in the measurement unit should not affect the dependence model. Such invariance is obtained by the probability integral transform, producing random variables $F_1(X_1), \ldots, F_d(X_d)$ that are uniformly distributed on the interval $(0, 1)$. Large values of X_j correspond to $F_j(X_j)$ being close to unity, whatever the original scale in which X_j was measured. In order to magnify large values, it is sometimes convenient to consider a further transformation to unit-Pareto margins via $X_j^* = 1/\{1 - F_j(X_j)\}$; note that $\mathbb{P}[X_j^* > z] = 1/z$ for $z \geq 1$.

In practice, the marginal distributions are unknown and need to be estimated. A simple, robust way to do so is via the empirical distribution functions. The data are then reduced to the vectors of ranks as in (17.1) and (17.2).

Starting point of the methodology is a mathematical description of the upper tail dependence of the vector $X^* = (X_1^*, \ldots, X_d^*)$ of standardized variables. There is a close connection with the classical theory on vectors of componentwise maxima, the multivariate version of the Fisher–Tippett–Gnedenko theorem (Fisher and Tippett, 1928; Gnedenko, 1943). In the literature, there is a plethora of objects available for describing the dependence component of the limiting multivariate extreme-value distribution. Out of these, we have singled out the stable tail dependence function ℓ and the spectral measure H (Section 17.2). The former is convenient as it yields a direct approximation of the joint tail of the distribution function of X^* and thus of X; see (17.6) below. The latter has the advantage of representing tail dependence as the distribution of the relative magnitudes of the components of X^* given that the vector itself is 'large'; see (17.11) below.

Nonparametric inference on tail dependence can then be based on the stable tail dependence function and the spectral measure (Section 17.3). For convenience and in line with most of the literature, the description of the spectral measure estimators is limited to the bivariate case, although the generalization to the general, multivariate case is straightforward.

A weak point of the classical theory is its treatment of asymptotic independence, that is, when the tail dependence coefficient in (17.25) is equal to zero. Classically, this case is assimilated to the situation where the variables are exactly independent. Since such a reduction applies to all bivariate normal distributions with pairwise correlation less than unity, it is clearly inadequate. More refined tail dependence coefficients as well as tests to distinguish asymptotic dependence from asymptotic independence are presented in Section 17.4.

The present chapter can serve as an introduction to nonparametric multivariate extreme value analysis. Readers looking for in-depth treatments of multivariate extreme-value distributions and their max-domains of attraction can consult, for instance, Chapter 5 in Resnick (1987) or Chapter 6 in de Haan and Ferreira (2006). Some other accessible introductions are the expository papers by de Haan and de Ronde (1998) and Segers (2012), Chapter 8 in Beirlant et al. (2004), and

Part II in Falk et al. (2011). The material on asymptotic independence is inspired by Coles et al. (1999).

17.2   Tail Dependence

17.2.1   The Stable Tail Dependence Function

Let $X = (X_1, \ldots, X_d)$ be a random vector with continuous marginal distribution functions $F_j(x_j) = \mathbb{P}(X_j \leq x_j)$ and joint distribution function $F(x) = \mathbb{P}(X_1 \leq x_1, \ldots, X_d \leq x_d)$. To study tail dependence, we zoom in on the joint distribution of $(F_1(X_1), \ldots, F_d(X_d))$ in the neighbourhood of its upper endpoint $(1, \ldots, 1)$. That is, we look at

$$1 - \mathbb{P}[F_1(X_1) \leq 1 - tx_1, \ldots, F_d(X_d) \leq 1 - tx_d]$$
$$= \mathbb{P}[F_1(X_1) > 1 - tx_1 \text{ or } \ldots \text{ or } F_d(X_d) > 1 - tx_d], \quad (17.3)$$

where $t > 0$ is small and where the numbers $x_1, \ldots, x_d \in [0, \infty)$ parametrize the relative distances to the upper endpoints of the d variables. The above probability converges to zero as $t \to 0$ and is in fact proportional to t:

$$\max(tx_1, \ldots, tx_d) \leqslant \mathbb{P}[F_1(X_1) > 1 - tx_1 \text{ or } \ldots \text{ or } F_d(X_d) > 1 - tx_d]$$
$$\leqslant tx_1 + \cdots + tx_d.$$

The *stable tail dependence function*, $\ell : [0, \infty)^d \to [0, \infty)$, is then defined by

$$\ell(x) := \lim_{t \downarrow 0} t^{-1} \mathbb{P}[F_1(X_1) > 1 - tx_1 \text{ or } \ldots \text{ or } F_d(X_d) > 1 - tx_d], \quad (17.4)$$

for $x \in [0, \infty)^d$ (Drees and Huang, 1998; Huang, 1992). The existence of the limit in (17.4) is an assumption that can be tested (Einmahl et al., 2006).

The probability in (17.3) represents the situation where *at least one* of the variables is large: for instance, the sea level exceeds a critical height at one or more locations. Alternatively, we might be interested in the situation where *all* variables are large simultaneously. Think of the prices of all stocks in a financial portfolio going down together. The *tail copula*, $R : [0, \infty)^d \to [0, \infty)$, is defined by

$$R(x) := \lim_{t \downarrow 0} t^{-1} \mathbb{P}[F_1(X_1) > 1 - tx_1, \ldots, F_d(X_d) > 1 - tx_d], \quad (17.5)$$

for $x \in [0, \infty]^d \setminus \{(\infty, \ldots, \infty)\}$. Again, existence of the limit is an assumption. In the bivariate case ($d = 2$), the functions ℓ and R are directly related by $R(x, y) = x + y - \ell(x, y)$. The difference between ℓ and R is visualized in Figure 17.2 for the log-returns of the stock prices of JP Morgan versus Citibank. From now on, we will focus on the function ℓ because of its direct link to the joint distribution function; see (17.6).

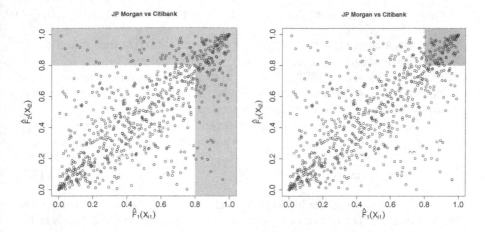

FIGURE 17.2
Scatterplots of negative weekly logreturns of stock prices of JP Morgan versus Citibank, transformed using ranks via (17.2). Left: the region where at least one variable is large, inspiring the definition of the stable tail dependence function ℓ in (17.4). Right: the region where both variables are large, inspiring the definition of the tail copula R in (17.5).

It is convenient to transform the random variables $X_1, \ldots, X_j$ to have the unit-Pareto distribution via the transformations $X_j^* = 1/\{1 - F_j(X_j)\}$ for $j = 1, \ldots, d$. The existence of ℓ in (17.4) is equivalent to the statement that the joint distribution function, H_0, of the random vector $\boldsymbol{X}^* = (X_1^*, \ldots, X_d^*)$ is in the max-domain of attraction of a d-variate extreme-value distribution, say G_0, with unit-Fréchet margins, i.e., $G_{0j}(z_j) = \exp(-1/z_j)$ for $z_j > 0$. The link between ℓ and G_0 is given by

$$\ell(\boldsymbol{x}) = \lim_{t \downarrow 0} t^{-1} \left[1 - \mathbb{P}\left\{ X_1^* \le 1/(x_1 t), \ldots, X_d^* \le 1/(x_d t) \right\} \right]$$
$$= \lim_{n \to \infty} -\log H_0^n(n/x_1, \ldots, n/x_d)$$
$$= -\log G_0(1/x_1, \ldots, 1/x_d),$$

so that $G_0(\boldsymbol{z}) = \exp\left\{ -\ell(1/z_1, \ldots, 1/z_d) \right\}$. Since, for "large" $z_1, \ldots, z_d$ we have

$$H_0(\boldsymbol{z}) \approx \exp\left\{ -\frac{1}{n} \ell\left(\frac{n}{x_1}, \ldots, \frac{n}{x_d} \right) \right\} = \exp\{-\ell(1/z_1, \ldots, 1/z_d)\}, \quad (17.6)$$

we see that estimating the stable tail dependence function ℓ is key to estimation of the tail of H_0, and thus, after marginal transformation, of F. If, in addition, the marginal distributions $F_1, \ldots, F_d$ are in the max-domains of attraction of the univariate extreme-value distributions $G_1, \ldots, G_d$, then F is in the max-domain of attraction of the d-variate extreme-value distribution $G(\boldsymbol{x}) = \exp\{-\ell(-\log G_1(x_1), \ldots, -\log G_d(x_d))\}$.

Every stable tail dependence function ℓ has the following properties:

1. $\max(x_1, \ldots, x_d) \leqslant \ell(\boldsymbol{x}) \leqslant x_1 + \cdots + x_d$, and in particular $\ell(0, \ldots, 0, x_j, 0, \ldots, 0) = x_j$ for $j = 1, \ldots, d$;
2. convexity, that is, $\ell(t\boldsymbol{x} + (1-t)\boldsymbol{y}) \leqslant t\,\ell(\boldsymbol{x}) + (1-t)\,\ell(\boldsymbol{y})$, for $t \in [0, 1]$;
3. order-one homogeneity: $\ell(ax_1, \ldots, ax_d) = a\,\ell(x_1, \ldots, x_d)$, for $a > 0$;

(Beirlant et al., 2004, page 257). When $d = 2$, these three properties characterize the class of stable tail dependence functions. When $d \geq 3$, a function satisfying these properties is not necessarily a stable tail dependence function.

The two boundary cases are complete dependence, $\ell(\boldsymbol{x}) = \max(x_1, \ldots, x_d)$, and independence, $\ell(\boldsymbol{x}) = x_1 + \cdots + x_d$. Another example is the *logistic model*,

$$\ell(\boldsymbol{x}; \theta) = \left(x_1^\theta + \cdots + x_d^\theta\right)^{1/\theta}, \qquad \theta \in [1, \infty). \tag{17.7}$$

As $\theta \to 1$, extremes become independent and as $\theta \to \infty$, extremes become completely dependent. The copula of the extreme-value distribution G_0 with this stable tail dependence function is the Gumbel–Hougaard copula.

The stable tail dependence function is related to two other functions:

*   The *exponent measure function* $V(\boldsymbol{z}) := -\log G_0(\boldsymbol{z}) = \ell(1/z_1, \ldots, 1/z_d)$, for $\boldsymbol{z} \in (0, \infty]^d$ (Coles and Tawn, 1991).
*   The *Pickands dependence function* A, which is just the restriction of ℓ to the unit simplex $\{\boldsymbol{x} \in [0, \infty)^d : x_1 + \cdots + x_d = 1\}$ (Pickands, 1981). In dimension $d = 2$, it is common to set $A(t) = \ell(1 - t, t)$ for $t \in [0, 1]$.

17.2.2   The Angular or Spectral Measure

Another insightful approach for modelling extremal dependence of a d-dimensional random vector is to look at the contribution of each of the d components to their sum conditionally on the sum being large.

Consider first the bivariate case. In Figure 17.3 we show pseudo-random samples from the Gumbel copula, which is related to the logistic model (17.7), at two different values of the dependence parameter θ: strong dependence ($\theta = 2.5$) at the top and weak dependence ($\theta = 1.25$) at the bottom. For convenience, the axes in the scatter plots on the left are on the logarithmic scale. The points for which the sum of the co-ordinates belong to the top 10% are shown in black. In the middle panels we plot the ratio of the first component to the sum (horizontal axis) against the sum itself (vertical axis). On the right, finally, histograms are given of these ratios for those points for which the sum exceeded the threshold in the plots on the left and in the middle. Strong extremal dependence leads to the ratios being close to 0.5 (top), whereas low extremal dependence leads to the ratios being close to either 0 or 1 (bottom).

We learn that the distribution of the ratios of the components to their sum, given that the sum is large, carries information about dependence between large values. This distribution is called the *angular* or *spectral measure* (de Haan and Resnick, 1977) and is denoted here by H. The reference lines in Figure 17.3 represent the spectral density associated with the Gumbel copula with parameter $\theta > 1$, that is,

$$\frac{\mathrm{d}H_\theta}{\mathrm{d}w}(w) = \frac{\theta - 1}{2} \{w(1-w)\}^{-1-\theta} \{w^{-\theta} + (1-w)^{-\theta}\}^{1/\theta - 2}, \qquad 0 < w < 1.$$

In Figure 17.4 we show the same plots as in Figure 17.3 but now for the pairs of negative log-returns of the stock prices of JP Morgan versus Citibank (top) and JP Morgan versus IBM (bottom). For each of the two pairs we do the following. Let (X_i, Y_i), $i = 1, \ldots, n$, denote a bivariate sample. First, we transform the data to the unit-Pareto scale using and (17.1) and (17.2), i.e.,

$$\widehat{X}_i^* = n/(n+1-R_{i,X}) \quad \text{and} \quad \widehat{Y}_i^* = n/(n+1-R_{i,Y}), \qquad i = 1, \ldots, n. \quad (17.8)$$

Here $R_{i,X}$ is the rank of X_i among $X_1, \ldots, X_n$ and $R_{i,Y}$ is the rank of Y_i among $Y_1, \ldots, Y_n$. Next, we construct an approximate sample from the spectral measure H by setting

$$\widehat{S}_i = \widehat{X}_i^* + \widehat{Y}_i^* \quad \text{and} \quad \widehat{W}_i = \widehat{X}_i^*/(\widehat{X}_i^* + \widehat{Y}_i^*), \qquad i = 1, \ldots, n. \quad (17.9)$$

The data in the pseudo-polar representation (17.9) are shown in the middle plots. The histograms of the ratios $\widehat{W}_i$ for those points for which the sum $\widehat{S}_i$ is above the 95% sample quantile are displayed on the right-hand side. The solid black lines show the smoothed spectral density estimator in (17.21) below.

For JP Morgan versus Citibank, the ratios are spread equally across the unit interval, suggesting extremal dependence. In contrast, for JP Morgan versus IBM, the spectral density has peaks at the boundaries 0 and 1, suggesting weak extremal dependence.

In the general, d-dimensional setting we transform the random vector $\boldsymbol{X}^* = (X_1^*, \ldots, X_d^*) = (1/\{1 - F_j(X_j)\} : j = 1, \ldots, d)$ as follows:

$$S = X_1^* + \cdots + X_d^* \quad \text{and} \quad \boldsymbol{W} = (X_1^*/S, \ldots, X_d^*/S). \quad (17.10)$$

The spectral measure is defined as the asymptotic distribution of the vector of ratios $\boldsymbol{W}$ given that the sum S is large:

$$\mathbb{P}\left[\boldsymbol{W} \in \cdot \mid S > u\right] \xrightarrow{d} H\left(\cdot\right), \qquad \text{as } u \to \infty. \quad (17.11)$$

The support of H is included in the unit simplex $\Delta_{d-1} = \{\boldsymbol{w} \in [0, \infty)^d : w_1 + \cdots + w_d = 1\}$. In case of asymptotic independence, the spectral measure is concentrated on the d vertices of Δ_{d-1}, whereas in case of complete asymptotic dependence it reduces to a unit point mass at the center of Δ_{d-1}.

The spectral measure is related to the stable tail dependence function via

$$\ell(\boldsymbol{x}) = d \int_{\Delta_{d-1}} \max_{j=1,\ldots,d} \{w_j x_j\} \, H(\mathrm{d}\boldsymbol{w}). \quad (17.12)$$

The property that $\ell(0, \ldots, 0, 1, 0, \ldots, 0) = 1$ implies the moment constraints

$$\int_{\Delta_{d-1}} w_j \, H(\mathrm{d}\boldsymbol{w}) = \frac{1}{d}, \qquad j = 1, \ldots, d. \quad (17.13)$$

Taking these constraints into account helps to improve the efficiency of nonparametric estimators of H (Section 17.3.1).

Any probability measure on the unit simplex satisfying (17.13) can be shown to be a valid spectral measure. The family of stable tail dependence functions is then characterized by (17.12) via a certain integral transform of such measures.

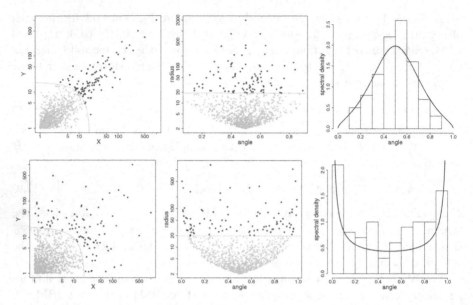

FIGURE 17.3
Pseudo-random samples from the Gumbel copula; $\theta = 2.5$ (top) and $\theta = 1.25$ (bottom). Left: the data transformed to unit-Pareto margins (log scale); middle: further transformation into pseudo-polar coordinates; right: histogram of the angles of the extreme pairs and the true spectral density (solid line).

17.3 Estimation

17.3.1 Estimating the Spectral Measure

Let (X_i, Y_i), $i = 1, \dots, n$, be independent copies of a random vector (X, Y) with distribution function F and continuous marginal distribution functions F_X and F_Y. Assume that F has a stable tail dependence function ℓ as in (17.24). The function ℓ can be represented by the spectral measure H via (17.12), which in the bivariate case specializes to

$$\ell(x, y) = 2 \int_0^1 \max\{wx, (1 - w)y\}\, H(\mathrm{d}w), \qquad (x, y) \in [0, \infty)^2, \qquad (17.14)$$

the unit simplex in $\mathbb{R}^2$ being identified with the unit interval, $[0, 1]$. The moment constraints in (17.13) simplify to

$$\int_0^1 w\, H(\mathrm{d}w) = \int_0^1 (1 - w)\, H(\mathrm{d}w) = \frac{1}{2}. \qquad (17.15)$$

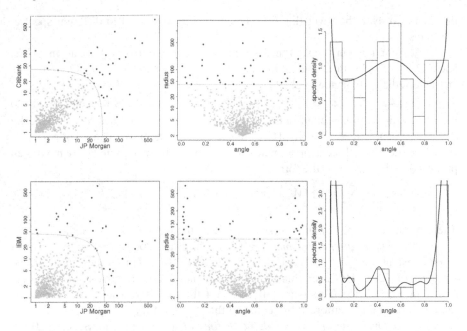

FIGURE 17.4
Log-returns of stock prices of JP Morgan versus Citibank (top) and JP Morgan versus IBM (bottom). Left: the data transformed to unit-Pareto margins (log scale); middle: further transformation into pseudo-polar coordinates; right: histogram of the angles of the extreme pairs and kernel density estimator (solid line).

The aim here is to estimate the spectral measure H. The starting point is the limit relation (17.11). Standardize the margins to the unit-Pareto distribution via $X^* = 1/\{1 - F_X(X)\}$ and $Y^* = 1/\{1 - F_Y(Y)\}$ and write

$$S = X^* + Y^*, \qquad W = X^*/(X^* + Y^*).$$

Equation (17.11) then becomes

$$\mathbb{P}[W \in \cdot \mid S > u] \overset{d}{\to} H(\cdot), \qquad \text{as } u \to \infty. \qquad (17.16)$$

The spectral measure H can be interpreted as the limit distribution of the angular component, W, given that the radial component, S, is large.

Mimicking the transformation from (X, Y) via (X^*, Y^*) to (S, W) in a given dataset, it is convenient to standardize the data to the unit-Pareto distribution via (17.8) and then switch to pseudo-polar coordinates using (17.9). Different choices for the angular components are possible; for example, Einmahl and Segers (2009) use $\arctan(X^*/Y^*)$. For the radial component one can take in general the L_p norm, $p \in [1, +\infty]$; Einmahl et al. (2001) take the L_∞ norm and de Carvalho et al. (2013) take the L_1 norm. Note however that different normalizations lead to different spectral measures.

For estimation purposes, transform pairs (X_i, Y_i) into $(\widehat{S}_i, \widehat{W}_i)$ as in (17.8) and (17.9). Set the threshold u in (17.16) to n/k, where $k = k_n \in (0, n]$ is an intermediate sequence such that $k \to \infty$ and $k/n \to 0$ as $n \to \infty$. Let $I_n = \{i = 1, \ldots, n : \widehat{S}_i \geq n/k\}$ denote the set of indices that correspond to the observations with pseudo-radial component $\widehat{S}_i$ above a high threshold, and let $N_n = |I_n|$ denote its cardinality. Here we specify u as the 95% quantile of $\widehat{S}_i$.

In the literature, three nonparametric estimators of the spectral measure have been proposed. All three are of the form

$$\widehat{H}_l(x) = \sum_{i \in I_n} \widehat{p}_{l,i} \, 1\{\widehat{W}_i \leq x\}, \qquad x \in [0, 1], \quad l = 1, 2, 3. \tag{17.17}$$

The estimators distinguish themselves in the way the weights $\widehat{p}_{l,i}$ are defined.

For the *empirical spectral measure* (Einmahl et al., 2001), the weights are set to be constant, i.e., $\widehat{p}_{1,i} = 1/N_n$. The estimator $\widehat{H}_1$ (17.17) becomes an empirical version of (17.16). However, this estimator does not necessarily satisfy the moment constraints in (17.15). This is a motivation for the two other estimators where the moment constraint is enforced by requiring that $\sum_{i \in I_n} \widehat{W}_i \widehat{p}_{l,i} = 1/2$.

The *maximum empirical likelihood estimator* (Einmahl and Segers, 2009), $\widehat{H}_2(x)$, has probability masses $\widehat{p}_{2,i}$ solving the optimization problem

$$\max_{mp \in \mathbb{R}_+^k} \quad \sum_{i \in I_n} \log p_i$$
$$\text{s.t.} \quad \sum_{i \in I_n} p_i = 1 \tag{17.18}$$
$$\sum_{i \in I_n} \widehat{W}_i p_i = 1/2.$$

By construction, the weights $\widehat{p}_{2,i}$ satisfy the moment constraints. The optimization problem in (17.18) can be solved by the method of Lagrange multipliers. In the non-trivial case where $1/2$ is in the convex hull of $\{W_i : i \in I_n\}$, the solution is given by

$$\widehat{p}_{2,i} = \frac{1}{N_n} \frac{1}{1 + \lambda(\widehat{W}_i - 1/2)}, \qquad i \in I_n,$$

where $\lambda \in \mathbb{R}$ is the Lagrange multiplier associated to the second equality constraint in (17.18), defined implicitly as the solution to the equation

$$\frac{1}{N_n} \sum_{i \in I_n} \frac{\widehat{W}_i - 1/2}{1 + \lambda(\widehat{W}_i - 1/2)} = 0.$$

Note that in (17.18), it is implicitly assumed that $\widehat{p}_{2,i} > 0$.

Another estimator that satisfies the moment constraints is the *maximum Euclidean likelihood estimator* (de Carvalho et al., 2013). The probability masses $\widehat{p}_{3,i}$ solve the following optimization problem:

$$\max_{mp \in \mathbb{R}^{I_n}} \quad -\tfrac{1}{2} \sum_{i \in I_n} (N_n p_i - 1)^2$$
$$\text{s.t.} \quad \sum_{i \in I_n} p_i = 1 \tag{17.19}$$
$$\sum_{i \in I_n} \widehat{W}_i p_i = 1/2.$$

This quadratic optimization problem with linear constraints can be solved explicitly with the method of Lagrange multipliers, yielding

$$\widehat{p}_{3,i} = \frac{1}{N_n}\{1 - (\overline{W} - 1/2)S_W^{-2}(\widehat{W}_i - \overline{W})\}, \qquad i \in I_n, \qquad (17.20)$$

where $\overline{W}$ and S_W^2 denote the sample mean and sample variance of $\widehat{W}_i, i \in I_n$, respectively, that is,

$$\overline{W} = \frac{1}{N_n}\sum_{i \in I_n}\widehat{W}_i, \qquad\qquad S_W^2 = \frac{1}{N_n}\sum_{i \in I_n}(\widehat{W}_i - \overline{W})^2.$$

The weights $\widehat{p}_{3,i}$ could be negative, but this does not usually occur as the weights are all nonnegative with probability tending to one.

It is shown in Einmahl and Segers (2009) and de Carvalho et al. (2013) that $\widehat{H}_2(x)$ and $\widehat{H}_3(x)$ are more efficient than $\widehat{H}_1(x)$. Moreover, asymptotically there is no difference between the maximum empirical likelihood or maximum Euclidean likelihood estimators. The maximum Euclidean likelihood estimator is especially convenient as the weights $\widehat{p}_{3,i}$ are given explicitly.

For the stock market data from JP Morgan, Citibank, and IBM, Figure 17.5 shows the empirical spectral measure $\widehat{H}_1(x)$ and the maximum Euclidean estimator $\widehat{H}_3(x)$ for the two pairs involving JP Morgan. In each case, the threshold u is set to be the 95% quantile of the sample of radii $\widehat{S}_i$. Enforcing the moment constraints makes a small but noticeable difference. Tail dependence is strongest for the pair JP Morgan versus Citibank, with spectral mass distributed approximately uniformly over $[0, 1]$. For the pair JP Morgan versus IBM, tail dependence is weaker, the spectral measure being concentrated mainly near the two endpoints, 0 and 1.

The nonparametric estimators of the spectral probability measure can be smoothed in such a way that the smoothed versions still obey the moment constraints. This can be done using kernel smoothing techniques, although some care is needed since the spectral measure is defined on a compact interval. In de Carvalho et al. (2013) the estimator is constructed by combining beta distributions with the weights, $\widehat{p}_{3,i}$, of the maximum Euclidean likelihood estimate. To ensure that the estimated measure obeys the marginal moment constraints, it is imposed that the mean of each smoother equals the observed pseudo-angle. The Euclidean spectral density estimator is defined as

$$\widetilde{h}_3(x) = \sum_{i \in I_n}\widehat{p}_{3,i}\,\beta(x; \widehat{W}_i\nu, (1 - \widehat{W}_i)\nu), \qquad x \in (0, 1), \qquad (17.21)$$

where $\nu > 0$ is the smoothing parameter, to be chosen via cross-validation, and

$$\beta(x; p, q) = \left\{\int_0^1 u^{p-1}(1 - u)^{q-1}\,\mathrm{d}u\right\}^{-1} x^{p-1}(1 - x)^{q-1}\,1_{[0,1]}(x),$$

is the beta density with parameters $p, q > 0$.

For the financial data, the realized spectral density estimator $\widetilde{h}_3(x)$ is plotted

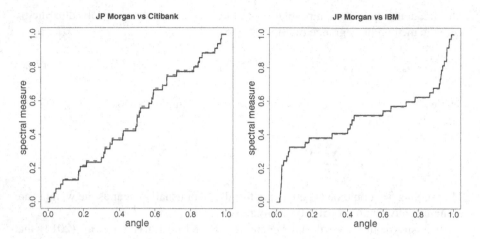

FIGURE 17.5
Estimated empirical spectral measures for JP Morgan versus Citibank (left) and JP Morgan versus IBM (right). The solid line and the dashed line correspond to $\widehat{H}_3(x)$ and $\widehat{H}_1(x)$, respectively.

on the right-hand panel of Figure 17.4. The picture is in agreement with the one in Figure 17.5. The distribution of the spectral mass points to asymptotic dependence for JP Morgan versus Citibank and to weaker asymptotic dependence for JP Morgan versus IBM.

Integrating out the estimated spectral density yields a smoothed version of the empirical spectral measure or its variants: for the maximum Euclidean likelihood estimator, we get

$$\widetilde{H}_3(x) = \int_0^x \widetilde{h}_3(v)\,\mathrm{d}v = \sum_{i \in I_n} \widehat{p}_{3,i}\,\mathcal{B}\big(x; \widehat{W}_i \nu, (1 - \widehat{W}_i)\nu\big), \qquad x \in [0,1], \quad (17.22)$$

with $\mathcal{B}(x; p, q) = \int_0^x \beta(y; p, q)\,\mathrm{d}y$ the regularized incomplete beta function. As

$$\int_0^1 x\,\widetilde{h}_3(x)\,\mathrm{d}x = \sum_{i \in I_n} \widehat{p}_{3,i}\,\frac{\nu \widehat{W}_i}{\nu \widehat{W}_i + \nu(1 - \widehat{W}_i)} = \sum_{i \in I_n} \widehat{p}_{3,i}\widehat{W}_i = 1/2, \quad (17.23)$$

the moment constraint (17.15) is satisfied.

17.3.2   Estimating the Stable Tail Dependence Function

Consider again the general, multivariate case. Given a random sample $X_i = (X_{i1}, \ldots, X_{id})$, $i = 1, \ldots, n$, the aim is to estimate the stable tail dependence function ℓ in (17.4) of the common distribution F with continuous margins $F_1, \ldots, F_d$. A straightforward nonparametric estimator can be defined as follows. Let $k = k_n \in$

$\{1, \ldots, n\}$ be such that $k \to \infty$ and $k/n \to 0$ as $n \to \infty$. By replacing $\mathbb{P}$ by the empirical distribution function, t by k/n, and $F_1, \ldots, F_d$ by $\widehat{F}_1, \ldots, \widehat{F}_d$ as defined in (17.2), we obtain the *empirical tail dependence function* (Drees and Huang, 1998; Huang, 1992)

$$\widehat{\ell}(\boldsymbol{x}) := \frac{n}{k} \frac{1}{n} \sum_{i=1}^{n} 1 \left\{ \widehat{F}_1(X_{i1}) > 1 - \frac{kx_1}{n} \text{ or } \ldots \text{ or } \widehat{F}_d(X_{id}) > 1 - \frac{kx_d}{n} \right\}$$

$$= \frac{1}{k} \sum_{i=1}^{n} 1 \{ R_{i1,n} > n + 1 - kx_1 \text{ or } \ldots \text{ or } R_{id,n} > n + 1 - kx_d \}.$$

Under minimal assumptions, the estimator is consistent and asymptotically normal with a convergence rate of $\sqrt{k}$ (Bücher et al., 2014; Einmahl et al., 2012). Alternatively, the marginal distributions might be estimated by $\widehat{F}_{j,2}(X_{ij}) = R_{ij,n}/n$ or $\widehat{F}_{j,3}(X_{ij}) = (R_{ij,n} - 1/2)/n$, resulting in estimators that are asymptotically equivalent to $\widehat{\ell}$.

Another estimator of ℓ can be defined by estimating the spectral measure, H, and applying the transformation in (17.12), an idea going back to Capéraà and Fougères (2000). In the bivariate case, we can replace H in (17.14) by one of the three nonparametric estimators $\widehat{H}_l$ studied in Section 17.3.1. Recall the notations in (17.8) and (17.9). If we modify the definition of the index set I_n to $I_n = \{i = 1, \ldots, n : \widehat{S}_i > \widehat{S}_{(k+1)}\}$, where $\widehat{S}_{(k+1)}$ denotes the $(k + 1)$-th largest observation of the $\widehat{S}_i$'s, then there are exactly k elements in the set I_n. Starting from the empirical spectral measure $\widehat{H}_1$, we obtain the estimator

$$\widehat{\ell}_{CF}(x, y) = \frac{1}{k} \sum_{i \in I_n} \max \left\{ \widehat{W}_i x, (1 - \widehat{W}_i) y \right\}.$$

For the maximum empirical or Euclidean likelihood estimators $\widehat{H}_2$ and $\widehat{H}_3$, one needs to replace the factor $1/k$ by the weights $\widehat{p}_{2,i}$ and $\widehat{p}_{3,i}$, respectively.

A way to visualize the function ℓ or an estimator thereof is via the level sets $\mathcal{D}_c := \{(x, y) : \ell(x, y) = c\}$ for a range of value of $c > 0$. Note that the level sets are equal to the lines $x + y = c$ in case of asymptotic independence and to the elbow curves $\max(x, y) = c$ in case of complete asymptotic dependence. Likewise, a plot of the level sets of an estimator of ℓ can be used as a graphical diagnostic of asymptotic (in)dependence; see de Haan and de Ronde (1998) or de Haan and Ferreira (2006, Section 7.2).

We plot the lines $\mathcal{D}_c$ for $c \in \{0.2, 0.4, 0.6, 0.8, 1\}$ and $k = 50$ of $\widehat{\ell}$ and $\widehat{\ell}_{CF}$ for the weekly log-returns of JP Morgan versus Citibank and JP Morgan versus IBM in Figure 17.6. The level sets for JP Morgan versus IBM resemble the straight lines $x + y = c$ much more closely than the level sets for JP Morgan versus Citibank do. The estimator based on the spectral measure, $\widehat{\ell}_{CF}$, acts as a smooth version of the empirical tail dependence function $\widehat{\ell}$.

Nonparametric estimators for ℓ can also serve as stepping stones for semiparametric inference. Assume that $\ell = \ell_{\theta_0} \in \{\ell_\theta : \theta \in \Theta\}$, a finite-dimensional parametric family. Then one could estimate θ_0 by minimizing a distance or discrepancy

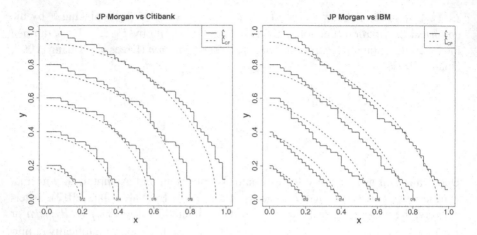

FIGURE 17.6
Level sets $\mathcal{D}_c$ of $\widehat{\ell}(x, y)$ and $\widehat{\ell}_{CF}$ for $c \in \{0.2, 0.4, 0.6, 0.8, 1\}$.

measure between $\widehat{\ell}$ and members of the parametric family (Einmahl et al., 2015, 2012).

17.4   Asymptotic Independence

In this section we will focus on the bivariate case. Let (X_i, Y_i), $i = 1, \dots, n$, be independent copies of a random vector (X, Y), with distribution function F having continuous margins F_X and F_Y. Recall from (17.4) that the stable tail dependence function (provided it exists) is given by

$$\ell(x, y) = \lim_{t \downarrow 0} t^{-1} \, \mathbb{P}[1 - F_X(X) \leqslant xt \text{ or } 1 - F_Y(Y) \leqslant yt], \qquad (x, y) \in [0, \infty)^2.$$
$$(17.24)$$

After marginal standardization, the pair of component-wise sample maxima converges weakly to the bivariate max-stable distribution G_0 given by $G_0(z_1, z_2) = \exp\{-\ell(1/z_1, 1/z_2)\}$ for $(z_1, z_2) \in (0, \infty)^2$; see (17.6). *Asymptotic independence* of the sample maxima, $G_0(z_1, z_2) = \exp(-1/z_1)\exp(-1/z_2)$ for all (z_1, z_2), is equivalent to $\ell(x, y) = x + y$ for all (x, y). The opposite case, $\ell(x, y) < x + y$ for some (x, y), is referred to as *asymptotic dependence*.

Sibuya (1960) already observed that bivariate normally distributed vectors are asymptotically independent as soon as the correlation is less than unity. We find that in case of asymptotic independence, we cannot rely on the function ℓ to quantify the amount of dependence left above high but finite thresholds.

Tail dependence coefficients serve to quantify the amount of tail dependence and to distinguish between asymptotic dependence versus independence (Section 17.4.1).

To decide which of the two situations applies for a dataset, a number of testing procedures are available (Section 17.4.2).

17.4.1   Tail Dependence Coefficients

A first measure for the strength of asymptotic dependence is the *tail dependence coefficient*

$$\chi := \lim_{u \uparrow 1} \mathbb{P}[F_X(X) > u \mid F_Y(Y) > u] = R(1,1) = 2 - \ell(1,1) \qquad (17.25)$$

(Coles et al., 1999). By the properties of ℓ (Section 17.2.1), we have $\ell(x,y) = x + y$ for all (x,y) if and only if $\ell(1,1) = 2$. That is, asymptotic independence is equivalent to $\chi = 0$, whereas asymptotic dependence is equivalent to $\chi \in (0,1]$.

To estimate χ, we will first write it as the limit of a function $\chi(u)$, so that $\lim_{u \uparrow 1} \chi(u) = \chi$; the function $\chi(u)$ is defined as

$$\chi(u) := 2 - \frac{1 - \mathbb{P}[F_X(X) < u, F_Y(Y) < u]}{1 - u}, \qquad 0 \leqslant u < 1.$$

To estimate $\chi(u)$ from a sample $(X_1, Y_1), \ldots, (X_n, Y_n)$, simply replace F_X and F_Y by their empirical counterparts in (17.2):

$$\widehat{\chi}(u) := 2 - \frac{1}{1-u}\left(1 - \frac{1}{n}\sum_{i=1}^{n} 1\left(\widehat{F}_X(X_i) < u, \, \widehat{F}_Y(Y_i) < u\right)\right).$$

For the stock price log-returns of JP Morgan, Citibank, and IBM, the estimated tail dependence coefficients $\widehat{\chi}(u)$ of the three possible pairs are shown in Figure 17.7 as a function of $u \geq 0.8$. The plot confirms the earlier finding that tail dependence is stronger between the two banking stocks, JP Morgan and Citibank, than between either of the two banks and IBM.

If the variables are asymptotically independent, $\chi = 0$, we need a second measure quantifying the extremal dependence. The *coefficient of tail dependence*, η, can be motivated as follows (Ledford and Tawn, 1996). Since $\mathbb{P}[1/\{1-F_X(X)\} > t] = 1/t$ for $t \geq 1$, we have

$$\mathbb{P}\left[\frac{1}{1-F_X(X)} > t, \frac{1}{1-F_Y(Y)} > t\right] \propto \begin{cases} t^{-1} & \text{for asymptotic dependence,} \\ t^{-2} & \text{for exact independence.} \end{cases}$$

A model that links these two situations is given by the assumption that the joint survival function above is regularly varying at ∞ with index $-1/\eta$, i.e.,

$$\mathbb{P}\left[\frac{1}{1-F_X(X)} > t, \frac{1}{1-F_Y(Y)} > t\right] = t^{-1/\eta} \mathcal{L}(t), \qquad (17.26)$$

where $\mathcal{L}$ is a slowly varying function, that is, $\mathcal{L}(tr)/\mathcal{L}(t) \to 1$ as $t \to \infty$ for fixed $r > 0$; see for example Resnick (1987).

For large t, if we treat $\mathcal{L}$ as constant, there are three cases of non-negative extremal association to distinguish:

1. $\eta = 1$ and $\mathcal{L}(t) \to c > 0$: asymptotic dependence;
2. $1/2 < \eta < 1$: positive association within asymptotic independence;
3. $\eta = 1/2$: (near) perfect independence.

Values of $\eta < 1/2$ occur when there is negative association between the two random variables at extreme levels.

In order to estimate η, we first define the structure variable

$$T := \min\left(\frac{1}{1 - F_X(X)}, \frac{1}{1 - F_Y(Y)}\right). \tag{17.27}$$

For a high threshold u and for $t \geq 0$, we have, by slow variation of $\mathcal{L}$,

$$\mathbb{P}[T > u + t \mid T > u] = \frac{\mathcal{L}(u+t)}{\mathcal{L}(u)}(1 + t/u)^{-1/\eta} \approx (1 + t/u)^{-1/\eta}.$$

On the right-hand side we recognize a generalized Pareto distribution with shape parameter η. Given the sample $(X_1, Y_1), \ldots, (X_n, Y_n)$, we can estimate η with techniques from univariate extreme value theory applied to the variables

$$\widehat{T}_i := \min\left(\frac{1}{1 - \widehat{F}_X(X_i)}, \frac{1}{1 - \widehat{F}_Y(Y_i)}\right), \qquad i = 1, \ldots, n.$$

One possibility consists of fitting the generalized Pareto distribution to the sample of excesses $\widehat{T}_i - u$ for those i for which $\widehat{T}_i > u$ (Ledford and Tawn, 1996, page 180). Alternatively, one can use the Hill estimator (Hill, 1975)

$$\widehat{\eta}_H := \frac{1}{k}\sum_{i=1}^{k} \log\left(\frac{\widehat{T}_{(i)}}{\widehat{T}_{(k+1)}}\right),$$

where $\widehat{T}_{(1)} \geqslant \cdots \geqslant \widehat{T}_{(n)}$ denote the order statistics of $\widehat{T}_1, \ldots, \widehat{T}_n$. Figure 17.7 shows the estimators $\widehat{\eta}_H$ of the coefficient of tail dependence between JP Morgan and Citibank, JP Morgan and IBM, and Citibank and IBM for a decreasing number of order statistics. The estimators suggest asymptotic dependence for the log-returns of JP Morgan versus Citibank.

Another frequently used measure in case of asymptotic independence can be obtained from (17.26). Taking logarithms on both sides and using slow variation of $\mathcal{L}$, we obtain

$$\frac{1}{\eta} = \lim_{u \uparrow 1} \frac{\log \mathbb{P}[F_X(X) > u, F_Y(Y) > u]}{\log \mathbb{P}[F_X(X) > u]}.$$

Coles et al. (1999) proposed the coefficient

$$\overline{\chi} := \lim_{u \uparrow 1} \overline{\chi}(u) = \lim_{u \uparrow 1} \frac{2 \log \mathbb{P}[F_X(X) > u]}{\log \mathbb{P}[F_X(X) > u, F_Y(y) > u]} - 1 = 2\eta - 1.$$

As $0 < \eta \leq 1$, we have $-1 < \overline{\chi} \leqslant 1$. For a bivariate normal dependence structure, $\overline{\chi}$ is equal to the correlation coefficient ρ. Together, the two coefficients χ and $\overline{\chi}$ contain complementary information on the amount of tail dependence:

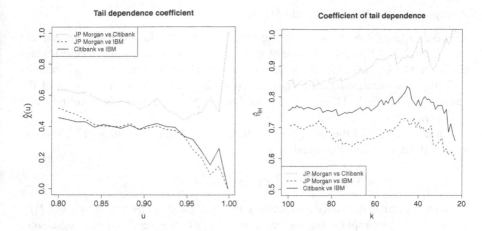

FIGURE 17.7
Estimators $\widehat{\chi}(u)$ (left) and $\widehat{\eta}_H$ (right) for the three pairs of log-returns.

- If $\chi = 0$ and $-1 < \overline{\chi} \leqslant 1$, the variables X and Y are asymptotically independent and $\overline{\chi}$ quantifies the degree of dependence. The cases $\overline{\chi} > 0$, $\overline{\chi} = 0$, and $\overline{\chi} < 0$ correspond to positive association, near independence, and negative association, respectively.
- If $\overline{\chi} = 1$ and $0 < \chi \leqslant 1$, the variables are asymptotically dependent, and χ quantifies the degree of dependence.

Although the graphical procedures in this section are easy to use, it is difficult to draw a conclusion from the plots, and it might be better to consider formal tests of asymptotic independence (Section 17.4.2). For an overview of more recent and advanced techniques in asymptotic independence, see for example de Carvalho and Ramos (2012) or Bacro and Toulemonde (2013).

17.4.2 Testing for Asymptotic (In)Dependence

Above we have called the variables X and Y asymptotically independent if $\ell(x, y) = x + y$. A sufficient condition for asymptotic independence is that the coefficient of tail dependence, η, in (17.26) is less than unity.

These observations yield two possible approaches to test for asymptotic (in)dependence, depending on the choice of the null hypothesis. Suppose the function $\mathcal{L}$ in (17.26) converges to a positive constant.

- Testing $H_0 : \eta = 1$ versus $H_1 : \eta < 1$ amounts to choosing asymptotic dependence as the null hypothesis.
- Testing $H_0 : \forall (x, y), \ \ell(x, y) = x + y$ versus $H_1 : \exists (x, y), \ \ell(x, y) < x + y$ means that asymptotic independence is taken as the null hypothesis.

Asymptotic dependence as the null hypothesis.

In order to test for $H_0 : \eta = 1$ versus $H_1 : \eta < 1$, the idea is to estimate η and to reject H_0 if the estimated value of η is below some critical value. The estimator chosen in Draisma et al. (2004) is the maximum likelihood estimator, $\widehat{\eta}_{\text{MLE}}$, obtained by fitting a generalized Pareto distribution to the k highest order statistics of the sample $\widetilde{T}_1, \ldots, \widetilde{T}_n$ in (17.27). The effective sample size, k, tends to infinity at a slower rate than n.

In order to derive the asymptotic distribution of $\widehat{\eta}_{\text{MLE}}$, condition (17.26) needs to be refined. Consider the function

$$q(x, y) := \mathbb{P}[1 - F_X(X) < x, 1 - F_Y(Y) < y].$$

Note that the tail dependence coefficient χ in (17.25) is given by $\chi = \lim_{t \downarrow 0} q(t, t)/t$. Equation (17.26) implies that the function $t \mapsto q(t, t)$ is regularly varying at zero with index $1/\eta$, i.e., $\lim_{t \downarrow 0} q(tx, tx)/q(t, t) = x^{1/\eta}$ for $x > 0$. In Draisma et al. (2004), this relation is refined to *bivariate regular variation*: it is assumed that

$$\lim_{t \downarrow 0} \frac{q(tx, ty)}{q(t, t)} = c(x, y), \qquad (x, y) \in (0, \infty)^2. \qquad (17.28)$$

The limit function c is homogeneous of order $1/\eta$, i.e., $c(ax, ay) = a^{1/\eta}c(x, y)$ for $a > 0$. Clearly, $c(1, 1) = 1$. To control the bias of $\widehat{\eta}_{\text{MLE}}$, a second-order refinement of (17.28) is needed: assume the existence of the limit

$$c_1(x, y) := \lim_{t \downarrow 0} \frac{\frac{q(tx, ty)}{q(t, t)} - c(x, y)}{q_1(t)},$$

for all $x, y \geqslant 0$ with $x + y > 0$. Here, $0 < q_1(t) \to t$ as $t \to 0$, while the limit function c_1 is neither constant nor a multiple of c. The convergence is assumed to be uniform on $\{(x, y) \in [0, \infty)^2 : x^2 + y^2 = 1\}$.

Suppose that the function c has partial derivatives $c_x := \partial c(x, y)/\partial x$ and $c_y := \partial c(x, y)/\partial y$. Let k be a sequence such that $\sqrt{k}\, q_1(q^{-1}(k/n, k/n)) \to 0$ as $n \to \infty$, where $q^{-1}(u, u) = t$ if and only if $q(t, t) = u$. Then $\widehat{\eta}_{\text{MLE}}$ is asymptotically normal,

$$\sqrt{k}(\widehat{\eta}_{\text{MLE}} - \eta) \xrightarrow{d} \mathcal{N}(0, \sigma^2(\eta)),$$

the asymptotic variance being

$$\sigma^2(\eta) = (1 + \eta)^2 \, (1 - \chi) \, (1 - 2\chi c_x(1, 1)c_y(1, 1)).$$

Under the null hypothesis $H_0 : \eta = 1$, consistent estimators of χ and $c_x(1, 1)$ are given by

$$\widehat{\chi} := \frac{k\widehat{T}_{(k+1)}}{n}, \qquad \widehat{c}_x(1, 1) := \frac{\widehat{p}^{5/4}}{n} \left[\widehat{T}^{(n, \widehat{p}^{-1/4})}_{(k+1)} - \widehat{T}_{(k+1)}\right],$$

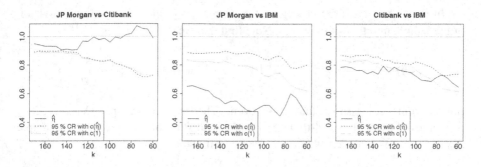

FIGURE 17.8
Estimates $\widehat{\eta}$ (solid lines) obtained by maximum likelihood based on excesses over the threshold $\widehat{T}_{(k+1)}$, together with pointwise 95% critical regions (CR).

where $\widehat{p} := k/\widehat{\chi}$ and $\widehat{T}_{(k+1)}^{(n,u)}$ is the $(k+1)$th largest observation of

$$\widehat{T}_i^{(n,u)} := \min\left(\frac{(1+u)}{1 - \widehat{F}_X(X_i)}, \frac{1}{1 - \widehat{F}_Y(Y_i)}\right), \qquad i = 1, \ldots, n.$$

The estimator $\widehat{c}_y(1,1)$ is defined analogously to $\widehat{c}_x(1,1)$. The null hypothesis $H_0 : \eta = 1$ is rejected at significance level α if

$$\widehat{\eta}_{\mathrm{MLE}} \leq 1 - \frac{\widehat{\sigma}}{\sqrt{k}}\Phi^{-1}(1 - \alpha), \tag{17.29}$$

where $\widehat{\sigma} = \widehat{\sigma}(1)$ or $\widehat{\sigma} = \widehat{\sigma}(\widehat{\eta}_{\mathrm{MLE}})$.

Figure 17.8 shows the estimates of $\widehat{\eta}_{\mathrm{MLE}}$ for a varying number of order statistics k, together with the lines defining the critical regions for the two tests in (17.29). We used the function gpd.fit from the ismev package (Heffernan and Stephenson, 2012). We see clearly that for JP Morgan versus Citibank, for every threshold, asymptotic dependence cannot be rejected; for JP Morgan versus IBM, asymptotic dependence is rejected at every threshold; and for Citibank versus IBM, results vary depending on the value of k.

Asymptotic independence as the null hypothesis.

Another approach for deciding between asymptotic dependence and asymptotic independence is to assume a null hypothesis of asymptotic independence, i.e., $H_0 : \forall (x,y),\ \ell(x,y) = x + y$. Recall the nonparametric estimator $\widehat{\ell}$ in Section 17.3.2. A natural approach would be to reject H_0 as soon as the difference between $\widehat{\ell}$ and the function $(x,y) \mapsto x + y$ is too large. However, from Einmahl et al. (2006) it follows that, under H_0,

$$\sup_{x,y \in [0,1]} \sqrt{k}\left|\widehat{\ell}(x,y) - (x+y)\right| \xrightarrow{\mathrm{Pr}} 0.$$

The limit being degenerate, it is not possible to compute critical values.

In Hüsler and Li (2009) another estimator of ℓ is proposed, based on a division of the sample into two sub-samples. Note first that we can write $\widehat{\ell}$ as

$$\widehat{\ell}(x, y) = \frac{1}{k} \sum_{i=1}^{n} 1\left\{ X_i > \widehat{F}_X^{-1}\left(1 - \frac{kx}{n}\right) \text{ or } Y_i > \widehat{F}_Y^{-1}\left(1 - \frac{ky}{n}\right) \right\}.$$

For convenience, assume that n is even. The first sub-sample $X_1, \ldots, X_{n/2}$ is compared with the $[kx]$-th largest order statistics of the second sub-sample $X_{n/2+1}, \ldots, X_n$; in other words, we use the estimator

$$\widetilde{\ell}_n(x, y) = \frac{1}{k} \sum_{i=1}^{n/2} 1\left\{ X_i > \widehat{\widetilde{F}}_X^{-1}\left(1 - \frac{2kx}{n}\right) \text{ or } Y_i > \widehat{\widetilde{F}}_Y^{-1}\left(1 - \frac{2ky}{n}\right) \right\}$$

$$= \frac{1}{k} \sum_{i=1}^{n/2} 1\left\{ \widetilde{R}_{i,X} > n/2 + 1 - kx \text{ or } \widetilde{R}_{i,Y} > n/2 + 1 - ky \right\},$$

where $\widetilde{R}_{i,X}$ for $i = 1, \ldots, n/2$ denotes the rank of X_i among $X_{n/2+1}, \ldots, X_n$. Under H_0, the estimator $\widetilde{\ell}_n(x, y)$ converges in probability to $x + y$, for fixed $x, y > 0$. Define

$$D_n(x, y) := \sqrt{k}\{\widetilde{L}_n(x, y) - (x + y)\}, \qquad x, y \in [0, 1].$$

Then under H_0, assuming certain regularity conditions on ℓ and k (Hüsler and Li, 2009, page 992), if $k = k_n \to \infty$ as $n \to \infty$,

$$\{D_n(x, y)\}_{x,y \in [0,1]} \xrightarrow{d} \{W_1(2x) + W_2(2y)\}_{x,y \in [0,1]}, \qquad \text{as } n \to \infty,$$

where W_1 and W_2 are two independent Brownian motions. By the continuous mapping theorem,

$$T_{I,n} := \int_{[0,1]^2} D_n(x, y)^2 \, dx \, dy \xrightarrow{d} \int_{[0,1]^2} \left(W_1(2x) + W_2(2y)\right)^2 dx \, dy =: T_I,$$

$$T_{S,n} := \sup_{x,y \in [0,1]} |D_n(x, y)| \xrightarrow{d} \sup_{x,y \in [0,1]} |W_1(2x) + W_2(2y)| =: T_S.$$

We reject the null hypothesis of asymptotic independence at significance level α if $T_{I,n} > Q_{T_I}(1 - \alpha)$ or if $T_{S,n} > Q_{T_S}(1 - \alpha)$, where Q_{T_I} and Q_{T_S} represent the quantile functions of T_I and T_S, respectively. Hüsler and Li (2009) compute $Q_{T_I}(0.95) = 6.237$ and $Q_{T_S}(0.95) = 4.956$.

Figure 17.9 shows the values of the test statistics $T_{I,n}$ and $T_{S,n}$ for the three pairs of stock returns, together with the pointwise critical regions for a range of k-values. The results are in agreement with Figure 17.8: asymptotic independence is rejected for JP Morgan versus Citibank, asymptotic independence cannot be rejected for JP Morgan versus IBM, and conclusions for Citibank versus IBM depend on the value of k.

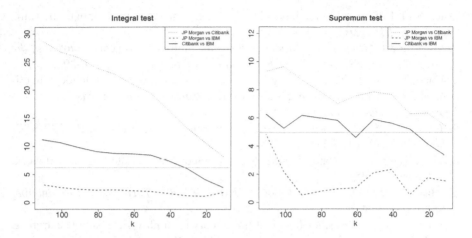

FIGURE 17.9
Test statistics $T_{I,n}$ (left) and $T_{S,n}$ (right) for a varying number of order statistics k. The horizontal lines define the critical regions $Q_{T_I}(0.95) = 6.237$ and $Q_{T_S}(0.95) = 4.956$.

Acknowledgments

The authors gratefully acknowledge funding by contract "Projet d'Actions de Recherche Concertées" No. 12/17-045 of the "Communauté française de Belgique" and by IAP research network Grant P7/06 of the Belgian government (Belgian Science Policy). Anna Kiriliouk and Michał Warchoł gratefully acknowledge funding from the Belgian Fund for Scientific Research (F.R.S. - FNRS).

References

Bacro, J.-N. and G. Toulemonde (2013). Measuring and modelling multivariate and spatial dependence of extremes. *Journal de la Société Française de Statistique 154*(2), 139–155.

Beirlant, J., Y. Goegebeur, J. Segers, and J. Teugels (2004). *Statistics of Extremes: Theory and Applications*. Wiley.

Bücher, A., J. Segers, and S. Volgushev (2014). When uniform weak convergence fails: Empirical processes for dependence functions and residuals via epi-and hypographs. *The Annals of Statistics 42*(4), 1568–1634.

Capéraà, P. and A.-L. Fougères (2000). Estimation of a bivariate extreme value distribution. *Extremes 3*(4), 311–329.

Coles, S., J. Heffernan, and J. Tawn (1999). Dependence measures for extreme value analyses. *Extremes 2*(4), 339–365.

Coles, S. and J. Tawn (1991). Modelling extreme multivariate events. *Journal of the Royal Statistical Society: Series B (Statistical Methodology) 53*, 377–392.

de Carvalho, M., B. Oumow, J. Segers, and M. Warchoł (2013). A Euclidean likelihood estimator for bivariate tail dependence. *Communications in Statistics: Theory and Methods 42*(7), 1176–1192.

de Carvalho, M. and A. Ramos (2012). Bivariate extreme statistics, II. *REVSTAT Statistical Journal 11*(1), 83–107.

de Haan, L. and J. de Ronde (1998). Sea and wind: Multivariate extremes at work. *Extremes 1*, 7–45.

de Haan, L. and A. Ferreira (2006). *Extreme Value Theory: An Introduction.* Springer-Verlag Inc.

de Haan, L. and S. I. Resnick (1977). Limit theory for multivariate sample extremes. *Z. Wahrscheinlichkeitstheorie und Verw. Gebiete 40*(4), 317–337.

Draisma, G., H. Drees, A. Ferreira, and L. De Haan (2004). Bivariate tail estimation: Dependence in asymptotic independence. *Bernoulli 10*(2), 251–280.

Drees, H. and X. Huang (1998). Best attainable rates of convergence for estimators of the stable tail dependence function. *Journal of Multivariate Analysis 64*(1), 25–47.

Einmahl, J. H. J., L. de Haan, and D. Li (2006). Weighted approximations of tail copula processes with application to testing the bivariate extreme value condition. *The Annals of Statistics 34*, 1987–2014.

Einmahl, J. H. J., L. de Haan, and V. I. Piterbarg (2001). Nonparametric estimation of the spectral measure of an extreme value distribution. *The Annals of Statistics 29*(5), 1401–1423.

Einmahl, J. H. J., A. Kiriliouk, A. Krajina, and J. Segers (2015). An M-estimator of spatial tail dependence. *Journal of the Royal Statistical Society: Series B (Statistical Methodology).* In press.

Einmahl, J. H. J., A. Krajina, and J. Segers (2012). An M-estimator for tail dependence in arbitrary dimensions. *The Annals of Statistics 40*, 1764–1793.

Einmahl, J. H. J. and J. Segers (2009). Maximum empirical likelihood estimation of the spectral measure of an extreme-value distribution. *The Annals of Statistics 37*(5B), 2953–2989.

Falk, M., J. Hüsler, and R.-D. Reiss (2011). *Laws of Small Numbers: Extremes and Rare Events* (extended ed.). Birkhäuser/Springer Basel AG, Basel.

Fisher, R. A. and L. H. C. Tippett (1928). Limiting forms of the frequency distribution of the largest or smallest member of a sample. *Mathematical Proceedings of the Cambridge Philosophical Society 24*(2), 180–190.

Gnedenko, B. (1943). Sur la distribution limite du terme maximum d'une serie aleatoire. *Annals of Mathematics 44*, 423–453.

Heffernan, J. E. and A. G. Stephenson (2012). *ismev: An Introduction to Statistical Modeling of Extreme Values.* R package version 1.38.

Hill, B. M. (1975). A simple general approach to inference about the tail of a distribution. *The Annals of Statistics 3*, 1163–1174.

Huang, X. (1992). *Statistics of Bivariate Extreme Values*. Ph. D. thesis, Tinbergen Institute Research Series.

Hüsler, J. and D. Li (2009). Testing asymptotic independence in bivariate extremes. *Journal of Statistical Planning and Inference 139*, 990–998.

Ledford, A. W. and J. A. Tawn (1996). Statistics for near independence in multivariate extreme values. *Biometrika 83*, 169–187.

Pickands, III, J. (1981). Multivariate extreme value distributions. In *Proceedings 43rd Session International Statistical Institute*, Volume 2, pp. 859–878. Amsterdam: International Statistical Institute.

Resnick, S. I. (1987). *Extreme Values, Regular Variation, and Point Processes*. Springer, New York.

Segers, J. (2012). Max-stable models for multivariate extremes. *REVSTAT Statistical Journal 10*, 61–82.

Sibuya, M. (1960). Bivariate extreme statistics, I. *Annals of the Institute of Statistical Mathematics 11*(2), 195–210.

18

An Overview of Nonparametric Tests of Extreme-Value Dependence and of Some Related Statistical Procedures

Axel Bücher

Ruhr-Universität Bochum, Germany

Ivan Kojadinovic

University of Pau, France

Abstract

An overview of existing nonparametric tests of extreme-value dependence is presented. Given an i.i.d. sample of random vectors from a continuous distribution, such tests aim at assessing whether the underlying unknown copula is of the *extreme-value* type or not. The existing approaches available in the literature are summarized according to how departure from extreme-value dependence is assessed. Related statistical procedures useful when modeling data with this type of dependence are briefly described next. Two illustrations on real datasets are then carried out using some of the statistical procedures under consideration implemented in the R package `copula`. Finally, the related problem of testing the *maximum domain of attraction* condition is discussed.

18.1 Introduction

By definition, the class of extreme-value copulas consists of all possible limit copulas of affinely normalized, componentwise maxima of a multivariate i.i.d. sample, or, more generally, of a multivariate stationary time series. As a consequence, extreme-value copulas can be seen as appropriately capturing the dependence between extreme or rare events. The famous extremal types theorem of multivariate extreme value theory leads to a rather simple characterization of extreme-value copulas: the class of extreme-value copulas merely coincides with the class of max-stable copulas (see Section 18.2 below for a precise definition). Other characterizations are possible, most of which are based on a parametrization by a lower-dimensional function or measure (see, e.g., Gudendorf and Segers, 2010, for an overview). Serial dependence of the underlying time series is explicitly allowed provided certain mixing conditions hold (Hsing, 1989; Hüsler, 1990).

The theory underlying extreme-value copulas motivates their use in combination with the famous *block maxima* method popularized in the univariate case in the monograph of Gumbel (1958): from a given time series, calculate (componentwise) monthly or annual or, more generally, block maxima, and consider the class of extreme-value copulas (or parametric subclasses thereof) as an appropriate model for the multivariate sample of block maxima. If the block size is sufficiently large, it is unlikely that the respective maxima within a block occur at the beginning or the end of the block, whence, even under weak serial dependence of the underlying time series, block maxima could be considered as approximately independent. In statistical practice, independence has usually been postulated hitherto. Applications of the block maxima method can also be found in contexts in which the underlying time series is not necessarily stationary, as is the case when seasonalities are present (for instance in some hydrological problems).

The use of extreme-value copulas is not restricted to the framework of multivariate extreme-value theory. These dependence structures can actually be a convenient choice to model any datasets with positive association. Moreover, many parametric submodels are available in the literature (e.g., Gudendorf and Segers, 2010).

Extreme-value copulas have been successfully applied in empirical finance and insurance (e.g., Cebrian et al., 2003; Longin and Solnik, 2001; McNeil et al., 2005), and environmental sciences (e.g., Salvadori et al., 2007; Tawn, 1988). They also arise in spatial statistics in connection with max-stable processes in which they determine the underlying spatial dependence (e.g., Davison et al., 2012; Ribatet, 2013; Ribatet and Sedki, 2013).

From a statistical point of view, it is important to test the hypothesis that the copula of a given sample is an extreme-value copula. When applied within the context of the block maxima method, a rejection of this hypothesis would indicate that the size of the blocks is too small and should be enlarged, or that the (broad) conditions of the extremal types theorem are not satisfied. When applied outside of the extremal types theoretical framework, tests of extreme-value dependence merely indicate whether the class of max-stable copulas is a plausible choice for modeling the cross-sectional dependence in the data at hand. If there is no evidence against this class, additional statistical procedures tailored for extreme-value copulas can be used to carry out the data analysis.

This chapter is organized as follows. A brief overview of the theory underlying extreme-value copulas is given in the second section. The third section provides a summary of the procedures available in the literature for testing whether the copula of a random sample from a continuous distribution can be considered of the extreme-value type or not. Rather detailed analyses of bivariate financial data and bivariate insurance data are presented next. They are accompanied by code for the R statistical system (R Development Core Team, 2014) from the `copula` package (Hofert et al., 2014). Finally, in the last section, the related issue of testing the maximum domain of attraction condition is discussed.

18.2 Mathematical Foundations

Consider a d-dimensional random vector $\mathbf{X} = (X_1, \ldots, X_d)$, $d \geq 2$, whose marginal cumulative distributions functions (c.d.f.s) $F_1, \ldots, F_d$ are assumed to be continuous. Then, by Sklar (1959)'s representation theorem, the c.d.f. F of $\mathbf{X}$ can be written in a unique way as

$$F(\mathbf{x}) = C\{F_1(x_1), \ldots, F_d(x_d)\}, \qquad \mathbf{x} = (x_1, \ldots, x_d) \in \mathbb{R}^d,$$

where the function $C : [0, 1]^d \to [0, 1]$ is a copula, i.e., the restriction of a multivariate c.d.f. with standard uniform margins to the unit hypercube. The above display is usually interpreted in the way that the copula C completely characterizes the stochastic dependence among the components of $\mathbf{X}$.

A d-dimensional copula C is an *extreme-value* copula if and only if there exists a copula C^* such that, for any $\mathbf{u} \in [0, 1]^d$,

$$\lim_{n \to \infty} \{C^*(u_1^{1/n}, \ldots, u_d^{1/n})\}^n = C(\mathbf{u}). \tag{18.1}$$

The copula C^* is then said to be in the *maximum domain of attraction* of C, which shall be denoted as $C^* \in D(C)$ in what follows.

Some algebra reveals that $\{C^*(u_1^{1/n}, \ldots, u_d^{1/n})\}^n$ is the copula, evaluated at $\mathbf{u} \in [0, 1]^d$, of the vector of componentwise maxima computed from an i.i.d. sample $\mathbf{Y}_1, \ldots, \mathbf{Y}_n$ with continuous marginal c.d.f.s and copula C^*. The latter fact motivates the terminology *extreme-value copula*. It is additionally very useful to note that C is an extreme-value copula if and only if it is *max-stable*, that is, if and only if, for any $\mathbf{u} \in [0, 1]^d$ and $r \in \mathbb{N}, r > 0$,

$$\{C(u_1^{1/r}, \ldots, u_d^{1/r})\}^r = C(\mathbf{u}). \tag{18.2}$$

The sufficiency follows by using, in combination with (18.1), the fact that, for any $\mathbf{u} \in [0, 1]^d$ and $r \in \mathbb{N}, r > 0$,

$$\left[C^*\{(u_1^{1/r})^{1/n}, \ldots, (u_d^{1/r})^{1/n}\}\right]^{1/n} = \left[\{C^*(u_1^{1/(nr)}, \ldots, u_d^{1/(nr)})\}^{1/(nr)}\right]^r.$$

The necessity is an immediate consequence of the fact $C \in D(C)$ for any max-stable copula C. Interestingly enough, it can be shown that a max-stable copula actually satisfies (18.2) for any real $r > 0$ (e.g., Galambos, 1978, Lemma 5.4.1).

An alternative, more complex characterization, essentially due to Pickands

(1981), is as follows: a copula C is of the extreme-value type if and only if there exists a function A such that, for any $\mathbf{u} \in (0, 1]^d \setminus \{(1, \ldots, 1)\}$,

$$C(\mathbf{u}) = \exp\left\{\left(\sum_{j=1}^{d} \log u_j\right) A\left(\frac{\log u_2}{\sum_{j=1}^{d} \log u_j}, \ldots, \frac{\log u_d}{\sum_{j=1}^{d} \log u_j}\right)\right\}, \quad (18.3)$$

where $A : \Delta_{d-1} \to [1/d, 1]$ is the *Pickands dependence function* and $\Delta_{d-1} = \{(w_1, \ldots, w_{d-1}) \in [0, 1]^{d-1} : w_1 + \cdots + w_{d-1} \leq 1\}$ is the unit simplex (see, e.g., Gudendorf and Segers, 2012, for more details). If relation (18.3) is met, then A is necessarily convex and satisfies the boundary condition $\max\{1 - \sum_{j=1}^{d-1} w_j, w_1, \ldots, w_{d-1}\} \leq A(\mathbf{w}) \leq 1$ for all $\mathbf{w} = (w_1, \ldots, w_{d-1}) \in \Delta_{d-1}$. The latter two conditions are, however, not sufficient to characterize the class of Pickands dependence functions unless $d = 2$ (see, e.g., Beirlant et al., 2004, for a counterexample).

Several other characterizations of extreme-value copulas are possible, for instance using the *spectral measure of C* (see, e.g., Gudendorf and Segers, 2012, for details) or the *stable tail dependence function* (Charpentier et al., 2014; Ressel, 2013).

18.3   Existing Tests of Extreme-Value Dependence

Let $\mathcal{EV}$ denote the class of extreme-value copulas. Given a random sample $\mathbf{X}_1, \ldots, \mathbf{X}_n$ from a c.d.f. $C\{F_1(x_1), \ldots, F_d(x_d)\}$ with $F_1, \ldots, F_d$ continuous and $C, F_1, \ldots, F_d$ unknown, tests of extreme-value dependence aim at testing

$$H_0 : C \in \mathcal{EV} \qquad \text{against} \qquad H_1 : C \notin \mathcal{EV}. \qquad (18.4)$$

The existing tests for H_0 available in the literature are all rank-based and therefore margin-free. They can be classified into three groups according to how departure from extreme-value dependence is assessed.

18.3.1   Approaches Based on Kendall's Distribution

The first class of approaches, which is also the oldest, finds its origin in the seminal work of Ghoudi et al. (1998) and is restricted to the case $d = 2$. Given a bivariate random vector $\mathbf{X} = (X_1, X_2)$ with c.d.f. F, continuous marginal c.d.f.s F_1 and F_2 and copula C, the tests in this class are based on the random variable

$$W = F(X_1, X_2) = C\{F_1(X_1), F_2(X_2)\}.$$

The c.d.f. of W is frequently referred to as *Kendall's distribution* and will be denoted by K subsequently. When $C \in \mathcal{EV}$, Ghoudi et al. (1998) showed that

$$K(w) = \Pr(W \leq w) = w - (1 - \tau)w \log w, \qquad w \in (0, 1], \qquad (18.5)$$

where τ denotes *Kendall's tau*. Whether C is of the extreme-value type or not, it is known since Schweizer and Wolff (1981) that

$$\tau = 4 \int_{[0,1]^2} C(u_1, u_2) \mathrm{d}C(u_1, u_2) - 1 = 4\mathrm{E}(W) - 1.$$

When $C \in \mathcal{EV}$, Ghoudi et al. (1998) also obtained from (18.5) that, for $k \in \mathbb{N}$, $\mu_k := E(W^k) = (k\tau + 1)/(k + 1)^2$, which for instance implies that

$$-1 + 8\mu_1 - 9\mu_2 = 0. \tag{18.6}$$

In order to test H_0 from a bivariate random sample $\mathbf{X}_1, \ldots, \mathbf{X}_n$ with c.d.f. $C\{F_1(x_1), F_2(x_2)\}$ where C, F_1, F_2 are unknown, Ghoudi et al. (1998) suggested to assess whether a sample version of the left-hand side of (18.6) is significantly different from zero or not. Specifically, they considered the statistic

$$S_{2n} = -1 + \frac{8}{n(n-1)} \sum_{i \neq j} I_{ij} - \frac{9}{n(n-1)(n-2)} \sum_{i \neq j \neq k} I_{ij} I_{kj}, \tag{18.7}$$

where $I_{ij} = \mathbf{1}(X_{i1} \leqslant X_{j1}, X_{i2} \leqslant X_{j2})$. As shown by Ghoudi et al. (1998), S_{2n} is a centered U-statistic which, under the null hypothesis, converges weakly to a normal random variable. To carry out the test, Ghoudi et al. (1998) proposed to estimate the variance of S_{2n} using a jackknife estimator. The test based on S_{2n} was revisited by Ben Ghorbal et al. (2009) who proposed two alternative strategies to compute approximate p-values for S_{2n}. The three versions of the test are implemented in the function evTestK of the R package copula.

The above approach was recently furthered by Du and Nešlchová (2013) who used the first three moments of Kendall's distribution and the theoretical relationship

$$-1 + 4\mu_1 + 9\mu_2 - 16\mu_3 = 0 \tag{18.8}$$

under the null instead of (18.6). The corresponding test statistic will subsequently be denoted by S_{3n}. An additional contribution of the latter authors was to find a counterexample to Ghoudi et al. (1998)'s conjecture that K has the form in (18.5) if and only if $C \in \mathcal{EV}$. The latter implies that tests in this class are not consistent. Despite that fact, the Monte Carlo experiments reported in Kojadinovic and Yan (2010b) and in Du and Nešlehová (2013) suggest that tests based on S_{2n} and its extension studied in Du and Nešlehová (2013) are among the most powerful procedures for testing bivariate extreme-value dependence.

Notice finally that additional extensions of the approach of Ghoudi et al. (1998) were studied in Quessy (2012) along with tests based on Cramér–von Mises-like statistics derived from the empirical process $\sqrt{n}(K_n - K_{\tau_n})$, where K_n is the empirical c.d.f. of $\widehat{W}_1, \ldots, \widehat{W}_n$ with $\widehat{W}_i = F_n(X_{i1}, X_{i2})$ and F_n the empirical c.d.f. of $\mathbf{X}_1, \ldots, \mathbf{X}_n$, and K_{τ_n} is defined as in (18.5) with τ replaced by its classical estimator denoted τ_n.

18.3.2 Approaches Based on Max-Stability

The second class of tests proposed in the literature consists of assessing empirically whether (18.2) holds or not. It was investigated in Kojadinovic et al. (2011) for $d \geqslant 2$. The key ingredient is a natural nonparametric estimator of the unknown copula C known as the *empirical copula* (e.g., Deheuvels, 1979, 1981; Rüschendorf, 1976).

Given a sample $\mathbf{X}_1, \ldots, \mathbf{X}_n$ from a c.d.f. $C\{F_1(x_1), \ldots, F_d(x_d)\}$ with $F_1, \ldots, F_d$ continuous and $C, F_1, \ldots, F_d$ unknown, let $\widehat{U}_{ij} = R_{ij}/(n+1)$ for all $i \in \{1, \ldots, n\}$ and $j \in \{1, \ldots, d\}$, where R_{ij} is the rank of X_{ij} among $X_{1j}, \ldots, X_{nj}$, and set $\widehat{\mathbf{U}}_i = (\widehat{U}_{i1}, \ldots, \widehat{U}_{id})$. It is worth noticing that the scaled ranks $\widehat{U}_{ij}$ can equivalently be rewritten as $\widehat{U}_{ij} = nF_{nj}(X_{ij})/(n+1)$, where F_{nj} is the empirical c.d.f. computed from $X_{1j}, \ldots, X_{nj}$, the scaling factor $n/(n+1)$ being classically introduced to avoid problems at the boundary of $[0, 1]^d$. The empirical copula of $\mathbf{X}_1, \ldots, \mathbf{X}_n$ is then frequently defined as the empirical c.d.f. computed from the *pseudo-observations* $\widehat{\mathbf{U}}_1, \ldots, \widehat{\mathbf{U}}_n$, i.e.,

$$C_n(\mathbf{u}) = \frac{1}{n} \sum_{i=1}^{n} \mathbf{1}(\widehat{\mathbf{U}}_i \leqslant \mathbf{u}), \qquad \mathbf{u} \in [0, 1]^d. \tag{18.9}$$

The inequalities between vectors in the above definition are to be understood componentwise.

To test (18.2) empirically, Kojadinovic et al. (2011) considered test statistics constructed from the empirical process

$$\mathbb{D}_{r,n}(\mathbf{u}) = \sqrt{n} \left[\{C_n(u_1^{1/r}, \ldots, u_d^{1/r})\}^r - C_n(\mathbf{u}) \right], \qquad \mathbf{u} \in [0, 1]^d, \tag{18.10}$$

for some strictly positive fixed values of r. The recommended test statistic is

$$T_{3,4,5,n} = T_{3,n} + T_{4,n} + T_{5,n}, \tag{18.11}$$

where $T_{r,n} = \int_{[0,1]^d} \{\mathbb{D}_{r,n}(\mathbf{u})\}^2 dC_n(\mathbf{u})$. Approximate p-values for the latter were computed using a *multiplier bootstrap*. The test based on $T_{3,4,5,n}$ is implemented in the function evTestC of the R package copula. It is not a consistent test either, because the validity of (18.2) is assessed only for a small number of r values.

18.3.3 Approaches Based on the Estimation of the Pickands Dependence Function

Recall that $\mathbf{X}_1, \ldots, \mathbf{X}_n$ is a random sample from a c.d.f. $C\{F_1(x_1), \ldots, F_d(x_d)\}$ with $F_1, \ldots, F_d$ continuous and $C, F_1, \ldots, F_d$ unknown. If $C \in \mathcal{EV}$, it can be expressed as in (18.3). The third class of tests exploits variations of the following idea: given a nonparametric estimator A_n of A and using the empirical copula C_n defined in (18.9), relationship (18.3) can be tested empirically.

The first test in this class is due to Kojadinovic and Yan (2010b) who, for $d = 2$ only, constructed test statistics from the empirical process

$$\mathbb{E}_n(u_1, u_2) = \sqrt{n} \left(C_n(u_1, u_2) - \exp\left[\log(u_1 u_2) A_n \left\{ \frac{\log(u_2)}{\log(u_1 u_2)} \right\} \right] \right),$$

for $(u_1, u_2) \in (0, 1]^2 \setminus \{(1, 1)\}$. The recommended statistic is

$$T_n^A = \int_{[0,1]^2} \mathbb{E}_n(u_1, u_2)^2 \mathrm{d}C_n(u_1, u_2), \qquad (18.12)$$

when A_n is the rank-based version of the Capéraà–Fougères–Genest (CFG) estimator of A studied in Genest and Segers (2009). The resulting test relies on a *multiplier bootstrap* and is implemented in the function `evTestA` of the R package `copula`. A multivariate version of this test was studied in Gudendorf (2012) using the multivariate extension of the rank-based CFG estimator of A investigated in Gudendorf and Segers (2012).

An alternative class of nonparametric multivariate rank-based estimators of A was proposed in Bücher et al. (2011) and Berghaus et al. (2013). These are based on the minimization of a weighted L^2-distance between the logarithms of the empirical and the unknown extreme-value copula. To derive multivariate tests of extreme-value dependence, the latter authors reused the aforementioned L^2-distance to measure the difference between the empirical copula in (18.9) and a plug-in nonparametric estimator of C under extreme-value dependence based on (18.3). The corresponding test statistic is subsequently denoted by $T_{L^2, n}$.

We end this subsection by briefly summarizing a recent graphical approach due to Cormier et al. (2014). Their idea, hitherto restricted to the bivariate case, is as follows: given a copula C, consider the transformation $T_C : (0, 1)^2 \to (0, 1) \times (0, \infty]$, defined by

$$T_C(u_1, u_2) = \left(\frac{\log(u_2)}{\log(u_1 u_2)}, \frac{\log\{C(u_1, u_2)\}}{\log(u_1 u_2)} \right), \qquad (u_1, u_2) \in (0, 1)^2.$$

If $C \in \mathcal{EV}$, representation (18.3) holds and we have $\log\{C(u_1, u_2)\} = \log(u_1 u_2) A\{\log(u_2) / \log(u_1 u_2)\}$ for all $(u_1, u_2) \in (0, 1)^2$, whence $\mathcal{S}_C = \{T_C(u, v) : (u, v) \in (0, 1)^2\}$ coincides with the graph of A, i.e., with the set $\{(t, A(t)) : t \in (0, 1)\}$. More generally, some thought reveals that H_0 is valid if and only if $\mathcal{S}_C$ is a convex curve. The latter observation suggests to test H_0 in (18.4) by estimating the set $\mathcal{S}_C$ and visually assessing the departure of that estimated set from a convex curve. The estimator defined in Cormier et al. (2014), called the *A-plot*, is given by

$$\widehat{\mathcal{S}}_n = \left\{ (\widehat{T}_i, \widehat{Z}_i) : \widehat{T}_i = \frac{\log(\widehat{U}_{i2})}{\log(\widehat{U}_{i1} \widehat{U}_{i2})}, \widehat{Z}_i = \frac{\log\{C_n(\widehat{U}_{i1}, \widehat{U}_{i2})\}}{\log(\widehat{U}_{i1} \widehat{U}_{i1})}, i \in \{1, \ldots, n\} \right\}.$$

Examples of A-plots when $C \in \mathcal{EV}$ and when $C \notin \mathcal{EV}$ can be found in Figure 1 of Cormier et al. (2014). When C is of the extreme-value type, the previous authors proposed a B-spline smoothing estimator for the Pickands dependence function A based on $\widehat{\mathcal{S}}_n$. The latter is subsequently denoted by A_n for simplicity (even though the estimator depends on several smoothing parameters). Additionally to a pure graphical check, the authors propose

$$T_n = \frac{1}{n} \sum_{i=1}^{n} \{\widehat{Z}_i - A_n(\widehat{T}_i)\}^2, \qquad (18.13)$$

a residual sum of squares, as a formal test statistic for H_0. The hypothesis is rejected for unlikely large values of T_n. Specifically, an approximate p-value for T_n is computed by means of a *parametric bootstrap* procedure based on simulating from a copula with Pickands dependence function A_n.

18.3.4   Finite-Sample Performance of Some of the Tests

The finite-sample performance of the tests reviewed in the preceding sections was investigated by various authors. Table 18.1, taken from Cormier et al. (2014), gathers those results from Cormier et al. (2014), Kojadinovic and Yan (2010b), Du and Nešlehová (2013), and Bücher et al. (2011) that were obtained under the same experimental settings (notice that the Gumbel–Hougaard copula is the only extreme-value copula among those considered in the table). As noted by Cormier et al. (2014), no test is uniformly better than the others: each test, except the one based on $T_{L^2,n}$ from Bücher et al. (2011), is favored for at least one of the considered scenarios under H_1. For high levels of dependence (as measured by Kendall's tau), the tests based on S_{2n} and S_{3n} described in Section 18.3.1 seem to yield the most accurate approximation of the nominal level (here 5%). The tests whose approximate p-values are computed by means of a multiplier bootstrap, i.e., the tests based on $T_{3,4,5,n}$ defined in (18.11) and on T_n^A and $T_{L^2,n}$ introduced in Section 18.3.3, are quite conservative for such scenarios. From a computational perspective, the test based on S_{2n} seems to be the fastest, while the one based on T_n^A defined in (18.12) is the most computationally intensive. Additional comparison of the tests based on S_{2n}, $T_{3,4,5,n}$ and T_n^A (resp. S_{2n} and S_{3n}) can be found in Kojadinovic and Yan (2010b, Tables 1–3) (resp. Du and Nešlehová, 2013, Table 5).

   Kojadinovic et al. (2011) and Berghaus et al. (2013) also present simulation results for $d > 2$, which are in favor of the test based on $T_{3,4,5,n}$ defined in (18.11). Preliminary results obtained in Gudendorf (2012) indicate that the multivariate extension of the test based on T_n^A defined in (18.12) is likely to outperform the test based on $T_{3,4,5,n}$ for several scenarios under H_1.

18.4   Some Related Statistical Inference Procedures

Once it has been decided to use an extreme-value copula to model dependence in a set of multivariate continuous i.i.d. observations, a typical next step is to choose a parametric family $\mathcal{C}$ in $\mathcal{EV}$ and estimate its unknown parameter(s) from the data. As many parametric families of extreme-value copulas are available (e.g., Gudendorf and Segers, 2010; Ribatet and Sedki, 2013), it is of strong practical interest to be able to test whether a given family $\mathcal{C}$ is a plausible model or not for the data at hand. In other words, tests for $H_0 : C \in \mathcal{C}$ against $H_1 : C \notin \mathcal{C}$ would be needed. Such goodness-of-fit procedures were investigated in the bivariate case by Genest et al. (2011) who considered Cramér–von Mises test statistics based on the difference be-

TABLE 18.1
Rejection rates of H_0 estimated from random samples of size $n = 200$ generated from a c.d.f. with copula C whose Kendall's tau is τ. All the tests were carried out at the 5% significance level. The table is taken from Cormier et al. (2014).

τ	C	T_n	S_{2n}	S_{3n}	T_n^A	$T_{3,4,5,n}$	$T_{L^2,n}$
0.25	Gumbel–Hougaard	4.7	5.4	5.3	3.8	5.0	4.5
	Clayton	97.7	98.0	96.6	**98.4**	94.6	87.4
	Frank	18.7	38.4	57.0	58.3	**66.1**	29.1
	Gaussian	25.5	37.3	**40.3**	36.5	38.7	16.8
	Student t with 4 d.f.	**37.7**	26.2	19.6	23.9	26.6	10.5
0.50	Gumbel–Hougaard	5.4	5.1	5.0	3.9	4.0	2.9
	Clayton	100.0	100.0	100.0	100.0	100.0	100.0
	Frank	87.8	59.4	84.4	95.7	**96.5**	73.0
	Gaussian	59.4	**62.6**	61.7	61.8	51.0	23.7
	Student t with 4 d.f.	**58.6**	56.0	45.3	50.1	52.7	15.8
0.75	Gumbel–Hougaard	6.2	4.9	5.3	3.2	2.3	2.5
	Clayton	100.0	100.0	100.0	100.0	100.0	100.0
	Frank	98.3	58.5	92.9	**99.9**	99.0	78.3
	Gaussian	56.5	**75.2**	71.1	66.5	46.7	8.4
	Student t with 4 d.f.	45.8	67.8	55.8	50.6	**69.2**	4.6

tween a nonparametric and a parametric estimator of the Pickands dependence function. The Monte Carlo experiments reported in the latter work highlighted the fact that, unless the amount of data is very large, there is hardly any practical difference among the existing bivariate symmetric parametric families of extreme-value copulas, and that an issue of more importance from a modeling perspective is whether a symmetric or asymmetric family should be used. For that purpose, the specific test of symmetry for bivariate extreme-value copulas investigated in Kojadinovic and Yan (2012) can be used as a complement to the goodness-of-fit test studied in Genest et al. (2011). Both tests are available in the `copula` R package.

When $d > 2$ but d remains reasonably small (say $d \leqslant 10$), generic goodness-of-fit tests (that is, developed for any parametric copula family, not necessarily of the extreme-value type) could be used (see, e.g., Genest et al., 2009; Kojadinovic and Yan, 2011, and the references therein). In a higher dimensional context, one possibility consists of using the specific approach for extreme-value copulas proposed by Smith (1990) in his seminal work on max-stable processes. It consists of comparing nonparametric and parametric estimators of the underlying *extremal coefficients* (which are functionals of the Pickands dependence function). The latter approach was recently revisited in Kojadinovic et al. (2015).

18.5   Illustrations and R Code from the copula Package

We provide two illustrations. The first one concerns bivariate financial logreturns and exemplifies the key theoretical connection between multivariate block maxima and extreme-value copulas briefly mentioned in the introduction and Section 18.2. The second illustration consists of a detailed analysis of the well-known LOSS/ALAE insurance data with particular emphasis on the effect and handling of ties.

18.5.1   Bivariate Financial Logreturns

As a first illustration, we considered daily logreturns computed from the closing values of the Dow Jones and the S&P 500 stock indexes for the period 1990–2004. The closing values are available in the QRM R package (Pfaff and McNeil, 2013) and can be loaded by entering the following commands into the R terminal:

```
> library(QRM)
> data(dji)
> data(sp500)
```

Daily logreturns for the period under consideration were computed using the timeSeries R package (Wuertz and Chalabi, 2013):

```
> d <- na.omit(cbind(dji,sp500))
> rd <- returns(d)
```

The statistical procedures mentioned in the previous sections should not, however, be directly applied on the resulting bivariate daily logreturns as the latter are strongly serially dependent. To obtain observations that might exhibit extreme-value dependence and could be considered approximately i.i.d., we first formed the bivariate series of componentwise monthly maxima. The last step was performed using functions from the timeSeries and timeDate R packages (Wuertz et al., 2013):

```
> library(timeSeries)
> by <- timeSequence(from=start(rd),  to=end(rd), by="month")
> mrd <- aggregate(rd, by, max)
```

The resulting component series do not contain ties, which is compatible with the implicit assumption of continuous margins:

```
> x <- series(mrd)
> nrow(x)
[1] 171
> apply(x, 2, function(x) length(unique(x)))
  DJI SP500
  171    171
```

After loading the copula package with the command library(copula) and setting the random seed by typing set.seed(123), the test of extreme-value dependence based on S_{2n} (resp. $T_{3,4,5,n}$, T_n^A) defined in (18.7) (resp. (18.11),

(18.12)) was applied using the command `evTestK(x)` (resp. `evTestC(x)`, `evTestA(x, derivatives="Cn")`) and returned an approximate p-value of 0.5737 (resp. 0.4191, 0.2423). In other words, none of the tests detected any evidence against extreme-value dependence thereby suggesting that the copula of componentwise block maxima, for blocks of length corresponding to a month, is sufficiently close to an extreme-value copula. Note that, as the tests are rank-based, they could have equivalently been called on the pseudo-observations computed from the monthly block maxima. The random seed was set (to ensure exact reproducibility) because the second and third tests involve random number generation as their p-values are computed using resampling.

For illustration purposes, we next formed monthly logreturns as follows:

```
> srd <- aggregate(rd, by, sum)
> x <- series(srd)
```

Proceeding as previously, it can be verified that the resulting component series do not contain ties, which is compatible with the implicit assumption of continuous margins. Monthly logreturns being merely sums of daily logreturns, the underlying unknown bivariate distribution should be far from exhibiting extreme-value dependence. The tests of extreme-value dependence based on S_{2n}, $T_{3,4,5,n}$, and T_n^A returned approximate p-values of 0.0003, 0.02, and 0.0005, respectively, confirming that there is strong evidence in the data against extreme-value dependence.

18.5.2 LOSS/ALAE Insurance Data

The well-known LOSS/ALAE insurance data are very frequently used for illustration purposes in copula modeling (e.g., Ben Ghorbal et al., 2009; Frees and Valdez, 1998; Kojadinovic and Yan, 2010a). The two variables of interest are LOSS, an indemnity payment, and ALAE, the corresponding allocated loss adjustment expense. They were observed for 1500 claims of an insurance company. Following Ben Ghorbal et al. (2009), the following study is restricted to the 1466 uncensored claims.

The data are available in the `copula` package, and can be loaded by typing `library(copula)` followed by `data(loss)`. The uncensored claims described in terms of LOSS and ALAE were obtained as follows:

```
> myLoss <- subset(loss, censored==0, select=c("loss", "alae"))
```

These data, consisting of 1466 bivariate observations, contain a non-negligible amount of ties, the variable LOSS being particularly affected:

```
> sapply(myLoss, function(x) length(unique(x)))
loss alae
 541 1401
```

The presence of ties is incompatible with the implicit assumption of continuous margins. Indeed, combined with the assumption that the data are i.i.d. observations, continuity of the margins implies that ties should not occur. Yet, ties are present here as in many other real datasets. The latter could be due either to the fact that the observed phenomena are truly discontinuous, or to precision/rounding issues. As far as the LOSS/ALAE data are concerned, the latter explanation applies.

Among the tests briefly described in Section 18.3, only that of Cormier et al. (2014) explicitly considers the case of discontinuous margins (see Section 6 in that reference). The remaining tests were all implemented under the assumption of continuous margins. For the test based on S_{2n} defined in (18.7), Genest et al. (2011) provide a heuristic explanation of the fact that, for discontinuous margins, S_{2n} is not necessarily centered anymore under the null.

Given the situation, there are roughly four possible courses of action: (i) stop the analysis, (ii) delete tied observations, (iii) use average ranks for ties, or (iv) break ties at random, sometimes referred to as *jittering* (which amounts to adding a small, continuous white noise term to all observations). Arguments for not considering solution (ii) are given in Genest et al. (2011, Section 2). To empirically study solutions (iii) and (iv), the latter authors carried out an experiment consisting of applying the test based on S_{2n} defined in (18.7) on binned observations from a bivariate Gumbel–Hougaard copula. More specifically, tied observations were obtained by dividing the unit square uniformly into bins of dimension 0.1 by 0.1 (resp. 0.2 by 0.2) resulting in at most 100 (resp. 25) different bivariate observations whatever the sample size. In such a setting, Genest et al. (2011) observed that both solutions (iii) and (iv) led to strongly inflated empirical levels for the test based on S_{2n}.

The situation in terms of ties in the LOSS/ALAE data is however far from being as extreme as in the experiment of Genest et al. (2011). In addition, ties mostly affect the LOSS variable. This prompted us first to consider solution (iv) as implemented in Kojadinovic and Yan (2010a).

Random ranks for ties

The idea consists of carrying out the analysis for many different randomizations (with the hope that this will result in many different configurations for the parts of the data affected by ties) and then looking at the empirical distributions (and not the averages) of the results (here the p-values of various tests).

For illustration purposes, we first detail the analysis for one randomization:

```
> set.seed(123)
> pseudoLoss <- sapply(myLoss, rank, ties.method="random") /
              (nrow(myLoss) + 1)
```

As a next step, the tests of extreme-value dependence based on S_{2n}, $T_{3,4,5,n}$ and T_n^A defined in (18.7), (18.11), and (18.12), respectively, were applied by successively typing evTestK(pseudoLoss), evTestC(pseudoLoss), and evTestA(pseudoLoss, derivatives="Cn"), resulting in approximate p-values of 0.8845, 0.468, and 0.4231, respectively.

```
> evTestK(pseudoLoss)

Test of bivariate extreme-value dependence based on Kendall's process
with argument 'method' set to fsample
```

```
data:  pseudoLoss
statistic = -0.1453, p-value = 0.8845

> evTestC(pseudoLoss)

Max-stability based test of extreme-value dependence for multivariate
copulas

data:  pseudoLoss
statistic = 0.1469, p-value = 0.468

> evTestA(pseudoLoss, derivatives="Cn")

Test of bivariate extreme-value dependence based on the CFG estimator
with argument 'derivatives' set to 'Cn'

data:  pseudoLoss
statistic = 0.0186, p-value = 0.4231
```

Hence, none of the three tests detected any evidence against extreme-value dependence. The following step consisted of fitting a parametric family of bivariate extreme-value copulas to the data. As discussed in Section 18.4, given the very strong similarities among the existing families of bivariate symmetric extreme-value copulas, the only issue of practical importance is to assess whether a symmetric or asymmetric family should be used. To do so, we applied the test developed in Kojadinovic and Yan (2012) by calling exchEVTest(pseudoLoss), with a resulting p-value of 0.1653.

```
> exchEVTest(pseudoLoss)

Test of exchangeability for bivariate extreme-value copulas with
argument 'estimator' set to 'CFG', argument 'derivatives' set to 'Cn'
and argument 'm' set to 100

data:  pseudoLoss
statistic = 0.0751, p-value = 0.1653
```

The previous result suggested to focus on an exchangeable family such as the Gumbel–Hougaard. We then ran the goodness-of-fit test proposed in Genest et al. (2011) by calling

```
> gofEVCopula(gumbelCopula(), pseudoLoss, method="itau", verbose=FALSE)

Parametric bootstrap based GOF test for EV copulas with argument
'method' set to 'itau' and argument 'estimator' set to 'CFG'

data:  pseudoLoss
statistic = 0.0377, parameter = 1.44, p-value = 0.2592
```

The resulting p-value of 0.2592 suggested to fit the Gumbel–Hougaard family:

```
> fitCopula(gumbelCopula(), pseudoLoss, method="itau")
fitCopula() estimation based on 'inversion of Kendall's tau'
```

and a sample of size 1466.
```
        Estimate Std. Error  z value Pr(>|z|)
param   1.44040    0.03327   43.29    <2e-16 ***
---
Signif. codes:  0 *** 0.001 ** 0.01 * 0.05 . 0.1   1
```

To assess how different randomizations of the ties affect the results, the above analysis was repeated 100 times using the following code:

```
> randomize <- function()
+ {
+     pseudoLoss <- sapply(myLoss, rank, ties.method="random") /
                          (nrow(myLoss) + 1)
+     evtK <- evTestK(pseudoLoss)$p.value
+     evtC <- evTestC(pseudoLoss)$p.value
+     evtA <- evTestA(pseudoLoss, derivatives="Cn")$p.value
+     exevt <- exchEVTest(pseudoLoss)$p.value
+     gofevGH <- gofEVCopula(gumbelCopula(), pseudoLoss, method="itau",
                             verbose=FALSE)$p.value
+     fitGH <- fitCopula(gumbelCopula(), pseudoLoss, method="itau")
+     c(evtK=evtK, evtC=evtC, evtA=evtA, exevt=exevt, gofevGH=gofevGH,
+         est=fitGH@estimate, se=sqrt(fitGH@var.est))
+ }
> reps <- t(replicate(100, randomize()))
> round(apply(reps, 2, summary), 3)
          evtK  evtC  evtA exevt gofevGH   est    se
Min.     0.868 0.430 0.353 0.092   0.191 1.441 0.033
1st Qu.  0.898 0.462 0.396 0.112   0.223 1.442 0.033
Median   0.914 0.475 0.411 0.120   0.235 1.442 0.033
Mean     0.913 0.474 0.411 0.122   0.236 1.442 0.033
3rd Qu.  0.928 0.489 0.425 0.129   0.248 1.443 0.033
Max.     0.955 0.525 0.464 0.162   0.292 1.444 0.033
```

The empirical distributions of the results show that the different randomizations did not affect the results qualitatively.

Average ranks for ties

We also considered solution (iii), that is, average ranks for ties. The p-values of the three tests of extreme-value dependence (applied in the same order as previously) were 0.6, 0.02, and 0, respectively. The p-values of the tests of exchangeability and goodness of fit were 0.12 and 0.18, respectively. The estimate of the parameter of the Gumbel–Hougaard copula was 1.446.

Random or average ranks for ties?

The previous computations illustrate that solutions (iii) and (iv) for dealing with ties can result in significantly different conclusions. To gain insight into which solution should be preferred, if any, we designed an experiment tailored to the LOSS/ALAE data. Specifically, we simulated a large number of samples of size $n = 1466$ from a Gumbel–Hougaard copula with parameter value 1.446, as suggested by the afore-mentioned parametric fit. We then modified each sample so that its marginal empirical c.d.f.s evaluated at the respective observations coincide with those of the

LOSS/ALAE data. For instance, in the original data, the 27th to the 49th smallest values of LOSS are equal. Each simulated sample was modified so that the 27th to the 49th smallest values of the first variable all get replaced by the 49th smallest observation. The same approach was used for the second variable of the generated samples. Solutions (iii) and (iv) were applied next to each modified sample prior to running the tests of extreme-value dependence, and the resulting p-values were compared with those obtained by applying the tests on the corresponding unmodified sample (that is, with no ties). The code used to carry out the experiment for the test based on S_{2n} defined in (18.7) is given below:

```
> mr.loss <- rank(myLoss[,1], ties.method="max")
> mr.alae <- rank(sort(myLoss[,2]), ties.method="max")
> test.func <- function(x) evTestK(x)$p.value
> do1 <- function()
+ {
+     x <- rCopula(1466, gumbelCopula(1.446))
+     y <- x[order(x[,1]),]
+     y[,1] <- y[mr.loss,1]
+     y <- y[order(y[,2]),]
+     y[,2] <- y[mr.alae,2]
+     z <- apply(y, 2, rank, ties.method="random")
+     c(test.func(x), test.func(y), test.func(z))
+ }
> res <- t(replicate(1000, do1()))
> summary(round(res[,1] - res[,3],3))
     Min.   1st Qu.   Median      Mean   3rd Qu.      Max.
-0.093000 -0.014000  0.001000  0.000033  0.014000  0.105000
> summary(round(res[,1] - res[,2],3))
    Min.  1st Qu.   Median     Mean  3rd Qu.     Max.
-0.54600 -0.25520  0.14000  0.06697  0.34750  0.52300
> apply(res, 2, function(x) mean(x <= 0.05))
0.046 0.107 0.047
```

For the test based on S_{2n}, the p-values computed from a continuous sample and the corresponding randomized sample are very close on average, the maximal deviation being relatively small. On the contrary, the p-values computed from a continuous sample are larger on average than the p-values computed from the corresponding sample involving average ranks, and the maximal deviation is very large. We also see that when solution (iii) is considered, the test based on S_{2n} is way too liberal, confirming the findings of Genest et al. (2011), while, when solution (iv) is used, the test holds its level well. A similar experiment was performed for the test based on $T_{3,4,5,n}$ (with 100 replications only) and the conclusions are of the same nature but more pronounced:

```
> summary(round(res[,1] - res[,3],3))
    Min.  1st Qu.   Median     Mean  3rd Qu.     Max.
-0.10400 -0.01425 -0.00150 -0.00050  0.01650  0.08700
> summary(round(res[,1] - res[,2],3))
   Min.  1st Qu.  Median    Mean  3rd Qu.    Max.
-0.0190  0.2700  0.4820  0.4536  0.6592  0.8730
> apply(res, 2, function(x) mean(x <= 0.05))
0.05 0.45 0.05
```

The previous experiment can be adapted to any dataset containing ties and suggests that, in the case of the LOSS/ALAE data, solution (iv) is meaningful while solution (iii) should be avoided.

18.6   Testing the Maximum Domain of Attraction Condition

The statistical framework considered in the three previous sections can be regarded as the "classical" setting of dependence modeling by copulas. As mentioned in the introduction, modeling a copula by an extreme-value copula, or testing extreme-value dependence within such a framework, is particularly sensible if there are reasons to assume that the data at hand are generated by some maxima-forming process. If this is not the case, or if the hypothesis of extreme-value dependence is rejected, it might still be reasonable to make the (mild) assumption that the copula of interest lies in the domain of attraction of some extreme-value copula. It is the aim of the present section to briefly discuss how the latter assumption could be tested.

A precise formulation of the problem is as follows: we observe a sample of d-dimensional i.i.d. vectors $\mathbf{Y}_1, \ldots, \mathbf{Y}_n$ with c.d.f. $C^*\{G_1(y_1), \ldots, G_d(y_d)\}$, where $G_1, \ldots, G_d$ are assumed continuous and $C^*, G_1, \ldots, G_d$ are unknown. We are interested in tests of

$$H_0 : C^* \in D(C) \text{ for some } C \in \mathcal{EV} \quad \text{against} \quad H_1 : C^* \notin D(C) \text{ for any } C \in \mathcal{EV},$$
(18.14)

where the notation $C^* \in D(C)$ is defined below (18.1). Notice that the analogue univariate problem (i.e., testing the null hypothesis that the underlying distribution of a given univariate i.i.d. sample lies in the maximum domain of attraction of some extreme-value distribution) was tackled in Dietrich et al. (2002), Drees et al. (2006), and Hüsler and Li (2006), while, in the multivariate case, only very few (validated) methods seem available.

A rejection of the null hypothesis in (18.14) gives indication that the stochastic dependence between componentwise block maxima formed from the $\mathbf{Y}_i$, no matter how large the blocks are, cannot be adequately described by an extreme-value copula. On the other hand, if the hypothesis is not rejected, it is promising to consider an extreme-value copula as a model provided the block size is sufficiently large. Also, in the latter case, one could make use of (18.1) to obtain the approximation that, for a sufficiently large r, $C^*(\mathbf{v}) \approx \{C(v_1^r, \ldots, v_d^r)\}^{1/r} = C(\mathbf{v})$ for all $\mathbf{v} \in [\mathbf{t}, \mathbf{1}]$, with $\mathbf{t} = (t_1, \ldots, t_d)$ close to $\mathbf{1}$. This would imply that, at least in the upper tail, the copula C^* can be well-approximated by an extreme-value copula C. A threshold model of that form was for instance considered in Ledford and Tawn (1996) in a bivariate setting with generalized Pareto marginals.

A first promising approach to test H_0 in (18.14) consists of comparing two estimators of C (or its characterizing objects) with different backgrounds. Under the null hypothesis, these estimators should not differ too much. Based on the peak-over-threshold method and in the bivariate case, Einmahl et al. (2006) developed a test based on an Anderson–Darling-type statistic between two estimators of the stable-tail dependence function $\ell : [0,1]^2 \to \mathbb{R}$ defined by $\ell(x,y) = |x+y|A(x/|x+y|)$, where A denotes the Pickands dependence function of C. Critical values for the test were obtained by approximately simulating from the limiting random variable. To the best of our knowledge, this testing procedure is the only validated method for testing the (bivariate) maximum domain of attraction condition.

A heuristic approach to test the null hypothesis in (18.14) in the bivariate case was described in Cormier et al. (2014). Their method consists of considering a trimmed A-plot (see also Section 18.3.3) defined by only including those points $(\widehat{T}_i, \widehat{Z}_i)$ in the set $\widehat{\mathcal{S}}_n = \widehat{\mathcal{S}}_n(\mathbf{t})$ for which $\widehat{\mathbf{U}}_i \in [\mathbf{t}, \mathbf{1}]$, with some suitable threshold parameter $\mathbf{t} = (t_1, t_2) \in [0,1]^2$ close to $\mathbf{1}$. Based on the trimmed A-plot, the approach briefly described in Section 18.3.3 can be followed to obtain a B-spline smoothing estimator of the Pickands dependence function corresponding to the limiting extreme-value copula C. Plotting the residual sum of squared errors defined in (18.13) against the threshold $\mathbf{t}$ serves as a data-driven method for the choice of the threshold. For that particular choice, the A-plot as well as the testing procedure described in Section 18.3.3 can be used to assess heuristically whether the maximum domain of attraction condition holds or not.

Finally, the tests described in Section 18.3 can be adapted to obtain simple heuristic procedures for testing H_0 in (18.14). Under the null hypothesis, given $\mathbf{Y}_1, \ldots, \mathbf{Y}_n$, if we form k (componentwise) block maxima from blocks of length m,

$$\mathbf{X}_i = (X_{i1}, \ldots, X_{id}), \quad X_{ij} = \max\{Y_{m(i-1)+1,j}, \ldots, Y_{mi,j}\},$$

$i \in \{1, \ldots, k\}, j \in \{1, \ldots d\}$, where $km = n$ and m is sufficiently large (if n is not an integer multiple of m, then a negligible remainder block of length strictly smaller than m occurs), then the copula of the block maxima $\mathbf{X}_i$ should (approximately) be an extreme-value copula. The tests described in Section 18.3 could next be applied to $\mathbf{X}_1, \ldots, \mathbf{X}_k$ to obtain an indication of whether the maximum domain of attraction condition holds or not. Another promising approach consists of adapting the approach in Section 18.3.2 by only testing max-stability in the upper tail $[\mathbf{t}, \mathbf{1}]$, with some suitable threshold parameter $\mathbf{t} = (t_1, t_2) \in [0,1]^2$ close to $\mathbf{1}$. This could be done by integrating the square of the process in (18.10) over the restricted set $[\mathbf{t}, \mathbf{1}]$.

Precise asymptotic validations of these methods are, however, not available. A treatment of occurring bias terms from an undersized choice of the block length or the threshold parameter would be necessary, as for instance carried out in Bücher and Segers (2014) in an estimation framework for time series based on block maxima. Also, data-driven methods to choose the block length m or the threshold parameter $\mathbf{t}$ would need to be developed.

18.7   Open Questions and Ignored Difficulties

Several issues dealt with in this chapter would need to be thoroughly investigated in future research. For instance, the suggested approach for handling ties in datasets for which it is actually reasonable to assume that the apparent discontinuities are only due to precision or rounding issues would need to be studied more in depth. While for the LOSS/ALAE dataset, it seemed reasonable to break ties at random a large number of times, this may not be the case for other datasets in which the proportion of ties is significantly larger (e.g., Genest et al., 2011, Section 4). Yet, even more difficult appears to be the problem of testing extreme-value dependence from truly discontinuous observations such as count data. A promising starting point for adapting some of the statistical procedures described in this work to such a context is the recent work of Genest et al. (2014) on the *multilinear empirical copula*.

With financial applications in mind in particular, tests of extreme-value dependence would also need to be adapted to multivariate stationary time series. The methods briefly described in this chapter all rely on the assumption that the observations at hand are serially independent which is hardly verified for many datasets of interest. Applying the discussed statistical procedures to (almost i.i.d.) standardized residuals from common time series models might be an option, but it is unclear whether the necessary additional estimation step affects the limiting null distribution of the test statistics or not. For that purpose, a starting point might be the work of Rémillard (2010) where the asymptotics of the empirical copula process of standardized residuals are investigated. If the tests are to be applied on the stationary raw time series data, then their empirical levels will most likely be affected by the serial dependence present in the observations. In such a situation, the dependent multiplier bootstrap studied in Bücher and Kojadinovic (2015) could be used to adapt some of the reviewed tests of extreme-value dependence.

Acknowledgments

The authors are grateful to Christian Genest, Johanna Nešlehová and an anonymous referee for their constructive comments on an earlier version of this chapter. This work has been supported in part by the Collaborative Research Center *Statistical Modeling of Nonlinear Dynamic Processes* (SFB 823, project A7) of the German Research Foundation (DFG).

References

Beirlant, J., Y. Goegebeur, J. Segers, and J. Teugels (2004). *Statistics of Extremes: Theory and Applications*. Wiley Series in Probability and Statistics. Chichester: John Wiley & Sons Ltd.

Ben Ghorbal, M., C. Genest, and J. Nešlehová (2009). On the test of Ghoudi, Khoudraji, and Rivest for extreme-value dependence. *The Canadian Journal of Statistics 37*(4), 534–552.

Berghaus, B., A. Bücher, and H. Dette (2013). Minimum distance estimators of the Pickands dependence function and related tests of multivariate extreme-value dependence. *Journal de la Société Française de Statistique 154*(1), 116–137.

Bücher, A., H. Dette, and S. Volgushev (2011). New estimators of the Pickands dependence function and a test for extreme-value dependence. *The Annals of Statistics 39*(4), 1963–2006.

Bücher, A. and I. Kojadinovic (2015). A dependent multiplier bootstrap for the sequential empirical copula process under strong mixing. *Bernoulli*. In press, available at arXiv:1306.3930.

Bücher, A. and J. Segers (2014). Extreme value copula estimation based on block maxima of a multivariate stationary time series. *Extremes 17*(3), 495–528.

Cebrian, A., M. Denuit, and P. Lambert (2003). Analysis of bivariate tail dependence using extreme values copulas: An application to the SOA medical large claims database. *Belgian Actuarial Journal 3*, 33–41.

Charpentier, A., A.-L. Fougères, C. Genest, and J. Nešlehová (2014). Multivariate Archimax copulas. *Journal of Multivariate Analysis 126*, 118 – 136.

Cormier, E., C. Genest, and J. Nešlehová (2014). Using B-splines for nonparametric inference on bivariate extreme-value copulas. *Extremes 17*(4), 633–659.

Davison, A., S. Padoan, and M. Ribatet (2012). Statistical modelling of spatial extremes (with discussion). *Statistical Science 27*(2), 161–186.

Deheuvels, P. (1979). La fonction de dépendance empirique et ses propriétés: un test non paramétrique d'indépendance. *Académie Royale de Belgique. Bulletin de la Classe des Sciences. 5e Série 65*, 274–292.

Deheuvels, P. (1981). A non parametric test for independence. *Publications de l'Institut de Statistique de l'Université de Paris 26*, 29–50.

Dietrich, D., L. de Haan, and J. Hüsler (2002). Testing extreme value conditions. *Extremes 5*(1), 71–85.

Drees, H., L. de Haan, and D. Li (2006). Approximations to the tail empirical distribution function with application to testing extreme value conditions. *Journal of Statistical Planning and Inference 136*(10), 3498–3538.

Du, Y. and J. Nešlehová (2013). A moment-based test for extreme-value dependence. *Metrika 76*, 673–695.

Einmahl, H. J., L. de Haan, and D. Li (2006). Weighted approximations of tail copula processes with application to testing the bivariate extreme value condition. *The Annals of Statistics 34*, 1987–2014.

Frees, E. and E. Valdez (1998). Understanding relationships using copulas. *North American Actuarial Journal 2*, 1–25.

Galambos, J. (1978). *The asymptotic theory of extreme order statistics*. Wiley Series in Probability and Mathematical Statistics. John Wiley & Sons, New York-Chichester-Brisbane.

Genest, C., I. Kojadinovic, J. Nešlehová, and J. Yan (2011). A goodness-of-fit test for bivariate extreme-value copulas. *Bernoulli 17*(1), 253–275.

Genest, C., J. Nešlehová, and M. Ruppert (2011). Comment on the paper by S. Haug, C. Klüppelberg and L. Peng entitled "Statistical models and methods for dependence in insurance data". *Journal of the Korean Statistical Society 40*, 141–148.

Genest, C., J. Nešlehová, and B. Rémillard (2014). On the multilinear empirical copula process. *Bernoulli 20*, 1344–1371.

Genest, C., B. Rémillard, and D. Beaudoin (2009). Goodness-of-fit tests for copulas: A review and a power study. *Insurance: Mathematics and Economics 44*, 199–213.

Genest, C. and J. Segers (2009). Rank-based inference for bivariate extreme-value copulas. *Annals of Statistics 37*, 2990–3022.

Ghoudi, K., A. Khoudraji, and L.-P. Rivest (1998). Propriétés statistiques des copules de valeurs extrêmes bidimensionnelles. *The Canadian Journal of Statistics 26*(1), 187–197.

Gudendorf, G. (2012). *Nonparametric estimation of multivariate extreme-value copulas*. Ph. D. thesis, Université catholique de Louvain.

Gudendorf, G. and J. Segers (2010). Extreme-value copulas. In P. Jaworski, F. Durante, W. Härdle, and W. Rychlik (Eds.), *Copula Theory and Its Applications (Warsaw, 2009)*, Lecture Notes in Statistics, pp. 127–146. Springer-Verlag.

Gudendorf, G. and J. Segers (2012). Nonparametric estimation of multivariate extreme-value copulas. *Journal of Statistical Planning and Inference 143*, 3073–3085.

Gumbel, E. J. (1958). *Statistics of extremes*. New York: Columbia University Press.

Hofert, M., I. Kojadinovic, M. Mächler, and J. Yan (2014). *copula: Multivariate Dependence with Copulas*. R package version 0.999-12.

Hsing, T. (1989). Extreme value theory for multivariate stationary sequences. *Journal of Multivariate Analysis 29*(2), 274 – 291.

Hüsler, J. (1990). Multivariate extreme values in stationary random sequences. *Stochastic Processes and their Applications 35*(1), 99 – 108.

Hüsler, J. and D. Li (2006). On testing extreme value conditions. *Extremes 9*(1), 69–86.

Kojadinovic, I., J. Segers, and J. Yan (2011). Large-sample tests of extreme-value dependence for multivariate copulas. *The Canadian Journal of Statistics 39*(4), 703–720.

Kojadinovic, I., H. Shang, and J. Yan (2015). A class of goodness-of-fit tests for spatial extremes models based on max-stable processes. *Statistics and Its Interface 8*(1), 45–62.

Kojadinovic, I. and J. Yan (2010a). Modeling multivariate distributions with continuous margins using the `copula` R package. *Journal of Statistical Software 34*(9), 1–20.

Kojadinovic, I. and J. Yan (2010b). Nonparametric rank-based tests of bivariate extreme-value dependence. *Journal of Multivariate Analysis 101*(9), 2234–2249.

Kojadinovic, I. and J. Yan (2011). A goodness-of-fit test for multivariate multiparameter copulas based on multiplier central limit theorems. *Statistics and Computing 21*(1), 17–30.

Kojadinovic, I. and J. Yan (2012). A nonparametric test of exchangeability for extreme-value and left-tail decreasing bivariate copulas. *Scandinavian Journal of Statistics 39*(3), 480–496.

Ledford, A. W. and J. A. Tawn (1996). Statistics for near independence in multivariate extreme values. *Biometrika 83*(1), 169–187.

Longin, F. and B. Solnik (2001). Extreme correlation of international equity markets. *The Journal of Finance 56*(2), 649–676.

McNeil, A., R. Frey, and P. Embrechts (2005). *Quantitative Risk Management*. New Jersey: Princeton University Press.

Pfaff, B. and A. McNeil (2013). *QRM: Provides R-language Code to Examine Quantitative Risk Management Concepts*. R package version 0.4-9.

Pickands, J. (1981). Multivariate extreme value distributions. With a discussion. Proceedings of the 43rd session of the International Statistical Institute. *Bulletin of the Institute of International Statistics 49*, 859–878, 894–902.

Quessy, J.-F. (2012). Testing for bivariate extreme-value dependence using Kendall's process. *Scandinavian Journal of Statistics 39*, 497–514.

R Development Core Team (2014). *R: A Language and Environment for Statistical Computing*. Vienna, Austria: R Foundation for Statistical Computing. ISBN 3-900051-07-0.

Rémillard, B. (2010). Goodness-of-fit tests for copulas of multivariate time series. *Social Science Research Network 1729982*, 1–32.

Ressel, P. (2013). Homogeneous distributions — And a spectral representation of classical mean values and stable tail dependence functions. *Journal of Multivariate Analysis 117*, 246–256.

Ribatet, M. (2013). Spatial extremes: Max-stable processes at work. *Journal de la Société Française de Statistique 154*(2), 156–177.

Ribatet, M. and M. Sedki (2013). Extreme value copulas and max-stable processes. *Journal de la Société Française de Statistique 154*(1), 138–150.

Rüschendorf, L. (1976). Asymptotic distributions of multivariate rank order statistics. *The Annals of Statistics 4*, 912–923.

Salvadori, G., C. D. Michele, N. Kottegoda, and R. Rosso (2007). *Extremes in Nature: An Approach Using Copulas*. Water Science and Technology Library, Vol. 56. Springer.

Schweizer, B. and E. Wolff (1981). On nonparametric measures of dependence for random variables. *The Annals of Statistics 9*(4), 879–885.

Sklar, A. (1959). Fonctions de répartition à n dimensions et leurs marges. *Publications de l'Institut de Statistique de l'Université de Paris 8*, 229–231.

Smith, R. L. (1990).   Max-stable processes and spatial extremes.   Unpublished manuscript, University of Surrey.

Tawn, J. A. (1988).   Bivariate extreme value theory: Models and estimation. *Biometrika 75*(3), 397–415.

Wuertz, D. and Y. Chalabi (2013). *timeSeries: Rmetrics — Financial Time Series Objects*. R package version 3010.97.

Wuertz, D., Y. Chalabi, M. Maechler, and J. W. Byers (2013). *timeDate: Rmetrics — Chronological and Calendar Objects*. R package version 3010.98.

19

Extreme Risks of Financial Investments

Ye Liu

JBA Risk Management, United Kingdom

Jonathan A. Tawn

Lancaster University, United Kingdom

Abstract

A range of statistical models for the joint distribution of different financial market returns has been developed. The statistical property of interest is the tail behaviour of these models and their abilities to capture features of extreme events in the financial markets, such as sharp falls in one or multiple markets within a short period of time. A conditional approach based on multivariate extreme value theory is considered and compared to a few other benchmark models commonly used in the industry. The conditional approach is extended to have hierarchically structured parameters with the aim to incorporate the underlying financial market factors. Analysis based on both simulated and empirical data shows that the proposed approaches are more suited for modelling the extreme events than the industrial benchmarks.

19.1 Introduction

Financial institutions such as pension funds or asset management companies, as well as wealthy individuals, tend to spread their investments over a wide range of assets from different financial markets, e.g., company shares and bonds, through a carefully managed portfolio. The ultimate goal for these practitioners is to minimise the risk of their investments while maintaining a satisfactory level of growth. This process, commonly referred to as portfolio management, can be put into the following mathematical form,

$$\min_{\mathbf{h}} f_{\text{Risk}} \left(\mathbf{h}^T \mathbf{X} \right) \quad \text{subject to} \quad f_{\text{Reward}} \left(\mathbf{h}^T \mathbf{X} \right) \geqslant r_T \quad \text{and} \quad \mathbf{h}^T \mathbf{1} = h_0,$$

399

where the control variable of this optimisation problem, $\mathbf{h} := (h_1, \ldots, h_n)'$, is a n-dimensional vector indicating the proportion of each asset held; integer $n \geqslant 2$ is commonly known as the breadth of a portfolio; variable $\mathbf{X} := (X_1, \ldots, X_n)'$ represents the corresponding asset returns; functions f_{Reward} and f_{Risk} are the quantitative interpretation of the investor's reward for making the investment and the investment's exposure to risk, respectively. The inequality constraint means that the reward must reach a target level of r_T whereas the equality constraint means that the total amount of funding is limited, typically $h_0 = 1$ for a fully funded portfolio or $h_0 = 0$ for a self-funded portfolio. An adequate risk metric f_{Risk} as well as a statistically sound model for the underlying asset returns $\mathbf{X}$ are two key aspects of a successful risk management system.

One of the popular candidates for f_{Risk} in the financial industry is the variance, as in the mean-variance analysis first proposed by Markowitz (1952). This approach makes redundant any effort to model the asset returns $\mathbf{X}$ beyond the complexity of a multivariate normal (MVN) distribution. The limitation of such a system is evident especially in abnormal market conditions where market returns exhibit fat tails and the dependence deviates from the linear correlation represented in an MVN distribution; see for example Embrechts et al. (1997). Research on downside risk of financial investments has focused on risk metrics for portfolio management such as value-at-risk (VaR) and conditional value-at-risk (CVaR); see for example Rockafellar and Uryasev (2000, 2002) and Jorion (2001). The disclosure of these measures is enforced by regulation in some countries as recommended in Basel II by the Basel Committee on Banking Supervision of Bank for International Settlements.

The choice of risk metric is difficult and in some contexts quite subjective. Even the use of popular downside risk metrics like VaR faces controversy; see Keating et al. (2001) for example. We assume that a sophisticated enough risk metric is used such that modelling of the joint tail behaviour of asset returns is important. Therefore the main aim of this study is to propose a multivariate distribution for financial market returns with focus on tail events so that it can be used efficiently alongside any risk metric that incorporates downside risks.

Extreme value theory deals with rare events and provides a viable approach for studying the tail of a distribution for financial data. It provides a scientific basis for extrapolation from the body of the distribution to rarely observed or unobserved levels. Extreme events in the financial market may refer to extreme returns of a single financial asset or an unusual joint behaviour across several markets. This article looks specifically at the joint behaviour, or the dependence structure, over a representative set of financial markets, under abnormal market regimes. Our objective may be summarised as the estimation of the tail probability $\mathbb{P}\left(\sum_{i=1}^{n} h_i X_i < -r_0\right)$ for an arbitrary portfolio $\mathbf{h}$ at a high level of loss $-r_0$ ($r_0 > 0$) based on realisations of $\mathbf{X}$.

Janse et al. (2000) and Hyung and de Vries (2007) discuss the tail probability of sums of random variables, i.e., $\sum_{i=1}^{n} h_i X_i$. These approaches avoid explicit assumptions about the extremal dependence and are potentially applicable to a portfolio with breadth $n \geqslant 3$. However they are highly sensitive to the marginal tail and dependence assumptions, which limits their applicability, as we explain in Section 19.2.1. In contrast, there are many approaches to model the joint distribution of $\mathbf{X}$ and then derive the associated distribution of $\sum_{i=1}^{n} h_i X_i$ for any portfolio $\mathbf{h}$. Copula theory and models (Joe, 1997) provide a framework for modelling separately the univariate marginal distributions and dependence structure. As it is the tail of the distribution of $\sum_{i=1}^{n} h_i X_i$ that is of interest the extremal properties of copulas are particularly important. General studies about extremal dependence include Coles and Tawn (1994), Ledford and Tawn (1996, 1997), Joe (1997), and E.Heffernan and A.Tawn (2004), whereas others are more targeted at the applications in finance, such as Embrechts et al. (1997), Poon et al. (2003), Demarta and McNeil (2005), and Luo and Shevchenko (2010). Methods proposed in the above studies have various limitations in practice. For example most models covered in Coles and Tawn (1994), Joe (1997), Demarta and McNeil (2005), and Luo and Shevchenko (2010) lack flexibility in terms of modelling different types of extremal dependence (see Section 19.2.2 for details), whereas the models of Ledford and Tawn (1996, 1997) and Poon et al. (2003) are confined to the joint tail of bivariate distributions.

In practice extreme value analysis based on the joint distribution of all asset returns is often preferred over that based on the univariate distribution of the portfolio return. The main reason is that an estimated joint distribution provides a consistent modelling approach (i.e., one set of assumptions) that is independent of the actual portfolio without the requirement of restrictive modelling assumptions.

The semiparametric approach of E.Heffernan and A.Tawn (2004) provides a flexible modelling approach for studying multivariate extreme events where at least one component, X_i, say, of the vector variable $\mathbf{X}$ is extreme. To extend the application to a typical portfolio of financial investments whose breadth n may grow to hundreds, the dimension of the parameter space may start to cause a problem. We propose a variation of the original Heffernan–Tawn model that incorporates a hierarchical parameter structure that is naturally linked with underlying financial market factors.

The rest of this chapter is structured as follows: Section 19.2 reviews the two existing methods for modelling extreme portfolio returns; Section 19.2.1 explains the theory behind the direct modelling approach, and Section 19.2.2 focuses on the copula-based approach. Section 19.3 summarises the original conditional model of E.Heffernan and A.Tawn (2004) and formulates the hierarchical factor model. Section 19.4 proposes diagnostic methods for evaluating the performance of a multivariate extreme value model. Section 19.5 compares the performance of the hierarchical factor model with some existing benchmark models, based on simulated and financial market returns, with the models illustrated in a passive investment strategy setting.

19.2 Existing Methods

19.2.1 Extended Feller Theorem

One approach is to study directly the extremes of the univariate variable $\sum_{i=1}^{n} h_i X_i$, using asymptotic theory to derive results that inform how this distribution changes over different portfolio choices. The most general available results are due to Embrechts et al. (2009) which extend the results of Feller (1971) derived under the strong assumption of independence. Under the assumption that the random variables X_i $(i = 1, \ldots, n)$ have Pareto form, with equal tail index $\alpha > 0$, i.e.,

$$\mathbb{P}(X_i > x) \sim c_i x^{-\alpha} \text{ as } x \to \infty \tag{19.1}$$

where $c_i > 0$ and $\mathbf{X}$ satisfies the conditions for multivariate regular variation (Resnick, 2007), then the tail probability of a nonnegatively weighted sum of X_i, i.e., $0 \leq h_i \leq 1$ for all i, is given by

$$\mathbb{P}\left(\sum_{i=1}^{n} h_i X_i > x\right) \sim K_n \sum_{i=1}^{n} h_i^{\alpha} \mathbb{P}(X_i > x) \text{ as } x \to \infty \tag{19.2}$$

for K_n a constant. Specifically, K_n is given by

$$K_n := \int_{S_n} \frac{\left[\sum_{i=1}^{n} h_i c_i^{1/\alpha} w_i^{1/\alpha}\right]^{\alpha}}{\sum_{i=1}^{n} h_i^{\alpha} c_i} dH_n(\mathbf{w})$$

where S_n the unit simplex subspace of $\mathbb{R}^n$ and H_n the limiting spectral distribution of $\mathbf{X}^*$, defined as the vector $\mathbf{X}$ standardised marginally to have standard identical unit Pareto margins, with for $\mathbf{w} \in S_n$

$$H_n(\mathbf{w}) = \lim_{r \to \infty} \mathbb{P}\left(\frac{\mathbf{X}^*}{\sum_{i=1}^{n} X_i^*} \leq \mathbf{w} \mid \sum_{i=1}^{n} X_i^* > r\right).$$

Liu (2013) shows that $1 \leq K_n \leq M_n$ and $M_n \leq K_n \leq 1$ for $\alpha \geq 1$ and $0 < \alpha < 1$, respectively, where $M_n = (\sum_{i=1}^{n} h_i c_i^{1/\alpha})^{\alpha} / \sum_{i=1}^{n} h_i^{\alpha} c_i$. The lower bound for $\alpha \geq 1$ and the upper bound for $0 < \alpha < 1$ are achieved when $\mathbf{X}$ are independent whereas the upper bound for $\alpha \geq 1$ and the lower bound for $0 < \alpha < 1$ are achieved when $\mathbf{X}$ are all perfectly dependent. In fact the case $K_n = 1$ holds much more broadly, specifically when the spectral measure H_n has all its mass on the vertex of the unit simplex S_n, a case termed pairwise asymptotic independence; see Section 19.2.2.

In a typical financial application, the assumption of equal tail indices $\alpha_i = \alpha$ for all i is very likely to be false due to the diversity of the financial products collected in an investment portfolio. When tail indices are not equal, the tail of the portfolio returns is dominated by the subset of X_i with the heaviest tails, i.e., X_i $i \in D$ where $D = \{i \in (1, \ldots, n) : \alpha_i = \min_{j=1,\ldots,n}\{\alpha_j\}\}$ hence limit (19.2) holds with K_n and the sum on the right-hand side applied to this particular subset. When only one variable has the smallest α_i value this result reduces to the portfolio tail being identical to the tail of this variable. Liu (2013) shows that both ignoring the lack of equality in the tail indices and reducing to set D generally leads to very poor approximations.

By estimating K_n directly assuming limit relationship (19.2) holds exactly for large x avoids the complication of modelling the dependence between financial returns explicitly as all the information about dependence is contained in the parameter K_n. However, Liu (2013) shows that if any pair of the variables are asymptotic independent (see Section 19.2.2), the estimation of K_n is typically poor as next-order terms bias its estimation, so asymptotic approximation (19.2) is often poor in this case.

Another issue with the use of the extended Feller result is its relatively limited application. In a situation where some other properties of the portfolio are the key interest, e.g., the conditional distribution of the portfolio return $\mathbf{h}^T\mathbf{X}|X_1 > x$ which is widely used in stress testing, this approach is not applicable. Lastly this method does not cover the case where the marginal distributions are all short-tailed, e.g., if the marginal tails decayed exponentially and hence were not of the form (19.1).

19.2.2 Copula-Based Models and Their Extremal Features

The most popular approach to capturing the tail behaviour of the weighted sum of random variables is through modelling the joint distribution of the individual random variables and then deriving the resulting distribution of the weighted sum through simulation from the model. Therefore we need to discuss models for the joint distribution and in particular explore their extremal properties given that it is the tail of the weighted sum that we will be interested in.

All continuous joint distribution functions can be written as

$$\mathbb{P}\left(X_1 < x_1, \ldots, X_n < x_n\right) = C\{F_1(x_1), \ldots, F_n(x_n)\}$$

where F_i is the marginal distribution function of X_i and $C : [0,1]^n \to [0,1]$ is the copula that describes the dependence structure. See Joe (1997) for properties of C. The copula describes the overall dependence structure.

We are typically interested in the dependence structure when one of the components of $(X_1, \ldots, X_n)$ is large, so it is helpful to quantify the extremal properties of the copula. Suppose two variables X_1 and X_2 have marginal distribution functions, F_1 and F_2, respectively; the most well-established way to quantify extremal dependence is through the limiting conditional tail probability χ, defined as

$$\chi := \lim_{u \to 1} \mathbb{P}\left(F_2(X_2) > u | F_1(X_1) > u\right) = \lim_{u \to 1} \frac{1 - 2u + C(u, u)}{1 - u}. \tag{19.3}$$

Two types of extremal dependence are defined according to this limiting value of χ: $\chi > 0$, variables X_1 and X_2 are said to be asymptotically dependent; $\chi = 0$ they are termed asymptotically independent. For an extensive list of copulas Heffernan (2000) identified whether they correspond to asymptotic independence or asymptotic dependence. Examples of pairwise asymptotically independent distributions include Gaussian copulas and inverted multivariate extreme value distributions (i.e., the lower tail). To capture the tendency for co-movements in the financial market, to be strongly related much of the finance literature has focused on copulas which are asymptotically dependent. The most widely studied classes of asymptotically dependent copulas are multivariate extreme value copulas (Joe, 1997) and various multivariate t-copulas (Demarta and McNeil, 2005). The latter have the benefit of giving asymptotic dependence in both the joint upper and lower tails, whereas the former class are asymptotically independent in the lower tail. Therefore most focus in modelling financial data has been on various t-copulas. We will review the main models and identify their extremal features.

The standard t copula is generated by a $[0, 1]$ uniform random variable U and a n-dimensional multivariate Gaussian random variable $\mathbf{Z} := (Z_1, \ldots, Z_n)^T$ with mean $\mathbf{0}$ and correlation matrix P in the following way:

$$\left\{ t_\nu \left[Z_1 \cdot G_\nu^{-1}(U) \right], \ldots, t_\nu \left[Z_n \cdot G_\nu^{-1}(U) \right] \right\} \tag{19.4}$$

where t_ν is the distribution function of the standard Student t distribution with degree of freedom $\nu > 0$; function G_ν^{-1} is the inverse distribution function of $\sqrt{\nu/\chi_\nu^2}$. Provided $\nu < \infty$, Demarta and McNeil (2005) showed that $\chi = 2t_{\nu+1}\left(\sqrt{\nu+1}\sqrt{1-\rho}/\sqrt{1+\rho}\right)$.

Shevchanko (2011) generated an extension of the standard t copula with distinct degrees of freedom on all margins, i.e.,

$$\left\{ t_{\nu_1} \left[Z_1 \cdot G_{\nu_1}^{-1}(U) \right], \ldots, t_{\nu_n} \left[Z_n \cdot G_{\nu_n}^{-1}(U) \right] \right\}. \tag{19.5}$$

The generalised t copula (19.4) provides more flexibility by allowing different dependence strength in separate regions of the joint tail. The cost of this generalisation is the increased number of parameters and hence complexity in inferences. We haven't found evidence that χ has been derived for the generalized t copula. We cannot derive its value other than to show $\chi > 0$; see Liu (2013) for a proof.

Daul et al. (2003) proposed the grouped t copula. Like the generalised t copula it also allows for different degrees of freedom on the margins. The grouped t copula is generated in the same way as the generalised t copula (19.5), except that

$$\nu_1 = \ldots = \nu_{s_1} = \nu_1^*, \ \nu_{s_1+1} = \ldots = \nu_{s_1+s_2} = \nu_2^*, \ \ldots \ , \nu_{n-s_m+1} = \ldots = \nu_n = \nu_m^*$$

where the n margins are divided into m groups with intragroup degree of freedom ν_j^* $(j = 1, \ldots, m)$ for the j-th group. This model captures the factor structure embedded in the financial data while maintaining a reasonable complexity for inferences. The grouped t copula is able to capture asymptotic independence within a group by allowing the corresponding degree of freedom to be infinity. However it is not possible to have both asymptotic independence and dependence within a group under the grouped t copula.

19.3   Conditional Approach for Multivariate Extremes

19.3.1   Motivation

When fitting copula models to data for inference for the tails of the distribution of a portfolio it is critical that the extremal properties implied by the model are in agreement with the data themselves, especially when we wish to extrapolate beyond observed levels. For example, assuming that the marginal distributions are known, if the bivariate normal dependence structure is fitted to a dataset with positively correlated bivariate Student t dependence structure, the upper tail of the sum $(X_1 + X_2)/2$, representing the return of an equally weighted portfolio, would to be under estimated. More generally fitting an asymptotically independent dependence structure to asymptotically dependent data leads to an under estimation of the extreme risk of this investment portfolio. Conversely, if an asymptotically dependent model is fitted to asymptotically independent data, the extreme risk will be over estimated which unnecessarily restrains portfolio weights and eliminates growth potential.

Therefore we require models which are flexible enough to cover both asymptotic independence and asymptotic dependence between all different pairs of variables. The copula models reviewed in Section 19.2.2 are restricted in this respect. An alternative approach is to focus on the form of the copula in the regions of interest, i.e., when at least one margin is very large, to see if any parsimonious model emerges. We will follow the approach of E.Heffernan and A.Tawn (2004) who proposed a conditional approach for studying the extremes which have the flexibility to describe a wide range of extremal dependence across all margins. Hilala et al. (2014) considers some implications of this model for finance. Other recent developments of this approach include methods to impose self-consistency of conditional distributions (Liu and Tawn, 2014) and additional constraints for parameter estimation (Keef et al., 2013). In this chapter we build on the original Heffernan–Tawn model to consider positive and negative tails and impose a factor structure similar to that of the grouped t copula.

19.3.2 The Heffernan–Tawn Model

Let $\mathbf{X}$ be an n-dimensional random variable and $\mathbf{Y}$ be the corresponding transformed random variable with standard Laplace marginal distribution, i.e., $Y_i = -F_L^{-1}[F_X(X_i)]$ for $i = 1, \ldots, n$ with F_L^{-1} the inverse distribution for the standard Laplace distribution with the change of sign to convert focus from the lower tail to the upper tail. E.Heffernan and A.Tawn (2004) and Keef et al. (2013) work under the assumption that there exist vector functions $\mathbf{a}$ and $\mathbf{b}$: $\mathbb{R} \to \mathbb{R}^{n-1}$ for the $(n-1)$-dimensional random variable $\mathbf{Y}_{-i} := (Y_1, \ldots, Y_{i-1}, Y_{i+1}, \ldots, Y_n)$ such that

$$\lim_{u \to \infty} \mathbb{P}\left(\frac{\mathbf{Y}_{-i} - \mathbf{a}(Y_i)}{\mathbf{b}(Y_i)} < \mathbf{z}, \; Y_i - u > y \,\middle|\, Y_i > u\right) = F_{|i}(\mathbf{z}) \exp(-y) \quad (19.6)$$

for $y > 0$, where $F_{|i}$ is an $(n-1)$-dimensional distribution function which is non-degenerate in each margin and the vector subtraction and division are performed componentwise. The normalisation is unique if we impose that $F_{|i}$ does not allow probability mass at $+\infty$ in any margin. They also proposed that the following choice of functions $\mathbf{a}$ and $\mathbf{b}$ in Equation (19.6), which was shown to cover a wide range of known dependence structures for $\mathbf{Y}$ including all the established copula models reviewed by Heffernan (2000):

$$
\begin{aligned}
\mathbf{a}(y) &= \left(\alpha_{1|i}y, \ldots, \alpha_{i-1|i}y, \alpha_{i+1|i}y, \ldots, \alpha_{n|i}y\right)^T \\
\mathbf{b}(y) &= \left(y^{\beta_{1|i}}, \ldots, y^{\beta_{i-1|i}}, y^{\beta_{i+1|i}}, \ldots, y^{\beta_{n|i}}\right)^T
\end{aligned} \tag{19.7}
$$

for $y > 0$; where $\alpha_{j|i} \in [-1, 1]$ and $\beta_{j|i} \in (-\infty, 1)$ $(j \neq i)$ are model parameters conditional on $Y_i > u$ as $u \to \infty$. Under the Heffernan–Tawn model, Equation (19.6) is assumed to hold approximately with the choice of $\mathbf{a}$ and $\mathbf{b}$ in Equations (19.7) for all sufficiently large u.

The estimation of the Heffernan–Tawn model parameters is performed on a pairwise basis, i.e., for $y > 0$ and u large,

$$
\mathbb{P}\left(\frac{Y_j - \alpha_{j|i}Y_i}{Y_i^{\beta_{j|i}}} < z, \ Y_i - u > y \ \middle|\ Y_i > u\right) \simeq F_{j|i}(z)\exp(-y).
$$

Distribution $F_{j|i}$ is the j-th marginal distribution of $F_{|i}$ of Equation (19.6). E.Heffernan and A.Tawn (2004) assumed that the first two moments exist for $F_{j|i}$ and proposed to use the Gaussian likelihood for $F_{j|i}$, i.e.,

$$
Y_j|Y_i = y \sim N\left(\alpha_{j|i}y + y^{\beta_{j|i}}\mu_Z, y^{2\beta_{j|i}}\sigma_Z^2\right),
$$

where μ_Z and σ_Z are the mean and standard deviation of the residual variable $Z_{j|i} := (Y_j - \alpha_{j|i}Y_i)/Y_i^{\beta_{j|i}}$, respectively, when $Y_i > u$. Note that μ_Z and σ_Z are only auxiliary parameters, such that when we make subsequent inferences of this pair we replace $\alpha_{j|i}$ and $\beta_{j|i}$ with their estimates $\widehat{\alpha}_{j|i}$, $\widehat{\beta}_{j|i}$ and use the empirical distribution of the residual $Z_{j|i}$ for $Y_i > u$ instead of assuming $F_{j|i}$ is Gaussian.

As illustrated in E.Heffernan and A.Tawn (2004), the goodness of fit of the Heffernan–Tawn model may be quickly tested by verifying the independence between the residual $Z_{j|i}$ and the conditioning margin Y_i. Another approach is discussed in Liu and Tawn (2014) where both the model self-consistency and model fit are verified based on the decay of conditional tail probability χ along the bivariate joint tail between two arbitrary margins Y_i and Y_j.

The pairwise estimation is then rotated through all combinations of (i, j) with $i \neq j$, or set to be $\alpha_{j|i} = 1$ and $\beta_{j|i} = 0$ if $i = j$ as Y_i is asymptotically dependent on itself. Note that the dependence may be asymmetric, e.g.. $\alpha_{j|i} \neq \alpha_{i|j}$, which requires both ways of each pairwise conditioning to be considered. The parameters can be put into a matrix form with the (i, j)th entry being $\alpha_{j|i}$ and $\beta_{j|i}$ in the corresponding matrix of αs and βs, respectively.

19.3.3 Handling Positive and Negative Returns

When fitting the Heffernan–Tawn model to asset returns, both the positive and negative extremes need to be accounted for. Denote $Y_i^{(+)} := Y_i$ and $Y_i^{(-)} := -Y_i$. Then there are four pairwise dependences, i.e., $Y_j^{(+)}|Y_i^{(+)} > u$, $Y_j^{(+)}|Y_i^{(-)} > u$,

$Y_j^{(-)}|Y_i^{(+)} > u$ and $Y_j^{(-)}|Y_i^{(-)} > u$. It appears that four sets of parameters A and B are required but as

$$\left(Y_j^{(+)}|Y_i^{(-)} > u\right) \sim -\left(Y_j^{(-)}|Y_i^{(-)} > u\right)$$

$$\text{and } \left(Y_j^{(-)}|Y_i^{(+)} > u\right) \sim -\left(Y_j^{(+)}|Y_i^{(+)} > u\right)$$

(19.8)

we can reduce the parameters matrices to (A_+, B_+) and (A_-, B_-) for parameters associated with pairwise dependence $Y_j^{(+)}|Y_i^{(+)} > u$ and $Y_j^{(-)}|Y_i^{(-)} > u$, respectively. These parameters can be estimated as described in Section 19.3.2. Therefore the final parameter set for the n-dimensional random variable $\mathbf{Y}$ consists of two $2n \times 2n$ matrices as follows,

$$A = \begin{pmatrix} A_+ & -A_- \\ -A_+ & A_- \end{pmatrix}, B = \begin{pmatrix} B_+ & B_- \\ B_+ & B_- \end{pmatrix},$$

where $\{A_+, A_-, B_+, B_-\}$ are all $n \times n$ matrices of α and β parameters.

The Heffernan–Tawn model does not provide a parametric model for the body of the distribution where none of the n margins of variable $\mathbf{Y}$ is large. This is a relatively minor issue since we are only concerned about extreme events and the empirical distribution of the nonextreme observations is sufficient to complete the model for the distribution of $\mathbf{Y}$. We follow the steps listed below to simulate random variables with standard Laplace margins for any given Heffernan–Tawn dependence parameter matrices A and B.

1. Fix a large positive value $u > 0$ such that $\mathbb{P}(L > u) = \mathbb{P}(L < -u)$ are small, where L has the standard Laplace distribution.
2. Draw at random $\mathbf{y} := (y_1, \ldots, y_n)^T$ from the original set of observations, i.e., from the empirical distribution of $\mathbf{Y}$.
3. If $\mathbf{y} \in (-u, u)^n$ then this sample $\mathbf{y}$ is accepted.
4. If at least one margin lies above u or below $-u$, i.e., an extreme event occurs. Denote the margin with largest absolute value as y_i, i.e., $|y_i| > |y_j|$ for all $j \neq i$. Then we follow these resimulation steps to generate a new sample using the Heffernan–Tawn model.

Case I - if $y_i > u$:

(a) Overwrite y_i with a realisation $y_i^{(+)}$ from the conditional distribution of $L|L > u$;

(b) Simulate the other margins $\mathbf{y}_{-i} := (y_1, \ldots, y_{i-1}, y_{i+1}, \ldots, y_n)^T$ using the conditional distribution that

$$\left(\mathbf{Y}_{-i}|Y_i = y_i^{(+)}\right) \sim \boldsymbol{\alpha}_{|i+}y_i^{(+)} + \left(y_i^{(+)}\right)^{\boldsymbol{\beta}_{|i+}} \otimes \mathbf{z}_{|i+}$$

where vector $\boldsymbol{\alpha}_{|i+}$ and $\boldsymbol{\beta}_{|i+}$ are the i-th column of submatrices A_+ and B_+, respectively, with the i-th entry removed; operation $\otimes$ means componentwise vector multiplication; vector $\mathbf{z}_{|i+}$ is a random sample drawn from the observed distribution of residuals $\mathbf{Z}_{|i} := (\mathbf{Y}_{-i} - \boldsymbol{\alpha}_{|i+}Y_i) \otimes Y_i^{-\beta_{|i+}}$ independently of $y_i^{(+)}$;

Case II - if $y_i < -u$:

(a) Overwrite y_i with a realisation $y_i^{(-)}$ from the conditional distribution of $L | L < -u$;

(b) Simulate the negative values of other margins, $\mathbf{y}_{-i}^{(-)}$, using the conditional distribution that

$$\left(\mathbf{Y}_{-i}^{(-)} | Y_i^{(-)} = y_i^{(-)}\right) \sim \boldsymbol{\alpha}_{|i-}y_i^{(-)} + \left(y_i^{(-)}\right)^{\beta_{|i-}} \otimes \mathbf{z}_{-|i-}$$

where vector $\boldsymbol{\alpha}_{|i-}$ and $\boldsymbol{\beta}_{|i-}$ are the i-th column of submatrices A_- and B_-, respectively, with the i-th entry removed; vector $\mathbf{z}_{|i-}$ is a random sample drawn from the observed distribution of residuals $\mathbf{Z}_{|i-} := (\mathbf{Y}_{-i}^{(-)} - \boldsymbol{\alpha}_{|i-}Y_i^{(-)}) \otimes Y_i^{(-)-\beta_{|i-}}$ independently of $y_i^{(-)}$;

5. If $|y_i| > |y_j|$ for all $j \neq i$, accept the new vector $\mathbf{y}$ or $\mathbf{y}^{(-)}$; otherwise reject $\mathbf{y}$ and repeat step (4).

19.3.4 Hierarchical Factor Model

In portfolio management variables are grouped into different types of asset classes. Even if all variables in $\mathbf{Y}$ from same financial market, e.g., commodities, groups are associated with a finer categorisation, e.g., metals, and agricultural products. Groupings are important as returns in different groups exhibit similar dependence to each other but quite different dependence to other groups. This is well recognised in the optimal estimation of the variance matrix where Disatnik and Katz (2012) exploit the block structure. The grouped t copula aims to capture this structure without being restricted to an assumption of MVN copula. We propose a hierarchical factor Heffernan–Tawn model, which imposes the factor structure on the tail regions of the copula, and capture a wide range of extremal dependence types through different combinations of the parameter matrices A and B given in Seciton 19.3.3.

As a motivational example, Figure 19.1 shows the estimated values of matrix A_+ of the Heffernan–Tawn model fitted to the returns of 19 futures indices returns, covering stocks, bonds, interest rates, metals and agricultural products. The data consist of daily returns of the above financial assets spanning from 01/2002 to 03/2010. We can observe roughly a block structure for the parameter matrix, e.g., the α parameters are relatively close to 1 within the "stocks × stocks" block whereas the parameters in the "stocks × bonds" block are generally smaller. The block-matrices are not symmetric, indicating the variables are pairwise nonexchangeable and the variance of the α parameters within each block can be quite different, which implies that a constant value for α in each block may be too strong an assumption.

Consider a portfolio of n assets classified into m groups $\{G_1, \ldots, G_m\}$ where the k-th group is of size s_k, i.e.,

$$G_1 = \{1, \ldots, s_1\}; G_2 = \{s_1 + 1, \ldots, s_1 + s_2\}; \ldots, G_m = \{n - s_m + 1, \ldots, n\}.$$

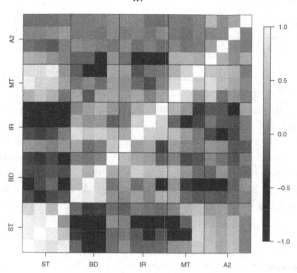

FIGURE 19.1
The estimated parameter $\alpha_{j|i+}$ for a portfolio of 19 future indices, grouped by the type of assets (ST-stocks, BD-bonds, IR-interest rates, MT-metals, AG-agricultural products) with the gray shade at the coordinates (i,j) corresponding to the size of the $\alpha_{j|i+}$.

If $i \in G_I$ and $j \in G_J$ the parameters of block-matrices A_* and B_*, where "$*$" represents the "$+$" or the "$-$" sign, are taken to be distributed independently over $i \neq j$ as $(A_*)_{i,j} \sim \Pi_{J|I}^{A_*}$ and $(B_*)_{i,j} \sim \Pi_{J|I}^{B_*}$ where

$$\frac{\Pi_{J|I}^{A_*} + 1}{2} \sim \text{Beta}\left(\theta_{J|I*}\right) \text{ and } 1 - \Pi_{J|I}^{B_*} \sim \text{Gamma}\left(\psi_{J|I*}\right) \quad (19.9)$$

with the bivariate vectors $\theta_{J|I*}$ and $\psi_{J|I*}$ being the beta and gamma distribution hyperparameters, respectively.

The hierarchical distribution hyperparameters for the I-th and J-th group are estimated by maximising the following pseudo-likelihood of (θ, ψ):

$$\prod_{i \neq j: \, i \in G_I, j \in G_J} \left\{ \iint_{\alpha, \beta} f_{Y_{j|i}^{(*)}}\left(y_j^{(*)} \mid y_i^{(*)}; \alpha, \beta\right) \cdot \pi_{J|I}^{A_*}(\alpha; \theta) \cdot \pi_{J|I}^{B_*}(\beta; \psi) \cdot d\alpha \, d\beta \right\}$$
$$(19.10)$$

where $f_{Y_{j|i}^{(*)}}(\cdot; \alpha, \beta)$ is the conditional density function for $Y_j^{(*)} | Y_i^{(*)}$ with the Heffernan–Tawn model parameters α and β. Efficient numerical methods may be necessary for the evaluation of the likelihood since the analytical form of the integration is non-attainable. The Heffernan–Tawn parameters A and B can then be estimated by maximising a conditional pseudo-likelihood with estimates $\left(\widehat{\alpha}_{j|i*}, \widehat{\beta}_{j|i*}\right)$

defined by

$$\arg\max_{\alpha,\beta} \left\{ f_{Y_{j|i}^{(*)}} \left(y_j^{(*)} \mid y_i^{(*)} ; \alpha, \beta \right) \cdot \pi_{J|I}^{A*} \left(\alpha; \widehat{\boldsymbol{\theta}}_{J|I*} \right) \cdot \pi_{J|I}^{B*} \left(\beta; \widehat{\boldsymbol{\psi}}_{J|I*} \right) \right\}$$

(19.11)

where $(\widehat{\boldsymbol{\theta}}_{J|I*}, \widehat{\boldsymbol{\psi}}_{J|I*})$ are the estimates of $(\boldsymbol{\theta}, \boldsymbol{\psi})$ obtained by maximising Equation (19.11). If the hierarchical parameter distributions Π^A and Π^B are narrowly (widely) spread the variance of the dependence parameters α and β in a group will be small (large), respectively. This approach describes the intrablock dependence with fewer parameters than the original Heffernan–Tawn model, i.e., using $(\boldsymbol{\theta}, \boldsymbol{\psi})$ instead of the full matrix of A and B, which is a desirable feature for high-dimensional portfolio management. The effective reduction in the number of parameters is at maximum when there is high consistency in intrablock dependence or minimum when there is no significant intrablock variation of dependence.

19.4   Diagnosis Methods for Portfolio Extremes

19.4.1   Pairwise Conditional Tail Test

For any two margins, i and j say, of the n-dimensional variable $\mathbf{X}$, we test the goodness of fit of our proposed multivariate dependence model by comparing the fitted distribution of $X_j|X_i > u$ (or $X_j|X_i < -u$) to the observed distribution. We perform the test using the Kolmogorov–Smirnov distance between the observed and modelled distributions. We simulate a sample from the fitted model; perform the Kolmogorov–Smirnov test between the pairwise conditional tails from the model and the data; and count the proportion of rejected tests. It is concluded that the model fits the observed data well, on a pairwise level, if the test has a similar proportion of rejections to the size of the test.

19.4.2   Generalized Pareto Distribution Test

This diagnostic is tailored for the extremes of weighted sums of margins of a multivariate distribution. It tests whether the observed tail and the fitted tail of the portfolio returns have the same distribution. In particular this method approximates both tails with the generalised Pareto distribution (GPD) and tests whether they have the same parameters over a range of different portfolio weights.

Balkema and de Haan (1974) showed that under suitable conditions, the only nondegenerate distribution that the scaled excess of a random variable X above a large threshold u, conditional on $X > u$, can converge to as u tends to the upper endpoint of X is the GPD. In practice this limiting distribution is assumed to hold above a large threshold u with the GPD fitted to observations above u, i.e.,

$$\mathbb{P}\left(X - u \leqslant x \mid X > u \right) \approx 1 - \left(1 + \xi \frac{x}{\sigma} \right)_+^{-\frac{1}{\xi}}$$

with scale parameter $\sigma > 0$, shape parameter ξ, and notation $(x)_+ = \max\{x, 0\}$; see Coles (2001).

Denoting the observed and modelled portfolio returns for given portfolio weights by X_O and X_M, the conditional excess distributions of X_O and X_M over a large threshold u are assumed to follow a GPD with parameters (σ_O, ξ_O) and (σ_M, ξ_M), respectively. We test the hypothesis

$$H_0 : \sigma_O = \sigma_M \text{ and } \xi_O = \xi_M$$
$$H_1 : \sigma_O \neq \sigma_M \text{ or } \xi_O \neq \xi_M$$

with a rejection of the null hypothesis implying the model does not resemble the data in the tails for this portfolio weight. In order to perform this diagnosis test on the financial data we preprocess the financial data, using a GARCH filter, by removing the volatility and standardizing the returns to achieve approximately independent and identically distributed residual returns. We then perform the hypothesis testing using the likelihood ratio test on these residuals.

19.5 Examples

Using the diagnostic methods outlined in Section 19.4 we assess the performance of some candidate modelling approaches mentioned earlier including the grouped t copula (GT), the Heffernan–Tawn model (HT) and the hierarchical Heffernan–Tawn model (HHT). The comparison is based on simulated data as well as observed financial returns.

19.5.1 Simulated Data

In the simulation study we simulate $n = 6$ variables with standard Gaussian marginal distributions and two different copulas: the grouped t copula (GT) and the asymmetric multivariate logistic (AL) extreme value copula (Tawn, 1990). Both copulas have a grouping of $G_1 = \{1, 2, 3\}$ and $G_2 = \{4, 5, 6\}$ with a different intragroup dependence.

The GT copula, given by expression (19.5) has parameters $\nu_1 = \nu_2 = \nu_3 = 5$ and $\nu_4 = \nu_5 = \nu_6 = 1000$; and with correlation matrix P of the MVN component with all off-diagonal terms being 0.5. Data are generated from this distribution using representation (19.5) (Daul et al., 2003). The AL distribution is given by

$$\mathbf{X}_{AL} \sim \left(\Phi^{-1} \left\{ \exp\left[-\exp(-Y_1) \right] \right\}, \ldots, \Phi^{-1} \left\{ \exp\left[-\exp(-Y_6) \right] \right\} \right)$$

where random variables Y_i $(i = 1, \ldots, 6)$ follow the asymmetric logistic multivariate extreme value distribution with joint distribution function $G_{AL}(\mathbf{y})$ equal to

$$\exp\left[-\lambda_1 \left\{ \sum_{i=1}^{3} e^{-y_i/r_1} \right\}^{r_1} - \lambda_2 \left\{ \sum_{i=4}^{6} e^{-y_i/r_2} \right\}^{r_2} - \lambda_3 \left\{ \sum_{i=1}^{6} e^{-y_i/r_3} \right\}^{r_3} \right],$$

TABLE 19.1
The proportion of GPD test rejected (under a 5% significance level) for simulated GT and AL datasets fitted under the GT, HT, and HHT modelling approach.

Data	Portfolio	Tail	GT	HT	HHT
GT	Long-only	Upper	0.004	0.003	0.001
GT	Long-only	Lower	0.000	0.000	0.000
GT	Long-short	Upper	0.107	0.087	0.068
GT	Long-short	Lower	0.112	0.081	0.074
	GT average		0.056	0.043	0.036
AL	Long-only	Upper	0.223	0.004	0.169
AL	Long-only	Lower	0.121	0.026	0.081
AL	Long-short	Upper	0.395	0.118	0.117
AL	Long-short	Lower	0.430	0.134	0.118
	AL average		0.292	0.071	0.121

where $\lambda_1 = \lambda_2 = 0.9$, $\lambda_3 = 0.1$, $r_1 = 0.8$, $r_2 = 0.2$ and $r_3 = 0.5$ and the marginal distributions are standard Gumbel distribution. The AL data are simulated using methods of Stephenson (2003).

This choice of copulas covers a good range of extremal dependence types, e.g., pairwise asymptotic independence or asymptotic dependence for the grouped margins in the GT if the degree of freedom $\nu \to \infty$ or $\nu < \infty$, respectively, so the G_1 margins of the GT data are pairwise asymptotically dependent and G_2 practically pairwise asymptotically independent. Similarly the lower or upper tail of the margins in the AL are asymptotically independent or asymptotically dependent, respectively, with this pattern holding for both groups. These copulas have small and large misspecification errors relative to the benchmark approach and cover an example with different dependence in the joint lower/upper tails and pairwise nonexchangeability.

We compare the performance of all three approaches (GT, HT, and HHT) on the GT and AL datasets, each of length 5000, based on the pairwise conditional tail test and the GPD test of Section 19.4. The former test is performed across all pairs of margins. Results given in Liu (2013) show that all three modelling approaches are reasonably good at capturing pairwise dependence. For the GPD diagnostic method we use the same 5000 GT and AL samples and perform the test on the corresponding portfolio returns for both the upper and lower tails (the top and bottom 5% of the distribution). We repeat the test over 4000 randomly generated portfolio weights (2000 long-only and 2000 long-short). A long-short portfolio has both positive and negative portfolio weights with the sum of weights equal to 0; whereas a long-only portfolio has only positive portfolio weights with the sum of weights equal to 1. Results are given in Table 19.1. For the GT data the results are similar across all approaches. However, the results for the AL data show that HT and HHT clearly outperform the GT benchmark approach. This confirms the disadvantage of the GT approach when applied to a relatively complex (asymmetric) dependence structure.

TABLE 19.2
Description of the financial data under study.

Type	Index name	Abbreviation
Stock	E-mini S& P 500	ESPC
Stock	FTSE 100	FTL
Stock	Dax Index	DXF
Stock	Tokyo Stock Exchange	TSJ
Bond	Treasury Bonds	TBC
Bond	Gilts	GTL
Bond	Euro-BUND	DBF
Bond	Japanese Bond	TSE

19.5.2 Financial Data

The financial data consists of four stock market indices and four bond market futures indices listed in Table 19.2. The return data are aggregated daily at the US market opening hour between 01/2002 and 03/2010 and contain 1792 complete datapoints for each market.

To ensure stationarity we first preprocess the data by removing the long-term mean and the volatility effect, the latter of which is achieved by taking the residual of the time series under a GARCH model. Figure 19.2 illustrates the effect of preprocessing on the TBC data. During the practice of portfolio construction, the GARCH modelling in the preprocessing may be replaced with more forecast-oriented approaches (e.g., the exponentially weighted moving average or the use of exogenous variables such as the implied volatility). The extreme value analysis based on the standardised returns remains approximately the same whatever the chosen method for the volatility model.

First we fit the standard HT model to the preprocessed data and examine the empirical distribution of the MLEs of α and β (as shown in Figure 19.3). These support the choice of beta and gamma distributions for the hierachical model. Figure 19.4 compares the A_+ and B_+ matrices under the HT and HHT approaches. While broadly similar, the parameters under the HHT model exhibit lower variance within each block (e.g., the inter-group blocks in matrix A_+) and yet allow occasional deviation from the block mean if supported by data (e.g., the bond-conditioning B_+ blocks).

We evaluate the performance of the three modelling approaches with the pairwise conditional tail test and the GPD test. The test settings used here are the 90-th or the 95-th quantile as the threshold for the first test; 4000 randomly generated portfolios of the standardised returns (including long-only and long-short) for the second test; both tests are performed at a 5% significance level. Tables 19.3 and 19.4 summarise the main findings. Table 19.3 shows superior performance from the benchmark GT approach, particularly at the lower threshold. But all three models perform equally well for the more extreme pairwise tails. In terms of the quality of fit to the full dependence structure the HT and HHT model again outperform the GT by a good

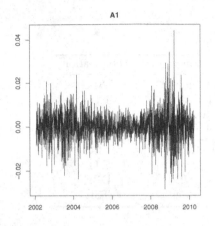
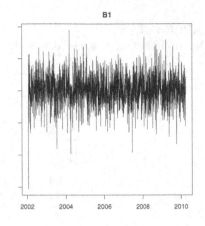

FIGURE 19.2
Time series plot of the raw (left) and standardised TBC daily returns between 01/2002 and 03/2010.

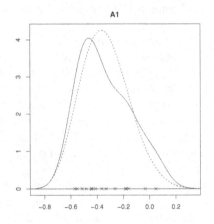
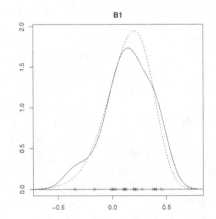

FIGURE 19.3
Illustration of the hierarchical parameter distribution choice using the "stock × bond" block with the kernel density (solid lines) of the $\alpha_{j|i+}$ (left) and $\beta_{j|i+}$ (right) MLEs (crosses) of the standard Heffernan–Tawn model and the corresponding best-fitting beta and gamma densities (dashed lines).

margin according to the GPD test results in Table 19.4. In addition the introduction of the hierarchical structure demonstrates its benefit by further lowering the rejection proportion of the original HT model from 0.180 to 0.115 while reducing the effective number of model parameters.

In practice tail dependence models can keep investors better informed about potential extreme risks in their investment portfolio. Through the following example,

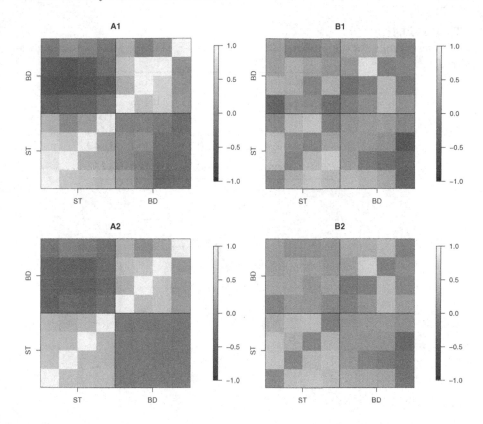

FIGURE 19.4
Comparison of the A_+ (left) and B_+ (right) matrices between the HT (upper) and HHT (lower) modelling approaches under the beta and gamma distribution choice for the hierarchical parameter distributions.

TABLE 19.3
Pairwise conditional tail test results for the stock-bond financial data under the GT, HT, and HHT modelling approach for two thresholds (the 90-th and 95-th quantiles).

Data-threshold	GT	HT	HHT
Financial-90	0.018	0.196	0.107
Financial-95	0.036	0.054	0.054

we can see the potential advantage of implementing a downside extreme risk indicator. To focus entirely on the tail dependence modelling, we avoid the use of any portfolio optimisation techniques and assume knowledge of the volatility of all assets.

The investment strategy is to split the portfolio equally between all eight assets; sell out all assets if the next day's portfolio return is prone to excessive downside

TABLE 19.4

GPD test results for the financial data under the GT, HT, and HHT modelling approach.

Portfolio	Tail	GT	HT	HHT
Long-only	Upper	0.271	0.168	0.124
Long-only	Lower	0.462	0.238	0.041
Long-short	Upper	0.206	0.161	0.144
Long-short	Lower	0.184	0.154	0.149
Average		0.281	0.180	0.115

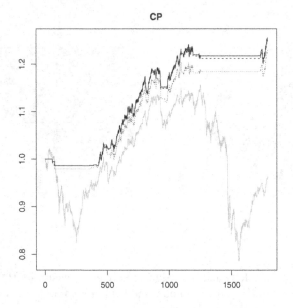

FIGURE 19.5

Performance of the parsimonious investment strategy, using GT (dotted line), HT (dashed line), or HHT (black solid line) as the tail dependence model, in comparison with the passive benchmark (gray solid line).

risk; buy back and start trading again when the downside risk returns to normal level. We use the CVaR as the indicator for downside risk and set the threshold to be -2% so that trading stops or resumes when the CVaR forecast for the next day is below or above -2%. We back-tested the performance of this investment strategy, using GT, HT or HHT as the tail dependence model, and compare the results with a passive benchmark (equally-weighted portfolio all the way through) over the duration of the original data (Figure 19.5).

Imposing a downside risk indicator for switch on/off portfolio positions proved to be reasonably successful, i.e., all three tail dependence models managed to provide a sufficient alert at major drawdowns. This allows the investment strategy to accumulate more than 20% return over the same period as the passive benchmark made a loss of −4%. Furthermore, HHT delivered the best performance among all three models, finishing with 125% of the initial portfolio value (0.5% and 2.7% more than HT and GT, respectively).

Note that this example does not reflect the reality since many other factors may affect the final performance. For example we found the volatility model, which we assumed to be known in advance, is also a key contributor to the effectiveness of the downside risk indicator. In the end, the performance of an investment strategy in the financial market will be as good as the modelling efforts for all aspects of the underlying financial assets.

Acknowledgment

We would like to thank Man Investments Ltd. for funding the research, Dr Anthony Ledford and Prof Ser-Huang Poon for helpful discussions.

References

Balkema, A. and L. de Haan (1974). Residual life time at great age. *Annals of Probability 2*, 792–804.

Coles, S. G. (2001). *An Introduction to Statistical Modeling of Extreme Values*. Springer.

Coles, S. G. and J. A. Tawn (1994). Statistical methods for multivariate extremes: An application to structural design (with discussion). *Applied Statistics 43*(1), 1–48.

Daul, S., E. De Giorgi, F. Linkskog, and A. J. McNeil (2003). The grouped t-copula with an application to credit risk. *RISK 16*, 73–76.

Demarta, S. and A. J. McNeil (2005). The t copula and related copulas. *International Statistical Review 73*, 111–129.

Disatnik, D. and S. Katz (2012). Portfolio optimization using a block structure for the covariance matrix. *Journal of Business Finance and Accounting 39*, 806–843.

E.Heffernan, J. and J. A.Tawn (2004). A conditional approach for multivariate extreme values (with discussion). *Journal of the Royal Statistical Society: Series B 66*(3), 497–546.

Embrechts, P., C. Klüppelberg, and T. Mikosch (1997). *Modelling Extremal Events: For Insurance and Finance*. Heidelberg: Springer-Verlag.

Embrechts, P., D. D. Lambrigger, and M. V. Wüthrich (2009). Multivariate extremes and aggregation of dependent risks: Examples and counter examples. *Extremes 12*, 107–127.

Feller, W. (1971). *An Introduction to Probability Theory and Its Applications.* John Wiley & Sons Inc.

Heffernan, J. E. (2000). A directory of coefficients of tail dependence. *Extremes 3,* 279–290.

Hilala, S., S.-H. Poon, and J. A. Tawn (2014). Portfolio risk assessment using multivariate extreme value methods. *Extremes 17,* 531–556.

Hyung, N. and C. G. de Vries (2007). Portfolio selection with heavy tails. *Journal of Empirical Finance 14*(3), 383–400.

Janse, D. W., K. G. Koedijk, and C. G. de Vries (2000). Portfolio selection with limited downside risk. *Journal of Empirical Finance 7*(3-4), 247–269.

Joe, H. (1997). *Multivariate Models and Dependence Concepts.* London: Chapman & Hall.

Jorion, P. (2001). *Value at Risk: The New Benchmark for Managing Financial Risk.* MacGraw-Hill international editions: Finance series. McGraw-Hill.

Keating, C., H. S. Shin, C. Goodhard, and J. Danielsson (2001, May). An academic response to Basel II. FMG Special Papers sp130, Financial Markets Group.

Keef, C., I. Papastathopoulos, and J. A. Tawn (2013). Estimation of the conditional distribution of a vector variable given that one of its components is large: Additional constraints for the Heffernan and Tawn model. *Journal of Multivariate Analysis 115,* 396–404.

Ledford, A. W. and J. A. Tawn (1996). Statistics for near independence in multivariate extreme values. *Biometrika 83,* 169–187.

Ledford, A. W. and J. A. Tawn (1997). Modelling dependence within joint tail regions. *Journal of Royal Statistical Society: Series B 59,* 475–499.

Liu, Y. (2013). *Extreme Value Theory in Finance.* Ph. D. thesis, Lancaster University.

Liu, Y. and J. A. Tawn (2014). Self-consistent estimation of conditional multivariate extreme distributions. *Journal of Multivariate Analysis 127,* 19–35.

Luo, X. and P. Shevchenko (2010). The t copula with multiple parameters of degrees of freedom: Bivariate characteristics and application to risk management. *Quantitative Finance 10*(9), 1039–1054.

Markowitz, H. (1952). Portfolio selection. *Journal of Finance 7*(1), 77–91.

Poon, S.-H., M. Rockinger, and J. Tawn (2003). Modelling extreme-value dependence in international stock. *Statistica Sinica 13,* 929–953.

Resnick, S. I. (2007). *Extreme Values, Regular Variation and Point Processes.* New York: Springer-Verlag.

Rockafellar, R. and S. Uryasev (2000). Optimization of conditional value-at-risk. *Journal of Risk 2,* 21–41.

Rockafellar, R. and S. Uryasev (2002). Conditional value-at-risk for general loss distributions. *Journal of Banking and Finance 26*(7), 1443–1471.

Shevchanko, P. V. (2011). *Modelling Operational Risk Using Bayesian Inference.* Springer.

Stephenson, A. (2003). Simulating multivariate extreme value distributions of logistic type. *Extremes 6,* 49–59.

Tawn, J. A. (1990). Modelling multivariate extreme value distributions. *Biometrika 77,* 245–253.

20

Interplay of Insurance and Financial Risks with Bivariate Regular Variation

Qihe Tang

University of Iowa, United States

Zhongyi Yuan

The Pennsylvania State University, United States

Abstract

It is known that for an insurer who invests in the financial market, the financial investments may affect its solvency as severely as do insurance claims. This conclusion is usually reached under an assumption of independence or asymptotic independence between insurance risk and financial risk. Such an assumption seems reasonable if the insurer focuses on the traditional insurance business that does not interact much with the capital market. However, we shall argue that at least for insurers who participate in financial guarantee insurance and as a result cause systemic risk, asymptotic dependence between insurance and financial risks needs to be considered. Under a bivariate regular variation structure, we investigate the interplay of insurance and financial risks, and show that the asymptotic dependence introduces extra risk for the insurer's solvency.

20.1 Introduction

In the aftermath of the 2008 financial crisis, the then Federal Reserve Chairman Ben Bernanke once commented on the bailout of AIG: "AIG ... was a hedge fund, basically, that was attached to a large and stable insurance company." By stressing its nature as a hedge fund, Mr. Bernanke was accusing AIG of making large amounts of aggressive investments and "irresponsible bets." In fact, many other insurance companies (the monolines, in particular) also engaged in such investments, which

419

greatly contributed to the financial crisis. Such aggressive investments were able to get around regulation due to the huge gap in the regulation system, which today's risk management is dedicated to fill. In the meantime, risk management also calls for better understanding of the risks the insurance industry faces. In this chapter, we shall discuss additional risk introduced to insurance companies by such an investment behavior, and shall pay particular attention to extreme risks that may cause bankruptcy for the insurance companies.

Typical risks that can affect an insurer's survival include large claims from catastrophic events and large losses from financial investments. For example, more than a dozen insurers became insolvent because of claims originating from the 2005 Hurricane Katrina (e.g., Muermann, 2008), and, as discussed above, a large number of insurers experienced financial distress or even collapsed due to their high exposure to mortgage-related securities (e.g., Harrington, 2009). Research also confirms that risky investments may impair the insurer's solvency just as severely as do large claims (e.g., Frolova et al., 2002; Kalashnikov and Norberg, 2002; Li and Tang, 2015; Norberg, 1999; Pergamenshchikov and Zeitouny, 2006; Tang and Tsitsiashvili, 2003). In the literature, the risk resulting from insurance claims and the risk from financial investments are generally referred to as insurance risk and financial risk, respectively, although general financial risk can be further categorized to market risk, credit risk, operational risk, and so on.

We shall describe the insurance business by a discrete-time risk model in which the two risks are quantified by concrete random variables. The probability of ruin in both finite-time and infinite-time horizons is then studied. This study has become particularly relevant for insurance because of modern regulatory frameworks (such as EU Solvency II) that require insurers to hold solvency capital so that the ruin probability is under control. Studies of the asymptotic behavior of the ruin probability of this model have been done by many authors (Chen, 2011; Li and Tang, 2015; Nyrhinen, 1999, 2001; Tang and Tsitsiashvili, 2003, 2004; Yang et al., 2012; Yang and Wang, 2013; Zhou et al., 2012).

A continuous-time analogy is the so-called bivariate Lévy-driven risk model, proposed by Paulsen (1993), in which the cash flow of premiums less claims is described as a Lévy process while the log price of the investment portfolio as another Lévy process. See a survey paper by Paulsen et al. (2008) for the study of the ruin probability of this risk model or one of its ramifications.

Thus far, it is commonly assumed that the insurance and financial risks are independent or asymptotically independent. Such an assumption is usually justified in the literature by virtue of the insufficient link between the traditional/core insurance business and the capital market. However, due to the recent changes in the insurance business, we have at least two reasons to believe that the two risks should be dependent or asymptotically dependent for certain insurers. First, to hedge against catastrophic risk, more and more insurers now securitize their insurance risk and export it to the capital market using insurance-linked securities, such as catastrophe bonds. Therefore, an insurer who invests in the capital market is likely to be exposed to the insurance risk exported by another insurer, yielding interconnectedness between his/her insurance and financial risks. Second, it has been argued in the recent literature of insurance and finance that, unlike the core insurance activities, noncore insurance activities such as financial guarantee insurance cause systemic risk, yielding another source of interconnectedness in extreme situations. For example, situations like the recent Great Recession would make insurers who have participated in both default protection insurance and the distressed financial market, such as AIG and the monolines, suffer losses from both sides. The more and more involvement of insurers in the noncore insurance business has resulted in significant systemic risk. Therefore, at least for insurers engaged in noncore activities, it is reasonable to assume that insurance claims and investment losses exhibit asymptotic dependence. For related discussions, we refer the reader to Acharya (2009), Bell and Keller (2009), Baluch et al. (2011), Billio et al. (2012), and Cummins and Weiss (2014).

To quantify the extreme insurance and financial risks impairing the insurer's solvency, it is necessary to model both their enormous sizes and extreme dependence. We shall show that this can be done in a unified framework of bivariate regular variation developed in multivariate extreme value theory. We shall characterize the insurance and financial risks through a bivariate regular variation structure, and derive a unified asymptotic approximation for the probability of ruin in both finite-time and infinite-time horizons. Our result reinforces the longstanding viewpoint that the asymptotic dependence between insurance claims and financial investment losses introduces substantial risk to the insurer's solvency.

The rest of this chapter is organized as follows: In Section 20.2, a discrete-time risk model is introduced as a platform for insurance and financial risks and some numerical studies are conducted to explore the asymptotic behavior of the ruin probability; the main result is presented in Section 20.3 after a brief review of bivariate regular variation; in Section 20.4, two more numerical examples are provided to examine the performance of the asymptotic formula; and finally, Section 20.5 completes the chapter with a proof of the main result.

20.2   Ruin under the Interplay of Insurance and Financial Risks

Throughout this chapter, the following notational conventions are in force. An operation $*$ between a scalar number c and a vector (x, y) yields a vector $(c * x, c * y)$. For a number $c \in \mathbb{R}$ and a set $A \subset \mathbb{R}^2$, write the positive part of c as $c^+ = c \vee 0$ and the set $\{cx : x \in A\}$ as cA. Moreover, for two positive functions $f(\cdot)$ and $g(\cdot)$, we write $f(\cdot) \sim g(\cdot)$ if $\lim f(\cdot)/g(\cdot) = 1$, write $f(\cdot) \lesssim g(\cdot)$ if $\limsup f(\cdot)/g(\cdot) \leqslant 1$, and write $f(\cdot) \gtrsim g(\cdot)$ if $\liminf f(\cdot)/g(\cdot) \geqslant 1$. For a nondecreasing function $h(\cdot)$, define its generalized inverse as $h^{\leftarrow}(y) = \inf \{x \in (-\infty, \infty) : h(x) \geqslant y\}$, where we follow the usual convention $\inf \varnothing = \infty$.

20.2.1 A Discrete-Time Risk Model

Denote by $W_0 = x > 0$ the deterministic initial wealth of the insurer, and by W_k the insurer's wealth at time $k \in \mathbb{N}$. The insurer's realized net profit from the insurance business is roughly equal to premiums collected, denoted by P_k, minus costs resulting from insurance claims and other expenses, denoted by Z_k. Suppose that, at the beginning of each period k, the insurer invests its wealth W_{k-1} into risk-free and risky assets that result in an overall return rate $R_k \in [-1, \infty)$, so that the insurer gets back $(1 + R_k)W_{k-1}$ at the end of the period. The wealth process $\{W_k, k \in \mathbb{N}\}$ evolves according to

$$W_k = (1 + R_k)W_{k-1} + (P_k - Z_k), \qquad k \in \mathbb{N}.$$

Iterating this yields

$$W_k = x \prod_{j=1}^{k}(1 + R_j) + \sum_{i=1}^{k}(P_i - Z_i) \prod_{j=i+1}^{k}(1 + R_j),$$

where multiplication over an empty index set produces value 1 by convention.

For $n \in \mathbb{N} \cup \{\infty\}$, we consider the probability of ruin in either finite-time or infinite-time horizon n, defined by

$$\psi(x; n) = \Pr\left(\min_{0 \leqslant k < n+1} W_k < 0 \,\middle|\, W_0 = x \right).$$

Following Tang and Tsitsiashvili (2003), introduce

$$X_i = Z_i - P_i \quad \text{and} \quad Y_i = \frac{1}{1 + R_i}, \qquad i \in \mathbb{N},$$

to quantify the insurance risk and financial risk, respectively. Then we can rewrite the ruin probability as

$$
\begin{aligned}
\psi(x; n) &= \Pr\left(\min_{0 \leqslant k < n+1} \prod_{j=1}^{k}(1 + R_j) \left(x + \sum_{i=1}^{k}(P_i - Z_i) \prod_{j=1}^{i} \frac{1}{1 + R_j} \right) < 0 \right) \\
&= \Pr\left(\max_{1 \leqslant k < n+1} \sum_{i=1}^{k} X_i \prod_{j=1}^{i} Y_j > x \right).
\end{aligned}
\tag{20.1}
$$

This way $\psi(x; n)$ reduces to the tail probability of the running maximum of a sequence of quantities with a sum-product structure.

20.2.2 Explorative Numerical Studies

To obtain a rough idea on how the asymptotic dependence between the insurance risk and financial risk would affect the insurer's solvency, we conduct two simple numerical studies to compare the ruin probability under the asymptotic dependence assumption with that under the independence assumption. For all our numerical studies in this chapter, we further assume that the nonnegative random vectors (Z_i, Y_i), $i \in \mathbb{N}$, form a sequence of independent and identically distributed (i.i.d.) random pairs with generic pair (Z, Y). For simplicity, the premium amount collected during each period is chosen to be a constant yielding a safety loading of about 10%, and the numerical studies are all conducted for the probability of ruin in time horizon $n = 3$ with initial capital x ranging from 50 to 4000.

Numerical study 1 Let Z and Y both follow a Pareto distribution F with shape parameter α and scale parameter θ; that is,

$$F(x) = 1 - \left(\frac{\theta}{x + \theta} \right)^{\alpha}, \qquad x > 0. \tag{20.2}$$

The parameters are set to be $\alpha = 2$ and $\theta = 1$, respectively. For the asymptotically dependent case, the dependence structure of Z and Y is described by a Gumbel copula.

$$C(u, v) = \exp\left\{ -((-\ln u)^r + (-\ln v)^r)^{1/r} \right\}, \qquad (u, v) \in (0, 1)^2. \tag{20.3}$$

It is known that a Gumbel copula with $r > 1$ produces asymptotic dependence with coefficient $2 - 2^{1/r}$, so a larger value of r means a stronger dependence in the tail area (e.g., McNeil et al., 2005, p. 209). The parameter r is set to be 2 or 10 in order to further demonstrate the effect of dependence strength. Samples of (Z, Y) following a Gumbel copula are generated in the R environment using the `copula` package (Yan, 2007).

Assume that $P_k \equiv (1 + 10\%)\mathrm{E}[Z]$, $k = 1, \ldots, n$. For both the asymptotically dependent case and the independent case, we simulate $N = 10^6$ samples of $((X_1, Y_1), \ldots, (X_n, Y_n))$ and use the empirical estimator to estimate the ruin probability in relation (20.1); that is, we estimate $\psi(x; n)$ by

$$\frac{1}{N} \sum_{m=1}^{N} 1_{\left(\max_{1 \leqslant k < n+1} \sum_{i=1}^{k} X_i^{(m)} \prod_{j=1}^{i} Y_j^{(m)} > x \right)}, \tag{20.4}$$

where $\left(\left(X_1^{(m)}, Y_1^{(m)} \right), \ldots, \left(X_n^{(m)}, Y_n^{(m)} \right) \right)$ for each $m = 1, \ldots, N$ is an independent copy of $((X_1, Y_1), \ldots, (X_n, Y_n))$, and $1_{(.)}$ is the indicator function. Figure 20.1 shows the ratios of the estimated ruin probability under the independent case to that under the asymptotically dependent case. We observe that the ratios are smaller than 1, and quickly decay to 0 as x increases, meaning that the ruin probability under the asymptotically dependent case is significantly higher.

Numerical study 2 Next we investigate the case in which the nonnegative generic pair (Z, Y) follows a mixture structure

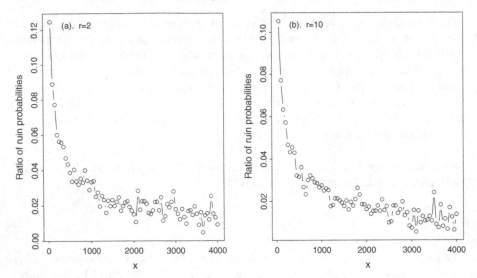

FIGURE 20.1
Independent case versus asymptotically dependent (Gumbel) case.

$$
\begin{cases}
Z &= \left[m_Z + \sqrt{W} \left(\sqrt{\rho}\eta_0 + \sqrt{1-\rho}\eta_Z \right) \right]^+, \\
Y &= \left[m_Y + \sqrt{W} \left(\sqrt{\rho}\eta_0 + \sqrt{1-\rho}\eta_Y \right) \right]^+,
\end{cases}
\tag{20.5}
$$

where $m_Z \in \mathbb{R}$, $m_Y \in \mathbb{R}$ and $0 < \rho < 1$ are nonrandom, η_0, η_Z, and η_Y are i.i.d. standard normal random variables, and W, independent of $\{\eta_0, \eta_Z, \eta_Y\}$, follows an inverse gamma distribution with both shape and scale parameters equal to $\alpha/2$; that is, W has probability density function

$$
f_W(w) = \frac{(\alpha/2)^{\alpha/2}}{\Gamma(\alpha/2)} w^{-(\alpha/2+1)} \exp\left\{ -\frac{\alpha}{2w} \right\}, \qquad w > 0.
$$

It is known that relation (20.5) yields an asymptotically dependent structure between Z and Y (e.g., McNeil et al., 2005, Section 5.3).

For comparison, we also consider the independent counterparty of (20.5) given by

$$
\begin{cases}
Z &= \left[m_Z + \sqrt{W_Z} \left(\sqrt{\rho}\eta_0 + \sqrt{1-\rho}\eta_Z \right) \right]^+, \\
Y &= \left[m_Y + \sqrt{W_Y} \left(\sqrt{\rho}\eta_0 + \sqrt{1-\rho}\eta_Y \right) \right]^+,
\end{cases}
\tag{20.6}
$$

where W_Z and W_Y are independent copies of W, and the other modeling components are specified the same as in relation (20.5).

We set the parameters to be $\alpha = 2$, $m_Z = 10$, $m_Y = 0.2$, and $\rho = 0.5$ or 0.9. The premium collected within each period is set to be $P_k \equiv 11$, $k = 1, \ldots, n$.

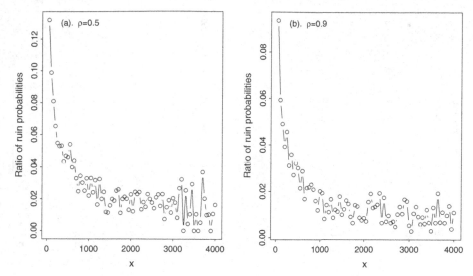

FIGURE 20.2
Independent case versus asymptotically dependent (mixture) case

Again, with $N = 10^6$ samples, we use the empirical estimator (20.4) to estimate the ruin probability $\psi(x; n)$ for both the independent case (20.6) and the asymptotically dependent case (20.5). The ratios of the ruin probability under the independent case to that under the asymptotically dependent case are demonstrated in Figure 20.2, which also shows rapid decays to 0.

The two explorative numerical studies above suggest that the decay rate of $\psi(x; n)$ with respect to x in the independent case is much faster than that in the asymptotically dependent case. More about this can be revealed via a rough quantitative analysis, for which we need the concept of regular variation. Recall that a positive measurable function $h(\cdot)$ on $\mathbb{R}^+ = [0, \infty)$ is said to be regularly varying at ∞ with regularity index $\alpha \in \mathbb{R}$ if

$$\lim_{x \to \infty} \frac{h(xy)}{h(x)} = y^\alpha, \qquad y > 0.$$

In this chapter, we shall often assume for a distribution function F that its tail $\overline{F} = 1 - F$ is regularly varying. The reader is referred to Bingham et al. (1987) or Resnick (1987) for a comprehensive treatment of regular variation.

Back to the ruin probability $\psi(x; n)$. For the independent case, by Corollary 2.1 of Chen and Xie (2005) we have

$$\psi(x; n) \sim \sum_{i=1}^{n} \Pr\left(X_i \prod_{j=1}^{i} Y_j > x\right).$$

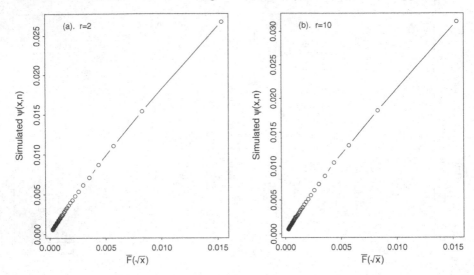

FIGURE 20.3
Comparison of the decaying rate of the estimated ruin probability and $\overline{F}(\sqrt{x})$.

Then applying the Corollary on Page 245 of Embrechts and Goldie (1980) we know that $\psi(\cdot; n)$ is regularly varying with index $-\alpha$. However, for the asymptotically dependent case we have

$$\psi(x; n) \geqslant \psi(x; 1) = \Pr(X_1 Y_1 > x).$$

For the special case where $Z = Y$, it is easy to show that $\Pr(X_1 Y_1 > x) \sim \overline{F}(\sqrt{x})$, and hence, $\psi(x; n)$ is bounded below by a positive function which is regularly varying with index $-\alpha/2$. This confirms what we have observed above and also motivates us to conduct the numerical study below.

Numerical study 3 Based on the setup of the two numerical studies above, we conduct two more simulations to compare $\psi(x; n)$ with $\overline{F}(\sqrt{x})$.

We see that the dots in Figure 20.3 and Figure 20.4 appear to be roughly on a straight line around the lower-left corner, which corresponds to larger values of x. This means that for x large, the ruin probability $\psi(x; n)$ decays roughly at rate $\overline{F}(\sqrt{x})$ under both the Gumbel case and the mixture case. Also, we observe that a stronger dependence between X and Y in the tail area leads to a larger ruin probability, which is no surprise. These observations will be theoretically verified in our main result below.

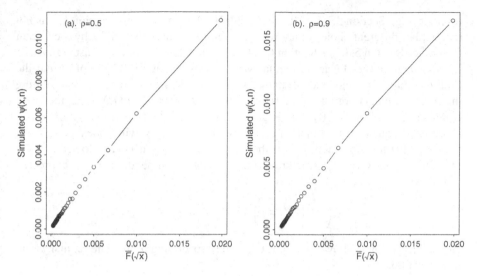

FIGURE 20.4
Comparison of the decaying rate of the estimated ruin probability and $\overline{F}(\sqrt{x})$.

20.3 Main Result

20.3.1 Bivariate Regular Variation

First we give a brief review of bivariate regular variation, which is the two-dimensional version of the concept of Multivariate!regular variation (MRV). Since its introduction by De Haan and Resnick (1981), the MRV framework has found its use in insurance, finance, and risk management by providing a tool for modeling extreme risks with both heavy tails and asymptotic (in)dependence. Recent works include Embrechts et al. (2009), Böcker and Klüppelberg (2010), Mainik and Rüschendorf (2010), Joe and Li (2011), Asimit et al. (2011), Part IV of Rüschendorf (2013), Tang and Yuan (2013), among many others. In particular, Fougères and Mercadier (2012) studied the ruin probability in a broader scope using an MRV structure with asymptotic dependence.

A random vector (ξ_1, ξ_2) taking values in $[0, \infty]^2 \backslash \{\mathbf{0}\}$ is said to follow a distribution with a bivariate regularly varying tail if there exist a positive normalizing function $b(\cdot)$ monotonically increasing to ∞ and a limit measure ν not identically 0, such that, as $x \to \infty$,

$$x \Pr\left(\frac{(\xi_1, \xi_2)}{b(x)} \in \cdot\right) \xrightarrow{v} \nu(\cdot) \qquad \text{on } [0, \infty]^2 \backslash \{\mathbf{0}\}, \tag{20.7}$$

where $\xrightarrow{v}$ denotes vague convergence. Discussions on vague convergence can be found in, e.g., Section 3.3.5 of Resnick (2007). The normalizing function $b(\cdot)$ is not unique, but different choices may result in limit measures that differ by a constant factor. See Section 5.4.2 or Section 6.1.4 of Resnick (2007) for more discussions.

The definition of bivariate regular variation by relation (20.7) implies that the limit measure ν is homogeneous; that is, there exists some $0 < \alpha < \infty$, representing the bivariate regular variation index, such that $\nu(tB) = t^{-\alpha}\nu(B)$ holds for every Borel set $B \subset [0, \infty]^2 \backslash \{\mathbf{0}\}$. See Page 178 of Resnick (2007) for the proof of this result. Consequently, the function $b(\cdot)$ is regularly varying with index $1/\alpha$. Thus, for some distribution function F with $\overline{F}(x) \sim 1/b^{\leftarrow}(x)$, it is easy to see that the bivariate regular variation structure in relation (20.7) can be alternatively expressed as follows:

$$\frac{1}{\overline{F}(x)} \Pr\left(\frac{(\xi_1, \xi_2)}{x} \in \cdot\right) \xrightarrow{v} \nu(\cdot) \qquad \text{on } [0, \infty]^2 \backslash \{\mathbf{0}\}. \tag{20.8}$$

Hereafter, we shall follow relation (20.8) when specifying a bivariate regular variation structure.

20.3.2   Main Result

We assume that (X_i, Y_i), $i \in \mathbb{N}$, form a sequence of i.i.d. random pairs with generic pair (X, Y), and that (X^+, Y) follows a bivariate regular variation structure given below:

Assumption 20.3.1. The vague convergence

$$\frac{1}{\overline{F}(x)} \Pr\left(\frac{(X^+, Y)}{x} \in \cdot\right) \xrightarrow{v} \nu(\cdot) \qquad \text{on } [0, \infty]^2 \backslash \{\mathbf{0}\} \tag{20.9}$$

holds for some auxiliary distribution function F with regularly varying tail with index $-\alpha$, and some limit measure ν with $\nu(1, \infty] > 0$.

Assumption 20.3.1 has some immediate implications. First, the assumption $\nu(1, \infty] > 0$ means that

$$\lim_{x \to \infty} \frac{1}{\overline{F}(x)} \Pr\left(X > x, Y > x\right) = \nu\left(1, \infty\right] > 0,$$

which indicates that X and Y exhibit large joint movements, and hence, are asymptotically dependent. Second, relation (20.9) also implies that

$$\lim_{x \to \infty} \frac{\Pr\left(X > x\right)}{\Pr\left(Y > x\right)} = \frac{\nu\left((1, \infty] \times [0, \infty]\right)}{\nu\left([0, \infty] \times (1, \infty]\right)},$$

indicating that the two risk variables X and Y have comparable tails. Third, introducing a set $A = \{(s, t) \in [0, \infty]^2 : st > 1\}$, it is easy to verify that $\nu(\partial A) = 0$ and that $\nu(A) > 0$. Thus, Assumption 20.3.1 leads to

$$\Pr\left(XY > x\right) = \Pr\left(\frac{(X, Y)}{\sqrt{x}} \in A\right) \sim \nu(A)\overline{F}\left(\sqrt{x}\right); \tag{20.10}$$

see also Proposition 7.6 of Resnick (2007) for a more general discussion. This shows that the product XY has a regularly varying tail with index $-\alpha/2$.

Theorem 20.3.1. *Under Assumption 20.3.1,*

   (a) *it holds for every* $n \in \mathbb{N}$ *that*

$$\psi(x; n) \sim \left(\sum_{i=1}^{n} \left(\mathrm{E} \left[Y^{\alpha/2} \right] \right)^{i-1} \right) \nu(A) \overline{F} \left(\sqrt{x} \right) ; \qquad (20.11)$$

   (b) *if further* $\mathrm{E} \left[Y^{\alpha/2} \right] < 1$, *then relation (20.11) holds uniformly for all* $n \in \mathbb{N}$ *and, hence,*

$$\psi(x; \infty) \sim \frac{\nu(A)}{1 - \mathrm{E} \left[Y^{\alpha/2} \right]} \overline{F} \left(\sqrt{x} \right) . \qquad (20.12)$$

It is meaningful to compare Theorem 20.3.1 with the results of Li and Tang (2015). In the latter paper, the authors have shown that for independent insurance and financial risks, each having a strongly regularly varying tail, the ruin probability $\psi(x; n)$ is equivalent to a linear combination of $\Pr(X > x)$ and $\Pr(Y > x)$. Thus, our Theorem 20.3.1 quantitatively captures the extra risk that the asymptotic dependence between the insurance and financial risks introduces to the insurer's solvency.

20.4   Accuracy Examination

In this section, we examine the accuracy of the asymptotic formula (20.11). Due to the infeasibility of simulating the infinite-time ruin probability, we skip formula (20.12). We shall demonstrate via two numerical studies that (20.11) can serve as a good approximation for the ruin probability when the initial surplus is reasonably large. Again, for simplicity we consider the probability of ruin in time horizon $n = 3$ with initial capital x ranging from 50 to 4000.

20.4.1   Gumbel Case

Similar to Section 20.2.2, we consider the case where Z and Y both follow a Pareto distribution given by (20.2) with $\alpha = 2$ and $\theta = 1$, and their dependence structure is described by the Gumbel copula (20.3) with $r = 2$ or 10. Note that in this case (X, Y) follows a bivariate regular variation structure on $[0, \infty]^2 \backslash \{0\}$ satisfying relation (20.9), in which the distribution function F is given by (20.2), and the limit measure ν is defined by

$$\nu [0, \mathbf{t}]^c = \left(t_1^{-\alpha r} + t_2^{-\alpha r} \right)^{1/r}, \qquad t > 0; \qquad (20.13)$$

see e.g., Lemma 5.2 of Tang and Yuan (2013).

We calculate the ratio of the left-hand side of (20.11) to the right-hand side. The left-hand side is estimated using the empirical estimator (20.4), while in order to evaluate the right-hand side, we need to estimate $\nu(A)$ for $A = \{(s,t) \in [0,\infty]^2 : st > 1\}$ and ν defined by (20.13). Note that the bivariate regular variation structure (20.9) implies that, as $N \uparrow \infty$,

$$\frac{1}{k} \sum_{i=1}^{N} \epsilon_{(X^{(i)},Y^{(i)})/F^{\leftarrow}(1-k/N)}(\cdot) \overset{w}{\to} \nu(\cdot),$$

for every sequence $k = k(N) \uparrow \infty$ with $k(N) \sim k(N+1)$ and $N/k \uparrow \infty$, where $\epsilon(\cdot)$ denotes the Dirac measure, $(X^{(i)}, Y^{(i)})$, $i = 1,\ldots,N$, are i.i.d. copies of (X,Y), and $\overset{w}{\to}$ denotes weak convergence; see Theorem 6.2 of Resnick (2007) for this assertion. Therefore, a natural estimator for $\nu(A)$ is given by

$$\frac{1}{k} \sum_{i=1}^{N} \mathbb{1}_{\left((X^{(i)},Y^{(i)})\in F^{\leftarrow}(1-k/N)A\right)} = \frac{1}{k} \sum_{i=1}^{N} \mathbb{1}_{\left(\sqrt{X^{(i)}Y^{(i)}}>F^{\leftarrow}(1-k/N)\right)};$$

see e.g., Section 9.2 of Resnick (2007) for discussions on the estimation of the limit measure of an MRV structure. The value of k is not a straightforward choice; in our numerical studies, with $N = 10^6$ samples, we use the algorithm described in Sections 9.2.4 and 11.2.2 of Resnick (2007) to choose a suitable value of k. We refer the reader to Nguyen and Samorodnitsky (2013) and references therein for recent discussions on this topic. Notice that although the limit measure still has to be estimated via a stochastic simulation, such a simulation is no longer a rare-event simulation, unlike the one we conducted for the ruin probability, and therefore, simulation efficiency will be less of an issue.

The graphs in Figure 20.5 show that the ratio approaches 1 as x becomes large. The fluctuation for larger values of x must be due to the variation increase of the empirical estimator for the ruin probability when large values of x lead to too small values of the ruin probability; see e.g., Tang and Yuan (2012) for a detailed discussion.

20.4.2   Mixture Case

In this subsection, we investigate the accuracy of the asymptotic approximation (20.11) under the mixture structure given by relation (20.5) in Section 20.2.2. Note that in this case (X,Y) follows a bivariate regular variation structure on $[0,\infty]^2 \setminus \{\mathbf{0}\}$ satisfying relation (20.9), where F is given by the distribution function of $\sqrt{W}$, and the limit measure is given by

$$\nu(\cdot) = \mathrm{E}\left[\nu_W\left((\eta^+)^{-1}\cdot\right)\right]$$

with $(\eta^+)^{-1} \cdot$ defined by the set of $(y_1,y_2) \in [0,\infty]^2 \setminus \{\mathbf{0}\}$ such that

$$\left(\left(\sqrt{\rho}\eta_0 + \sqrt{1-\rho}\eta_X\right)^+ y_1, \left(\sqrt{\rho}\eta_0 + \sqrt{1-\rho}\eta_Y\right)^+ y_2\right) \in \cdot,$$

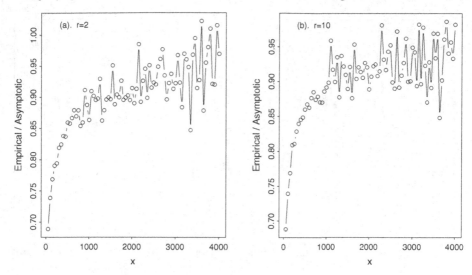

FIGURE 20.5
The ratio of empirical estimate to asymptotic approximation of the ruin probability — the Gumbel case.

and the limit measure ν_W on $[0, \infty]^2 \backslash \{\mathbf{0}\}$ defined by

$$\nu_W [\mathbf{0}, \mathbf{t}]^c = \left(\bigwedge_{i=1}^{n} t_i \right)^{-\alpha}, \qquad \mathbf{t} > 0;$$

see e.g., Section 5.2 of Tang and Yuan (2013) for this claim. Thus, $\nu(A)$ in relation (20.10) can be calculated as

$$
\begin{aligned}
&\nu(A) \\
=\ & \mathrm{E}\left[\nu_W \left(\left(\eta^+ \right)^{-1} A \right) \right] \\
=\ & \mathrm{E}\left[\nu_W \left\{ \mathbf{y} > \mathbf{0} : \left(\sqrt{\rho}\eta_0 + \sqrt{1-\rho}\eta_X \right)^+ \left(\sqrt{\rho}\eta_0 + \sqrt{1-\rho}\eta_Y \right)^+ y_1 y_2 > 1 \right\} \right] \\
=\ & \mathrm{E}\left[\left(\left(\sqrt{\rho}\eta_0 + \sqrt{1-\rho}\eta_X \right)^+ \left(\sqrt{\rho}\eta_0 + \sqrt{1-\rho}\eta_Y \right)^+ \right)^{\alpha/2} \right].
\end{aligned}
$$

The expectation above is estimated via simulation.

Again, we estimate the ruin probability $\psi(x; n)$ using the empirical estimator with $N = 10^6$ simulation samples, and compare it with the asymptotic approximation given on the right-hand side of (20.11). The parameters are set the same as in Section 20.2.2. The graphs in Figure 20.6 also indicate that the ratio approaches 1 as x becomes large.

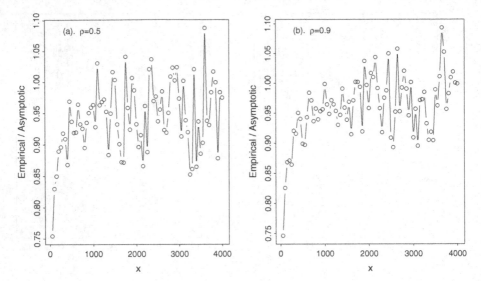

FIGURE 20.6
The ratio of empirical estimate to asymptotic approximation of the ruin probability
— the mixture case.

Therefore, we may conclude that for x reasonably large our asymptotic formula
(20.11) can provide a good approximation for $\psi(x; n)$. Although we do have to carefully choose the parameters to present good graphs, our purpose is mainly to ensure
that the ruin probability is not too small, so that the reliability of simulation with a
sample size of 10^6 is not an issue.

20.5   Proof of Theorem 20.3.1

The following lemma will be needed for the proof of Theorem 20.3.1:

Lemma 20.5.1. Under Assumption 20.3.1, the products $X_i \prod_{j=1}^{i} Y_j$, $i \in \mathbb{N}$, appearing in (20.1) are pairwise asymptotically independent.

Proof. Since XY has a regularly varying tail with index $-\alpha/2$ as shown in (20.10),
by Breiman's theorem (see Breiman, 1965; Cline and Samorodnitsky, 1994) we have

$$\Pr\left(X_i \prod_{j=1}^{i} Y_j > x\right) \sim \left(\mathrm{E}\left[Y^{\alpha/2}\right]\right)^{i-1} \Pr\left(X_i Y_i > x\right)$$

$$\sim \left(\mathrm{E}\left[Y^{\alpha/2}\right]\right)^{i-1} \nu(A)\overline{F}\left(\sqrt{x}\right). \qquad (20.14)$$

For arbitrarily chosen $1 \leqslant i_1 < i_2 < \infty$, by Lemma 6.2 of Tang and Yuan (2012)

$$\Pr\left(X_{i_1}\prod_{j=1}^{i_1}Y_j > x, X_{i_2}\prod_{j=1}^{i_2}Y_j > x\right)$$

$$= \Pr\left(X_{i_1}\prod_{j=1}^{i_1}Y_j > x, (X_{i_2}Y_{i_2})\prod_{j=1}^{i_2-1}Y_j > x\right)$$

$$= o(1)\Pr\left(X_{i_2}Y_{i_2} > x\right)$$

$$= o(1)\overline{F}\left(\sqrt{x}\right).$$

By relation (20.14), this joint tail probability is $o(1)$ of each individual tail probability, indicating asymptotic independence between $X_{i_1}\prod_{j=1}^{i_1}Y_j$ and $X_{i_2}\prod_{j=1}^{i_2}Y_j$. This concludes the proof of Lemma 20.5.1.   □

Proof of Theorem 20.3.1. (a) By Lemma 20.5.1 above, Theorem 3.1 of Chen and Yuen (2009), and relation (20.14),

$$\psi(x; n) \leqslant \Pr\left(\sum_{i=1}^{n}X_i^{+}\prod_{j=1}^{i}Y_j > x\right)$$

$$\sim \sum_{i=1}^{n}\Pr\left(X_i\prod_{j=1}^{i}Y_j > x\right)$$

$$\sim \left(\sum_{i=1}^{n}\left(\mathrm{E}\left[Y^{\alpha/2}\right]\right)^{i-1}\right)\nu(A)\overline{F}\left(\sqrt{x}\right).$$

To derive a corresponding asymptotic lower bound, introduce

$$\tau(x) = \inf\left\{k \in \mathbb{N}: \sum_{i=1}^{k}X_i\prod_{j=1}^{i}Y_j > x\right\},$$

which is a stopping time with respect to the natural filtration generated by $\{(X_i, Y_i), i \in \mathbb{N}\}$. Then

$$\psi(x; n) = \sum_{k=1}^{n}\Pr\left(\tau(x) = k\right). \qquad (20.15)$$

For arbitrarily fixed $\varepsilon \in (0, 1)$, we have

$$\Pr\left(\tau(x) = k\right)$$

$$= \Pr\left(\max_{1 \leqslant l \leqslant k-1}\sum_{i=1}^{l}X_i\prod_{j=1}^{i}Y_j \leqslant x, \sum_{i=1}^{k}X_i\prod_{j=1}^{i}Y_j > x\right)$$

$$\geqslant \Pr\left(\max_{1\leqslant l\leqslant k-1} \sum_{i=1}^{l} X_i \prod_{j=1}^{i} Y_j \leqslant x, \ \sum_{i=1}^{k-1} X_i \prod_{j=1}^{i} Y_j > -\varepsilon x, \ X_k \prod_{j=1}^{k} Y_j > (1+\varepsilon)x \right)$$

$$\geqslant \Pr\left(X_k \prod_{j=1}^{k} Y_j > (1+\varepsilon)x \right)$$

$$- \Pr\left(\max_{1\leqslant l\leqslant k-1} \sum_{i=1}^{l} X_i \prod_{j=1}^{i} Y_j > x, \ X_k \prod_{j=1}^{k} Y_j > (1+\varepsilon)x \right)$$

$$- \Pr\left(\sum_{i=1}^{k-1} X_i \prod_{j=1}^{i} Y_j \leqslant -\varepsilon x, \ X_k \prod_{j=1}^{k} Y_j > (1+\varepsilon)x \right)$$

$$\sim \ \left(\mathrm{E}\left[Y^{\alpha/2} \right] \right)^{k-1} \nu(A)\overline{F}\left(\sqrt{(1+\varepsilon)x} \right) + o(1)\Pr\left(X_k Y_k > (1+\varepsilon)x \right)$$

$$\sim \ (1+\varepsilon)^{-\alpha/2} \left(\mathrm{E}\left[Y^{\alpha/2} \right] \right)^{k-1} \nu(A)\overline{F}\left(\sqrt{x} \right),$$

where in the second last step we applied relation (20.14) to deal with the first probability and applied Lemma 6.2 of Tang and Yuan (2012) to deal with the last two probabilities. It follows that

$$\Pr\left(\tau(x) = k \right) \gtrsim \left(\mathrm{E}\left[Y^{\alpha/2} \right] \right)^{k-1} \nu(A)\overline{F}\left(\sqrt{x} \right).$$

Substituting this into (20.15) yields the desired asymptotic lower bound for $\psi(x;n)$.

(b) Note that the uniformity over $n \in \mathbb{N}$ of the asymptotics in (20.11) is an immediate consequence of relations (20.11) and (20.12) since the sequence $\sum_{i=1}^{n} \left(\mathrm{E}\left[Y^{\alpha/2} \right] \right)^{i-1}$, $n \in \mathbb{N}$, starts with 1 and increases to $\left(1 - \mathrm{E}\left[Y^{\alpha/2} \right] \right)^{-1}$. Thus, it suffices to prove relation (20.12). In addition, by Lemma 1.7 of Vervaat (1979), both $\sum_{i=1}^{\infty} X_i \prod_{j=1}^{i} Y_j$ and $\sum_{i=1}^{\infty} X_i^{+} \prod_{j=1}^{i} Y_j$ converge almost surely.

Since $\psi(x;\infty) \geqslant \psi(x;n)$ for every $n \in \mathbb{N}$, it follows from Theorem 20.3.1(a) that

$$\psi(x;\infty) \gtrsim \frac{\nu(A)}{1 - \mathrm{E}\left[Y^{\alpha/2} \right]} \overline{F}\left(\sqrt{x} \right).$$

Now we derive the corresponding asymptotic upper bound for $\psi(x;\infty)$. For arbitrarily fixed $n \in \mathbb{N}$ and $0 < \varepsilon < 1$, we derive

$$\psi(x;\infty) \ \leqslant \ \Pr\left(\sum_{i=1}^{\infty} X_i^{+} \prod_{j=1}^{i} Y_j > x \right)$$

$$\leqslant \ \Pr\left(\sum_{i=1}^{n} X_i^{+} \prod_{j=1}^{i} Y_j > (1-\varepsilon)x \right) + \Pr\left(\sum_{i=n+1}^{\infty} X_i^{+} \prod_{j=1}^{i} Y_j > \varepsilon x \right)$$

$$= \ I_1(x,n) + I_2(x,n).$$

By Theorem 20.3.1(a),

$$I_1(x, n) \sim (1 - \varepsilon)^{-\alpha/2} \left(\sum_{i=1}^{n} \left(\mathrm{E} \left[Y^{\alpha/2} \right] \right)^{i-1} \right) \nu(A) \overline{F} \left(\sqrt{x} \right).$$

To deal with $I_2(x, n)$, choose some $0 < \delta < 1$ such that the function

$$m(s) = \mathrm{E} \left[\left(\frac{Y}{1 - \delta} \right)^{s} \right]$$

is smaller than 1 in a neighborhood of $s = \alpha/2$. Then choose some large n such that $\sum_{i=n+1}^{\infty} (1 - \delta)^i < 1$ and choose $0 < \alpha_1 < \alpha/2 < \alpha_2 < \alpha$ such that both $m(\alpha_1)$ and $m(\alpha_2)$ are strictly smaller than 1. We derive

$$
\begin{aligned}
I_2(x, n) &\leqslant \Pr \left(\sum_{i=n+1}^{\infty} X_i^+ \prod_{j=1}^{i} Y_j > \varepsilon x \sum_{i=n+1}^{\infty} (1 - \delta)^i \right) \\
&\leqslant \sum_{i=n+1}^{\infty} \Pr \left(X_i \prod_{j=1}^{i} Y_j > \varepsilon (1 - \delta)^i x \right) \\
&= \sum_{i=n+1}^{\infty} \Pr \left(X_i Y_i \prod_{j=1}^{i-1} \frac{Y_j}{1 - \delta} > \varepsilon (1 - \delta) x \right).
\end{aligned}
$$

Notice that in each term on the right-hand side above, $X_i Y_i$ has a regularly varying tail with index $-\alpha/2$ and is independent of $\prod_{j=1}^{i-1} \frac{Y_j}{1-\delta}$. By Lemma 3.2 of Hao et al. (2012), for some $c > 0$, $x_0 > 0$, and all $x > x_0$,

$$
\begin{aligned}
\frac{I_2(x, n)}{\Pr \left(XY > \varepsilon(1 - \delta)x \right)} &\leqslant c \sum_{i=n+1}^{\infty} \mathrm{E} \left[\left(\prod_{j=1}^{i-1} \frac{Y_j}{1 - \delta} \right)^{\alpha_1} \vee \left(\prod_{j=1}^{i-1} \frac{Y_j}{1 - \delta} \right)^{\alpha_2} \right] \\
&\leqslant c \sum_{i=n+1}^{\infty} \left[m(\alpha_1)^{i-1} + m(\alpha_2)^{i-1} \right].
\end{aligned}
$$

The right-hand side above converges and, hence, it tends to 0 as $n \to \infty$. Thus, $I_2(x, n)$ is negligible in comparison to $I_1(x, n)$ for large n. It follows that

$$\psi(x; \infty) \lesssim \frac{\nu(A)}{1 - \mathrm{E} \left[Y^{\alpha/2} \right]} \overline{F} \left(\sqrt{x} \right).$$

This concludes the proof of Theorem 20.3.1. $\qquad\square$

Acknowledgments

This work was supported by an NSF grant (CMMI-1435864).

References

Acharya, V. V. (2009). *On the Financial Regulation of Insurance Companies*. Ph. D. thesis, NYU Stern School of Business.

Asimit, A. V., E. Furman, Q. Tang, and R. Vernic (2011). Asymptotics for risk capital allocations based on conditional tail expectation. *Insurance: Mathematics and Economics 49*(3), 310–324.

Baluch, F., S. Mutenga, and C. Parsons (2011). Insurance, systemic risk and the financial crisis. *The Geneva Papers on Risk and Insurance-Issues and Practice 36*(1), 126–163.

Bell, M. and B. Keller (2009). Insurance and stability: The reform of insurance regulation. Technical report, Zurich Financial Services Group.

Billio, M., M. Getmansky, A. W. Lo, and L. Pelizzon (2012). Econometric measures of connectedness and systemic risk in the finance and insurance sectors. *Journal of Financial Economics 104*(3), 535–559.

Bingham, N. H., C. M. Goldie, and J. L. Teugels (1987). *Regular Variation*, Volume 27. Cambridge University Press.

Böcker, K. and C. Klüppelberg (2010). Multivariate models for operational risk. *Quantitative Finance 10*(8), 855–869.

Breiman, L. (1965). On some limit theorems similar to the arc-sin law. *Theory of Probability & Its Applications 10*(2), 323–331.

Chen, Y. (2011). The finite-time ruin probability with dependent insurance and financial risks. *Journal of Applied Probability 48*(4), 1035–1048.

Chen, Y. and X. Xie (2005). The finite time ruin probability with the same heavy-tailed insurance and financial risks. *Acta Mathematicae Applicatae Sinica 21*(1), 153–156.

Chen, Y. and K. C. Yuen (2009). Sums of pairwise quasi-asymptotically independent random variables with consistent variation. *Stochastic Models 25*(1), 76–89.

Cline, D. B. and G. Samorodnitsky (1994). Subexponentiality of the product of independent random variables. *Stochastic Processes and Their Applications 49*(1), 75–98.

Cummins, J. D. and M. A. Weiss (2014). Systemic risk and the US insurance sector. *Journal of Risk and Insurance 81*(3), 489–527.

De Haan, L. and S. I. Resnick (1981). On the observation closest to the origin. *Stochastic Processes and Their Applications 11*(3), 301–308.

Embrechts, P. and C. M. Goldie (1980). On closure and factorization properties of subexponential and related distributions. *Journal of the Australian Mathematical Society (Series A) 29*(02), 243–256.

Embrechts, P., D. D. Lambrigger, and M. V. Wüthrich (2009). Multivariate extremes and the aggregation of dependent risks: Examples and counter-examples. *Extremes 12*(2), 107–127.

Fougères, A.-L. and C. Mercadier (2012). Risk measures and multivariate extensions of Breiman's theorem. *Journal of Applied Probability 49*(2), 364–384.

Frolova, A., Y. Kabanov, and S. Pergamenshchikov (2002). In the insurance business risky investments are dangerous. *Finance and Stochastics 6*(2), 227–235.

Hao, X., Q. Tang, et al. (2012). Asymptotic ruin probabilities for a bivariate lévy-driven risk model with heavy-tailed claims and risky investments. *Journal of Applied Probability 49*(4), 939–953.

Harrington, S. E. (2009). The financial crisis, systemic risk, and the future of insurance regulation. *Journal of Risk and Insurance 76*(4), 785–819.

Joe, H. and H. Li (2011). Tail risk of multivariate regular variation. *Methodology and Computing in Applied Probability 13*(4), 671–693.

Kalashnikov, V. and R. Norberg (2002). Power tailed ruin probabilities in the presence of risky investments. *Stochastic Processes and Their Applications 98*(2), 211–228.

Li, J. and Q. Tang (2015). Interplay of insurance and financial risks in a discrete-time model with strongly regular variation. *Bernoulli 21*(3), 1800–1823.

Mainik, G. and L. Rüschendorf (2010). On optimal portfolio diversification with respect to extreme risks. *Finance and Stochastics 14*(4), 593–623.

McNeil, A. J., R. Frey, and P. Embrechts (2005). *Quantitative Risk Management: Concepts, Techniques, and Tools.* Princeton University Press.

Muermann, A. (2008, 07). Market price of insurance risk implied by catastrophe derivatives. *North American Actuarial Journal 12*(3), 221–227.

Nguyen, T. and G. Samorodnitsky (2013). Multivariate tail estimation with application to analysis of COVAR. *Astin Bulletin 43*(02), 245–270.

Norberg, R. (1999). Ruin problems with assets and liabilities of diffusion type. *Stochastic Processes and Their Applications 81*(2), 255–269.

Nyrhinen, H. (1999). On the ruin probabilities in a general economic environment. *Stochastic Processes and Their Applications 83*(2), 319–330.

Nyrhinen, H. (2001). Finite and infinite time ruin probabilities in a stochastic economic environment. *Stochastic Processes and Their Applications 92*(2), 265–285.

Paulsen, J. (1993). Risk theory in a stochastic economic environment. *Stochastic processes and Their applications 46*(2), 327–361.

Paulsen, J. et al. (2008). Ruin models with investment income. *Probability Surveys 5*, 416–434.

Pergamenshchikov, S. and O. Zeitouny (2006). Ruin probability in the presence of risky investments. *Stochastic Processes and Their Applications 116*(2), 267–278.

Resnick, S. I. (1987). *Extreme Values, Regular Variation, and Point Processes.* Springer.

Resnick, S. I. (2007). *Heavy-Tail Phenomena: Probabilistic and Statistical Modeling.* Springer.

Rüschendorf, L. (2013). *Mathematical Risk Analysis. Dependence, Risk Bounds, Optimal Allocations and Portfolios.* Springer, Heidelberg.

Tang, Q. and G. Tsitsiashvili (2003). Precise estimates for the ruin probability in finite horizon in a discrete-time model with heavy-tailed insurance and financial risks. *Stochastic Processes and Their Applications 108*(2), 299–325.

Tang, Q. and G. Tsitsiashvili (2004). Finite-and infinite-time ruin probabilities in the presence of stochastic returns on investments. *Advances in Applied Probability 36*(4), 1278–1299.

Tang, Q. and Z. Yuan (2012). A hybrid estimate for the finite-time ruin probability in a bivariate autoregressive risk model with application to portfolio optimization. *North American Actuarial Journal 16*(3), 378.

Tang, Q. and Z. Yuan (2013). Asymptotic analysis of the loss given default in the presence of multivariate regular variation. *North American Actuarial Journal 17*(3), 253–271.

Vervaat, W. (1979). On a stochastic difference equation and a representation of non-negative infinitely divisible random variables. *Advances in Applied Probability 11*(4), 750–783.

Yan, J. (2007). Enjoy the joy of copulas: With a package copula. *Journal of Statistical Software 21*(4), 1–21.

Yang, Y., R. Leipus, and J. Šiaulys (2012). On the ruin probability in a dependent discrete time risk model with insurance and financial risks. *Journal of Computational and Applied Mathematics 236*(13), 3286–3295.

Yang, Y. and Y. Wang (2013). Tail behavior of the product of two dependent random variables with applications to risk theory. *Extremes 16*(1), 55–74.

Zhou, M., K.-y. Wang, and Y.-b. Wang (2012). Estimates for the finite-time ruin probability with insurance and financial risks. *Acta Mathematicae Applicatae Sinica, English Series 28*(4), 795–806.

21

Weather and Climate Disasters

Richard W. Katz

National Center for Atmospheric Research, United States

Abstract

Weather and climate disasters receive much public attention, as well as dominating the portfolios in catastrophe insurance. Public policy concerns include the questions of whether such disasters are becoming more frequent and/or more intense. Addressing such questions is complicated because not only is the climate possibly changing, but societal vulnerability as well. Extreme value theory provides a natural framework for addressing these questions, yet statistical methods based on this theory have been only rarely applied in this context. In the present chapter, first the extremal characteristics of extreme weather and climate events that may result in disasters are described, including possible trends and relationships to physically based covariates such as the El Niño-Southern Oscillation phenomenon. Then the extremal characteristics of economic damage are described, including the random sum representation for annual total damage. Finally, damage functions converting the intensity of extreme events into the corresponding economic damage are considered. In an attempt to attribute the upper tail behavior of economic damage to that of the underlying weather or climate phenomenon, penultimate approximations in extreme value theory are invoked. Applications include the US billion-dollar weather and climate disasters, with the main focus being on the economic damage caused by hurricanes.

21.1 Introduction

Much of the economic impact of weather and climate is realized through extreme events. As an example, Figure 21.1 shows the time series of the annual total economic damage from US billion-dollar weather and climate disasters, adjusted for inflation to constant 2011 dollars, for the time period 1980–2011 (Smith and Katz,

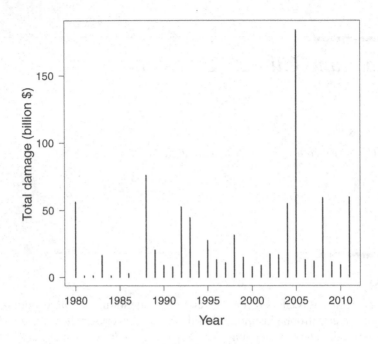

FIGURE 21.1

Annual total economic damage from US billion-dollar weather and climate disasters (adjusted for inflation to constant 2011 dollars) for time period 1980–2011.

2013). The total damage for the year 2005 stands out; still no long-term trend is necessarily evident. The total damage shown in the figure can be viewed as a random sum, with sources of variation including both the annual number of events and the damage from individual events (Embrechts et al., 1997; Katz, 2002). In general, the extreme upper tail of the distribution of the economic damage caused by weather or climate phenomena is of concern, especially to the insurance and reinsurance industries (Embrechts et al., 1997; Murnane, 2004).

The US billion-dollar weather and climate disasters dataset (on which Figure 21.1 is based) is maintained by the National Climatic Data Center (NCDC) of the US National Oceanic and Atmospheric Administration (NOAA). All types of weather and climate disasters are covered, including hurricanes, droughts, and floods, with damages being adjusted for inflation using the US Consumer Price Index. The annual frequency of such disasters and the annual total damage from these disasters are available to the public. For more information about this dataset, see the Appendix and Smith and Katz (2013).

In trend analysis of time series of economic damage, one complication is isolating the influence of weather and climate extremes. Trends in damage are also

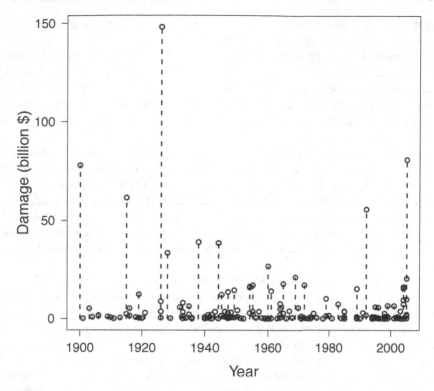

FIGURE 21.2

Economic damage caused by individual hurricanes making landfall along US coast for time period 1900–2005, adjusted for inflation and changes in societal vulnerability to year 2005. For multiple events during the same year, the points are superimposed.

associated with shifts in societal vulnerability, including higher population and more valuable capital. Thus it is difficult to necessarily attribute any observed trend to climate change. For some extreme weather and climate phenomena, especially hurricanes, attempts have been made to adjust the damage data accounting for these shifts in societal vulnerability (e.g., Pielke et al., 2008).

Hurricanes are the single most prevalent type of weather or climate phenomenon causing billion-dollar events (Smith and Katz, 2013), with these damage estimates being of higher quality than those for other extreme phenomena (e.g., floods; Pielke and Downton, 2000). As can be seen from Figure 21.2, only a few of the damaging hurricanes making landfall along the Gulf and Atlantic coasts of the US caused much of the total economic damage during the time period 1900–2005 (based on a dataset adjusted for both inflation and shifts in societal vulnerability to the year 2005; Pielke et al. (2008)). The highest adjusted damage, about $150 billion (US$ for year 2005), is associated with the Great Miami hurricane in 1926; that is, the hypothetical damage if this storm had occurred in 2005 instead. For more information on the hurricane damage dataset, see the Appendix and Pielke et al. (2008).

An indirect approach to the quantification of the economic impact of extreme weather and climate events involves what is termed a "damage function" (Pielke, 2007); typically, assumed to be in the form of a power transformation (e.g., Nordhaus, 2010). It might be anticipated that, depending on the form of damage function, the upper tail behavior of economic damage is inherited from that of the underlying weather or climate variable.

It would seem natural to make use of the statistical theory of extreme values (e.g., Coles, 2001) to study the statistical characteristics of the economic damage caused by extreme weather and climate. Nevertheless, the majority of the statistical analyses so far have relied on distributions, such as the lognormal (Katz, 2002; Nordhaus, 2010), whose properties are not necessarily flexible enough for the extreme upper tail. In the case of economic damage caused by hurricanes, exceptions in which methods based on extreme value theory have been used include Jagger et al. (2011) and Katz (2010, 2015).

In this chapter, first the statistical characteristics of extreme weather and climate events are reviewed, including trends in frequency and intensity and relationships to physically based covariates such as the El Niño-Southern Oscillation phenomenon (Section 21.2). As an application, the economic damage caused by hurricanes striking the US is analyzed as well as, to a lesser extent, the US billion-dollar disaster data (Section 21.3). Then the approximate upper tail behavior of economic damage is related to that of the underlying weather or climate variable through a damage function, in the form of a power transformation, in combination with a penultimate approximation (Section 21.4). Finally, Section 21.5 consists of a discussion. Details about the billion-dollar weather and climate disasters and economic damage from hurricanes datasets are relegated to the Appendix.

21.2   Extreme Weather and Climate Events

21.2.1   Extremal Characteristics

To characterize the statistics of extreme weather and climate events, first the statistical theory of extreme values is briefly reviewed.

21.2.1.1   Block Maxima

The fundamental result in the classical statistical theory of extreme values is the Extremal Types Theorem (Coles, 2001). Suppose that the limiting distribution of the maximum of a stationary sequence of random variables (each with cumulative distribution function (cdf) F, sometimes called the "parent" distribution), suitably normalized, has a limiting distribution, say with cdf G. Then G must be in the form of the generalized extreme value (GEV) distribution; that is,

$$G(x; \mu, \sigma, \xi) = \exp\{-[1 + \xi(x - \mu)/\sigma]^{-1/\xi}\}, \tag{21.1}$$

$1 + \xi(x - \mu)/\sigma > 0$, $\sigma > 0$. Here μ, σ, and ξ denote the location, scale, and shape parameters, respectively. Note that the Extremal Types Theorem does not require that the sequence of random variables necessarily be temporally independent (e.g., Chapter 5 of Coles, 2001).

The GEV distribution has three types: (i) the Weibull, unbounded below and with a finite upper bound of $\mu - \sigma/\xi$ ($\xi < 0$); (ii) the Fréchet, with a finite lower bound of $\mu - \sigma/\xi$ and an unbounded, heavy upper tail ($\xi > 0$); and (iii) the Gumbel, unbounded below and above with a light upper tail ($\xi = 0$), formally obtained by taking the limit as $\xi \to 0$ in (21.1). If the limiting distribution of the suitably normalized maximum is the Gumbel type of GEV, then the parent cdf F is said to be in the "domain of attraction" of the Gumbel (with analogous terminology for the other two types of GEV).

Weather or climate variables such as precipitation totaled over short time periods (e.g., an hour or a day) and stream flow typically possess an apparent heavy tail (e.g., Katz et al., 2002; Koutsoyiannis, 2004). That is, block maxima (e.g., annual maximum of daily precipitation amount or annual peak stream flow) are usually approximated by a GEV distribution with shape parameter $\xi > 0$. Other weather and climate variables, such as temperature or wind speed, have an apparent bounded upper tail (i.e., a GEV distribution for block maxima with shape parameter $\xi < 0$; (e.g., Furrer et al., 2010; Katz, 2015)).

21.2.1.2 Excess over High Threshold

More modern extreme value theory has focused on the limiting behavior of the upper tail of distributions. Consider a high threshold u and suppose that the random variable X exceeds this threshold (i.e., $X > u$). Denote the "excess" over the threshold by $Y = X - u$. Analogous to the Extremal Types Theorem, for sufficiently high threshold u, the excess has an approximate generalized Pareto (GP) distribution (Coles, 2001) with cdf

$$H[y; \sigma(u), \xi] = 1 - \{1 + \xi[y/\sigma(u)]\}^{-1/\xi}, y > 0, \tag{21.2}$$

$1 + \xi[y/\sigma(u)] > 0$. Here $\sigma(u) > 0$ and ξ denote the scale and shape parameters, respectively. The interpretation of the shape parameter is identical to that for the GEV distribution (termed the beta ($\xi < 0$), Pareto ($\xi > 0$), and exponential ($\xi = 0$) types). As the notation suggests, the scale parameter depends on the threshold u, with the excess over a higher threshold having an approximate GP distribution with the same shape parameter, but requiring an adjustment to the scale parameter.

Several techniques are commonly used to estimate the parameters of the GP distribution, including maximum likelihood and probability-weighted moments. Here maximum likelihood is adopted, in part because of the ease in introducing trends and other covariates into the extreme value models. For shape parameter $\xi > -0.5$, the maximum likelihood estimators have the usual asymptotic properties, including

TABLE 21.1
Maximum likelihood estimates and standard errors for parameters of GP and conventional Weibull distributions fit to hurricane wind speed.

Distribution	Parameter	Estimate*
(i) GP ($u = 107.5$ kt, $n = 34$)	$\sigma(u)$	17.203 (3.702)
	ξ	-0.188 (0.134)
(ii) Conventional Weibull ($u = 27.5$ kt, $n = 160$)	σ	63.920 (2.391)
	c	2.214 (0.141)
(iii) Conventional Weibull ($u = 82.5$ kt, $n = 78$)	σ	26.956 (2.214)
	c	1.444 (0.133)

*Standard error given in parentheses.

approximate expressions for the standard errors (Smith, 1985). This constraint on the shape parameter is satisfied in practice for the extremes of weather and climate variables and of the economic damage caused by these variables.

As an example, part i of Table 21.1 gives the parameter estimates and standard errors for a fit by maximum likelihood of the GP distribution to the excess in hurricane wind speed at landfall (a common measure of hurricane intensity) over a threshold of 107.5 kt. This threshold value was obtained by trial and error, first fitting GP distributions for a number of possible thresholds, next identifying a tentative value based on stability of the shape parameter estimates, and finally using a Q-Q plot to verify the goodness-of-fit of the GP distribution for the tentative value (diagnostics described in Chapter 4 of Coles, 2001). As anticipated for wind speed, there is an apparent bounded upper tail with a shape parameter estimate of about -0.19. The wind speed data are part of the US hurricane damage dataset, and will be analyzed in more detail in Section 21.4 (also see Appendix).

21.2.1.3   Poisson-GP Model

Besides the excess over a high threshold, the rate at which the threshold is exceeded also needs to be modeled. Given that exceeding a high threshold is a rare event, it is natural to model the sequence of exceedances as a Poisson process, say with rate parameter $\lambda > 0$. Consequently, the total number of threshold exceedances within a time interval of length T would have an approximate Poisson distribution with rate parameter (or mean) λT. The combination of these two components (i.e., a Poisson process governing the rate of exceedance of a high threshold and a GP distribution for the excess over the threshold) is sometimes called a Poisson-GP model (e.g., Katz et al., 2002). Such a model is consistent with using the GEV distribution as an approximation for block maxima (e.g., Smith, 1989).

For the billion-dollar disaster dataset, 133 events occurred over the 32-yr time period, or a rate of about 4.2 events per year. For the hurricane damage dataset, 160 events occurred over the 106-yr time period (excluding events with damage less than \$0.1 billion, see the Appendix), or a rate of about 1.5 events per year. In the next two

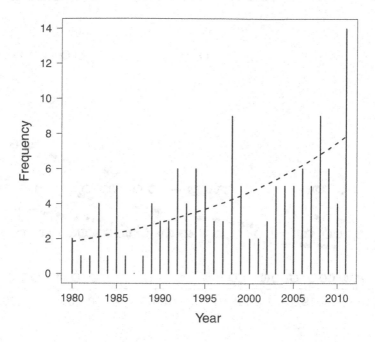

FIGURE 21.3
Time series of annual frequency of US billion-dollar disasters, along with trend (dashed line) fit by Poisson regression for time period 1980–2011.

subsections, possible trends in these rates as well as modulation by physically based covariates will be considered.

21.2.2 Trends in Frequency and Intensity

Concerns about possible climate change call for the introduction of nonstationarity into extremal models. One straightforward approach is to express one or more of the parameters of an extremal distribution as a function of covariates such as time (e.g., Coles, 2001). Parameters can still be readily estimated by maximum likelihood, with model selection being based on a likelihood ratio test (e.g., Coles, 2001) or model selection criterion such as Akaike's information criterion (AIC) or the Bayesian information criterion (BIC) (e.g., Venables and Ripley, 2002).

As an example, Figure 21.3 shows the time series of annual frequency of US billion-dollar disasters. Also included is the fitted trend curve, expressing the log-transformed rate parameter λ of the Poisson distribution for the annual frequency of

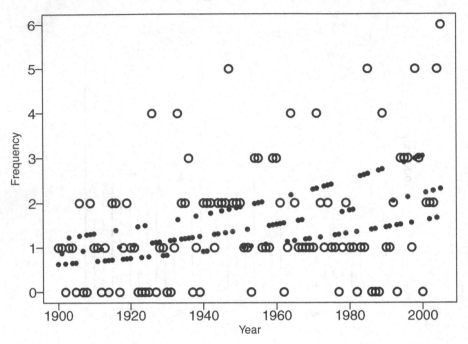

FIGURE 21.4
Trend in annual frequency of hurricanes fit by Poisson regression, conditional on ENSO state, for time period 1900–2005 (observed values indicated by open circles, fitted values by solid circles; fitted points on highest curve correspond to La Niña events, etc.).

disasters as a linear function of the year t (i.e., a form of Poisson regression):

$$\log \lambda(t) = \lambda_0 + \lambda_1 t. \tag{21.3}$$

The estimated slope parameter λ_1 corresponds to a relative increase in the frequency of disasters of about 4.8% per year. This apparent trend is overwhelmingly statistically significant (P-value $< 10^{-5}$ for likelihood ratio test). For more details, see Smith and Katz (2013).

Similarly, trends in extreme event intensity can be modeled by expressing one or both of the parameters of the GP distribution for the excess in intensity over a high threshold as a function of time (e.g., the log-transformed scale parameter as a linear function of time). Nevertheless, it is difficult to detect trends in, for example, hurricane intensity (Kunkel et al., 2013), notwithstanding physical arguments suggesting an increase in intensity with global warming (Emanuel, 2005).

21.2.3   Relationship to Covariates

In addition to trends, covariates in extremal models can be physically based, such as indices of large-scale atmospheric or oceanic circulation. As an example, Figure 21.4 shows the time series of annual frequency of hurricanes damaging the US. Also included is the fitted trend curve, modulated by an index of the El Niño-Southern Oscillation (ENSO) phenomenon denoted by Z_t (i.e., $Z_t = -1, 0, 1$ for La Niña, neutral, and El Niño events, respectively). In this case (again a form of Poisson regression), the log-transformed rate parameter of the Poisson distribution for the annual frequency of hurricanes is expressed as a linear combination of both the ENSO state and the year t:

$$\log \lambda(t) = \lambda_0 + \lambda_1 Z_t + \lambda_2 t. \tag{21.4}$$

Controlling for the effect of ENSO (with damaging hurricanes being less frequent during El Niño events; Katz (2002)), there is fairly strong evidence for a trend (P-value about 0.0004), with the estimated trend parameter λ_2 corresponding to a relative increase in the frequency of hurricanes of about 0.9% per year (holding the ENSO state fixed). For more details about the ENSO data, see the Appendix and Katz (2010).

As with trends, it is difficult to detect any dependence of extreme high hurricane intensity on physically based covariates. Nevertheless, Mestre and Hallegatte (2009) found an influence of sea surface temperature on the upper tail of the distribution of hurricane intensity.

21.3   Economic Damage

In this section, the statistical properties of economic damage from extreme weather and climate events are considered.

21.3.1   Damage from Individual Extreme Events

The economic damage caused by hurricanes striking the US coast along the Gulf of Mexico and Atlantic Ocean exhibits a rapid increase over the past century, even if corrected for inflation (Pielke et al., 2008). Yet most, if not all of this increase is attributable to shifts in societal vulnerability, as opposed to any change in climate. To remedy this problem, Pielke et al. (2008) developed a hurricane damage dataset adjusted for wealth and population in addition to inflation (see the Appendix).

Part i of Table 21.2 includes the parameter estimates and standard errors based on maximum likelihood for the GP distribution fit to the excess in damage over a high threshold. The value of the threshold $u = \$7.5$ billion was obtained using the same diagnostics as described for the hurricane wind speed data (see Section 21.2.1.2), leaving a total of 31 excesses. In particular, a Q-Q plot (not shown) looks satisfactory.

TABLE 21.2
Maximum likelihood estimates and standard errors for parameters of GP distribution fit to excess in hurricane damage over high threshold.

Distribution	Parameter	Estimate*
(i) All excesses ($n = 31, u = 7.5$)	$\sigma(u)$	11.603 (3.677)
	ξ	0.476 (0.275)
(ii) Trend ($n = 31, u = 7.5$)	σ_0	3.326 (0.543)
	σ_1	-0.013 (0.008)**
	ξ	0.382 (0.244)
(iii) Covariate V ($n = 31, u = 7.5$)	σ_0	1.258 (1.061)
	σ_1	0.011 (0.009)***
	ξ	0.452 (0.287)
(iv) $V > 82.5$ kt ($n = 25, u = 7.5$)	$\sigma(u)$	12.793 (4.616)
	ξ	0.496 (0.316)

*Standard error given in parentheses.
**P-value ≈ 0.094 for likelihood ratio test of $\sigma_1 = 0$ in (21.7).
***P-value ≈ 0.228 for likelihood ratio test of $\sigma_1 = 0$ in (21.11).

There is evidence of a heavy tail, with the estimated shape parameter being 0.476, consistent with other analyses (Chavas et al., 2013; Jagger et al., 2011; Katz, 2010).

To assess the fit of the GP distribution to excess damage, it is convenient to consider the survival function. From (21.2), it follows that

$$\log\{1 - H[y; \sigma(u), \xi]\} = -(1/\xi)\log\{1 + \xi[y/\sigma(u)]\}. \qquad (21.5)$$

In other words, the survival function of the GP distribution is approximately linear for large excesses on a log-log scale. Further, the slope of the approximate straight line is negative for a heavy-tailed distribution (i.e., $\xi > 0$).

Figure 21.5 shows a plot of the log-transformed survival function versus the log-transformed excess damage for both the empirical and fitted GP distributions. It indicates that the GP distribution provides a reasonable fit, with the approximate linearity of the empirical survival function for high excesses being evident. A Q-Q plot uses essentially the same information as shown in Figure 21.5, but in a linearized form.

21.3.2   Total Damage as Random Sum

It is common to aggregate damage (e.g., over a year) from extreme weather and climate events. Such an aggregation is especially convenient for the insurance and reinsurance industry, with premiums typically being tied to an annual time scale (recall Figure 21.1 for the billion-dollar disaster data). As mentioned in the Introduction, total damage can be viewed as a random sum. Because random sums are the "bread and butter of insurance mathematics" (Embrechts et al., 1997, p. 96), much is known about their statistical properties. In particular, the upper tail of the distribution of a

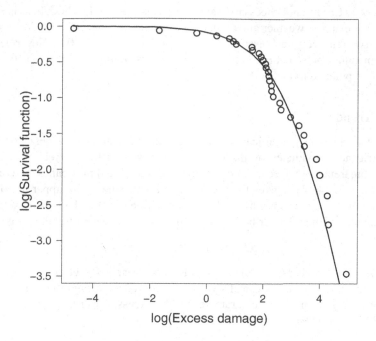

FIGURE 21.5
Log-transformed empirical survival function versus log-transformed excess in hurricane damage over $7.5 billion, along with survival function for fitted GP distribution (solid line).

random sum closely resembles the corresponding one for damage from individual events under broad conditions ("the tail of the maximum determines the tail of the sum," (Embrechts et al., 1997, p. 38)). Nevertheless, it is generally more informative to directly model the two component processes of a random sum: (i) the frequency of events; and (ii) the individual damage associated with each event.

Let the random variable $N(t)$ $(= 0, 1, \ldots)$ denote the number of events in year t. Conditional on $N(t) > 0$, let L_k denote the damage from the kth event in year t, $k = 1, \ldots, N(t)$. Then the total annual damage in year t, say $S(t)$, can be expressed as

$$S(t) = L_1 + L_2 + \ldots + L_{N(t)}. \tag{21.6}$$

As already discussed in 21.2.1.3, it is natural to assume that the annual frequency of events has a Poisson distribution. Additional common assumptions are that, given the number of events in year t, say $N(t) = k$, $k > 0$, the damages from individual events, $L_1, \ldots, L_k$, are independent and identically distributed, as well as being independent of $N(t)$. For the damage from extreme weather and climate events, at least some seasonality would be anticipated, suggesting that the terms on the right-hand side of (21.6) are not necessarily identically distributed. Further, there is some evidence that extreme weather and climate events tend to "cluster," with $N(t)$ being overdispersed relative to a Poisson distribution (Vitolo et al., 2009). Nevertheless, the random sum representation with these common assumptions still seems to work fairly well in practice (Katz, 2002).

21.3.3   Trends

Recall that a statistically significant increasing trend in the annual frequency of damaging hurricanes, modulated by the ENSO state, has been found (Section 21.2.3). Now possible trends in the economic damage from individual hurricanes are considered. Part ii of Table 21.2 gives the results of fitting a trend to the upper tail of the distribution of damage from individual hurricanes. Specifically, the log-transformed scale parameter of the GP distribution is assumed to have a linear trend; that is,

$$\log \sigma(t, u) = \sigma_0 + \sigma_1 t, \tag{21.7}$$

where t denotes the year ($t = 1$ corresponds to the year 1900, etc.). A likelihood-ratio test yields borderline statistical significance for a decreasing trend estimated at about 1.3% per year in the scale parameter for the excesses over a threshold of \$7.5 billion (P-value = 0.094).

21.4   Damage Functions

The economic damage from hurricanes has an apparent heavy tail (see Section 21.3); yet hurricane intensity apparently does not (see Section 21.2). So, at first glance, it would seem futile to attempt to relate the upper tail behavior of damage to that of event intensity. But more refined extreme value theory, based on so-called penultimate approximations, provides an avenue to make sense of this result for hurricanes.

21.4.1   Functional Form

21.4.1.1   Power Transform

It is common to assume that the function converting the intensity of the extreme weather or climate event into the corresponding economic damage is in the form of a power transformation (e.g., Nordhaus, 2010). This functional form has been justified on the basis of physical principles, empirical evidence, and mathematical convenience.

Suppose that the intensity of the weather or climate variable is denoted by V and the corresponding economic damage (or loss) by L. Then it is assumed that L is related to V by the power transformation

$$L = aV^b, a > 0, b > 0. \tag{21.8}$$

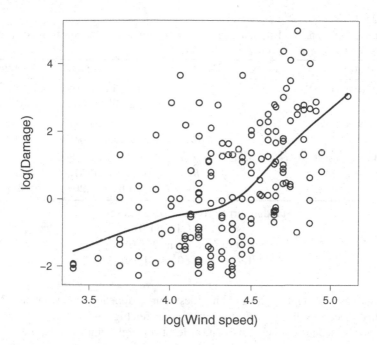

FIGURE 21.6
Scatterplot of log-transformed hurricane damage versus log-transformed maximum wind speed at landfall, along with smoother based on loess (solid line).

For hurricanes, with V being the maximum wind speed at landfall, it has been argued on a physical basis that the exponent $b = 3$ in (21.8), as consistent with the hurricane power dissipation index used by Emanuel (2005). Otherwise, both the parameters a and b in (21.8) would need to be estimated.

The power transformation (21.8) implies log-log linearity for the relationship between the damage L and the intensity V of the extreme weather or climate event. So it is straightforward to estimate the exponent b as the slope in the regression of $\log L$ versus $\log V$. That is, applying the logarithmic transformation to both sides of (21.8) gives

$$\log L = \log a + b \log V. \tag{21.9}$$

Some empirical evidence obtained from such regression analyses indicates that $b > 3$ for hurricane damage (Nordhaus, 2010; Pielke, 2007).

The maximum wind speed at landfall, as already analyzed in Table 21.1, is used as a measure of hurricane intensity (see Appendix for more details about this dataset), even though ideally other storm features such as size should be included as well (Chavas et al., 2013). Figure 21.6 shows a scatterplot of log-transformed hurricane damage versus log-transformed wind speed, along with a smoothed curve based on

TABLE 21.3

Least squares estimates and standard errors for parameters of power transform damage function fit to hurricane wind speed and damage.

Distribution	Parameter	Estimate*	P-Value
(i) All data ($n = 160$)	$\log a$	$-11.153\ (1.503)$	
	b	$2.635\ (0.342)$	≈ 0**
(ii) $V > 82.5$ kt ($n = 78$)	$\log a$	$-18.509\ (5.045)$	
	b	$4.239\ (1.082)$	< 0.001**
(iii) Piecewise linear ($n = 160$)	$\log a$	$-5.439\ (2.403)$	
	b_1	$1.230\ (0.575)$	0.034***
	b_2	$3.624\ (1.207)$	0.003****

*Standard error given in parentheses.
** Test of $b = 0$ in (21.9).
*** Test of $b_1 = 0$ in (21.10).
**** Test of $b_2 = 0$ in (21.10).

loess (Venables and Ripley, 2002). The loess curve appears somewhat nonlinear, but roughly piecewise linear with a steeper slope for high wind speeds (say $V > 82.5$ kt). So the scatterplot suggests that, at least for high wind speeds, the damage function may be approximately in the form of a power transformation (21.8). Note that Murnane and Elsner (2012) also found a lack of log-log linearity in the damage function for hurricanes.

Table 21.3 gives the estimated parameters and standard errors from the regression of log-transformed damage on log-transformed wind speed, both for all the data (part i) and for only the subset with wind speed $V > 82.5$ kt (part ii, reducing the size of the dataset from 160 to 78). By (21.9), this slope can be viewed as an estimate of the exponent of the power transformation b. For all the data, the estimated slope is about 2.64 or roughly one standard error below the physically based value of $b = 3$; for only the data with $V > 82.5$ kt, the estimated slope is about 4.24 or roughly one standard error above $b = 3$. In both cases, a t-test of $b = 0$ gives overwhelming statistical significance for a linear relationship between log-transformed wind speed and log-transformed damage.

Table 21.3 (part iii) also includes the results of a piecewise linear regression of $\log L$ versus $\log V$, with a join point at 82.5 kt. That is, a damage function of the following form is fitted:

$$\log L = \log a + b_1 \log V + b_2 \log(V/82.5) I_{\{V > 82.5\}}, \qquad (21.10)$$

where I denotes the indicator function. A piecewise linear function is a better fit to the scatterplot than a linear one (P-value ≈ 0.003 for t-test of $b_2 = 0$ in (21.10)). For this model, the slope for $V > 82.5$ kt is given by $b_1 + b_2$, with an estimate being about 4.85 or somewhat higher than that based on only the data for which $V > 82.5$ kt (i.e., part ii of Table 21.3). As already mentioned, a few recent studies have found estimates of b considerably higher than three. For instance, Nordhaus

(2010) concluded that $b \approx 9$, whereas Bouwer and Botzen (2011) claimed that $b \approx 8$ is a better choice.

21.4.1.2 Extreme Value Approach

An alternative approach to obtaining a hurricane damage function would be based on extreme value theory, considering only the relationship with wind speed for the most damaging hurricanes (Chavas et al., 2013). Instead of fitting an unconditional GP distribution to the excess in damage above a high threshold u (as in part i of Table 21.2), one or more of the parameters of the GP could depend on wind speed. In particular, it is assumed that the log-transformed scale parameter of the GP can be modeled as a linear function of the wind speed V; that is,

$$\log \sigma(V, u) = \sigma_0 + \sigma_1 V, \tag{21.11}$$

with the shape parameter ξ being held constant.

Part iii of Table 21.2 includes the results of fitting such a conditional GP distribution with the same threshold of $u = \$7.5$ billion, again using maximum likelihood to estimate the parameters. A likelihood ratio test can be used to compare the fit of this conditional GP distribution to that of the unconditional GP (Coles, 2001). This test of the slope parameter $\sigma_1 = 0$ in (21.11) is not statistically significant, with a P-value about 0.228. Most likely, this lack of statistical significance is attributable to the relatively small sample size involved (i.e., only 31 excesses). In fact, if a lower threshold of $u = \$5$ billion is used instead as in Chavas et al. (2013), a P-value of about 0.05 is obtained or at least borderline statistical significance.

21.4.2 Induced Damage Distribution

Suppose that the weather or climate variable V has a conventional Weibull distribution (a reverse or negated version of the Weibull type of GEV distribution (1)); that is, with cdf given by

$$F(x; u, \sigma, c) = 1 - \exp\left\{-[(x - u)/\sigma]^c\right\}, x > u. \tag{21.12}$$

Here $\sigma > 0$ and $c > 0$ denote the scale and shape parameters, respectively, and u the lower bound or threshold (often $u = 0$). This distribution is commonly fitted to wind speed (e.g., Cook and Harris, 2004). Then the power transform relationship (21.8) implies that the damage L would also have a Weibull distribution, but with different scale and shape parameters, σ^* and c^*, related to the original scale and shape parameters, σ and c, by

$$c^* = c/b, \sigma^* = a\sigma^b \tag{21.13}$$

(Johnson and Kotz, 1970, Chapter 20).

Table 21.1 also includes the maximum likelihood parameter estimates and standard errors of Weibull distributions fit to the hurricane wind speed data. The estimated shape parameter $c = 2.214$ for all the data (part ii of Table 21.1) and $c = 1.444$ for only wind speed greater than 82.5 kt (i.e., the same threshold as used for the damage function estimation, part ii of Table 21.3). These estimates are reasonably consistent with the typical values of c being between 1 and 3 for wind speeds in general (Cook and Harris, 2004). So, using the estimates of c and b based on only the wind speed greater than 82.5 kt (part iii of Table 21.1 and part ii of Table 21.3), (21.13) implies that damage would have a Weibull distribution with shape parameter of $c^* = 0.341$. As will be seen later in this section, this shift to a smaller Weibull shape parameter would correspond to a heavier upper tail for damage than for wind speed (in the sense of penultimate extreme value theory).

21.4.2.1 Penultimate Approximations

One way to determine the domain of attraction for the parent cdf F is based on a concept known as the "hazard rate"; that is,

$$h_F(x) = F'(x)/[1 - F(x)], \tag{21.14}$$

where F' (i.e., the derivative of F) denotes the corresponding probability density function (pdf) (Reiss and Thomas, 2007). For simplicity, assume that the upper tail of the cdf F is unbounded. A sufficient condition for F to be in the domain of attraction of the Gumbel type of GEV distribution (known as the von Mises condition, Reiss and Thomas (2007)) can be expressed in terms of the limiting behavior of the derivative of the reciprocal of the hazard rate as follows:

$$(1/h_F)'(x) \to 0 \tag{21.15}$$

as $x \to \infty$.

The extreme value theory approximations described in Section 21.2 can be thought of as "ultimate" (or first-order), in the sense of only holding in the limit (i.e., as the block size n or the threshold u trends to infinity). More refined penultimate (or second-order) approximations are also available, with potentially increased accuracy in practice.

For simplicity, only the case in which the cdf F is in the domain of attraction of the Gumbel is considered, with it further being assumed that the von Mises condition (21.15) holds. The basic idea is to make use of the approximate behavior of the quantity (i.e., derivative of the reciprocal of the hazard rate) on the left-hand side of (21.15) for large x, rather than simply taking the limit as $x \to \infty$. Instead of the Gumbel, the GEV distribution is used with shape parameter ξ_n depending on the block size n as follows:

$$\xi_n \approx (1/h_F)'(x)|_{x=u(n)} \tag{21.16}$$

(Reiss and Thomas, 2007). Here

$$u(n) = F^{-1}(1 - 1/n) \tag{21.17}$$

denotes the $(1 - 1/n)$th quantile of the cdf F (sometimes called the "characteristic largest value"). By (21.15), $\xi_n \to 0$ as $n \to \infty$, so this penultimate approximation converges toward the ultimate approximation as the block size increases.

If F is the conventional Weibull distribution (21.12), then the hazard rate and the characteristic largest value can be expressed in closed form as:

$$h_F(x; \sigma, c) = (c/\sigma)(x/\sigma)^{c-1}, u(n) = \sigma(\log n)^{1/c}. \tag{21.18}$$

So by (21.16) and (21.18), the shape parameter of the penultimate GEV distribution is given by:

$$\xi_n = (1-c)/(c \log n) \tag{21.19}$$

(Furrer and Katz, 2008).

Because the conventional Weibull distribution is in the domain of attraction of the Gumbel, the penultimate shape parameter (21.19) does converge to zero as the block size n increases. Nevertheless, for finite block size n, (21.19) implies that: (i) $\xi_n < 0$ (i.e., Weibull type of GEV) for $c > 1$; and (ii) $\xi_n > 0$ (i.e., Fréchet type of GEV) for $c < 1$, a penultimate approximation used for precipitation extremes (Furrer and Katz, 2008; Wilson and Toumi, 2005).

Katz (2015) compared the penultimate approximation (21.19) to that for the GEV distribution directly fitted to block maxima obtained from pseudorandom numbers simulated from conventional Weibull distributions. For a plausible range of block sizes and of shape parameters for the Weibull, the GEV shape parameters based on the penultimate approximation (21.19) are only slightly too large in absolute magnitude as compared to the simulated ones.

21.4.2.2 Application to Hurricane Damage

Now inferences about the upper tail behavior of the distribution of economic damage caused by hurricanes can be made on the basis of these penultimate approximations in extreme value theory. Making use of (21.13), the penultimate approximation (21.19) for the shape parameter of the upper tail of the distribution of damage can be expressed as

$$\xi_n = [1 - (c/b)]/[(c/b) \log n]. \tag{21.20}$$

Because the shape parameter of damage has been estimated for the excess in damage over a high threshold (i.e., not directly in terms of block maxima), the value of the block size n to substitute into (21.20) is somewhat ambiguous. Nevertheless, because the expression for ξ_n depends on the block size only through $\log n$, it is not very sensitive to the exact choice.

If the estimates of c and b based on only the data for which the wind speed $V > 82.5$ kt are used, then damage would have a Weibull distribution with shape parameter $c^* = c/b = 0.341$. So (21.20) gives a GEV (or equivalently GP) shape parameter $\xi_n \approx 0.495$ for the upper tail of damage if the block size $n = 50$ and 0.420 if $n = 100$. Both these shape parameter estimates correspond to rather heavy tails, with the value for $n = 50$ being close to the actual estimate of 0.476 for the direct fit of a GP distribution to damage (part i of Table 21.2) and virtually identical to the estimate of 0.496 based on only damage for which the wind speed exceeds 82.5 kt (part iv of Table 21.2).

21.5 Discussion

It has been demonstrated that statistical methods based on extreme value theory provide a natural framework for quantifying the risk of high damage from extreme weather and climate events. In particular, it has been shown that economic damage from hurricanes has a distribution with an apparent heavy upper tail. Further, increasing trends in the frequency of damaging hurricanes, as well as in the frequency of all billion-dollar disasters, are evident but not necessarily in extreme high damage from individual hurricanes.

One chance mechanism has been proposed concerning how the wind speed at landfall, a measure of hurricane intensity, could have a distribution with an apparent bounded upper tail, yet the economic damage from hurricanes has a distribution with an apparent heavy upper tail. This chance mechanism is based on the assumption of a conventional form of Weibull distribution for wind speed and a power transformation relating wind speed to economic damage. Under these assumptions, penultimate approximations in extreme value theory can be invoked. Still it should be noted that, strictly speaking, both the upper tail of the distribution of wind speed and of damage could be light in the sense of ultimate extreme value theory.

Nevertheless, there are other possible explanations for the heavy upper tail of the distribution of hurricane damage. It may be that this heavy tail arises, at least in part, as a consequence of damage being aggregated over a population, irrespective of the underlying weather or climate phenomenon. In fact, Vilfredo Pareto originally derived the Pareto distribution (or "Pareto's law") as a model for how income or capital should be distributed over a population (Arnold, 1983).

Other statistical challenges remain in the analysis of economic damage from extreme weather and climate events. In particular, the current procedures for adjusting damage to take into account shifts in societal vulnerability may introduce systematic biases through ignoring the small-scale granularity of the adjustment data, such as the rate of insurance coverage (Smith and Katz, 2013). It may be that temporal shifts in these biases also affect trend analysis, further complicating any effort to attribute any trend in damage data to climate change.

21.6 Appendix

Billion-dollar weather and climate disasters

The US billion-dollar weather and climate disasters dataset is available on the NOAA/NCDC website www.ncdc.noaa.gov/billions. Because the damage from individual disasters is considered proprietary (being based in part on estimates from the reinsurance industry), only the annual frequency and annual total damage are publicly available. The damage estimates have been adjusted for inflation using the Consumer Price Index, but not for any shifts in societal vulnerability. For convenience, the total damage for the year 1987, in which no billion-dollar disasters occurred, is treated as "missing" (i.e., coded as "NA" in R).

The billion-dollar disasters dataset is updated annually, both to incorporate additional weather and climate disasters and to adjust for inflation in terms of the most recent year. As such, the data values presently posted on the NCDC website necessarily differ somewhat from those analyzed here. For more information about this dataset, see the NCDC website and Smith and Katz (2013).

Economic damage from hurricanes

The US economic damage data from hurricanes (strictly speaking, including some tropical cyclones which caused damage despite not attaining official "hurricane" status), adjusted for inflation and shifts in societal vulnerability (including population and wealth), are available at sciencepolicy.colorado.edu/publications/special/ normalized_hurricane_damages.htm. The dataset termed "PL05" is used, with the adjustment process being described in Pielke et al. (2008). To avoid bias against relatively low damage events early in the record, only storms causing greater than $0.1 billion are analyzed, reducing the number of events from 208 to 160 over the time period of 1900–2005.

As a measure of hurricane intensity, the maximum wind speed at landfall was obtained from the NOAA Atlantic basin hurricane database (called HURDAT and available at www.aoml.noaa.gov/hdr/hurdat), following Chavas et al. (2013). Hurricanes that made multiple landfalls were treated as separate events. Because of ambiguity about landfall, a few events were eliminated. Covering tropical cyclones (as well as full-fledged hurricanes), the lowest wind speed value in the dataset is 30 kt, with a resolution of 5 kt. So 27.5 kt is treated as the effective lower bound.

The classification of the ENSO state (i.e., La Niña, neutral, or El Niño event) follows that in Trenberth (1997), as adapted by Pielke and Landsea (1999). It is based on the average sea surface temperature over the Niño 3.4 region of the Pacific, with these data being available at www.cgd.ucar.edu/cas/catalog/climind/TNI_N34/ index.html#Sec5. But it should be noted that the numerical values now posted on the web page differ slightly from those used in this chapter because of changes in computational methods.

Acknowledgments

I thank an anonymous reviewer for helpful comments, and Christina Karamperidou for providing the hurricane maximum wind speed data extracted from the HURDAT dataset. The statistical analysis made use of the `ismev` package in the open source statistical programming language R (www.r-project.org). Research was partially supported by the NCAR Weather and Climate Assessment Science Program.

The National Center for Atmospheric Research is sponsored by the National Science Foundation.

References

Arnold, B. C. (1983). *Pareto Distributions*. Burtonsville, MD: International Cooperative Publishing House.

Bouwer, L. M. and W. J. W. Botzen (2011). How sensitive are US hurricane damages to climate? Comment on a paper by W. D. Nordhaus. *Climate Change Economics 2*(1), 1–7.

Chavas, D., E. Yonekura, C. Karamperidou, N. Cavanaugh, and K. Serafin (2013). US hurricanes and economic damage: Extreme value perspective. *Natural Hazards Review 14*(4), 237–246.

Coles, S. G. (2001). *An Introduction to Statistical Modeling of Extreme Values*. Springer.

Cook, N. J. and R. I. Harris (2004). Exact and general FT1 penultimate distributions of extreme wind speeds drawn from tail-equivalent Weibull parents. *Structural Safety 26*(4), 391–420.

Emanuel, K. (2005). Increasing destructiveness of tropical cyclones over the past 30 years. *Nature 436*(7051), 686–688.

Embrechts, P., C. Klüppelberg, and T. Mikosch (1997). *Modelling Extremal Events*. Springer.

Furrer, E. M. and R. W. Katz (2008). Improving the simulation of extreme precipitation events by stochastic weather generators. *Water Resources Research 44*(12), W12439. W12439.

Furrer, E. M., R. W. Katz, M. D. Walter, and R. Furrer (2010). Statistical modeling of hot spells and heat waves. *Climate Research 43*(3), 191–205.

Jagger, T. H., J. B. Elsner, and R. K. Burch (2011). Climate and solar signals in property damage losses from hurricanes affecting the United States. *Natural Hazards 58*(1), 541–557.

Johnson, N. L. and S. Kotz (1970). *Distributions in Statistics: Continuous Univariate Distributions: Vol. 1*. Houghton Mifflin.

Katz, R. W. (2002). Stochastic modeling of hurricane damage. *Journal of Applied Meteorology 41*(7), 754–762.

Katz, R. W. (2010). Discussion on 'Predicting losses of residential structures in the state of Florida by the public hurricane loss evaluation model' by S. Hamid et al. *Statistical Methodology 7*(5), 592–595.

Katz, R. W. (2015). Economic impact of extreme events: An approach based on extreme value theory. In *Extreme Events: Observations, Modeling and Economics*. Wiley-Blackwell. M. Chavez, M. Ghil, and J. Urrutia-Fucugauchi (eds.).

Katz, R. W., M. B. Parlange, and P. Naveau (2002). Statistics of extremes in hydrology. *Advances in Water Resources 25*(8), 1287–1304.

Koutsoyiannis, D. (2004). Statistics of extremes and estimation of extreme rainfall: II. Empirical investigation of long rainfall records. *Hydrological Sciences Journal 49*(4), 591–610.

Kunkel, K. E., T. R. Karl, H. Brooks, J. Kossin, J. H. Lawrimore, D. Arndt, L. Bosart, D. Changnon, S. L. Cutter, N. Doesken, et al. (2013). Monitoring and understanding trends in extreme storms: State of knowledge. *Bulletin of the American Meteorological Society 94*(4), 499–514.

Mestre, O. and S. Hallegatte (2009). Predictors of tropical cyclone numbers and extreme hurricane intensities over the North Atlantic using generalized additive and linear models. *Journal of Climate 22*(3), 633–648.

Murnane, R. J. (2004). Climate research and reinsurance. *Bulletin of the American Meteorological Society 85*(5), 697–707.

Murnane, R. J. and J. B. Elsner (2012). Maximum wind speeds and US hurricane losses. *Geophysical Research Letters 39*(16), L16707. L16707.

Nordhaus, W. D. (2010). The economics of hurricanes and implications of global warming. *Climate Change Economics 1*(01), 1–20.

Pielke, Jr., R. A. (2007). Future economic damage from tropical cyclones: Sensitivities to societal and climate changes. *Philosophical Transactions of the Royal Society A: Mathematical, Physical and Engineering Sciences 365*(1860), 2717–2729.

Pielke, Jr., R. A. and M. W. Downton (2000). Precipitation and damaging floods: Trends in the United States, 1932–97. *Journal of Climate 13*(20), 3625–3637.

Pielke, Jr., R. A., J. Gratz, C. W. Landsea, D. Collins, M. A. Saunders, and R. Musulin (2008). Normalized hurricane damage in the United States: 1900–2005. *Natural Hazards Review 9*(1), 29–42.

Pielke, Jr., R. A. and C. N. Landsea (1999). La Niña, El Niño and Atlantic hurricane damages in the United States. *Bulletin of the American Meteorological Society 80*(10), 2027–2033.

Reiss, R.-D. and M. Thomas (2007). *Statistical Analysis of Extreme Values with Applications to Insurance, Finance, Hydrology and Other Fields*. Springer Science & Business Media.

Smith, A. B. and R. W. Katz (2013). US billion-dollar weather and climate disasters: Data sources, trends, accuracy and biases. *Natural Hazards 67*(2), 387–410.

Smith, R. L. (1985). Maximum likelihood estimation in a class of nonregular cases. *Biometrika 72*(1), 67–90.

Smith, R. L. (1989). Extreme value analysis of environmental time series: An application to trend detection in ground-level ozone (with discussion). *Statistical Science 4*, 367–393.

Trenberth, K. E. (1997). The definition of El Niño. *Bulletin of the American Meteorological Society 78*(12), 2771–2777.

Venables, W. N. and B. D. Ripley (2002). *Modern Applied Statistics with S* (4 ed.). Springer.

Vitolo, R., D. B. Stephenson, I. M. Cook, and K. Mitchell-Wallace (2009). Serial clustering of intense European storms. *Meteorologische Zeitschrift 18*(4), 411–424.

Wilson, P. S. and R. Toumi (2005). A fundamental probability distribution for heavy rainfall. *Geophysical Research Letters 32*(14), L14812. L14812.

22

The Analysis of Safety Data from Clinical Trials

Ioannis Papastathopoulos

University of Edinburgh and Maxwell Institute, United Kingdom

Harry Southworth

Data Clarity Consulting Ltd., United Kingdom

Abstract

The analysis of the efficacy of a drug in a clinical trial amounts to quantifying the expected response of patients to the drug. The analysis of safety data, though, amounts to characterizing the unexpected, or extreme, response of a minority of patients. As such, the statistical methods used to analyse efficacy data are inappropriate to the analysis of safety data. We illustrate how univariate and multivariate approaches to the analysis of clinical trial safety data can identify and characterize potential toxicities.

22.1 Safety Data from Clinical Trials

22.1.1 Introduction

Clinical trials are usually designed to address a specific question about the effectiveness, or efficacy, of a drug in treating a particular condition. For example, a trial of a new antihypertension drug would measure how much the drug reduces blood pressure relative to a placebo or another drug. A trial of a new drug for treating cancer would study survival times between the new drug and a comparator. Since such questions of efficacy are usually well-formed and since the clinical trials are designed to answer them, the analysis of the resulting data is usually a straightforward prespecified analysis, using a standard statistical technique to compare, say, the mean values in the treatment groups, or the duration of survival of the patients.

However, the bulk of the data collected in clinical trials relates not to the *efficacy* of the drug, but to its *safety*. Safety data will typically include data on the patients' vital signs (blood pressure, heart rate, etc.), adverse events reported by patients to their doctors, other medications that patients might have been taking, and many measurements resulting from sending patients' blood samples to laboratories. Most adverse events, whether or not related to the drug, tend to be uncommon so that the data tend to be sparse. The laboratory data often have skewed distributions and are subject to outliers. For safety data then, unlike efficacy, the questions are typically not well-defined, some of them will not be known until the data have been studied in some detail, and the data are messy, usually of mixed types that come from different labs. For these reasons, statisticians have tended to shy away from the analysis and particularly modelling of safety data, leaving interpretation to clinicians. Some attempts have been made and these include graphical and visualisation techniques for adverse events, nonparametric survival models and Bayesian hierarchical mixture modelling (Enas and Goldstein, 1995; Siddiqui, 2009; Xia et al., 2011).

Published statistical guidance for clinical trials (International Conference on Harmonisation, 1998) summarizes this position: "The investigation of safety and tolerability is a multidimensional problem. Although some specific adverse effects can usually be anticipated and specifically monitored for any drug, the range of possible adverse effects is very large, and new and unforeseeable effects are always possible. Further, an adverse event experienced after a protocol violation, such as use of an excluded medication, may introduce a bias. This background underlies the statistical difficulties associated with the analytical evaluation of safety and tolerability of drugs, and means that conclusive information from confirmatory clinical trials is the exception rather than the rule."

22.1.2   The Importance of Extreme Values

Typically, the data tend to be far from normally distributed and subject to outliers. In fact, it is usually the outliers that indicate safety issues, as supported by the following observations.

(i) The U.S. Food and Drug Administration (FDA) has published a guidance document (U.S. Food and Drug Administration, 2008) which advises that extremely high values of alanine aminotransferase (ALT) and aspartate aminotransferase (AST), laboratory values that are recorded as standard in most clinical trials, can be indicative of drug-induced liver injury.

(ii) Rhabdomyolysis, an adverse event that has been linked to certain drugs in the past, involves the breakdown of muscles and the release of their contents into the blood. Typically, if the muscles have begun to break down, levels of creatine kinase in the blood rise dramatically.

(iii) In the U.S., the National Institutes of Health and the National Cancer Institute jointly publish the Common Toxicity Criteria (CTC) (DCTD et al., 2003), a guideline listing many laboratory variables and categorizing abnormally high (or sometimes low) levels by severity. The guideline is used by clinicians to aid interpretation of safety data.

(*iv*) Various guidelines exist for the diagnosis of kidney failure, usually relying on high values of creatinine in the blood.

The clinical trial statistician concerned with efficacy is concerned with characterizing the *expected* response patients have to the drug. When it comes to safety, what is important is the characterization of the *unexpected* responses, i.e., the extreme values. Hence, measures of central tendency such as means and medians, and the familiar tools in the clinical trial statistician's toolbox, i.e., analysis of variance, regression, tests for equality of means, etc., are practically of no benefit in the interpretation of clinical safety data. Since it is the extreme values of laboratory variables that tend to indicate toxicity, it is usual for them to have an associated *upper limit of normal* (ULN) and *lower limit of normal* (LLN). The way in which these quantities are calculated can vary from laboratory to laboratory, but they represent high and low quantiles, such as the 5th and 95th percentiles, of the distribution. Thus any observation that is above the ULN, or some multiple of it, is considered to be abnormal and potentially relating to a safety issue.

To illustrate the problems that can arise, consider the case of troglitazone, a drug used for the treatment of diabetes, and liver toxicity. The U.S. Food and Drug Administration (1999) review of the available data states: "Mean ALT and AST levels fell in patients receiving troglitazone in phase 3 trials... It was also stated that 2.2% of patients in phase 3 trials had a transaminase (ALT or AST) level exceeding $3\times$ULN... What was not appreciated by FDA was that many of the patients classified as ALT $> 3\times$ULN actually had ALT values that were VERY much higher than $3\times$ULN... 23 patients had treatment-emergent ALT values over $3\times$ULN... In 14 of these 23 patients, the ALT value exceeded $8\times$ULN... and in 5/23 patients the ALT value exceeded $30\times$ULN." The review goes on to describe how, once troglitazone was on the market, cases of "frank liver failure" were reported in association with the drug. Further, in a postmarketing clinical trial, a patient developed irreversible liver damage and ultimately died. The drug was subsequently withdrawn from the market.

The apparent toxicity of troglitazone was not at all predicted by the central tendency of the distributions of ALT and AST — the means and medians went *down*, intuitively suggesting a beneficial effect of the drug on the liver. It was the extreme values of ALT and AST that contained the useful information. Making inferences about the potential for liver toxicity by looking at the mean would be misleading and dangerous.

In this chapter, we summarise the methodology and findings of Southworth and Heffernan (2012a,b) for modelling laboratory variables of an experimental drug that has been linked to liver toxicity. We specifically focus our attention on potentially alternative signals of liver toxicity than those suggested by the U.S. Food and Drug Administration (2008). In Section 22.2 we introduce methodology that is used in the analysis of safety data, focusing on univariate and multivariate threshold methods. Subsequently, we illustrate in Section 22.3 the statistical analysis of safety data from the phase 3 clinical trial. Our conclusions are given in Section 22.4.

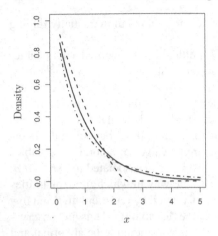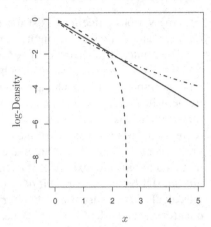

FIGURE 22.1

Left: The GP density function $h(x) = H'(x) = \sigma^{-1}\left(1 + \xi x/\sigma\right)_+^{-1-1/\xi}$ for $\sigma = 1$ and ξ equal to -0.4 (dashed), 0 (solid) and 0.4 (dotted-dashed). Right: Logarithm of the GP density function using the same configuration of parameters. Negative values of ξ result in short-tailed distributions and positive values of ξ result in heavy tails and thus greater probability of outliers.

22.2 Extreme Value Methods

22.2.1 Univariate Threshold Models

Different extreme value models exist for dealing with data arising in a variety of settings. In the clinical trial setting, we typically have a fairly small number of observed values from each of several dozen or a few hundred patients who are categorized in treatment groups. For each patient, several laboratory variables are usually recorded. Since we are interested in the extremes of the laboratory data, methods that allow the characterization of the upper tail of the distribution of each of the variables in the dataset are appropriate. The most appropriate tool is known as the *generalized Pareto* (GP) distribution function

$$H(x) = \left(1 + \xi x/\sigma\right)_+^{-1/\xi} \qquad x > 0, \tag{22.1}$$

where $\sigma > 0$, $\xi \in (-\infty, \infty)$ and $x_+ = \max(x, 0)$. This distribution was first given by Pickands (1975) and the statistical approach is due to Davison and Smith (1990). Let $x_1, \ldots, x_n$ be n observed measurements that are assumed to have arisen from a random sample $X_1, \ldots, X_n$ of a random variable X. In this approach, extreme events are defined by threshold exceedances, i.e., $\{x_i : x_i > u\}$, where

$u < \max(x_1, \ldots, x_n)$ is a sufficiently large threshold above which the GP distribution (22.1) is assumed to be a reasonable approximation to the distribution of $X - u \mid X > u$. Subsequently, the GP model is fitted, using maximum likelihood, to the data that exceed this large threshold and the model is extrapolated to levels above which no data are observed. For example, a multiple of ULN or a high sample quantile above which there is a sufficient number of recorded observations needed for statistical estimation and extrapolation can be chosen in the analysis. Asymptotic theory suggests that if u_0 is a valid threshold for excesses to follow the GP distribution then estimates of the shape parameter ξ and the modified scale parameter $\sigma - \xi u$ should be relatively stable above u_0. Hence, standard practice is to choose as low a threshold as possible which does not violate the suitability of model (22.1) for the distribution of the excesses.

The parameter σ controls the scale of the distribution whereas the shape parameter ξ controls how heavy the right tail of the distribution is. Plots of the GP density function and its logarithm are shown in Figure 22.1 for different values of ξ. By examining how the values of σ or (more importantly) ξ depend on the drug and dose, we can form an idea for the strength of evidence for treatment and dose-response effects, and for the nature of those relationships. For example, a positive association of ξ with dose level would imply that the higher the dose, the more likely to observe extremes in the sample.

However, the parameters in the GP distribution do not have a straightforward physical interpretation and in order to communicate the analysis to people working in the clinical industry, it is more useful to work with predictions from the fitted models. For example, fitted models can be used to predict how large the values that we see are likely to get if we expose larger numbers of patients to the experimental drug. This is usually done via the T-patient return level

$$x_T = u + (\sigma/\xi)\left\{(Tp_u)^\xi - 1\right\}, \tag{22.2}$$

which is the level that is expected to be exceeded on average once in T patients, where $p_u = \Pr(X > u)$. In practice, u is the threshold used to fit the GP distribution and the quantities p_u, σ and ξ are replaced by their estimates. The uncertainty of the return level estimates is usually illustrated with approximate confidence intervals obtained either from the delta method or more preferably, the profile likelihood of the maximum likelihood estimators; see Coles (2001). Last, quantities such as the probability of any particular variable being above ULN, $3\times$ULN or some other multiple, can be predicted.

22.2.2   Dependence Modelling

In example (i) of Section 22.1.2, the U.S. Food and Drug Administration (2008) also states that a finding of ALT, usually substantial and greater than thrice its ULN, seen concurrently with bilirubin level (TBL), greater than twice its ULN, identifies a drug likely to cause severe drug-induced liver injury. Moreover, *alkaline phosphatase* (ALP) should not be elevated so as to explain the elevation of TBL. Univariate extreme value methods are appropriate for understanding and predicting the behaviour of extremes of each laboratory variable. However, as this example illustrates, it is often the case that the joint occurrence of extremes of several laboratory variables is indicative of toxicity.

Models such as the multivariate normal, despite being highly useful for modelling the main body of joint distributions, usually underestimate the frequency of joint extremes. As such, multivariate threshold models that are applied to upper fractions of the data are more appropriate. Let (X, Y) be a bivariate random variable that has double unit-exponential margins. There is no loss in assuming double unit-exponential margins since in practice the margins can be standardized. Following the approach of Heffernan and Tawn (2004) and Keef et al. (2013), the *conditional dependence model* assumes that the variable Y, conditionally on X exceeding a large threshold v, can be well approximated by

$$Y = \alpha X + X^\beta Z, \qquad (22.3)$$

where $\alpha \in [-1, 1]$ and $\beta < 1$ are parameters that describe the dependence of Y with large X, and Z is a random variable that is independent of X. The parameter α is a slope parameter with positive (negative) values indicating positive (negative) association. The parameter β controls how the variance of Y changes with X, with positive (negative) values indicating increasing (decreasing) variance as X increases. The case $\alpha = \beta = 0$ corresponds to independence. This model is motivated by a broad range of dependence structures for (X, Y) (Heffernan and Tawn, 2004), covering all dependence models studied by Nelsen (1999). These include the bivariate normal, the bivariate max-stable distribution, and the bivariate inverted max-stable distribution (Papastathopoulos and Tawn, 2015a). However, different dependence structures for X and Y can result in different distributions for Z so unlike the univariate threshold models, the distribution of Z cannot be expressed in a simple closed parametric form.

Regarding estimation, a simple approach is to use the false working assumption that Z has a normal distribution with parameters μ and σ^2. Under this assumption, maximisation of the induced pseudolikelihood yields consistent point estimates $\widehat{\alpha}$ and $\widehat{\beta}$. Subsequently, the distribution of Z is estimated nonparametrically via the empirical distribution function of

$$\widehat{Z} = \frac{Y - \widehat{\alpha}X}{X^{\widehat{\beta}}}, \quad \text{for } X > v.$$

Similarly with the univariate threshold exceedances model, asymptotic theory suggests that if v_0 is a valid threshold for model (22.3) to hold, then the parameter estimates of α and β should remain constant above v_0. Hence, standard practice is to choose as low a threshold as possible that satisfies the stability property.

Given the parameter estimates $\widehat{\alpha}$ and $\widehat{\beta}$, and the empirical distribution function of $\widehat{Z}$, the following algorithm allows the generation of N datapoints from the fitted bivariate model (22.3).

1. Let $i = 1$.
2. Simulate e from an exponential distribution and set $x_i = v + e$.

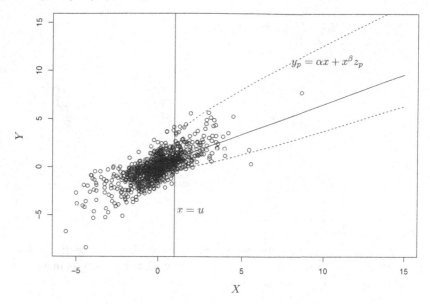

FIGURE 22.2
Simulated points from a standard bivariate normal distribution with correlation 0.75 shown in double exponential margins. The conditional dependence model (22.3) is fitted above the 90% quantile of the double exponential distribution, denoted by u, and the $0.025, 0.5$, and 0.975 model-based conditional quantile estimates are shown.

3. Sample z, independently of x_i, from the empirical distribution function of $\widehat{Z}$.
4. Set $y_i = \widehat{\alpha} x_i + x_i^{\widehat{\beta}} z$.
5. If $i < N$ go to step 1, otherwise output $(x_1, y_1), \ldots, (x_N, y_N)$.

As such, the functional $\Pr(Y \in C \mid X > v)$ can be estimated by repeating steps 1–5 and evaluating the estimate as the long-run proportion of the sample that falls in the set $C \in \mathbb{R}$. As far as confidence intervals are concerned, these can be obtained for the estimate of any functional using the following bootstrap algorithm: data generation under the fitted model, estimation of model parameters, and derivation of the estimate of the functional linked to extrapolation (Heffernan and Tawn, 2004).

Figure 22.2 shows conditional quantiles estimates from model (22.3) fitted to simulated data from the standard bivariate normal distribution with correlation 0.75. The data have been transformed to double exponential margins. Both estimates $\widehat{\alpha}$ and $\widehat{\beta}$ are greater than 0, indicating positive dependence and increasing variance as X increases. The model based conditional quantiles are given by $y_p = \alpha x + x^\beta z_p$, for $x > u$, where z_p is the pth quantile of Z.

Similarly with the univariate threshold model in Section 22.2.1, model (22.3) can be used to answer important questions such as: i) Is there a dose-response relationship in the extremes of ALT and TBL? ii) What is the probability of a patient exceeding 3 ULN$_{ALT}$ and 2 ULN$_{TBL}$?

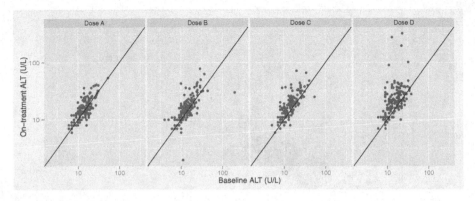

FIGURE 22.3
Plots of pretreatment values versus on-treatment values in each dose group. The reference lines are lines of identity, i.e., a point above the line indicates an increase from baseline and a point below the line indicates a decrease.

22.3   An Illustrative Analysis

A clinical trial collected data from patients receiving one of four different doses of an experimental drug. ALT, AST, ALP, and TBL measurements were collected from all patients at baseline (before treatment) and post baseline (after treatment). The data can be obtained from the R package texmex (Southworth and Heffernan, 2010). There were approximately 160 patients in each dose group. The doses are ordered such that Dose A is the lowest and Dose D is the highest.

The ULN for ALT, say ULN_{ALT}, is 36 international units per litre (U/L). A detailed analysis of this example is provided by Southworth and Heffernan (2012a). Figure 22.3 shows baseline (i.e., pre-treatment) values of ALT against the on-treatment values. It can be seen that there are some outliers in the on-treatment values, particularly in the patients receiving Dose D. However, there is only a handful of large observations, and it is not clear if there is a tendency for outliers to occur in some of the lower dose groups. How concerned should we be about the large ALT values in group D?

The baseline and on-treatment values are clearly related to each other, i.e., higher baselines values are associated with higher on-treatment values. So it makes sense to eliminate the systematic effect of baseline and apply extreme value modelling to any remaining variation. Before doing so, it is important to mention that as dose increases, there appears to be a considerable number of outliers in the response variable. As such, ordinary least squares estimates of the linear regression would be severely affected. As mentioned in Section 22.1.1, drug-induced changes in laboratory variables related to safety are expected to occur in the extremes and not in the location measures of the sampling distributions. It is for this reason that we regard

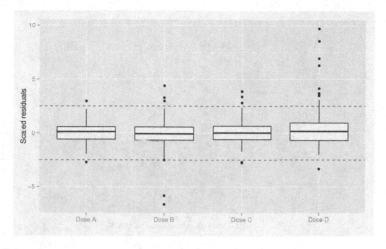

FIGURE 22.4
Boxplots of residuals from using robust regression to eliminate the baseline effect. If the residuals were from a normal distribution, we would expect them to mostly fall between about -2.5 and 2.5 (within the dashed lines). Clearly there are outliers indicating that the residuals come from a heavy-tailed distribution. A dose-response relationship now becomes more clear.

the positive association shown in Figure 22.3 as systematic and resort to more robust-to-outliers alternatives than ordinary least squares for adjusting it. In particular, we use the class of MM-estimators for linear regression with Tukey's biweight function (Yohai, 1987) which can be implemented in R using package MASS (Venables and Ripley, 2002). Plots of the residuals from the model, i.e., what remains after the baseline effect has been removed, are displayed by dose group in Figure 22.4. The same robust model was also applied to TBL and ALP.

Based on standard diagnostic techniques (Coles, 2001) on threshold selection and in fitting the generalized Pareto distribution (22.1) to the residuals displayed in Figure 22.4, it was found that i) the 70% quantile of the pooled residual ALT was a stable level for the analysis of the threshold exceedances for all dose levels and ii) the model that allowed the shape parameter to depend on log-dose was the best fit to the data. In order to interpret the model, we simulate values from it and then re-apply the baseline effect. In general, we will likely be interested in just how high ALT is likely to get on each dose if we treat increasing numbers of patients. To address this question, we compute the T-patient return level (22.2). Figure 22.5 displays the estimated 1000-patient return levels for ALT, i.e., the values of ALT likely to be exceeded only once every 1000 patients. It is important to note here that the uncertainty in the parameter estimation in the robust regression is masked by the uncertainty in the extreme value analysis.

In Figure 22.5, the dose-response effect is clearly visible and the model indicates that if 1000 patients were to take Dose D, we should expect one of them to have an ALT greater than $50 \times$ULN. Such high values of ALT can be observed in practice.

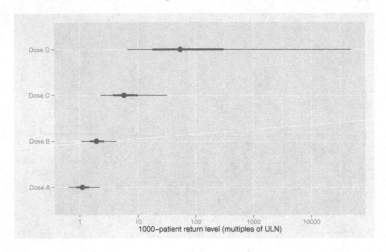

FIGURE 22.5
Estimated 1000-patient return levels for ALT at each dose. The heavy line segments are 50% confidence intervals, and the lighter segments are 90% intervals.

TABLE 22.1
Guideline toxicity grading scheme for ALT.

Threshold	Grade	Toxicity
ULN	1	Mild
$2\frac{1}{2} \times$ ULN	2	Moderate
$5 \times$ ULN	3	Severe
$20 \times$ ULN	4	Life threatening

However, the 90% interval estimate for Dose D contains values of ALT so high that it is unlikely they ever could be observed because either the patient would experience symptoms and stop taking the drug, the patient's doctor would notice the change in ALT and stop the drug, or the patient would die. Another way to assess the nature of the ALT effect that the drug appears to have is to consider the probability of breaches of certain thresholds. U.S. Food and Drug Administration (2008) provides the toxicity grading displayed in Table 22.1, so we use our extreme value and robust regression models to estimate the probabilities of these thresholds being exceeded. In Figure 22.6, the dose-response relationship that the drug appears to have on ALT is evident. Of particular note, the model suggests that Dose D could result in an $ALT > 20 \times$ ULN in approximately 1 in 400 patients, though there is no obvious cause for concern for Dose A.

In practice, it is known that the incidence of severe liver-related adverse events

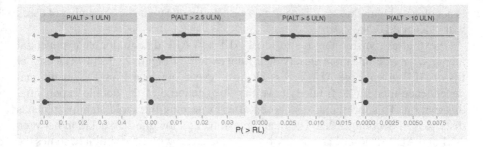

FIGURE 22.6
Estimated probabilities of ALT exceeding the thresholds in Table 22.1 with 90% confidence intervals. Note that the horizontal axes vary from plot to plot.

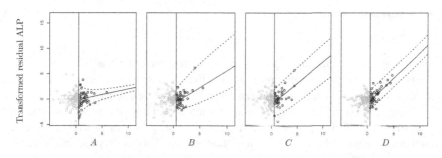

FIGURE 22.7
Conditional quantile estimates from conditional dependence model fitted to residual ALP given residual ALT exceeding the 70% quantile of the double exponential distribution. Plots show the 0.025 (dashed), 0.5 (solid), and 0.975 (dashed) conditional quantiles for doses A, B, C and D (left to right).

such as jaundice and hepatitis that occur with Dose D is approximately 1 in 500 patients whereas there is no evidence that such events occur with Dose A.

Published guidance (DCTD et al., 2003) suggests that liver toxicity is triggered by a combination of extreme ALT and TBL, provided that ALP remains at moderate levels so that it does not explain the elevation of TBL. Hence, a natural next step is to analyse the joint extremes of ALT and TBL. However, due to the limited sample size and the insufficient duration of the clinical trial (6 weeks only), extrapolation to the tail area that identifies drug-induced liver injury, i.e., ALT>3 ULN_{ALT} and TBL>2 ULN_{TBL}, is not feasible for the laboratory data. Southworth and Heffernan (2012b) and Papastathopoulos and Tawn (2015b) found some weak dose-response relationship for the probability of joint extreme elevations of ALT and TBL. However, TBL is not elevated with dose so this pattern is essentially due to the large number of cases with ALT$>$ 3 ULN$_{ALT}$.

Next, after transforming the residual ALT and ALP data to double exponential margins, we examine more closely their joint extremes. Figure 22.7 shows the model fits of ALP given large ALT, across all dose levels, obtained from the conditional dependence model (22.3) fitted above the 70% quantile of the double exponential distribution. This threshold was chosen after examining threshold stability plots as discussed in Section 22.2.2. The quantile estimates show mild association between ALT and ALP at Dose A and this association increases with dose until it becomes the strongest possible, i.e., At dose D, $(\alpha, \beta) = (1, 0)$, a case that is known as *asymptotic dependence* (Ledford and Tawn, 1996). A dose-response effect in the extremes of ALT and ALP is apparent.

22.4   Discussion

Clinical trial safety data are typically complicated, and the traditional statistical methods that are useful for the analysis of efficacy data are of little or no use in the context of safety. These observations have led to the belief that statisticians have little of value to add to the interpretation of such data. However, recent experiences suggest that the field is ripe for more statistical input, and clinical trial statisticians would benefit from looking further afield than the set of tools they are taught in typical medical statistics courses, or in the clinical trials textbooks.

Section 22.3 illustrates how extreme value techniques can lead to identification of interesting dose-response effects and patterns. In particular, ALT and ALP appear to be elevated with dose level whereas TBL remains moderate. Current biological understanding (U.S. Food and Drug Administration, 2008), however, suggests that a different combination of joint extremes is responsible for drug-induced liver injury, i.e., elevated ALT and TBL seen with moderate ALP, also known as *Hy's law*. Hence, the interesting finding of dose-response relationship between the extremes of ALT and ALP could possibly indicate that the drug is not toxic since elevated ALP might explain a subsequent TBL elevation. However, the drug analysed in Section 22.3 is known to be liver toxic in dose D and therefore, it is natural to ask whether this combination could also be an indication of drug-related altered liver behaviour.

Acknowledgments

The first author acknowledges funding from the SuSTaIn program Engineering and Physical Sciences Research Council grant EP/D063485/1 at the School of Mathematics, University of Bristol. This work was partially funded by AstraZeneca.

References

Coles, S. G. (2001). *An Introduction to Statistical Modelling of Extreme Values.* Springer.

Davison, A. C. and R. L. Smith (1990). Models for exceedances over high thresholds. *Journal of the Royal Statistical Society, Series B 53*, 393–442.

DCTD, NCI, NIH, and DHHS (2003). Cancer therapy evaluation program, common terminology criteria for adverse events, version 3.0. http://ctep.cancer.gov/protocolDevelopment/electronic_ applications/docs/ctcaev3.pdf.

Enas, G. G. and D. J. Goldstein (1995). Defining, monitoring and combining safety information in clinical trials. *Statistics in Medicine 14*(9), 1099–1111.

Heffernan, J. E. and J. Tawn (2004). A conditional approach for multivariate extreme values. *Journal of the Royal Statistical Society, Series B 56*, 497–546.

International Conference on Harmonisation (1998). Statistical principles for clinical trials, ICH topic E9. http://www.ich.org/ products/guidelines/efficacy/efficacy-single/article/ statistical-principles-for-clinical-trials.html.

Keef, C., I. Papastathopoulos, and J. A. Tawn (2013). Estimation of the conditional distribution of a multivariate variable given that one of its components is large: Additional constraints for the Heffernan and Tawn model. *Journal of Multivariate Analysis 115*, 396–404.

Ledford, A. W. and J. A. Tawn (1996). Statistics for near independence in multivariate extreme values. *Biometrika 83*(1), 169–187.

Nelsen, R. B. (1999). *An Introduction to Copulas.* Springer.

Papastathopoulos, I. and J. A. Tawn (2015a). Conditioned limit laws for inverted max-stable processes. arXiv:1402.1908.

Papastathopoulos, I. and J. A. Tawn (2015b). Stochastic ordering under conditional modelling of extreme values: Drug-induced liver injury. *Journal of the Royal Statistical Society, Series C 64*(2), 299–317.

Pickands, J. (1975). Statistical inference using extreme order statistics. *The Annals of Statistics 3*, 119–131.

Siddiqui, O. (2009). Statistical methods to analyze adverse events data of randomized clinical trials. *Journal of Biopharmaceutical Statistics 19*, 889–899.

Southworth, H. and J. E. Heffernan (2010). *texmex: Threshold Exceedances and Multivariate Extremes.* R package version 1.0.

Southworth, H. and J. E. Heffernan (2012a). Extreme value modelling of laboratory safety data from clinical studies. *Pharmaceutical Statistics 11*, 361–366.

Southworth, H. and J. E. Heffernan (2012b). Multivariate extreme value modelling of laboratory safety data from clinical studies. *Pharmaceutical Statistics 11*, 367–372.

U.S. Food and Drug Administration (1999). Medical reviewer's final report on troglitazone. http://www.accessdata.fda.gov/drugsatfda_docs/nda/99/20720S12S14_Rezulin_medr_P2.pdf.

U.S. Food and Drug Administration (2008). Guidance for industry drug-induced liver injury: Premarketing clinical evaluation. http://www.fda.gov/cder/guidance/7507dft.htm.

Venables, W. N. and B. D. Ripley (2002). *Modern Applied Statistics with S* (Fourth ed.). New York: Springer.

Xia, H. A., H. Ma, and B. P. Carlin (2011). Bayesian hierarchical modeling for detecting safety signals in clinical trials. *Journal of Biopharmaceutical Statistics 21*, 1006–1029.

Yohai, V. J. (1987). High breakdown-point and high efficiency robust estimates for regression. *The Annals of Statistics 15*(20), 642–656.

23

Analysis of Bivariate Survival Data Based on Copulas with Log-Generalized Extreme Value Marginals

Dooti Roy

University of Connecticut, United States

Vivekananda Roy

Iowa State University, United States

Dipak Dey

University of Connecticut, United States

Abstract

This chapter introduces a novel copula-based methodology to analyze right censored bivariate survival data using flexible log-generalized extreme value (GEV) marginals. Copulas are quite popular in high-dimensional data modeling as they allow modeling and estimating the parameters of the marginal distributions separately from the dependence parameter of the copula. The Clayton copula structure has been used to represent the dependency between the log individual survival times for this chapter. We propose an empirical Bayes (EB) method for estimating the dependency parameter which has several benefits over the other existing methodologies. The EB methodology is implemented using an efficient importance sampling scheme.

23.1 Introduction

During the last two decades, there has been a growing interest in modeling paired data or bivariate survival data which arises primarily from the field of medicine. For example, it has been medically observed that incidence of one disease often increases the risk of another in a patient affected by human immunodeficiency virus. Diabetic retinopathy is the leading cause of blindness in the United States. It has been observed that once one eye gets affected, the chance of the other eye contracting the disease increases many fold. Clayton (1978) and Oakes (1982) were the first to develop a fully parametric approach to model bivariate survival data. Huster et al. (1989) extended the fully parametric approach to include censored bivariate paired data with covariates. Since then, from year 1985 to year 1999, several methods using both parametric and semiparametric modeling techniques have been introduced and developed. Nelson (1986) in his landmark paper discussed simulation from a copula given uniform marginals. Nelson also derived the connection between the three well-known nonparametric measures of association, namely Kendall's τ, Pearson's correlation coefficient, and Spearman's rank correlation coefficient, with the copula parameter of association. Oakes (1986) provides an explicit formula for the limiting variance of the score function derived in Clayton (1978) and then extends the result for the case where the survival data is censored. Genest and Rivest (1993) discusses the issue of copula selection within the Archimedean family using a semiparametric method of moments based estimator. Wang and Wells (2000) later modified the model selection estimator proposed by Genest and Rivest (1993) for censored failure time data. The authors also adopted the graphical approach of goodness of fit proposed by Genest and Rivest (1993) for censored data. Shih and Louis (1995) wrote another flagship paper comparing a semiparametric with a fully parametric method of estimating the association parameter of the copula and the parameters concerning the respective marginal distribution of the survival times. We will discuss the estimation process using copula in greater detail later. Employing copula to model dependency between marginal survival data has been particularly popular since the univariate marginals do not depend on the choice of the dependency structure and can be consequently estimated separately from the dependency parameters.

Bayesian methods to model bivariate survival data started developing in the early 2000s. In Sahu and Dey (2000), the authors used a Bayesian frailty model to model bivariate survival data with covariates. Chen et al. (2002) introduced a Bayesian model for right censored bivariate survival data with cure fraction. Another interesting work is by Shemyakin and Youn (2001), where the authors develop a Bayesian joint survival model for life insurance data. More recently, Romeo and Tanaka (2006) used a copula-based Bayesian framework to model bivariate survival data.

In the analysis of bivariate survival data using a copula-based model it is important to estimate the copula dependence parameter as it measures the strength of association between the individual survival times. There are currently two available approaches to the estimation of these copula association parameters. Firstly, there is a two-step procedure as prescribed in Shih and Louis (1995). In the first step, independence between individual failure times is assumed and marginal survival parameters are estimated. In the second step, the dependence parameter is estimated assuming the estimated marginals are fixed. The other approach is the full Bayesian analysis, where a prior is assumed on the association parameter and all parameters including the marginal survival parameters are estimated from the joint posterior distribution. The problem with the first approach is the unnatural independence assumption. The set of values that the copula dependence parameter can take varies with the copula family and choosing an appropriate prior for the dependence parameter as required in the second approach is not intuitive. Since finding an appropriate prior directly for the copula association parameter is a difficult task, an alternative is to use a beta prior on the Kendall's τ as done in Romeo and Tanaka (2006).

Recently Roy (2014) used an empirical Bayes (EB) methodology with an efficient importance sampling procedure for selecting the link function parameter in a generalized linear model for binomial data. Following Roy (2014), in this chapter the copula association parameter is selected by maximizing its marginal likelihood function. This proposed EB approach avoids the difficulty of specifying a prior on the copula dependence parameters, which, as mentioned before, can be difficult. Also in the proposed methodology, there is no need to sample from the nonstandard conditional distributions of the copula dependence parameters, which is needed for obtaining Markov chain samples in the full Bayesian analysis. Further, the mixing of a Markov chain with updates on the association parameters can be slow.

Thus in this chapter, we propose a novel empirical Bayes method of estimating the dependence parameter for copula-based bivariate survival models. In particular, we use the Clayton copula structure for illustration of our methodology. We also propose new bivariate survival models where the flexible generalized extreme value (GEV) distribution is used to model the marginal distributions. As shown in Roy et al. (2013), many commonly used lifetime distributions such as the exponential, the Rayleigh, and the Weibull distributions are all special cases of the GEV distribution. The rest of this chapter is organized as follows: In Section 23.2, the concept of copula, its properties, and different types of copulas are discussed. In Section 23.3, the model settings are described. Our proposed method is outlined in detail in Section 23.4. Section 23.5 provides background and results of simulation studies. A real data analysis is provided in Section 23.6. The chapter concludes with limitations and future work directions.

23.2 A Brief Overview of Copula

Copula "couples" a joint cumulative density function to its marginal distributions and hence the name. Essentially a copula is a bivariate distribution with uniform marginals. The Sklar's theorem (Nelson, 1999, page 41), defines a copula structure in the following way:

Given a p-dimensional joint cumulative distribution function (cdf) F with marginal cdfs $F_1, F_2, \ldots, F_p$, there exists a p-copula C such that,

$$F(x_1, x_2, \ldots, x_p) = C(F_1(x_1), F_2(x_2), \ldots, F_p(x_p)) \qquad (23.1)$$

for all $(x_1, x_2, \ldots, x_p) \in \mathbb{R}^p$. If the marginal distributions are continuous, then C is unique, otherwise C is uniquely determined on Range $F_1 \times \cdots \times$ Range F_p. Conversely, if C is a p-copula and $F_1, F_2, \ldots, F_p$ are cdfs, then F defined by (23.1) is a p-dimensional cdf with marginal cdfs $F_1, F_2, \ldots, F_p$.

Nelson in his book *Introduction to Copulas (1999)* defined the properties of a p-dimensional copula on page 40:

1. For every $u = (u_1, u_2, \ldots, u_p) \in [0,1]^p$, $C(u) = 0$ if at least one element $u_i = 0$.
2. If all coordinates of u are 1, except u_i, then $C(u) = u_i$.
3. For every $a = (a_1, a_2, \ldots, a_p), b = (b_1, b_2, \ldots, b_p) \in [0,1]^p$, such that $a_i \leq b_i$ for all $i = 1, 2, \ldots, p$,

$$\Delta_{a_p}^{b_p} \Delta_{a_{p-1}}^{b_{p-1}} \cdots \Delta_{a_1}^{b_1} C(u) \geq 0, \tag{23.2}$$

where $\Delta_{a_k}^{b_k}$ defines the first-order differences as

$$\Delta_{a_k}^{b_k} C(u) = C(u_1, \ldots, u_{p-1}, b_k, u_{k+1}, \ldots, u_p) - C(u_1, \ldots, u_{k-1}, a_k, u_{k+1}, \ldots, u_p).$$

The first and third properties are satisfied if $C(u)$ is a distribution function. The second property is satisfied if the marginal distributions of C are uniform. Using copulas to model dependent bivariate data is advantageous for many reasons. Copula structure being extremely flexible, allows nonlinear dependence between the two associated marginal distributions, or to measure dependence for heavy tailed distributions. It can be used under fully parametric, semiparametric, or nonparametric settings and also allows for faster computations. Several families of copula have been studied in detail, for example: Gaussian copula, Clayton copula, Frank copula, Gumbel copula, etc. Copulas are chosen depending on the tail concentrations. In this chapter, Clayton copula is chosen to develop the framework and model the Diabetes Retinopathy Study data, since this copula has a nice relationship with the Kendall's τ. Further, Romeo and Tanaka (2006) have shown that Clayton copula performs reasonably well on the same real data.

23.3 Marginal Distributions and Bivariate Survival Model

23.3.1 Generalized Extreme Value Distribution as Marginal

Roy et al. (2013) introduced modeling univariate right censored survival data with a cure fraction using generalized extreme value (GEV) distribution (see also Roy and Dey, 2014). The GEV distribution characterized by three parameters is extremely flexible. As mentioned in the introduction, many commonly used lifetime distributions such as the Exponential, the Rayleigh and the Weibull are all special cases of the minima generalized extreme value distribution.

If we assume a GEV distribution for $\log T$, where T is a marginal survival time, i.e., $\log T \sim \text{GEV}(\mu, \sigma, \xi)$, then the corresponding density and survival functions can be defined respectively as:

$$f(t|\mu,\sigma,\xi) = \begin{cases} \frac{1}{\sigma t}(1 + \xi\frac{(\log t-\mu)}{\sigma})_+^{\frac{1}{\xi}-1} \exp[-(1 + \xi\frac{(\log t-\mu)}{\sigma})_+^{\frac{1}{\xi}}] & \text{if } \xi \neq 0 \\ \frac{1}{\sigma t}\exp(\frac{\log t-\mu}{\sigma})\exp[-\exp(\frac{\log t-\mu}{\sigma})] & 0 < t < \infty; \text{if } \xi = 0, \end{cases}$$

$$S(t|\mu,\sigma,\xi) = \begin{cases} \exp[-(1 + \xi\frac{(\log t-\mu)}{\sigma})_+^{\frac{1}{\xi}}] & \text{if } \xi \neq 0 \\ \exp[-\exp(\frac{\log t-\mu}{\sigma})] & \text{if } \xi = 0, \end{cases}$$

where $\mu \in \mathbb{R}$, $\sigma \in \mathbb{R}^+$, and $\xi \in \mathbb{R}$ are the location, scale, and shape parameters, respectively, and the function $(a)_+$ is defined to be a when $a > 0$ and 0 otherwise.

23.3.2 Bivariate Model

Let (T_1, T_2) denote failure times of two events for each subject or failure times of members of each group. Marginally we assume that

$$\log T_i \sim \text{GEV}(\mu_i, \sigma_i, \xi_i) \tag{23.3}$$

for $i = 1, 2$. The joint survival function based on a copula $C_\phi, \phi \in G$ is given by

$$S(t_1, t_2) = C_\phi(S_1(t_1), S_2(t_2)), \tag{23.4}$$

where $S_i(t_i) = S(t_i|\mu_i, \sigma_i, \xi_i)$ is the marginal survival function obtained from Section (23.3.1). The parameter ϕ measures the "intensity" of dependence between the individual failure times. We use the popular bivariate copula from the Clayton family which is given by

$$C_\phi(u_1, u_2) = \max[(u_1^{-\phi} + u_2^{-\phi} - 1)^{-1/\phi}, 0]. \tag{23.5}$$

In this case $G = (-1, \infty) \setminus \{0\}$. One of the major goals while studying bivariate data is to gauge the direction and extent of the association between the two variables. When linear correlation coefficient is not a valid measure of association due to nonlinearity of the association, Kendall's τ and Spearman's ρ are two most popular measures of pairwise concordance (Nelson, 1999, Chapter 5). Certain families of copula have very nice relationship with Kendall's τ. For example, in case of Clayton copula family, the Kendall's τ measure is given by

$$\tau_\phi = \frac{\phi}{\phi + 2} \in (-1, 1).$$

Due to the easy computation and interpretability, Clayton copula remains one of the widely used copula families. For proof of the above relationship, see Nelson (1999), pages 162–163.

23.3.3 Model Settings

Let T_{ij} (W_{ij}) be the survival (censoring) time of the jth event for the ith subject, $j = 1, 2; i = 1, 2, \ldots, n$. We assume (T_{i1}, T_{i2}) and (W_{i1}, W_{i2}) are independent for $i = 1, 2, \ldots, n$. We also assume that (T_{i1}, T_{i2}), are independently and identically distributed (i.i.d.) with joint probability density function (pdf) $f(t_1, t_2)$ and joint survival time $S(t_1, t_2)$, as defined in (23.4), $i = 1, 2, \ldots, n$. The observed data is $(\boldsymbol{y}_1, \boldsymbol{y}_2, \boldsymbol{\delta}_1, \boldsymbol{\delta}_2)$, where $\boldsymbol{y}_1 = (y_{11}, y_{21}, \ldots, y_{n1}), \boldsymbol{y}_2 = (y_{12}, y_{22}, \ldots, y_{n2}), \boldsymbol{\delta}_1 = (\delta_{11}, \delta_{21}, \ldots, \delta_{n1})$, and $\boldsymbol{\delta}_2 = (\delta_{12}, \delta_{22}, \ldots, \delta_{n2})$, with $y_{ij} = \min(T_{ij}, W_{ij})$ and $\delta_{ij} = 1(y_{ij} = T_{ij})$ for $j = 1, 2; i = 1, 2, \ldots, n$. Let $\boldsymbol{y} = (\boldsymbol{y}_1, \boldsymbol{y}_2)$ and $\boldsymbol{\delta} = (\boldsymbol{\delta}_1, \boldsymbol{\delta}_2)$. Let (S_1, S_2) and (f_1, f_2) be the marginal survival and density functions of GEV distribution as given in Section 3.2. Let $\boldsymbol{\theta}_1$ and $\boldsymbol{\theta}_2$ be the parameters associated with each of the marginal distribution. So, $\boldsymbol{\theta}_i = (\mu_i, \sigma_i, \xi_i), i = 1, 2$. where μ, σ, and ξ are location, scale, and shape parameters, respectively, of the marginal GEV distributions.

Then the complete data likelihood function (joint distribution) of the parameters $(\boldsymbol{\theta}_1, \boldsymbol{\theta}_2, \phi)$ can be written as (see also Chen, 2012):

$$
\begin{aligned}
L(\boldsymbol{\theta}_1, \boldsymbol{\theta}_2, \phi | \boldsymbol{y}, \boldsymbol{\delta}) = {} & \prod_{i=1}^{n} \{ f(y_{i1}, y_{i2}) \}^{\delta_{i1}\delta_{i2}} \left(\frac{\partial S(y_{i1}, y_{i2})}{\partial y_{i1}} \right)^{\delta_{i1}(1-\delta_{i2})} \\
& \left(\frac{\partial S(y_{i1}, y_{i2})}{\partial y_{i2}} \right)^{(1-\delta_{i1})\delta_{i2}} \{ S(y_{i1}, y_{i2}) \}^{(1-\delta_{i1})(1-\delta_{i2})} \\
= {} & \prod_{i=1}^{n} [c_\phi \{ S_{1\boldsymbol{\theta}_1}(y_{i1}), S_{2\boldsymbol{\theta}_2}(y_{i2}) \} f_{1\boldsymbol{\theta}_1}(y_{i1}), f_{2\boldsymbol{\theta}_2}(y_{i2}]^{\delta_{i1}\delta_{i2}} \\
& \left(-\frac{\partial C_\phi \{ S_{1\boldsymbol{\theta}_1}(y_{i1}), S_{2\boldsymbol{\theta}_2}(y_{i2}) \}}{\partial S_{1\boldsymbol{\theta}_1}(y_{i1})} \{ -f_{1\boldsymbol{\theta}_1}(y_{i1}) \} \right)^{\delta_{i1}(1-\delta_{i2})} \\
& \left(-\frac{\partial C_\phi (S_{1\boldsymbol{\theta}_1}(y_{i1}), S_{2\boldsymbol{\theta}_2}(y_{i2}))}{\partial S_{2\boldsymbol{\theta}_2}(y_{i2})} (-f_{2\boldsymbol{\theta}_2}(y_{i2})) \right)^{(1-\delta_{i1})\delta_{i2}} \\
& [C_\phi \{ S_{1\boldsymbol{\theta}_1}(y_{i1}), S_{2\boldsymbol{\theta}_2}(y_{i2}) \}]^{(1-\delta_{i1})(1-\delta_{i2})}, \quad (23.6)
\end{aligned}
$$

where $c_\phi(., .)$ is the second derivative of the copula function $C_\phi(., .)$ with respect to both $S_{1\boldsymbol{\theta}_1}(y_{i1})$ and $S_{2\boldsymbol{\theta}_2}(y_{i2})$ defined in (23.5).

23.4 Proposed Methodology

As mentioned in the introduction, there are problems with the current available approaches to the estimation of copula association parameters. In this section we develop an EB methodology based on an efficient importance sampling scheme for the estimation of ϕ.

Let $\pi(\boldsymbol{\theta}_1)$ and $\pi(\boldsymbol{\theta}_2)$ be the priors on $\boldsymbol{\theta}_1$ and $\boldsymbol{\theta}_2$. Then for fixed ϕ, the joint posterior density of θ_1 and θ_2 is given by,

$$\pi_\phi(\theta_1, \theta_2 | y, \delta) = \frac{L(\theta_1, \theta_2, \phi | y, \delta) \pi(\theta_1) \pi(\theta_2)}{m_\phi(y, \delta)}, \quad (\theta_1, \theta_2) \in \Theta \qquad (23.7)$$

where $m_\phi(y, \delta)$ is the normalizing constant given by

$$m_\phi(y, \delta) = \int_\Theta L(\theta_1, \theta_2, \phi | y, \delta) \pi(\theta_1) \pi(\theta_2) d\theta_1 d\theta_2,$$

Following Roy (2014), we select that value of $\widehat{\phi} \in G$ which maximizes the marginal likelihood of the data $m_\phi(y, \delta)$. Note that $m_\phi(y, \delta)$ is not available in closed form and needs to be estimated. Instead of maximizing $m_\phi(y, \delta)$, we choose to maximize $a \times m_\phi(y, \delta)$ as it is often easier to estimate. We choose "a" as $1/m_{\phi_1}(y, \delta)$ for a pre fixed value ϕ_1. Then by ergodic theorem, we have a simple importance sampling consistent estimator of $B_{\phi, \phi_1} := m_\phi(y, \delta)/m_{\phi_1}(y, \delta)$:

$$\frac{1}{N} \sum_{l=1}^{N} \frac{L(\theta_1^{(l)}, \theta_2^{(l)}, \phi | y, \delta)}{L(\theta_1^{(l)}, \theta_2^{(l)}, \phi_1 | y, \delta)} \xrightarrow{a.s.} \int_\Theta \frac{L(\theta_1, \theta_2, \phi | y, \delta)}{L(\theta_1, \theta_2, \phi_1 | y, \delta)} \pi_{\phi_1}(\theta_1, \theta_2 | y, \delta) d\theta_1 d\theta_2$$

$$= \frac{m_\phi(y, \delta)}{m_{\phi_1}(y, \delta)}, \qquad (23.8)$$

as $N \to \infty$ where $\{\theta_1^{(l)}, \theta_2^{(l)}\}_{l=1}^N$ is a single Harris ergodic Markov chain with stationary density $\pi_{\phi_1}(\theta_1, \theta_2 | y, \delta)$. So one *single* MCMC sample $\{\theta_1^{(l)}, \theta_2^{(l)}\}_{l=1}^N$ from $\pi_{\phi_1}(\theta_1, \theta_2 | y, \delta)$ is used to estimate B_{ϕ, ϕ_1} for all $\phi \in G$. Once the association parameter $\widehat{\phi}$ is estimated, θ_1 and θ_2 are estimated using Markov chain (Gibb's sampler) samples from $\pi_{\widehat{\phi}}(\theta_1, \theta_2 | y, \delta)$.

23.4.1 A Two-Stage Importance Sampling Procedure

The above estimate of B_{ϕ, ϕ_1} although simple, is often unstable, in particular, when ϕ is not close to ϕ_1 since then the ratio $L(\theta_1^{(l)}, \theta_2^{(l)}, \phi | y, \delta)/L(\theta_1^{(l)}, \theta_2^{(l)}, \phi_1 | y, \delta)$ can take on very large values for some $(\theta_1^{(l)}, \theta_2^{(l)})$ (Geyer, 1994). To remove this instability, following Geyer (1994), Doss (2010), and Roy (2014) here we consider a generalized importance sampling method. Let $\phi_1, \phi_2, \ldots, \phi_m \in G$ be m appropriately chosen skeleton points. Let $\{\theta_1^{(j;l)}, \theta_2^{(j;l)}\}_{l=1}^{N_j}$ be a Markov chain with stationary density $\pi_{\phi_j}(\theta_1, \theta_2 | y, \delta)$ for $j = 1 \ldots, m$. Define $r_k = m_{\phi_k}(y, \delta)/m_{\phi_1}(y, \delta)$ for $k = 2, 3, \ldots, m$, with $r_1 = 1$. Then B_{ϕ, ϕ_1} is consistently estimated by

$$\widehat{B}_{\phi, \phi_1} = \sum_{j=1}^{m} \sum_{l=1}^{N_j} \frac{L(\theta_1^{(j;l)}, \theta_2^{(j;l)}, \phi | y, \delta)}{\sum_{k=1}^{m} N_k L(\theta_1^{(j;l)}, \theta_2^{(j;l)}, \phi_k | y, \delta)/\widehat{r}_k}, \qquad (23.9)$$

where $\widehat{r}_1 = 1$, $\widehat{r}_k$, $k = 2, 3, \ldots, m$ are consistent estimators of r_k's obtained by the "reverse logistic regression" method proposed by Geyer (1994) described in detail in the next section. See Roy et al. (2014) for a discussion on how to choose the skeleton points.

23.4.1.1   Estimation of r_k

Following Geyer (1994), $\widehat{r}_k$ are estimated by maximizing the quasi-likelihood function,

$$l_n(r) = \sum_{j=1}^{m} \sum_{l=1}^{N_j} \log p_j(X_{jl}, r), \qquad (23.10)$$

where $X_{jl} = (\boldsymbol{\theta_1}^{(j;l)}, \boldsymbol{\theta_2}^{(j;l)})$, $r = (r_1, r_2, \ldots, r_m)$, N_j is the number of MCMC samples with stationary density $\pi_{\phi_j}(\boldsymbol{\theta_1}, \boldsymbol{\theta_2}|\boldsymbol{y}, \boldsymbol{\delta})$ corresponding to the skeleton point $\phi_j, j = 1, 2, \ldots, m$ and the function $p_j(\cdot, \cdot)$ is defined below. To avoid identifiability issues, $\widehat{r}_1$ is assumed to be 1. The function $p_j(\cdot, \cdot)$ is defined as,

$$p_j(X_{jl}, r) = \frac{L(\boldsymbol{\theta_1}^{(j;l)}, \boldsymbol{\theta_2}^{(j;l)}, \phi_j|\boldsymbol{y}, \boldsymbol{\delta})e^{r_j}}{\sum\limits_{k=1}^{m} L(\boldsymbol{\theta_1}^{(j;l)}, \boldsymbol{\theta_2}^{(j;l)}, \phi_k|\boldsymbol{y}, \boldsymbol{\delta})e^{r_k}}, \qquad (23.11)$$

where ϕ_k is the kth skeleton point, $\theta_i^{(j;l)} = (\mu_i^{(j;l)}, \sigma_i^{(j;l)}, \xi_i^{(j;l)})$, for $i = 1, 2$ and $(\theta_1^{(j;l)}, \theta_2^{(j;l)})$ is the lth MCMC sample of the chain with stationary density $\pi_{\phi_j}(\boldsymbol{\theta_1}, \boldsymbol{\theta_2}|\boldsymbol{y}, \boldsymbol{\delta})$ corresponding to the skeleton point $\phi_j, j = 1, 2, \ldots, m$ and $l = 1, \ldots, N_j$.

Hence the two-stage procedure for estimating B_{ϕ, ϕ_1} and $\widehat{\phi}$ are as follows. In the first-stage, we run MCMC chains $\{\boldsymbol{\theta_1}^{(j;l)}, \boldsymbol{\theta_2}^{(j;l)}\}_{l=1}^{N_j}$ with invariant density $\pi_{\phi_j}(\boldsymbol{\theta_1}, \boldsymbol{\theta_2}|\boldsymbol{y}, \boldsymbol{\delta})$, $j = 1, 2, \ldots, m$. We estimate r using these first stage posterior samples and the reverse logistic method as described in Section 23.4.1.1. In the second stage, independent of the first-stage samples, we obtain *new* samples $\{\boldsymbol{\theta_1}^{(j;l)}, \boldsymbol{\theta_2}^{(j;l)}\}_{l=1}^{N_j}$ from $\pi_{\phi_j}(\boldsymbol{\theta_1}, \boldsymbol{\theta_2}|\boldsymbol{y}, \boldsymbol{\delta})$, $j = 1, 2, \ldots, m$ to estimate B_{ϕ, ϕ_1} using Equation (23.9). Finally, the estimate $\widehat{\phi}$ where $\widehat{B}_{\phi, \phi_1}$ attains maximum, can be obtained by an optimization procedure or simply by looking at the plot of $\widehat{B}_{\phi, \phi_1}$ against ϕ. As mentioned before, once $\widehat{\phi}$ is obtained, inference on $\boldsymbol{\theta_1}$ and $\boldsymbol{\theta_2}$ is made by getting samples from $\pi_{\widehat{\phi}}(\boldsymbol{\theta_1}, \boldsymbol{\theta_2}|\boldsymbol{y}, \boldsymbol{\delta})$.

23.5   Simulation Study

23.5.1   Generating Data

R package "copula" was used to generate samples from bivariate distributions with uniform$(0, 1)$ marginals. Let (u_1, u_2) denote the two random variables where marginally $u_i \sim$ uniform$(0, 1), i = 1, 2$. The data has dependency according to the previously mentioned Clayton copula structure. Next we used "inverse survival function" approach mentioned below to get survival times generated from the Clayton copula-based bivariate distribution with GEV$(\mu, 1, \xi_i), i = 1, 2$ marginals. Let

(t_1, t_2) denote the generated paired survival times. A single covariate X, was introduced through the mean parameter $\mu = \beta_0 + \beta_1 * X$.

Case I: $\xi \neq 0$

$$1 - u_i = S(t_i)$$

$$= \exp[-(1 + \xi(\log t_i - \mu_i))_+^{\frac{1}{\xi}}] \tag{23.12}$$

$$\Rightarrow t_i = \exp[\frac{-1 + (-\log(1 - u_i))^{\xi}}{\xi} + \mu_i]. \tag{23.13}$$

Case II: $\xi = 0$

$$1 - u_i = S(t_i)$$

$$= \exp[-\exp(\log(t_i) - \mu_i)/\sigma)] \tag{23.14}$$

$$\Rightarrow t_i = \exp[\sigma \log(-\log(1 - u_i)) + \mu_i]. \tag{23.15}$$

A random sample of 200 observations was generated following the above method. The true values of the parameters that we used were $\xi_1 = \xi_2 = 0.1$, $\beta_{01} = \beta_{02} = 2$, $\beta_{11} = \beta_{12} = 0.5$, and $\phi = 1$. These choices of ξ_1 and ξ_2 were made keeping in mind that in practical scenarios ξ rarely exceeds $[-0.5, 0.5]$ (see Castillo and Hadi, 1997; Smith, 1985). The survival times t_1 and t_2 were censored at random so that the censoring percentages in both t_1 and t_2 were about 11%. It was observed through simulation that higher censoring sometimes leads to biased estimates of the coefficients of the covariates. Further study is required to address this issue.

23.5.2 Estimation of Parameters

To estimate ϕ, we first used the naive importance sampling method (23.8) mentioned in Section 23.4 with several values of ϕ_1. A uniform prior of $(-1, 1)$ was considered for ξ. For the intercept and the slope parameters, diffused normal priors were considered with 0 mean and 400 variance. MCMC samples of size 20,000 were generated from the posterior density $\pi_{\phi_1}(\xi_1, \xi_2 | y, \delta)$ and the first 10,000 were discarded as burn-in. The results of the simulation are displayed in Table 23.1. It is observed for values of ϕ_1 close to the true value of ϕ, the naive estimate is a good approximation. However, when the value of ϕ_1 is chosen far away from the true value of ϕ, the estimates are not robust. The real data analysis in Section 23.6 also shows problems with the naive importance sampling method. Table 23.2 provides Bayesian estimates of the marginal parameters for certain values of ϕ_1 along with their standard errors and highest posterior density intervals. As the value of ϕ_1 moves away from the true value of ϕ, the marginal estimates also deteriorate.

Next, we use the two-stage importance sampling scheme mentioned in Section 23.4.1. In order to use our multichain importance sampling method, seven diverse skeleton points (0.01, 0.1, 0.3, 0.8, 1.5, 3, 4.5) were selected keeping the true value of ϕ. For each of the four ϕ values in the skeleton set, 20,000 MCMC samples were generated and the first 10,000 were discarded as burn-in and then the chain was thinned by every 10th sample to adjust for the autocorrelation. Using these MCMC

TABLE 23.1
Naive importance sampling: Estimate of ϕ for simulated data.

ϕ_1	$\widehat{\phi}$	$\widehat{\tau}$
0.3	0.845	0.297
0.8	0.843	0.296
1.5	0.863	0.301
3.0	1.234	0.382
4.5	1.822	0.477
6.0	2.546	0.560
8.0	3.276	0.621

samples, following Geyer (1994), r_k, $k = 2, 3, 4, 5, 6, 7$ were estimated (we assume $r_1 = 1$). Then using new MCMC samples and the $\widehat{r}_k$ values obtained in the first stage, estimates of B_{ϕ, ϕ_1} and hence $\widehat{\phi}$ were obtained. Confidence intervals were empirically obtained for $\widehat{\phi}$ and $\widehat{\tau}$ from 100 independent MCMC simulations. The final $\widehat{\phi}$ is the average of the 100 estimated $\widehat{\phi}$s. Finally fixing ϕ as $\widehat{\phi}$, marginal parameters ξ_1, ξ_2, and the coefficients of the covariate X were estimated with fresh MCMC samples from $\pi_{\widehat{\phi}}(\beta_{01}, \beta_{11}, \beta_{02}, \beta_{12}, \xi_1, \xi_2 | \boldsymbol{y}, \boldsymbol{\delta})$. Table 23.3 and Table 23.4 provide the details.

23.5.3   Note on Monte Carlo Simulation

Since the extreme value distributions have interdependent parameters, i.e., the criterion: $1 + \xi(\log t - \mu)/\sigma > 0$ must always hold, when performing Monte Carlo simulations, there is a need to specify bounds for each parameter contained in the likelihood function. When conducting a real data analysis we assume that both the marginal GEV's have three unknown parameters: μ, σ, and ξ in order to allow the data to have maximum flexibility. Also more often than not there is one or more associated covariate. Let us assume that there is one covariate Z for simplicity. We introduce the covariate through the parameter μ as $\mu = \beta_0 + \beta_1 Z$. So overall we will have 8 parameters in the likelihood function: $\sigma_1, \xi_1, \sigma_2, \xi_2, \beta_{01}$ (intercept for log of censored survival time y_1), β_{11} (slope for log of censored survival time y_1), β_{02} (intercept for log of censored survival time y_2), β_{12} (slope for log of censored survival time y_2).

We define the bounds for each of the parameters in the following way:

For ξ, define $\nu_i = \log y_i - \mu_i$, $i = 1, .., n$. We assume n is the sample size and the standard assumption that $\sigma > 0$ also holds. The parameters ξ_1 and ξ_2 will have similar bounds. Then,

$$\max(-1, \max_{1 \le i \le n} [-\frac{\sigma}{\nu_i} 1_{\nu_i > 0}]) < \xi < \min(1, \min_{1 \le i \le n} [-\frac{\sigma}{\nu_i} 1_{\nu_i < 0}]).$$

For σ, define $\zeta_i = \xi(\log y_i - \mu_i)$, $i = 1, ..., n$. As before, parameters σ_1 and σ_2 will

TABLE 23.2
Naive importance sampling: Posterior estimates of marginal parameters for simulated data.

ϕ_1	Parameter	Estimate [S.E.]	95% HPD
1.5	ξ_1	0.148[0.061]	(0.024, 0.255)
	ξ_2	0.146[0.060]	(0.028, 0.259)
	β_{01}	1.725[1.464]	(1.417, 1.975)
	β_{11}	0.219[0.261]	(-0.289, 0.715)
	β_{02}	1.672[0.148]	(1.396, 1.965)
	β_{12}	0.262[0.263]	(-0.258, 0.778)
4.50	ξ_1	0.122[0.057]	(0.008,0.226)
	ξ_2	0.138[0.052]	(0.037,0.231)
	β_{01}	1.954[0.135]	(1.694, 2.219)
	β_{11}	0.218[0.243]	(-0.293, 0.653)
	β_{02}	1.872[0.132]	(1.622, 2.144)
	β_{12}	0.295[0.244]	(-0.157, 0.796)
6.00	ξ_1	0.087[0.055]	(-0.018,0.197)
	ξ_2	0.119[0.049]	(0.017,0.208)
	β_{01}	2.127[0.131]	(1.872, 2.384)
	β_{11}	0.228[0.232]	(-0.224, 0.654)
	β_{02}	2.037[0.132]	(1.775, 2.227)
	β_{12}	0.303[0.235]	(-0.159, 0.754)
8.00	ξ_1	0.053[0.056]	(-0.064,0.153)
	ξ_2	0.087[0.051]	(-0.009,0.184)
	β_{01}	2.293[0.125]	(2.051, 2.519)
	β_{11}	0.213[0.227]	(-0.197, 0.679)
	β_{02}	2.193[0.124]	(1.964, 2.430)
	β_{12}	0.313[0.226]	(-0.154, 0.725)

TABLE 23.3
Generalized importance sampling: Estimate of ϕ for simulated data.

Parameter	Estimate[S.E.]	95% CI
ϕ	1.444[0.207]	(1.312, 2.004)
τ	0.419[0.031]	(0.396, 0.500)

have similar bounds. Then,

$$\sigma > \max(0, \max_{1 \le i \le n}(-\zeta_i)).$$

For the intercept β_0, we define $A_i = \log y_i + \sigma/\xi - \beta_1 Z_i$. For $\xi > 0$, $\beta_0 \in (-\infty, \min A_i)$. For $\xi < 0$, $\beta_0 \in (\max A_i, \infty)$. For the slope β_1, we define $B_i = (\log y_i + \sigma/\xi - \beta_0)/Z_i$. For $\xi > 0$, $\beta_1 \in (-\infty, \min B_i)$. For $\xi < 0$,

TABLE 23.4
Generalized importance sampling: Posterior estimates of marginal parameters for simulated data.

Parameter	Estimate[S.E.]	95% HPD
ξ_1	0.135[0.062]	(0.016, 0.252)
ξ_2	0.152[0.051]	(0.057, 0.248)
β_{01}	1.864[0.146]	(1.582, 2.155)
β_{11}	0.203[0.254]	(-0.270, 0.724)
β_{02}	1.795[0.145]	(1.509, 2.077)
β_{12}	0.259[0.254]	(-0.242, 0.763)

$\beta_1 \in (\max B_i, \infty)$. Note that these bounds hold for positive Z_i, $i = 1, ..., n$. Similar bounds can be obtained in case the covariate takes negative values.

23.5.4   Note on Implementation of the Algorithm

In practice, often log likelihood function is used instead of likelihood for computational issues. While estimating r_k or B_{ϕ,ϕ_1} function in the estimation procedure, using likelihood can be a deterrent due to the large magnitude of the evaluated likelihood. One way around this problem is to deal with log likelihood and consider maximizing:

$$l_n(r) = \sum_{j=1}^{m} \sum_{l=1}^{N_j} \log p_j(X_{jl}, r) \tag{23.16}$$

where,

$$
\begin{aligned}
p_j(X_{jl}, r) &= \cfrac{1}{\sum\limits_{k=1}^{m} \cfrac{e^{L^*(\theta_1{}^{(j;l)}, \theta_2{}^{(j;l)}, \phi_k | y, \delta)} e^{r_k}}{e^{L^*(\theta_1{}^{(j;l)}, \theta_2{}^{(j;l)}, \phi_j | y, \delta)} e^{r_j}}} \\[2mm]
&= \cfrac{1}{\sum\limits_{k=1}^{m} \cfrac{e^{L^*(\theta_1{}^{(j;l)}, \theta_2{}^{(j;l)}, \phi_k | y, \delta)} e^{r_k}}{e^{L^*(\theta_1{}^{(j;l)}, \theta_2{}^{(j;l)}, \phi_j | y, \delta)} e^{r_j}}} \tag{23.17} \\[2mm]
&= \left[\sum_{k=1}^{m} e^{L_k^* - L_j^*} e^{r_k - r_j} \right]^{-1},
\end{aligned}
$$

$L_k^* = L^*(\theta_1{}^{(j;l)}, \theta_2{}^{(j;l)}, \phi_k | y, \delta) = \log L(\theta_1{}^{(j;l)}, \theta_2{}^{(j;l)}, \phi_k | y, \delta)$. Depending on the converged MCMC chains, the quantity $L_k^* - L_j^*$ should not be overly large and hence the maximization can be carried out. Similar techniques can be applied to the estimation and subsequent maximization of the function B_{ϕ,ϕ_1} while implementing the second stage of the proposed estimation procedure of ϕ.

23.6   Application to Diabetes Retinopathy Study Data

We consider the well-known Diabetes Retinopathy Study (DRS) data for our analysis. This data was first introduced by Huster et al. (1989) in their seminal paper on modeling bivariate data with covariates. Diabetes retinopathy is the leading cause of blindness in American adults, accounting for approximately 12% of all new cases. It has been shown that 90% of the patients who have been diabetic for more than a decade will eventually develop retinopathy, or an ocular manifestation of blindness. The Diabetes Retinopathy Study (DRS) data consists of 197 patients with severe retinopathy affecting the eye. For each patient, the data records a treated eye and a control, untreated eye. Time from detection until blindness is the response variable. One or both eyes may be censored. The censoring happens in approximately 79% of the treated eyes and 49% of the untreated eyes. Age at the onset of diabetes is considered as a covariate.

Huster et al. (1989) analyzed the data using a frequentist approach with the Clayton copula family and Weibull marginal distributions. Therneau and Grambsch (2000) considered a random effects frailty model and Sahu and Dey (2000) considered exponential and Weibull bivariate distributions with a Bayesian approach. Romeo and Tanaka (2006) considered copulas to model the dependency of the data and follow the same two-step procedures outlined by Shih and Louis (1995).

Let us assume the time to blindness for the jth eye of the ith patient, $i = 1, ..., 197$, is t_{ij}. Then $\log t_{ij} \sim \text{GEV}(\mu_{ij}, \sigma_j, \xi_j)$, $j = 1, 2$. Note that we do not directly observe t_{ij}; instead we observe $y_{ij} = \min(t_{ij}, w_{ij})$, where w_{ij} is the censoring time. The covariate "age at the time of diagnosis" (Z) is included through $\mu_{ij} = \beta_{0j} + \beta_{1j} Z_i$, $j = 1, 2$. For simplicity, independent priors are considered on σ, ξ and β. Since we know from Section 23.5.3, that the values of the marginal parameters σ, ξ, and β are all interdependent, we adjust for the dependency by restricting the proposed parameter space within the Monte Carlo simulations. Also, we take $\sigma^2 \sim \text{Inv Gamma}(a_\sigma, b_\sigma)$ and $\xi \sim \text{uniform}(-a, a)$. We pick $a = 1$, $a_\sigma = 0.01, b_\sigma = 2$. We put diffuse priors for β_0's and β_1's as N(0, 100). The choices of the hyper parameters are such that the posterior distribution is proper.

We first use the simple importance sampling method outlined in Section 23.4, with $\phi_1 = 3$ for estimating ϕ. We use the Gibbs sampler with Metropolis–Hastings steps to get the samples from the posterior density $\pi_{\phi_1}(\theta_1, \theta_2 | y, \delta)$. We run the MCMC chain 10,000 iterations. Convergence was checked using the trace plots, ergodic mean plots, and the autocorrelation plots for all the parameters. Consequently, we find that 4,000 iterations are adequate as burn-in and every 5th sample was chosen after thinning. Using these 1,200 posterior samples, B_{ϕ,ϕ_1} values are estimated and subsequently maximized to obtain $\widehat{\phi}$. Table 23.5 provides the estimates for the dependence parameter and the concordance measure. However, as Figure 23.1 demonstrates the estimates of ϕ obtained from the naive importance sampling are not stable and the function B_{ϕ,ϕ_1} achieves its maximum at different points for different values of ϕ_1. This instability becomes more pronounced as higher values of ϕ_1 are used.

TABLE 23.5
Naive importance sampling: Estimates of ϕ and τ for DRS data.

Parameter	Estimate
ϕ	1.065
τ	0.347

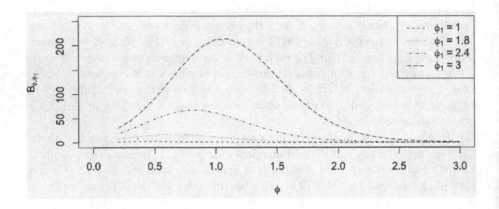

FIGURE 23.1
Plot of B_{ϕ,ϕ_1} against ϕ.

TABLE 23.6
Estimated $\hat{r}_j$ for DRS data.

r	r_1	r_2	r_3	r_4
$\hat{r}$	1	0.679	1.445	7.867

   In order to obtain stable estimates of the model parameters, following the method outlined in Section 23.4.1, four skeleton points were chosen as $\phi_1 = 0.2$, $\phi_2 = 0.5$, $\phi_3 = 1$, $\phi_4 = 3$. MCMC samples of size 10,000 were generated for each skeleton point and 4,000 were considered as burn-in after checking for convergence and the chains were thinned every 5th sample. The maximum likelihood estimates of the marginal parameters given the value of skeleton point ϕ were estimated by maximizing the log likelihood function and were considered as initial starting values for the MCMC simulations. Once all four MCMC chains were obtained, the function (23.10) was maximized to obtain estimated $\hat{r}_j$'s. The estimated values of $\hat{r}_j$ were quite robust to the change in initial starting values of the MCMC chains. Table 23.6 provides the values of $\hat{r}_j$.

   Using the estimated $\hat{r}_j$ s, $\widehat{B}_{\phi,\phi_1}$ was maximized as a function of ϕ and the final estimate $\hat{\phi}$ was obtained; 100 independent MCMC simulations were conducted and

TABLE 23.7

Generalized importance sampling: Empirical estimate of ϕ and τ for DRS data.

Parameter	Estimate[S.E.]	95% CI
ϕ	2.017[0.042]	(1.945, 2.098)
τ	0.502[0.005]	(0.493, 0.512)

TABLE 23.8

Generalized importance sampling: Posterior estimates of marginal parameters for DRS data.

Parameter	Estimate[S.E.]	95% HPD
σ_1	1.176[0.196]	(0.830, 1.599)
ξ_1	0.202[0.160]	(-0.139, 0.455)
σ_2	1.524[0.289]	(1.032, 2.082)
ξ_2	0.179[0.188]	(-0.229, 0.486)
β_{01}	2.726[0.351]	(1.983, 3.404)
β_{11}	0.022 [0.015]	(-0.005, 0.054)
β_{02}	2.712[0.454]	(1.754, 3.571)
β_{12}	0.017[0.020]	(-0.017, 0.0561)

the final $\widehat{\phi}$ is the average of the 100 estimated $\widehat{\phi}$s. Finally, using the estimated $\widehat{\phi}$, the marginal parameters were determined using Gibb's sampling algorithm as done before in Section 23.5. Table 23.7 provides the estimates of the dependence parameter ϕ and the concordance measure Kendall's τ while Table 23.8 provides the final estimates of the marginal parameters of the model.

The final estimate $\widehat{\phi}$ is quite different from the simple importance sampling estimator obtained in Table 23.5. The stability of the final estimated ϕ using our two-stage importance sampling scheme was tested by including additional $\phi_1 = 2.5, 0.8, 3.5, 4, 4.5$ as skeleton points and dropping one or more ϕ_1 as a skeleton point. The final estimate of ϕ obtained in the process was quite close to the estimated $\widehat{\phi}$. The standard error of the estimated $\widehat{\phi}$ and $\widehat{\tau}$ is calculated empirically from 100 independent MCMC simulations. As seen from Table 23.8, the final estimates of ϕ are quite robust. In Romeo and Tanaka (2006), the authors found the estimated $\widehat{\phi}$ to be approximately 1.061. Our estimated $\widehat{\phi}$ is higher but in the same direction. The estimated $\widehat{\tau}$ is 0.502. It can be thus concluded that the marginal time to blindness for the patients' eyes are indeed positively associated. The estimates of slopes of predictor Age are not significant for both the marginals. This finding agrees with Romeo and Tanaka (2006).

23.7   Discussion

A new bivariate survival model using copula with flexible log-generalized extreme value marginals is proposed in this chapter. Since the construction of appropriate priors for the association parameter of the copula may be difficult, an empirical Bayes methodology is developed for making inference on these parameters by maximizing the marginal likelihood function. The proposed method will work on any bivariate data and is not restricted to survival data or to a particular family of copulas. A future direction of work includes extending the proposed methodology for multivariate responses. If the multivariate copula has many association parameters, use of the proposed methodology for estimating these parameters seems challenging. Detailed goodness of fit tests are also needed to assess the model performance. We consider this in our related future project.

Acknowledgment

The authors would like to thank the editor and two anonymous reviewers for their careful reading and helpful comments that have improved the chapter.

References

Castillo, E. and A. Hadi (1997). Fitting generalized Pareto distribution to data. *Journal of American Statistical Association 92*, 1609–1620.

Chen, M. H., J. Ibrahim, and D. Sinha (2002). Bayesian inference for multivariate survival data with a cure fraction. *Journal of Multivariate Analysis 80*, 101–126.

Chen, Z. (2012). *A Flexible Copula Model for Bivariate Survival Data*. Ph. D. thesis, Department of Biostatistics and Computational Biology, School of Medicine and Dentistry, University of Rochester.

Clayton, D. G. (1978). A model for association in bivariate life tables and its application in epidemiological studies of familial tendency in chronic disease coincidence. *Biometrika 65*, 141–151.

Doss, H. (2010). Estimation of large families of Bayes factors from Markov chains output. *Statistica Sinica 20*, 537–560.

Genest, C. and L. Rivest (1993). Statistical inference procedures for bivariate Archimedean copulas. *Journal of the American Statistical Association 88*, 1034–1043.

Geyer, C. J. (1994). Estimating normalizing constants and reweighting mixtures in Markov chain Monte Carlo. Technical Report 568, School of Statistics, University of Minnesota.

Huster, W. J., R. Brookmeyer, and S. G. Self (1989). Modelling paired survival data with covariates. *Biometrics 45*, 145–156.

Nelson, R. B. (1986). Properties of a one-parameter family of bivariate distributions with specified marginals. *Communications in Statistics, Part A 153*, 3277–3285.

Nelson, R. B. (1999). *An Introduction to Copulas*. Springer.

Oakes, D. (1982). A model for association in bivariate survival data. *Journal of Royal Statistical Society B 44*, 414–428.

Oakes, D. (1986). Semiparametric inference in a model for association in bivariate survival data. *Biometrika 73*, 353–361.

Romeo, J. and N. Tanaka (2006). Bivariate survival modeling: A Bayesian approach based on copulas. *Lifetime Data Analysis 12*, 205–222.

Roy, D., V. Roy, and D. Dey (2013). Analysis of survival data with a cure fraction under generalized extreme value distribution. Techical report, Department of Statistics, Iowa State University.

Roy, V. (2014). Efficient estimation of the link function parameter in a robust Bayesian binary regression model. *Computational Statistics and Data Analysis 73*, 87–102.

Roy, V. and D. Dey (2014). Propriety of posterior distributions arising in catgorical and survival models under generalized extreme value distribution. *Statistica Sinica 2*, 699–722.

Roy, V., E. Evangelou, and Z. Zhu (2014). Efficient estimation and prediction for the Bayesian binary spatial model with flexible link functions. Techical report, Department of Statistics, Iowa State University.

Sahu, S. K. and D. Dey (2000). A comparison of frailty and other models for bivariate survival data. *Lifetime Data Analysis 6*, 207–228.

Shemyakin, A. and H. Youn (2001). Bayesian estimation of joint survival functions in life insurance. In P. Nanopoulos and G. I. Edward (Eds.), *Monographs of Official Statistics. Bayesian Methods with Applications to Science, Policy and Official Statistics, European Communities*, Volume 489–496. Office for Official Publ. of the European Communities.

Shih, J. H. and T. A. Louis (1995). Inferences on the association parameter in copula models for bivariate survival data. *Biometrics 51*, 1384–1399.

Smith, R. L. (1985). Maximum likelihood estimation in a class of non-regular cases. *Biometrika 72*, 67–90.

Therneau, T. M. and P. Grambsch (2000). *Modeling Survival Data: Extending the Cox Model*. New York: Springer.

Wang, W. and M. Wells (2000). Model selection and semiparametric inference for bivariate failure-time data. *Journal of the American Statistical Association 95*, 62–72.

24

Change Point Analysis of Top Batting Average

Sy Han Chiou
Harvard University, United States

Sangwook Kang
Yonsei University, Korea

Jun Yan
University of Connecticut, United States

Abstract

In modern times of professional baseball, a season batting average higher than 0.400 is considered a nearly unachievable goal, but it was more frequently seen in the early years before the 1940s. It is tempting to suggest that the disappearance of 0.400 hitters indicates decline in the skills of the players, but Gould (1997) makes a case for the contrary. According to Gould, the variance of the batting average among individual players has decreased as professional baseball gets better, and with the mean level remaining unchanged, it caused the extreme value in batting average to decrease. In that case, the change would have happened gradually over time. We approach the phenomenon with a change point analysis of extreme batting average using the top batting average data every year from Major League baseball in the United States. A likelihood ratio test is proposed to test the change point with a profile likelihood method, and the p-value of the observed testing statistic is obtained from a parametric bootstrap procedure. The test procedure is extended to test a smoothly changing model versus a model with a change point, either one of which could be set as the null hypothesis with the other one as the alternative hypothesis. A change point was detected in the 1940s, and the change point model provided better fit than a smoothly changing model in model comparison. The results call for further, alternative explanation of the disappearance of 0.400 hitters.

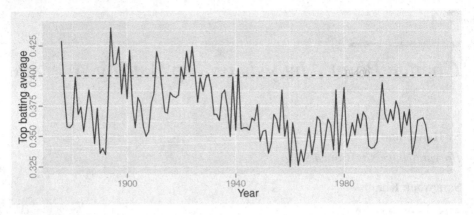

FIGURE 24.1
Top baseball batting average in the MLB from 1876 to 2013.

24.1 Introduction

Batting average (BA) of a player in baseball is the number of successful hits divided by number of plate appearances. The seasonal BA of 0.406 that Ted Williams of the Boston Red Sox recorded in 1941 is the most recent one above 0.400 in Major League baseball (MLB) of the United States. Figure 24.1 shows the time series of the top BA from these who met the minimum plate appearance requirement in MLB from its initiation in 1876 to 2013. In the MLB history of more than 120 years, there have been only seven other players who have achieved the milestone of seasonal BA of 0.400 or above, but all those happened before Ted Williams. No one has reached the landmark since, for more than 70 years. The closest one was 0.394 by Tony Gwynn of the San Diego Padres in 1994. Similar stories can be told in the professional leagues of other countries. In the Japanese (or Nippon) Professional Baseball (NPB), the MLB equivalent in Japan, no one has ever hit 0.400 in its more than 70-year history; 0.389 by Randy Bass in 1986 was the highest one. In the Korean Professional Baseball League (KPL), only one player has achieved this; 0.412 by In-Cheon Baek in 1982, the year that the KPL officially launched.

This disappearance of 0.400 hitters in the MLB has naturally been a heated issue for the last several decades among baseball fans. Among many arguments to explain the extinction of 0.400 hitters, the one by Stephen Jay Gould (1986, 1997) stands out. Unlike the conventional arguments focusing either on the declining hitting ability of individual players or toughening baseball environment for batters, Gould, an evolutionary biologist, approached the phenomenon from the perspective of evolution of systems. He claimed that the sport of baseball had "matured" over time, as evidenced from the declining of the standard deviation and the unchanged mean of BAs among all players. As variation shrinks around a constant BA, 0.400 hitters disappear — it is as simple as that. Decreasing variability has been studied as an index of improvement of the sport by other authors too (e.g., Ahn et al., 2014; Chatterjee and Hawkes, 1995; Leonard, 1995; Schell, 1999). One of the works, Leonard (1995), statistically tested Gould's theory with team statistics on 12 hitting performance variables. The results supported Gould's proposition in most of the variables, but interestingly for BA, statistically significant difference was found in means but not in variance across time, in contrast to Gould's claim.

We approach the phenomenon by fitting the top BA over time with generalized extreme value (GEV) distribution and a change point analysis to detect changes in the parameters in the GEV distribution. Extreme value theory has been applied to assess how much of an outlier an exceptional track performance could be (Barao and Tawn, 1999; Robinson and Tawn, 1995; Smith, 1997) and to predict the ultimate world record (Einmahl and Magnus, 2008; Henriques-Rodrigues et al., 2011). We are unaware of any extreme value analysis with the top baseball BA. To evaluate Gould's claim, we focus on the following two questions: First, did the top performance in the BA category change, or more specifically decrease, over time? Second, if so, did it decrease smoothly over time or did it change suddenly at some time point? These questions are answered with change point analysis for extremes. Similar to detection of changes in mean (e.g., Hawkins, 1977; Sen and Srivastava, 1975) or in variance (e.g., Chen and Gupta, 1997; Hsu, 1977), a likelihood ratio test has been applied to detection of changes in extremes (Dierckx and Teugels, 2010; Jarušková and Rencova, 2008). The likelihood ratio test can be extended to compare nonnested models which provide alternative interpretation to the data. In particular, we are interested in comparing the change point model with a model where the GEV parameters are changing smoothly with trend captured by monotone splines (Ramsay, 1988). Unlike existing work using the asymptotic distribution of the likelihood ratio statistic under the null hypothesis, we use parametric bootstrap to obtain an approximate null distribution of the testing statistic to assess the significance or the p-value of the observed testing statistic. Our results detected a change point in the 1940s and, more interestingly, the change point model provided better fit than a smoothly changing model in model comparison for the nonnested models. A simulation was conducted to assess the size and power of the proposed testing procedure.

The rest of the chapter is organized as follows: In Section 24.2, the testing procedures for detecting change point in extremes and for comparing nonnested models are described. The proposed methods are applied to the top BA from the MLB, with results explained and insights highlighted in Section 24.3. A confirmatory simulation study that mimics the application is reported in Section 24.4 to check the validity of the methods. A discussion concludes in Section 24.5.

24.2 Methods

Let $X_1, \ldots, X_n$ be a sequence of independent random variables with density $f(x; \theta_1, \eta), \ldots, f(x; \theta_n, \eta)$, respectively, where θ_t's, $t = 1, \ldots, n$, are individual specific parameters while η is shared across individuals. Both θ_t's and η can be vectors. We are interested in testing if some of the parameters θ_t's changed over time, and if so, whether the change occurred smoothly over time or abruptly.

24.2.1 Change Point Test

The statistical hypotheses for detecting a single change point τ are formulated as:

$$H_0: \quad \theta_t = \theta,$$
$$H_1: \quad \theta_t = \theta I(t \leq \tau) + \gamma I(t > \tau), \tag{24.1}$$

where $t = 1, \ldots, n$, $1 < \tau \leq n$, θ is the pre-change parameter, γ is the post-change parameter, and η does not change. The unknown change point τ under H_1 can be estimated along with other parameters by the maximum likelihood (ML) approach. The likelihood for each given change point τ can be maximized first for a range of candidate values of τ, and then the MLE of τ is obtained by maximizing the profile likelihood with respect to τ. Let $\mathbf{X} = (X_1, \ldots, X_n)$ be the observed data. Let $L_0(\mathbf{X})$ and $L_1(\mathbf{X})$ be the maximized likelihood under H_0 and H_1, respectively. That is,

$$L_0(\mathbf{X}) = \sup_{\theta, \eta} \prod_{t=1}^{n} f(X_t; \theta, \eta),$$

$$L_1(\mathbf{X}) = \sup_{\tau, \theta, \gamma, \eta} \prod_{t=1}^{\tau-1} f(X_t; \theta, \eta) \prod_{t=\tau}^{n} f(X_t; \gamma, \eta).$$

Define the likelihood ratio statistic

$$T_n = -2 \log \frac{L_0(\mathbf{X})}{L_1(\mathbf{X})}. \tag{24.2}$$

This statistic is generally used in change point detection (e.g., Csörgö and Horváth, 1997), with large value of T_n indicating an evidence against H_0.

To assess the significance of the observed T_n or its p-value, we need the distribution of T_n under H_0. Dierckx and Teugels (2010) adapted the asymptotic results of Csörgö and Horváth (1997) to the context of extreme values with f being the density of a generalized Pareto distribution. The critical value based on the asymptotic null distribution of T_n, however, did not perform satisfactorily with finite sample sizes; the tests were too liberal, not maintaining their levels in simulation studies (Dierckx and Teugels, 2010, Table 2). We propose a general parametric bootstrap procedure that is applicable for any f as long as one can generate data from f under H_0. The procedure generates a bootstrap sample of T_n under H_0 and approximates the p-value of the observed T_n by the proportion in the bootstrap sample that are at least as extreme as the observed T_n.

Specifically, the parametric bootstrap testing procedure is summarized as follows for some large bootstrap sample size B.

1. Compute $(\widehat{\theta}_n, \widehat{\eta}_n)$ and $(\widehat{\tau}_n, \widehat{\theta}_n, \widehat{\gamma}_n, \widehat{\eta}_n)$ as the ML estimates under H_0 and H_1, respectively.

2. Compute testing statistic T_n as in (24.2).
3. For every $k \in \{1, \dots, B\}$, repeat:
   (a) Generate a sample $\mathbf{X}^{(k)} = \left(X_1^{(k)}, \dots, X_n^{(k)} \right)$ from the null model with parameters $(\widehat{\theta}_n, \widehat{\eta}_n)$.
   (b) Compute the bootstrap version of the ML estimates in Step 1 with data $\mathbf{X}^{(k)}$.
   (c) Compute the testing statistic $T_n^{(k)} = -2 \log \left[L_0(\mathbf{X}^{(k)})/L_1(\mathbf{X}^{(k)}) \right]$.
4. Return an approximate p-value of T_n as $B^{-1} \sum_{k=1}^{B} 1(T_n^{(k)} \geq T_n)$.

24.2.2 Smooth Change versus Jump Change

In a more general setting, one could consider θ_t as a function of time t with some parameter vector β, $h(t; \beta)$. This formulation covers both models of θ_t in (24.1) as special cases: under H_0 we have $h(t; \beta) = \theta$ with $\beta = \theta$ while under H_1, we have $h(t; \beta) = \theta I(t \leqslant \tau) + \gamma I(t > \tau)$ with $\beta = (\tau, \theta, \gamma)$. Motivated by Gould's reasoning based on reduced variation among individual players, a smooth monotone function can be specified for θ_t in the GEV model for the top BA data. The smooth monotone change model can be compared with the jump change model with a single change point in (24.1). Note that neither of the two models is nested in the other model. Nevertheless, the likelihood ratio can still be used as a testing statistic, with either one of the models as the null model and the other one as the alternative model.

To incorporate monotone smooth change in a model parameter θ_t over time, we use monotone splines and express the parameter as a linear combination of I-spline basis (Ramsay, 1988). We give a brief description for completeness; for more details, see Ramsay (1988). I-splines are monotone splines defined as integrals of M-splines, which are nonnegative. Consider splines basis of order k (degree $k - 1$) in time window $[L, U]$, with m degrees of freedom. Let $\mathbf{t} = \{t_1, \dots, t_{m+k}\}$ be a knot sequence such that $t_1 = \cdots = t_k = L$, $t_m = \cdots = t_{m+k} = U$, and $t_i \leq t_{i+k}$ for all $i \in \{1, \dots, m\}$. M-spline basis with order 1 are defined as, for $i = 1, \dots, m$,

$$M_i(x|1, \mathbf{t}) = \frac{1}{t_{i+1} - t_i}, \qquad t_i \leq x < t_{i+1}, \quad \text{and 0 otherwise.} \qquad (24.3)$$

M-spline bases with order $k > 1$ can be defined recursively as, for $i = 1, \dots, m$,

$$M_i(x|k, \mathbf{t}) = \frac{k[(x - t_i)M_i(x|k - 1, \mathbf{t}) + (t_{i+k} - x)M_{i+1}(x|k - 1, \mathbf{t})]}{(k - 1)(t_{i+k} - t_i)}, \qquad (24.4)$$

for $t_i \leqslant x < t_{i+k}$, and 0 otherwise. Each basis M_i has the normalization $\int M_i(x)\mathrm{d}x = 1$. The nonnegativity of $\sum a_i M_i$ is ensured by $a_i \geq 0$. I-spline basis are obtained by integrating the M-spline basis

$$I_i(x|k, \mathbf{t}) = \int_L^x M_i(u|k, \mathbf{t})\mathrm{d}u, \qquad (24.5)$$

for $i = 1, \dots, m$. Spline function $\sum b_i I_i$ is monotone nondecreasing if $b_i \geq 0$ for all i, and monotone nonincreasing if $b_i \leq 0$ for all i.

We present a monotone model for a scalar θ; a set of I-spline basis coefficients is needed for each component of θ. Given degrees of freedom m with order k, the number of interior knots is determined and we place them equally spaced in (L, U). The model for θ is then $\theta_t = \sum_{i=1}^{m} b_i I_i(t|k, \mathbf{t})$ with regression coefficients $\beta = (b_1, \ldots, b_m)^\top$, which controls the monotonicity of θ_t, $\beta \geq 0$ for non-decreasing and $\beta \leq 0$ for non-increasing, respectively. The degrees of freedom m can be chosen with ad hoc method such as Akaike information criterion (AIC), which penalizes the likelihood by model complexity as $2m$. The smooth monotone model and the jump change model are not nested one way or the other, and both of them can be tested as the null hypothesis with the other as the alternative hypothesis.

In a general framework, we may consider the following hypotheses:

$$H_0 : \theta_t = h_0(t; \beta_0), \quad \text{versus} \quad H_1 : \theta_t = h_1(t; \beta_1), \tag{24.6}$$

where $h_0(t; \beta_0)$ and $h_1(t; \beta_1)$ are two models that are not necessarily nested one way or the other. The corresponding parameters under each model can still be estimated with the ML approach, and let L_j be the maximized likelihood under $H_j, j \in \{0, 1\}$. The likelihood ratio statistic $T_n = -2 \log [L_0(\mathbf{X})/L_1(\mathbf{X})]$ can be computed, which is still good to test (24.6). The only difficulty is the null distribution of T_n, but it can be approximated by a parametric bootstrap procedure too.

In summary, the parametric bootstrap testing procedure goes as follows.

1. Compute $(\widehat{\beta}_{0,n}, \widehat{\eta}_n)$ and $(\widehat{\beta}_{1,n}, \widehat{\eta}_n)$ as the ML estimates under H_0 and H_1, respectively.
2. Compute testing statistic T_n.
3. For every $k \in \{1, \ldots, B\}$, repeat:
   (a) Generate a sample $\mathbf{X}^{(k)} = \left(X_1^{(k)}, \ldots, X_n^{(k)} \right)$ from the null model with parameters $(\widehat{\beta}_{0,n}, \widehat{\eta}_n)$.
   (b) Compute the bootstrap version of the ML estimates in Step 1. with data $\mathbf{X}^{(k)}$.
   (c) Compute the testing statistic $T_n^{(k)} = -2 \log \left[L_0(\mathbf{X}^{(k)})/L_1(\mathbf{X}^{(k)}) \right]$.
4. Return an approximate p-value of T_n as $B^{-1} \sum_{k=1}^{B} 1(T_n^{(k)} \geq T_n)$.

Since the two hypotheses are not nested, we may switch their position and consider test with the other model as the null hypothesis. The only change in the parametric bootstrap procedure is that the bootstrap sample $\mathbf{X}^{(k)}$ is generated from the fitted hypothesized model.

24.3   Data Analysis

24.3.1   Change Point Test

In GEV modeling of the top BA, f is the GEV density (Equation (1.6) in Chapter 1) and (θ_t, η) is a partition of the three GEV parameters (μ, σ, ξ), where μ is the location, σ is the scale, and ξ is the shape. Even for a single change point, there are many

TABLE 24.1

Log-likelihood (LL) and AIC comparisons for selected jump change models and monotone smooth change models.

Model	$\theta_t = \mu_t$		$\theta_t = \sigma_t$		$\theta_t = (\mu_t, \sigma_t)$	
	LL	AIC	LL	AIC	LL	AIC
Jump change	347.1	-684.2	337.9	-665.9	355.0	-698.0
Smooth change						
1 knot	339.4	-666.8	335.4	-658.7	346.9	-675.8
2 knots	338.2	-662.5	335.8	-657.5	347.7	-673.3
3 knots	344.4	-672.7	336.4	-656.8	353.8	-681.7
4 knots	342.5	-667.0	337.6	-657.2	353.0	-676.0

possible partitions for (μ, σ, ξ): one, or two, or all three parameters may be put in θ_i and allowed to change after the change point. In GEV modeling, it is the most difficult to estimate ξ, followed by σ, and μ is the easiest to estimate, often incorporating covariates. Therefore, we considered change point models with change in μ only, in σ only, and in (μ, σ). That is, the changing part of the parameter θ_t takes 1) $\theta_t = \mu_t$, 2) $\theta_t = \sigma_t$ and 3) $\theta_t = (\mu_t, \sigma_t)$. The best one among these models in terms of AIC will be used to test the change point model against no change.

Table 24.1 (upper panel) summarizes the AIC and log-likelihood (LL) for the three change point models, with AIC $= -2\text{LL} + 2p$, where p is the number of model parameters. Note that the change point τ is included as a model parameter. The results indicate that the model with changes in both μ and σ yielded the lowest AIC; the two additional parameters in comparison to the other two models were well worth it, increasing the log-likelihood much more than the penalty they brought. The model with change in μ came next in terms of AIC, followed by the model with change in σ only, meaning that if one parameter were allowed to change, a change in μ would give a better model than a change in σ. Interestingly, in all three models, the estimated change point was $\hat{\tau} = 1941$.

We consider testing the hypothesis (24.1) with $\theta_t = (\mu_t, \sigma_t)$ and $\eta = \xi$. The parametric bootstrap procedure was carried out with $B = 1000$, and the resulting approximate p-value was 0.015. At a significant level of 0.05, we reject the null hypothesis of no change; the evidence supports that (μ, σ) changed simultaneously around 1941.

The common shape parameter ξ was estimated to be -0.231. Combined with the estimated μ and σ, the ultimate limit of the BA is estimated to be 0.472 before 1941 and 0.413 after 1941. These purely statistical results need to be interpreted in corroboration with substantive changes in baseball.

24.3.2 Smooth Change versus Jump Change

Now that the null hypothesis of no change in the GEV distribution has been rejected, we consider the question of how the change happened. Did the change happen abruptly as implied by the change point model, or gradually and smoothly as suggested by Gould's evolution theory? We use I-splines to model μ_t, σ_t, or (μ_t, σ_t) such that these parameters change over time monotonically. With quadratic splines (order $k = 3$), we fitted with the number of interior knots $\nu \in \{1, 2, 3, 4\}$ for each parameter. The log-likelihood was maximized such that all the basis coefficients constrained were nonpositive to ensure that the parameters were monotone decreasing over time. The best smooth change model in terms of AIC will then be compared with the change point model.

The log-likelihood and AIC for all the smooth change models we fitted are summarized in Table 24.1 (lower panel). For each number of interior knots, the μ-change model is better than the σ-change model, but inferior to the model with simultaneous changes in (μ, σ). The log-likelihood increases as ν increases but levels off at $\nu = 3$, and the AIC prefers the model with $\nu = 3$. Therefore, the model with $\nu = 3$ and $\theta_t = (\mu_t, \sigma_t)$ will be used as the smooth model in the following comparison.

Figure 24.2 presents the estimates of μ and σ when fitted with constant model, jump model, and smooth model. It is interesting to note that the estimated μ_t in the smooth model is very similar to that in the jump model: it is flat before 1915 and after 1975, with the smooth drop occurring between 1915 and 1975. On the other hand, the smooth estimates of σ_t decreases before 1935 and after 1985, and remains constant between 1935 and 1985. The smooth changes in μ_t and σ_t appear to compensate each other over a big portion of the time, jointly controlling the changes in the GEV distribution of the top BA over time.

The jump model and the smooth model were compared in hypothesis tests. Two sets of hypotheses are:

$$H_0 : \text{smooth model} \qquad \text{versus} \qquad H_1 : \text{jump model}, \tag{24.7}$$

and

$$H_0 : \text{jump model} \qquad \text{versus} \qquad H_1 : \text{smooth model}. \tag{24.8}$$

Parametric bootstrap procedures with $B = 1000$ gave approximate p-values of 0.013 for testing (24.7) and 0.188 for testing (24.8). At a significant level of 0.05, we reject the null hypothesis of smooth change model in (24.7) and fail to reject the null hypothesis of jump change model in (24.8). The results suggest that the change in (μ, σ) is rather a sudden change with a jump change at $\hat{\tau} = 1942$. From a model comparison perspective, the jump model is also preferred to the smooth model in terms of AIC.

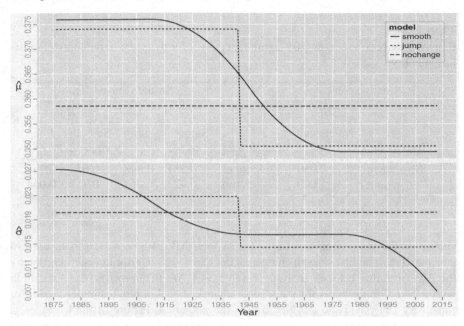

FIGURE 24.2
Estimates of μ (top) and σ (bottom) under constant model (- - -), jump model (– –), and smooth model (—).

24.4 Simulation

To assess the performance of the proposed testing procedures, we conducted simulations with data generated from the fitted models. In particular, the size and power of the tests were studied by the empirical rejection rates for the test of change point and the testing of jump model versus smooth change model.

The first study focused on the change point test (24.1). Using the ML estimates as parameter values, the fitted models under both H_0 and H_1 were used as the data generating model. We generated 1000 datasets, each with length 138 to mimic real data. For each dataset, the parametric bootstrap procedure was carried out with bootstrap sample size $B = 1000$. The empirical rejection rates for the tests at significance level 0.05 are summarized in Table 24.2. When H_0 was the true model, the empirical rejection rate was 0.050, suggesting that the test maintains its level at this sample size. When H_1 was the true model, all empirical rejection rate was 1.000, suggesting that the test is very powerful in detecting the difference between the no change model and the single change point model with the fitted parameter values from the real data.

We then examined the comparison of the smooth change model and the jump change model in testing (24.7) and (24.8). The smooth change model and the jump

TABLE 24.2
Empirical rejection rates at 0.05 significance level for the parametric bootstrap tests in simulation.

Change point			Smooth versus Jump			
H_0			H_0		H_0	
No change			Smooth	Jump	Smooth	Jump
True model		True model	$\nu = 1$		$\nu = 3$	
No change	0.050	Smooth	0.037	0.898	0.056	0.934
Jump	1.000	Jump	0.922	0.051	0.788	0.054

change model with parameter values set at their ML estimates were both used to generate data. The first smooth change model had $\nu = 3$ interior knots, as chosen by the AIC from the real data analysis in Table 24.1. We then repeated the simulation with $\nu = 1$ interior knots for the smooth change model, a much simpler model for sensitivity analysis. Note that because of the nonnested feature of the hypotheses, the null distribution of the testing statistic depends on H_1 too. Again, we had 1000 replicates for each scenario and bootstrap sample size $B = 1000$ for each test. The empirical rejection rates for the tests with level 0.05 are summarized in Table 24.2. When H_0 is the true model, the tests appear to maintain their levels, with all empirical rejection rates very close to the nominal level 0.05. When H_1 is the true model, the tests show substantial power in rejecting H_0, regardless of the number of interior knots in the smooth change model. These results suggest that in the real data analysis setting, the rejection of smooth change model in favor of the jump change model is a quite reliable decision; the fitted models are sufficiently different.

24.5   Discussion

By fitting GEV models to the top seasonal BA in the MLB, we detected a single change point and compared the jump change model with smooth change models implied by Gould (1997). Although our results are statistical so far instead of substantive, they might provide some quantitative support for a new perspective in explaining the disappearance of the 0.400 hitters. The rejection of the no-change model is obvious; one may point out a change point around 1942 by visual inspection of Figure 24.1. What is more interesting is that the jump change model with a single change point fitted the data better than smooth change models in AIC and in comparison of nonnested models based on hypothesis testing. This finding cannot be completely explained by Gould's reasoning, under which the changes occur smoothly and monotone over time. In fact, the reduction in variation (Gould, 1997) was concluded by visual observation of a declining trend in the range of the BA of all players. Suspicions have been raised from a series of statistical tests on the homogeneity of variance and mean over time in Leonard (1995). Our results further suggest that there might be something else around 1942 that triggered the disappearance of the 0.400 hitters, which was not captured by Gould's theory.

For a more substantive explanation, we investigated the MLB in the 1940s. World War II caused a shortage of professional baseball players. More than 500 players from the MLB were drafted to serve in the military during the war, and many MLB teams played without their best players (Anton et al., 2008). Further, the 1942 season was nearly canceled due to the wartime blackout. Fewer players and fewer games might be the key factor to cause the tail of the distribution of the BA among players to shrink. Following the war, another important event was the breaking of the color barrier in baseball by Jackie Robinson with the Brooklyn Dodgers in 1947. The new recruitment strategy improved the quality of batting as well as fielding and pitching, the dynamic interaction of which affected the BA, as a measure of batting skill, in relation to fielding and pitching skills. Over time, rule makers have sought to balance advantages for pitching and fielding with advantages for batting, which makes 0.400 hitters really rare. Our finding is not necessarily contradictory to Gould's claim; instead, it suggests that the stabilization of baseball as a system happened after some triggering events rather than in a smooth fashion.

Our analysis only considered jump model with a single change point. More complicated models with multiple change points are possible. For example, one may consider a model where some parameters are modeled as smooth functions while others are modeled as step functions. The current analysis used only the top BA in each season. It might be worth including the highest r BAs to use more data. We assumed that the top BAs were independent from season to season. A more comprehensive analysis needs to account for the temporal dependence. Finally, the statistical analysis is not intended to replace, but to stimulate, support and supplement substantive analysis for the intriguing phenomenon of the disappearance of 0.400 hitters.

References

Ahn, J. Y., G. J. Byun, H. Byun, Y. Chekal, G. Cheon, B. H. Cheon, S. H. Cho, S. Choi, K. H. Chung, B. Hong, S. Jang, W. C. Jang, S. M. Jeon, J. Jeong, K. Jun, Y. Jung, M. Kang, K. M. Kim, K. Kim, D. J. Kim, D. S. Kim, L. Y. Kim, S. M. Kim, S. W. Kim, Y. J. Kim, Y. N. Kim, Y. K. Kim, J. H. Kim, T. H. Kim, H. Kim, H.-I. Kim, J. H. Kwon, K. Lee, S. H. Lee, S. K. Lee, J. S. Lee, J. S. Lee, C.-H. Lee, H. K. Lee, S. Lim, S. Lim, S. Nam, S. W. Nam, N. H. Noh, J. m. Noh, W. Oh, M. Pak, S. H. Park, S. Park, S. Park, J. H. Park, C. E. Park, H. J. Park, B. G. Shin, E. Shin, E. J. Song, M. Yi, and S.-y. Yoon (2014). Why 0.400 hitters have disappeared: 30-year evolution of the Korean Professional Baseball League. Baek In-Cheon Project.

Anton, T. W., B. Nowlin, and C. Schilling (2008). *When Baseball Went to War.* Chicago: Triumph Books.

Barao, M. I. and J. A. Tawn (1999). Extremal analysis of short series with outliers: Sea-levels and athletics records. *Journal of the Royal Statistical Society: Series C (Applied Statistics) 48*(4), 469–487.

Chatterjee, S. and J. Hawkes (1995). Apparent decline as a sign of improvement? Or, can less be more? *Teaching Statistics 17*(3), 82–83.

Chen, J. and A. Gupta (1997). Testing and locating variance changepoints with application to stock prices. *Journal of the American Statistical Association 92*(438), 739–747.

Csörgö, M. and L. Horváth (1997). *Limit Theorems in Change-Point Analysis*, Volume 18. New York: Wiley.

Dierckx, G. and J. L. Teugels (2010). Change point analysis of extreme values. *Environmetrics 21*(7–8), 661–686.

Einmahl, J. H. J. and J. R. Magnus (2008). Records in athletics through extreme-value theory. *Journal of the American Statistical Association 103*(484), 1382–1391.

Gould, S. J. (1986, August). Entropic homogeneity isn't why no one hits .400 any more. *Discover 7*(8), 60–66.

Gould, S. J. (1997). *Full House: The Spread of Excellence from Plato to Darwin.* Three Rivers Press.

Hawkins, D. M. (1977). Testing a sequence of observations for a shift in location. *Journal of the American Statistical Association 72*(357), 180–186.

Henriques-Rodrigues, L., M. I. Gomes, and D. Pestana (2011). Statistics of extremes in athletics. *REVSTAT–Statistical Journal 9*(2), 127–153.

Hsu, D.-A. (1977). Tests for variance shift at an unknown time point. *Applied Statistics 26*(3), 279–284.

Jarušková, D. and M. Rencova (2008). Analysis of annual maximal and minimal temperatures for some European cities by change point methods. *Environmetrics 19*(3), 221–233.

Leonard, W. M. I. (1995). The decline of the .400 hitter: An explanation and a test. *Journal of Sport Behavior 18*, 226–236.

Ramsay, J. O. (1988). Monotone regression splines in action. *Statistical Science 3*(4), 425–441.

Robinson, M. E. and J. A. Tawn (1995). Statistics for exceptional athletics records. *Applied Statistics 44*(4), 499–511.

Schell, M. J. (1999). *Baseball's All-Time Best Hitters: How Statistics Can Level the Playing Field.* Princeton University Press.

Sen, A. and M. S. Srivastava (1975). On tests for detecting change in mean. *The Annals of Statistics 3*(1), 98–108.

Smith, R. L. (1997). Statistics for exceptional athletics records. *Journal of the Royal Statistical Society: Series C (Applied Statistics) 46*(1), 123–128.

25

Computing Software

Eric Gilleland

National Center for Atmospheric Research, United States

Abstract

Software aimed at facilitating statistical extreme value analysis is increasingly abundant, as well as becoming increasingly refined. This chapter provides a summary of available software tools with an emphasis on the numerous packages of R because it is a freely available open-source statistical language and environment that has become the language most used by academic statisticians.

25.1 Introduction

Software development for extreme value analysis (EVA) has been rapidly increasing over the last several years. Software packages are being created and refined very often; for example, much has changed since the recent software review by Gilleland et al. (2013), and much had changed at that time since an earlier review by Stephenson and Gilleland (2005). It is possible that many of the packages will have changed considerably by the time this current text is published. Therefore, the hope is to give the reader an overall flavor of the types of software tools that are available. The primary emphasis, here, is on packages in the R programming language, but some attempt to summarize other software tools is also given.

TABLE 25.1
Primary general extreme value analysis packages in R.

---

evd (Stephenson, 2002)
evdbayes (Stephenson and Ribatet, 2014)
evir (Pfaff and McNeil, 2012)
extRemes (Gilleland and Katz, 2011a, 2015)
fExtremes (Wuertz et al., 2013)
ismev (Heffernan and Stephenson, 2012)
lmom (Hosking, 2014a)
lmomRFA (Hosking, 2014b)
lmomco (Asquith, 2014)
SpatialExtremes (Ribatet et al., 2013)
texmex (Southworth and Heffernan, 2013)
VGAM (Yee, 2013; Yee and Stephenson, 2007)

25.2   R Packages

A host of R packages are available with capabilities for carrying out univariate EVA (Table 25.1). Nearly all of the packages listed in Table 25.1 utilize maximum likelihood estimation (MLE) for parameter estimation of the generalized extreme value (GEV) distribution, except evdbayes, which as the name suggests employs Bayesian estimation, as well as lmom, lmomRFA and lmomco, all of which use L-moments. Versions $\geqslant 2.0$ of extRemes has some Bayesian estimation capability, generalized MLE (Martins and Stedinger, 2000, 2001) as well as L-moments in addition to the usual MLE method. Package fExtremes also contains probability weighted moments for parameter estimation. Package SpatialExtremes also implements maximum composite likelihood estimation and Bayesian techniques. Package texmex also employs penalized MLE and Bayesian methods. Finally, VGAM provides vector generalized linear and additive extreme value models (Yee and Stephenson, 2007).

The packages in Table 25.1 to include functionality for nonstationary regression for the GEV parameters are extRemes, ismev, SpatialExtremes, texmex and VGAM. Packages evd and evdbayes include limited functionality for nonstationary regression.

Package in2extRemes (Gilleland and Katz, 2011b) contains a graphical user interface for some of the functionality available in extRemes, and is primarily aimed at shortening the learning curve to performing EVA and using extRemes. Originally, extRemes < 2.0 included a graphical user interface to ismev with some additional command-line functionality, and in2extRemes replaces all of the graphical user interfaces so that they are separate from the command-line codes. In this way, a user wanting only the command-line code will not need in2extRemes at all. For the most part, extRemes $\geqslant 2.0$ includes the same functionality as ismev,

TABLE 25.2
R packages with multivariate extreme value analysis capabilities.

copula (Hofert et al., 2014; Hofert and Maechler, 2011; Kojadinovic and Yan, 2010; Yan, 2007)
evd (Stephenson, 2002)
evir (Pfaff and McNeil, 2012)
lmomRFA (Hosking, 2014b)
lmomco (Asquith, 2014)
SpatialExtremes (Ribatet et al., 2013)
texmex (Southworth and Heffernan, 2013)

but much more. Exceptions are that it does not allow for the r-th largest order statistic model, and only allows for a log-link function in the scale parameter. Package ismev uses matrices with column identifiers for incorporation of parameter covariates where extRemes copies texmex by using R formulas for this capacity. Which is better depends on user preferences, but the former can be more efficient while the latter more flexible and intuitive.

Table 25.2 lists the primary R packages for carrying out multivariate EVA. Package copula includes five extreme value copula models, among other choices. Package evd contains parametric and nonparametric estimation for the class of bivariate extreme value distributions, and diagnostic plotting methods are also included. Package lmomco contains L-comoments for multivariate analysis. As spatial analysis is inherently multivariate, it is no surprise that SpatialExtremes contains much multivariate functionality, including max-stable processes, Bayesian hierarchical models (requires a conditional independence assumption), as well as copula approaches. Package texmex employs the conditional extreme value modeling approach introduced by Heffernan and Tawn (2004).

Package extRemes $\geqslant$ 2.0 contains a function (taildep) to calculate tail dependence parameters for investigating whether or not two variables are dependent in their extreme values, as well as a function (taildep.test) to conduct a hypothesis test about tail dependence, as described in Reiss and Thomas (2007). However, no formal multivariate modeling is included at the time of this writing.

Various other R packages also have utility when interest is in extreme values (Table 25.3). A good many of the packages listed in Table 25.3 contain functionality for volatility modeling and/or quantitative financial risk management (e.g., ccgarch, CreditMetrics, fGarch, gogarch, QRM, rmgarch and rugarch, and tseries). Package QRM is an R port of the S-Plus package QRMlib (McNeil et al., 2005).

Package inla has utility for estimating a class of models via the novel inferential procedure proposed by Rue et al. (2009), and includes models based on the GEV distribution.

Package maxstat contains functions for maximally selected rank statistics with p-value approximations.

TABLE 25.3

Additional R packages.

`ABCExtremes` (Erhardt, 2013)
`acer` (Haug, 2012)
`actuar` (Dutang et al., 2008)
`bgeva` (Marra et al., 2013)
`BMAmevt` (Sabourin, 2013)
`BSquare` (Smith and Reich, 2013)
`ccgarch` (Nakatani, 2014)
`CreditMetrics` (Wittman, 2007)
`eventstudies` (Shah et al., 2013)
`evmix` (Hu and Scarrott, 2013)
`extremevalues` (van der Loo, 2010a,b)
`extWeibQuant` (Liu, 2014)
`fgac` (Gonzalez-Lopez, 2012)
`fGarch` (Chausse and others, 2013)
`gamlss` (Rigby and Stasinopoulos, 2005)
`gogarch` (Pfaff, 2012)
`inla` (Rue et al., 2009)
`maxstat` (Hothorn, 2014)
`nacopula` (Hoefert and Maechler, 2012)
`QRM` (Pfaff and McNeil, 2014)
`quantreg` (Koenker, 2013)
`rugarch` (Ghalanos, 2014b)
`rmgarch` (Ghalanos, 2014a)
`Renext` (Deville and IRSN, 2013)
`Runuran` (Leydold and Hörmann, 2012)
`RandomFields` (Schlather et al., 2014)
`spatial.gev.bma` (Lenkoski, 2014)
`TestEVC1d.r` (Hüsler and Li, 2006)
`tseries` (Trapletti and Hornik, 2013)

Quantile regression can be carried out with functions in `quantreg`, as well as a novel Bayesian spatial approach proposed by Reich et al. (2011) incorporating quantile regression with `BSquare`.

Package `extremevalues` contains functions for distribution-based outlier detection, and comes with a graphical user interface.

Package `Renext` provides an alternative to the usual peaks over threshold (POT) approach that is apparently popular among francophone hydrologists called the *méthode de renouvellement* (renewal method).

Several packages have been uploaded to CRAN since the review papers by Gilleland et al. (2013); Stephenson and Gilleland (2005), including the following. Package `ABCExtremes` is a newer package that constructs the summary statistic needed to implement approximate Bayesian computing (ABC) for spatial EVA, as well as running the ABC rejection method to obtain a set of draws from the ABC posterior distribution. The `acer` package implements the method proposed by Naess and Gaidai (2009) for estimating return levels. Package `bgeva` contains regression models for binary rare events with linear and nonlinear covariate effects when using the quantile function of the GEV distribution. Package `BMAmevt` allows for Bayesian estimation of the dependence structure in parametric multivariate extreme value models. Package `eventstudies` provides some analysis methods for extreme event studies. The `evmix` package provides univariate extreme value mixture modelling (Chapter 3) and threshold estimation (Chapter 4) diagnostics; see the Supplementary Materials for Chapter 3 for details. Package `extWeibQuant` contains functionality for estimating the lower quantile by way of the subjectively censored Weibull MLE and censored Weibull mixture model. Package `spatial.gev.bma` is a package that was designed for a specific project that uses spatial Bayesian hierarchical modeling to fit the GEV distribution to data with Bayesian model averaging.

25.3 Other Software

25.3.1 MATLAB®

Package `EVIM` is a free MATLAB [1] package (Gençay et al., 2001, available at: `http://www.sfu.ca/~rgencay/`) that contains univariate EVA routines similar to those in R package `evir`. In addition to fitting GEV distributions to block maxima and POT, it includes functionality for declustering data, making hill plots, and estimating the extremal index.

Another MATLAB package is `NEVA` (Cheng et al., 2014) which provides univariate estimation for the GEV distribution with the possibility of including a covariate in the location parameter as a linear function. Bayesian estimation is carried out by way of differential evolution Markov chain (DE-MC) simulation.

The `WAFO` (Brodtkorb et al., 2000) package is another freely available MATLAB and Python package focused on wave analysis for oceanography, sea modeling, and fatigue analysis. It also contains the customary block maxima and POT approaches allowing for uncertainty estimation through bootstrapping, the profile likelihood, as well as the maximum product of spacings method proposed by Wong and Li (2006).

25.3.2 S-Plus

S-Plus Package `ismev` is available on the Comprehensive S-PLUS Archive Network (CSAN, `http://csan.insightful.com/Default.aspx`) with functions supporting the text by Coles (2001).

Package `QRMlib` is a freely available (`http://www.macs.hw.ac.uk/`

---

[1] MATLAB® is a registered trademark of The Mathworks, Inc. For product information please contact: The Mathworks Inc., 3 Apple Hill Drive, Natick, MA, 01760-2098 USA. Tel: 508-647-7000. Fax: 508-647-8001. Email: info@mathworks.com. Web: `www.mathworks.com`.

˜mcneil/book/QRMlib.html) that accompanies the text by McNeil et al. (2005). The package contains functionality for stationary univariate EVA, as well as copula modeling.

Package S+FinMetrics (Carmona, 2004) is an S-Plus module (and, therefore, available at additional cost) intended for econometric analysis. The module includes functionality for time series analysis, econometric system estimation, volatility modeling (e.g., GARCH processes), copulas, and state space modeling in addition to the functionality formally available from the EVIS package for implementing univariate EVA.

25.3.3 Standalone Packages

Package EXTREMES (available at: http://extremes.gforge.inria.fr/) is a point-and-click software package with functionality for executing most univariate EVA analyses (Diebolt et al., 2003), as well as many standard statistical methods, such as sample simulation, Anderson–Darling test, parametric quantile estimation, etc. The software includes a test for comparing the parametric estimate of an extreme quantile with its semiparametric estimate obtained by EVT (Diebolt et al., 2007). Additionally, EXTREMES contains a regularization procedure that preserves the general form of a parametric distribution allowing for an improved fit in the tail of the distribution (Diebolt et al., 2003).

Package GLSnet and peakFQ are two freely available Fortran packages available from the U.S. Geological Survey's Water Resources Applications Software (http://water.usgs.gov/software). They are aimed at modeling flow characteristics, but do not contain any extreme value distribution functionality. Package GLSnet performs predictions of low and high flows at unobserved sites by way of regression and generalized least squares techniques. Package peakFQ estimates annual peak flows for different return periods by fitting the Pearson Type III (gamma) distribution using the logarithmic sample moments.

Package HYFRAN and HYFRAN-PLUS (El Aldouni et al., 2008) are commercial products developed by researchers at INRS University, Canada. The package was originally designed for hydrological frequency analysis, but has grown to include tools for fitting numerous distributions with different tail weights to i.i.d. data. Package HYFRAN-PLUS additionally contains graphical diagnostics for choosing between different models.

Wallingford HydroSolutions provides various commercial software products, such as WINFAP-FEH 3, which provides flood frequency analysis techniques for both annual maxima and peaks over threshold methods.

A standalone Windows package, called Xtremes Version 3.1, accompanies the EVT text (Reiss and Thomas, 2007). It constitutes a relatively thorough package containing functionality for performing most EVA tasks. The STABLE routines written by Nolan (2007) are included with Xtremes.

25.4   Discussion

Much changed in the R software landscape between the review papers of Stephenson and Gilleland (2005) and Gilleland et al. (2013), and since the latter, many more software tools are available while existing packages have undergone tremendous modifications and improvements. Others are no longer available or have been merged into other packages. Therefore, the above guide should be useful for finding what might be out there, even if numerous further changes take place before the present text is published.

In terms of R packages, Gilleland et al. (2013) suggested that it would be beneficial for the various R packages, especially those aiming toward a global goal (e.g., extRemes, SpatialExtremes, texmex), to become more consistent with each other in order to make a comprehensive set of tools that can be readily applied. It is unclear how likely such an endeavor would be. Moreover, it is clear that as new methods have come to the fore, new packages seem to follow, rather than new functions within existing packages. As a developer, it can be difficult to discern what new methods are worth adding to existing packages, as it can require considerable effort on the part of the maintainer who may not have time to handle numerous such requests. Version control software can be useful there, but still requires vigilance for maintainers to keep functions consistent and control the quality.

Regardless of what the future holds, a good number of high-quality software packages are available for conducting extreme value analysis, and it is hoped that this summary chapter will at least give some guidance about which tools might be useful for particular purposes. Here, it is useful to summarize R packages in a more task-oriented manner.

25.4.1   General Univariate Extreme Value Analysis

For fitting univariate extreme value distributions to data using MLE, a few R packages are fairly complete: evd, extRemes, and ismev. Other packages perform similar analyses, but are aimed at specific estimation methods (e.g., evdbayes for Bayes estimation, lmom for L-moments estimation). If it is desired to incorporate covariates into the parameters, then the best choices are either extRemes or ismev, although other options exist for specific models, and VGAM allows for more flexibility in model choices, but is less specific to handling extremes.

25.4.2   Multivariate Extreme Value Analysis

Classical multivariate extreme value analysis is perhaps best handled by evd and SpatialExtremes. Copula modeling can be handled by numerous packages (cf. Section 25.2 and, in particular, Table 25.2). Heffernan and Tawn (2004) introduced a new approach to multivariate analysis that utilizes a conditional approach, and the texmex package is excellent for implementing this procedure.

25.4.3   Finance

Many packages have functionality for handling financial data, or for methods often used in finance circles. These include fairly standard time series packages (e.g., tseries, actuar, gogarch, fGarch, QRM), as well as more specific packages aimed at the extreme values, such as fExtremes and evir.

References

Asquith, W. H. (2014). *lmomco—L-moments, Trimmed L-moments, L-comoments, Censored L-moments, and Many Distributions.* R package version 2.1.1.

Brodtkorb, P., P. Johannesson, G. Lindgren, I. Rychlik, J. Rydén, and E. Sjö (2000). WAFO: A MATLAB toolbox for the analysis of random waves and loads. *Proc. 10'th Int. Offshore and Polar Eng. Conf., ISOP, Seattle, WA, U.S.A. 3*, 343–350.

Carmona, R. A. (2004). *Statistical Analysis of Financial Data in S-Plus.* London, UK, 455 pp.: Springer-Verlag.

Chausse, P. and others (2013). *fGarch: Rmetrics—Autoregressive Conditional Heteroskedastic Modelling.* R package version 3010.82.

Cheng, L., A. AghaKouchak, E. Gilleland, and R. Katz (2014). Non-stationary extreme value analysis in a changing climate. *Climatic Change 127*(2), 353–369.

Coles, S. (2001). *An Introduction to Statistical Modeling of Extreme Values.* Springer-Verlag, London, UK, 208 pp.

Deville, I. and IRSN (2013). *Renext: Renewal Method for Extreme Values Extrapolation.* R package version 2.1-0.

Diebolt, J., J. Ecarnot, M. Garrido, S. Girard, and D. Lagrange (2003). Le logiciel extremes, un outil pour l'étude des queues de distributions. *Revue Modulad 30*, 53–60.

Diebolt, J., M. Garrido, and S. Girard (2007). *Extreme Value Distributions*, Chapter A Goodness-of-Fit Test for the Distribution Tail, pp. 95–109. New Nork: Ahsanulah, M. and S. Kirmani (eds.) Nova Science.

Diebolt, J., M. Garrido, and C. Trottier (2003). Improving extremal fit: A Bayesian regularization procedure. *Reliability Engineering & System Safety 82*(1), 21–31.

Dutang, C., V. Goulet, and M. Pigeon (2008). actuar: An R package for actuarial science. *Journal of Statistical Software 25*(7), 1–37.

El Aldouni, S., B. Bobée, and T. B. M. J. Ouarda (2008). On the tails of extreme event distributions in hydrology. *Journal of Hydrology 355*, 16–33.

Erhardt, R. (2013). *ABCExtremes: ABC Extremes.* R package version 1.0.

Gençay, R., F. Selçuk, and A. Ulugülyağci (2001). EVIM: A software package for extreme value analysis in MATLAB. *Studies in Nonlinear Dynamics and Econometrics 5*(3), 213–239.

Ghalanos, A. (2014a). *rmgarch: Multivariate GARCH Models.* R package version 1.2-6.

Ghalanos, A. (2014b). *rugarch: Univariate GARCH Models*. R package version 1.3-3.

Gilleland, E. and R. W. Katz (2011a, 11 January). New software to analyze how extremes change over time. *Eos 92*(2), 13–14.

Gilleland, E. and R. W. Katz (2011b). New software to analyze how extremes change over time. *EOS 92*(2), 13–14.

Gilleland, E. and R. W. Katz (2015). extRemes 2.0: An extreme value analysis package in R. *Journal of Statistical Software*. Forthcoming.

Gilleland, E., M. Ribatet, and A. G. Stephenson (2013). A software review for extreme value analysis. *Extremes 16*(1), 103–119.

Gonzalez-Lopez, V. A. (2012). *fgac: Generalized Archimedean Copula*. R package version 0.6-1.

Haug, E. (2012). *acer: The ACER Method for Extreme Value Estimation*. R package version 0.1.2.

Heffernan, J. E. and A. G. Stephenson (2012). *ismev: An Introduction to Statistical Modeling of Extreme Values*. R package version 1.39.

Heffernan, J. E. and J. A. Tawn (2004). A conditional approach for multivariate extreme values (with discussion). *Journal of the Royal Statistical Society: Series B (Statistical Methodology) 66*(3), 497–546.

Hoefert, M. and M. Maechler (2012). *nacopula: Nested Archimedean copulas*. R package version 0.8-1.

Hofert, M., I. Kojadinovic, M. Maechler, and J. Yan (2014). *copula: Multivariate Dependence with Copulas*. R package version 0.999-9.

Hofert, M. and M. Maechler (2011). Nested Archimedean copulas meet R: The nacopula package. *Journal of Statistical Software 39*(9), 1–20.

Hosking, J. R. M. (2014a). *L-moments*. R package version 2.4.

Hosking, J. R. M. (2014b). *Regional Frequency Analysis Using L-Moments*. R package version 3.0.

Hothorn, T. (2014). *maxstat: Maximally Selected Rank Statistics*. R package version 0.7-20.

Hu, Y. and C. J. Scarrott (2013). evmix: Extreme value mixture modelling, threshold estimation and boundary corrected kernel density estimation. Available on CRAN.

Hüsler, J. and D. Li (2006). *How to Use the Package TestEVC1d.r*. Department of Statistics, School of Management, Fudan University, 3pp.

Koenker, R. (2013). *quantreg: Quantile Regression*. R package version 5.05.

Kojadinovic, I. and J. Yan (2010). Modeling multivariate distributions with continuous margins using the copula R package. *Journal of Statistical Software 34*(9), 1–20.

Lenkoski, A. (2014). *spatial.gev.bma: Hierarchical Spatial Generalized Extreme Value (GEV) Modeling with Bayesian Model Averaging (BMA)*. R package version 1.0.

Leydold, J. and W. Hörmann (2012). *Runuran: R Interface to the UNU.RAN Random Variate Generators*. R package version 0.20.0.

Liu, Y. (2014). *extWeibQuant: Estimate the Lower Extreme Quantile with the Censored Weibull MLE and Censored Weibull Mixture*. R package version 1.0.

Marra, G., R. Calabrese, and S. A. Osmetti (2013). *bgeva: Generalized Extreme Value Additive Modelling for Binary Rare Events Data*. R package version 0.2.

Martins, E. S. and J. R. Stedinger (2000). Generalized maximum-likelihood generalized extreme-value quantile estimators for hydrologic data. *Water Resources Res. 36*(3), 737–744.

Martins, E. S. and J. R. Stedinger (2001). Generalized maximum-likelihood Pareto-Poisson estimators for partial duration series. *Water Resources Research 37*(10), 2551–2557.

McNeil, A. J., R. Frey, and P. Embrechts (2005). *Quantitative Risk Management: Concepts, Techniques, and Tools*. Princeton University Press, Princeton, New Jersey, U.S.A., 544 pp.

Naess, A. and O. Gaidai (2009). Estimation of extreme values from sampled time series. *Structural Safety 31*, 325–334.

Nakatani, T. (2014). *ccgarch: An R Package for Modelling Multivariate GARCH Models with Conditional Correlations*. R package version 0.2.3.

Nolan, J. P. (2007). *Stable Distribution: Models for Heavy Tailed Data*. Birkhäuser, Boston, U.S.A., 352pp.

Pfaff, B. (2012). *gogarch: Generalized Orthogonal GARCH (GO-GARCH) Models*. R package version 0.7-2.

Pfaff, B. and A. J. McNeil (2012). *evir: Extreme Values in R*. R package version 1.7-3.

Pfaff, B. and A. J. McNeil (2014). *QRM: Provides R-language Code to Examine Quantitative Risk Management Concepts*. R package version 0.4-10.

Reich, B. J., M. Fuentes, and D. B. Dunson (2011). Bayesian spatial quantile regression. *Journal of the American Statistical Association 106*(493), 6–20.

Reiss, R. and M. Thomas (2007). *Statistical Analysis of Extreme Values: With Applications to Insurance, Finance, Hydrology and Other Fields* (3rd ed.). Birkhäuser, 530 pp.

Ribatet, M., R. Singleton, and R Core Team (2013). *SpatialExtremes: Modelling Spatial Extremes*. R package version 2.0-0.

Rigby, R. A. and D. M. Stasinopoulos (2005). Generalized additive models for location, scale and shape. *Journal of the Royal Statistical Society: Series C (Applied Statistics) 54*(3), 507–554.

Rue, H., S. Martino, and N. Chopin (2009). Approximate Bayesian inference for latent Gaussian models using integrated nested Laplace approximations (with discussion). *Journal of the Royal Statistical Society: Series B (Statistical methodology) 71*(2), 319–392.

Sabourin, A. (2013). *BMAmevt: Multivariate Extremes: Bayesian Estimation of the Spectral Measure*. R package version 1.0.

Schlather, M., A. Malinowski, M. Oesting, D. Boecker, K. Strokorb, S. Engelke, J. Martini, P. Menck, S. Gross, K. Burmeister, J. Manitz, R. Singleton, and R Core Team (2014). *RandomFields: Simulation and Analysis of Random Fields*. R package version 3.0.10.

Shah, A., V. Balasubramaniam, and V. Bahure (2013). *eventstudies: Event Study and Extreme Event Analysis*. R package version 1.1.

Smith, L. and B. J. Reich (2013). *BSquare: Bayesian Simultaneous Quantile Regression*. R package version 1.1.

Southworth, H. and J. E. Heffernan (2013). *texmex: Statistical Modelling of Extreme Values*. R package version 2.1.

Stephenson, A. G. (2002). evd: Extreme value distributions. *R News 2*(2), 31–32.

Stephenson, A. G. and E. Gilleland (2005). Software for the analysis of extreme events: The current state and future directions. *Extremes 8*, 87–109.

Stephenson, A. G. and M. Ribatet (2014). *evdbayes: Bayesian Analysis in Extreme Value Theory*. R package version 1.1-1.

Trapletti, A. and K. Hornik (2013). *tseries: Time Series Analysis and Computational Finance*. R package version 0.10-32.

van der Loo, M. P. J. (2010a). Distribution based outlier detection for univariate data. Technical Report 10003, Statistics Netherlands.

van der Loo, M. P. J. (2010b). *extremevalues, an R Package for Outlier Detection in Univariate Data*. R package version 2.0.

Wittman, A. (2007). *CreditMetrics: Functions for Calculating the CreditMetrics Risk Model*. R package version 0.0-2.

Wong, T. S. T. and W. K. Li (2006). A note on the estimation of extreme value distributions using maximum product of spacings. *IMS Lecture Notes 52*, 272–283.

Wuertz, D. et al. (2013). *fExtremes: Rmetrics—Extreme Financial Market Data*. R package version 3010.81.

Yan, J. (2007). Enjoy the joy of copulas: With a package copula. *Journal of Statistical Software 21*(4), 1–21.

Yee, T. W. (2013). *VGAM: Vector Generalized Linear and Additive Models*. R package version 0.9-1.

Yee, T. W. and A. G. Stephenson (2007). Vector generalized linear and additive extreme value models. *Extremes 10*(1–2), 1–19.

Index

Printed in the United States
by Baker & Taylor Publisher Services